A climate of arctic severity prevailed in Britain throughout much of the last two million years, resulting in the widespread formation of a great range of periglacial landforms and deposits. Many of these features provide key evidence for understanding the evolution of the present landscape and reconstructing former climate. Appreciation of the significance of periglacial deposits and structures is also important in many engineering operations and in understanding the development of present-day soils.

This book provides a synthesis of current theory in periglacial geomorphology and applies this to the study of periglacial phenomena in Great Britain. The first part of the book introduces the chronological and environmental background to periglaciation in Britain. The second and third parts deal respectively with the periglaciation of lowland Britain and upland Britain. The book concludes by considering the implications of periglacial phenomena for environmental reconstruction.

The Periglaciation of Great Britain can be used as a text for geography and geology undergraduates studying cold-climate geomorphology, and is also an important and unique reference for all whose teaching or research encompasses a knowledge of periglacial phenomena, including geomorphologists, geologists, Quaternary scientists, civil engineers and pedologists.

D1439414

The Periglaciation of Great Britain

The Periglaciation of Great Britain

COLIN K. BALLANTYNE
Senior Lecturer in Geography and Geology, University of St Andrews

CHARLES HARRIS
Senior Lecturer in Geology, University of Wales, Cardiff

CAMBRIDGE
UNIVERSITY PRESS

Published by the Press Syndicate of the University of Cambridge
The Pitt Building, Trumpington Street, Cambridge CB2 1RP
40 West 20th Street, New York, NY 10011-4211, USA
10 Stamford Road, Oakleigh, Melbourne 3166, Australia

First published 1994

Printed in Great Britain at the University Press, Cambridge

A catalogue record for this book is available from the British Library

Library of Congress cataloguing in publication data

Ballantyne, C. K.
The Periglaciation of Great Britain / Colin K. Ballantyne, Charles Harris.
   p.   cm.
includes bibliographical references and index.
ISBN 0-521-32459-9.
1. Glacial landforms–Great Britain. I. Harris, Charles, 1947–   II. Title.
GB588.43.B35 1993
551.3'15'0941–dc20   92-43477   CIP

ISBN 0 521 32459 9 hardback
ISBN 0 521 31016 4 paperback

Contents

Preface

One of the most interesting paradoxes of geomorphology is that few of the earth's landscapes can be explained in terms of the processes currently operating on them. The explanation for this paradox lies in the dramatic changes that have affected global climate over the last two million years or so. During much of this time, Great Britain experienced conditions significantly colder than those of the present, and such cooling introduced a wide range of geomorphic effects that are now confined to arctic, subarctic and cold montane environments. The most dramatic of these was the formation of glaciers, which carved out distinctive erosional forms in upland areas and deposited a mantle of drift over much low ground. More subtle, but more widespread, was the influence of periglaciation, the modification of landscape under nonglacial cold-climate conditions. Periglacial processes have dominated the recent evolution of those parts of the British landscape that lay beyond the reach of the ice sheets, and have modified – in some areas extensively – those areas that formerly lay under glacier ice. Even when (as at present) warmer interglacial conditions interrupted the prevailing cold, the zone of periglacial activity merely retreated to the highest parts of British mountains, awaiting a climatic summons to return to low ground.

This book synthesises current knowledge concerning the effects of periglaciation on the British landscape. It is set out in four parts. The first provides an introduction to the concept of periglaciation, and establishes the chronological and environmental context necessary for the understanding of later material. Part 2 is devoted to the periglaciation of lowland Britain, which comprises those areas that lie below about 400 m in altitude, are underlain mainly by sedimentary rocks and are extensively mantled by drift deposits. The third part describes the periglaciation of upland Britain, an area of generally more resistant rocks that have been subject to repeated episodes of glacial erosion, and which is consequently characterised by steep slopes with only a thin, discontinuous drift cover. This upland/lowland distinction reflects not only the rather different assemblages of periglacial phenomena encountered in these two zones, but also the fact that parts of upland Britain continue to experience periglacial activity at present, whereas the periglacial landscapes of lowland Britain are entirely relict. The final part of the book brings together many of the phenomena previously considered individually by outlining the characteristics of three contrasting periglacial environments: that of the Dimlington Stadial of *c.* 26 000–13 000 years BP (before present), when glacier ice advanced to cover two-thirds of the present land surface; that of the Loch Lomond Stadial of *c.* 11 000–10 000 years BP, when glacier ice remained confined to upland areas; and that of the present, when glacier ice is absent and periglacial activity is restricted to high ground.

The aim of this book, however, is not merely to summarise current knowledge concerning periglacial features in Great Britain, but also to demonstrate how such features illustrate the wider corpus of knowledge that constitutes periglacial geomorphology. With this end in view, the chapters that constitute the core of the book (Chapters 4–13) are prefaced by general introductions that summarise current understanding of the formation and environmental significance of particular periglacial phenomena. Each of these introductions draws widely from a range of theoretical, laboratory and field research, and incorporates findings from a wide range of present periglacial environments, such as the arctic wastes of Canada, Alaska and Siberia, or the mountains of the Alps, the North American Cordillera and Scandinavia. In this way the volume attempts to demonstrate how current theory in periglacial research can be applied to enhance understanding of the periglacial geomorphology of Great Britain, and equally how periglacial phenomena may be interpreted to help reconstruct past episodes in the overall evolution of the British landscape. The book thus represents a case study in the interpretation of periglacial phenomena in a landscape of varying terrain and complex climatic history. It is designed to be of relevance to all earth scientists whose research interests include cold-climate geomorphology and the reconstruction of Quaternary palaeoenvironments.

Acknowledgements

Although this book was some four years in actual preparation, our interests in periglacial geomorphology and in particular the periglaciation of Great Britain go back to our undergraduate days over 20 years ago. In the intervening period our interest in periglacial geomorphology and the British Quaternary has been stimulated by many friends, colleagues, students and postgraduate students. Amongst the numerous individuals who have (wittingly or unwittingly) played a major part in shaping our outlook we owe a particular debt to the following: Dr Douglas Benn, Professor Hugh French, Professor Brian Luckman, Professor Ross Mackay, Professor Brian McCann, Dr Robert Price, Professor Jim Rose, Dr Alison Sandeman, Dr Brian Sissons, Professor Ming-ko Woo and Professor Peter Worsley. In addition, we acknowledge with thanks and affection the eccentric, enthusiastic and frequently exasperating influence of our mutual friend Dr John Matthews.

We are also more directly indebted to numerous individuals for advice and assistance during the writing of this book. Particular thanks are due to Doug Benn, Ali Dawson, Jim Rose and Peter Worsley, all of whom read parts of the manuscript and offered advice concerning its improvement. We wish to record in particular our gratitude to Janet Mykura of Geography

Graphics and Cartography at St Andrews University, who drafted the majority of the figures; and to Graeme Sandeman, Bert Bremner and Margaret Millen who contributed the remainder. Various individuals also kindly supplied us with photographs and are acknowledged in the captions to these.

Every effort has been made to trace the holders of copyright to the figures that we have reproduced, and we offer our apologies for any errors or omissions that may have occurred. We are grateful to the following for permission to reproduce copyright material: Pergamon Press for Figures 2.1, 2.4, 9.7, 9.27, 9.28, 9.29, 9.30, 13.32 and 14.5; Macmillan Magazines Ltd for Figures 2.2, 5.22, 8.22, 13.33 and 14.1 (right); Scandinavian University Press and Dr Peter Thorp for Figure 2.3; University of Toronto Press for Figure 3.4; Regents of the University of Colorado for Figures 3.5, 6.7(a) and 6.15; Professor A.H. Lachenbruch for Figure 4.1; *American Journal of Science* for Figure 4.5; the Geologists' Association for Figures 4.8, 4.13, 4.15, 4.16, 6.19, 6.27(b) and 9.26; the Geological Society Publishing House for Figures 4.11(a), 7.22, and 7.31; Svenska Sällskapet för Antropologi och Geografi for Figures 4.11(c), 5.9, 5.17(b), 5.21, 8.1(a) and 13.8; John Wiley and Sons Ltd for Figures 4.18, 4.24, 6.10, 7.27, 9.25, 12.5, 12.6, 12.14, 12.17, 12.20, 12.21 and 12.22; Dr K.S. Richards and Professor J. Rose for Figures 4.22 and 6.20; Cambridge University Press for Figures 4.23, 4.26(a), 5.23, 7.13, 7.25, 7.26, 8.11, 9.31, 10.6, 10.7, 11.4 and 14.7; The Royal Society for Figures 4.26(b), 5.16, 7.21, 7.24, 7.28, 8.16, and 14.1 (left); *Scottish Journal of Geology* for Figures 4.27, 11.13, 12.23 and 12.27; Norsk Polaristitutt for Figure 5.7; the National Research Council of Canada for Figures 5.8, 6.7(b), 7.1, 7.3(b) and 8.1(c); Professor A. Pissart for Figures 5.13 and 11.9; Mrs S. Watson for Figures 5.17(a), 5.19, 5.20, 11.3 and 13.2; Gerbrüder Borntraeger for Figure 5.17(c); L.A. Vierek and D.J. Lev for Figure 6.1(a); Tapir Forlag, University of Trondheim, for Figure 6.1(b); Longman Group UK Ltd for Figure 6.8; Academic Press Inc. for Figure 6.9(a); R.J. Ray, W.B. Kranz, N. Caine and R.D. Gunn for Figure 6.9(b); Antiquity, Cambridge, for Figure 6.22(a); the Yorkshire Geological Society for Figures 6.27(a) and 9.21 (right); Dr D.N. Mottershead for Figure 7.16; L.J. Onesti and S.A. Watt for Figure 8.1(d); Elsevier Science Publishers for Figures 8.6 and 14.3(b); Dr Ian Bryant for Figure 8.7; Dr R.A. Shakesby for Figure 8.8; *Catena* for Figure 8.9; the Quaternary Research Association for Figure 8.10; the Institute of British Geographers for Figures 8.13, 8.14 and 12.4; the *Geographical Journal* for Figures 9.21 (left) and 12.19; the Royal Scottish Geographical Society for Figures 12.10 and 13.5; the Royal Society of Edinburgh for Figures 12.16, 13.13, 13.15 and 13.16; Technischen Hochschule, Zurich, for Figure 12.26; V.H. Winston and Sons, inc. for Figure 13.3; Dr J.L. McArthur for Figure 13.4; Professor Rob Ferguson for Figure 13.9; The Arctic Institute of North America for Figure 14.2(a); Dr B. Van Vliet-Lanoë for Figure 14.2(b); and Dr R.B.G. Williams for Figure 14.3(a).

On a more personal note, we wish to thank Catherine Flack of Cambridge University Press for her advice and encouragement, and for guiding the book through the production stage. CKB thanks Rebecca Trengove, Rusty Mooney and Roger Mason for providing sympathy, distraction and whisky (though not necessarily in that order of priority) when things looked bleak. CH acknowledges with affection and gratitude the constant support and unending patience of Sue and the forbearance of the rest of the Harris family during the writing of this book.

PART 1
Introduction and context

Throughout much of the last two million years, the landscape of Great Britain evolved under climatic conditions very much colder than those of the present. The most dramatic consequence of such cooling was the development of glaciers in the mountains of Britain and their subsequent expansion across the adjacent lowlands. Though these glaciers sometimes achieved remarkable dimensions, even overrunning most of the English Midlands and East Anglia, on no occasion was all of the present land surface completely buried under ice. Always there remained a zone, the *periglacial zone*, that experienced prolonged exposure to intensely cold nonglacial conditions. Often, indeed, the periglacial zone was much more extensive than the area covered by glacier ice, and even when periods of warmer climate interrupted the prevailing cold, periglacial conditions continued to affect the highest ground. This volume examines the effects of such periglacial conditions in modifying the British landscape, and describes the various landforms, deposits and sedimentary structures that have survived from the frigid past, as well as those still actively developing on British mountains during the milder present. The first three chapters provide the context for this examination. The opening chapter defines the nature of periglaciation, traces the historical development of periglacial research in Britain and introduces the aims and rationale of the volume as a whole. Chapter 2 sets the study of periglaciation in chronological context by tracing the broad outline of environmental changes that have affected Britain over the last million years or so. The third chapter introduces some of the principal characteristics of present-day periglacial environments, particularly those of the arctic and subarctic, which are similar in many respects to those that prevailed in Britain in the recent geological past.

1
Introduction

The term *periglacial* refers to 'the conditions, processes and landforms associated with cold, nonglacial environments' (Harris *et al.*, 1988) and is used in this sense throughout this book. *Periglacial geomorphology* is concerned with our understanding of the landforms, deposits and processes of cold nonglacial environments. *Periglacial environments*, past or present, are those in which cold-climate nonglacial processes have produced distinctive landforms and deposits, often (but not always) as a result of ground freezing. The concept of *periglaciation* is perhaps less familiar, and is used here to describe the collective and cumulative effects of periglacial processes in modifying the landscape, much as 'glaciation' describes the general effects of glacial action.

A volume devoted to the periglaciation of Great Britain may appear something of a paradox. This island lies far to the south of the arctic tundra, supports no mountains of alpine stature and presently experiences a maritime temperate climate with mild winters. Over much of the country the only 'periglacial' effect that may be evident under present conditions is the growth of needle ice in fallow land during an exceptional frost. Heavy snowfall in lowland areas is sufficiently unusual to be elevated by the weather-conscious populace to the status of a rather enjoyable natural catastrophe. Moreover, to the professional eyes of geomorphologists, much of the northern two-thirds of the island are dominated by the inherited effects of Pleistocene glaciation. Glacier ice sculpted the British uplands into a landscape of corries, arêtes and troughs, and the landscapes of all but the southernmost lowlands are dominated by a cover of glacial drift. The role of periglaciation in modifying the British landscape has been more subtle but more widespread. In southern Britain, beyond the limits of maximum glaciation, cold climates recurred during successive glacial episodes, creating what is essentially a relict periglacial landscape. Moreover, since the waning of the last great ice sheet all parts of northern Britain have experienced periglacial conditions, and these linger on even today in the highest mountains. The legacy of periglaciation across the country is a rich variety of landforms, deposits and sedimentary structures, many of which provide insights into the nature and severity of the sweeping climatic changes that affected Great Britain during the final millennia of the Pleistocene Epoch.

The general aim of this volume is to present a comprehensive picture of current knowledge concerning the periglaciation of Great Britain. Subsumed within this aim are four particular objectives. The first concerns evaluation of the geomorphological and environmental implications of relict periglacial phenomena in the British landscape. A basic tenet of much geological and geomorphological research is James Hutton's dictum that the present represents the key to understanding the past. To permit accurate interpretation of the significance of relict periglacial phenomena in Britain, reference must be made to modern analogues in high latitude and high altitude environments where periglacial conditions currently prevail. Over the past two decades, much research has been devoted to establishing a deeper understanding of the physical, chemical and hydrological processes that operate in such environments (cf. Clark, 1988a; Williams & Smith, 1989). Many areas of uncertainty and contention remain unresolved, but such work has nevertheless placed our understanding of cryogenic (freeze–thaw) activity on increasingly secure theoretical foundations. A particular aim of this book is to integrate recent findings on the mechanisms of periglacial processes and the significance of periglacial phenomena with field observations of such phenomena in Great Britain. For this reason, most chapters are prefaced with a review of recent theoretical developments, against which the British evidence may be assessed. Inevitably, this sometimes involves reinterpretation of the significance of certain sites and landforms.

Our second objective in writing this book has been to attempt a synthesis of evidence recorded from different parts of the country. Periglacial research in Great Britain has often proceeded on a site-by-site basis; great attention has been lavished on particular areas, individual mountains and isolated critical sections. Whilst this approach is of great value in elucidating detail and establishing stratigraphic or geomorphological relationships, it inevitably militates against the establishment of regional patterns. A number of syntheses have demonstrated the value of collating evidence from wider areas within Britain (e.g. Williams, 1965, 1975; Kelletat, 1970a; Watson, 1977a; Worsley, 1977, 1987; Ballantyne, 1984, 1987a; Bryant & Carpenter, 1987; Harris, 1987a), and this approach is extended and developed here. Conversely, we have also tried to identify sites that are outstanding in representing particular phenomena, and have where possible described and illustrated these in detail. In periglacial research, as in other branches of earth science, field experience is indispensable; our aim has been to highlight sites that will both stimulate interest and deepen understanding. A fourth objective has been to identify some priorities for future research. Our quarrying of the literature has revealed numerous gaps in understanding as well as opportunities for future work. Should this volume stimulate fellow researchers to take up the challenge of hitherto unresolved problems or unrecognised possibilities, so much the better.

The overriding aim of this book, however, is to demonstrate the importance of periglaciation in modifying the British landscape and the immense diversity of its manifestations. Despite the relatively limited compass of this island, few periglacial phenomena are unrepresented in relict or active form. For this reason, we believe that synthesis and reassessment of current knowledge will be of interest not only to researchers within Great Britain, but to all earth scientists whose work encompasses the interpretation of periglacial phenomena. There is much that remains to be learnt about – and from – the periglaciation of Great Britain.

The development of periglacial research in Great Britain

The study of periglacial phenomena in Great Britain has a long but somewhat erratic history. Identification of significant milestones in its progress is inevitably arbitrary, but for convenience we have grouped developments into three broad periods. The earliest involved the growing awareness amongst geographers and geologists of the existence of deposits and landforms of periglacial origin in the British landscape, and extended from the subject's hesitant beginnings in the nineteenth century until around 1940. The following quarter century, from 1940 until 1965, witnessed a steady increase in appreciation of the effects of periglaciation, an increase that reflects the quickening pace of Quaternary research in this country. Such developments have reached fruition in the present phase, in which periglacial phenomena have become of primary concern not only to geomorphologists, but also to a wide range of earth scientists whose interests encompass such diverse fields as Quaternary geology and stratigraphy, present-day geomorphic processes, soil science, engineering geology and climatic change.

Early beginnings

The first periglacial phenomenon to have excited the interest of British geologists was the presence of widespread sheets of 'rubble drift' or *head* deposits in southern England. The earliest account of head deposits is attributable to Borlase (1758, p.76) who described '...a rough yellow clay, charged here and there with small and large stones, all with their angles on...' overlying raised beach deposits in the coastal cliffs of Cornwall. Chalky head or *coombe rock* in SE England was subsequently described by Mantell (1833) and considered by him to represent a manifestation of the 'diluvium' supposedly deposited by the biblical flood. The local quarryman's term 'head' was given to the rubble drift by De la Beche (1839), and such deposits were first attributed to periglacial processes by Austen (1851). In a paper remarkable for its astuteness of observation, Austen attributed the head deposits of Devon and Cornwall to rock decomposition and mass movement over low-gradient slopes. 'They are', he noted, 'so totally beyond the power of any present agencies, that it seems necessary to call in the operation of cold to adequately account for them' (p.130). Unhappy with the then novel idea of ice ages, however, Austen invoked massive uplift and then subsidence to explain the existence of such periglacial deposits at present sea level.

Not all Victorian geologists, however, were satisfied with Austen's explanation. Head and coombe rock were variously attributed to earth tremors (Murchison, 1851), torrential snowmelt floods (Ussher, 1879) and even catastrophic pulsed uplift (Prestwich, 1892), a hypothesis judged by an eminent contemporary to be 'more ingenious than convincing' (Geikie, 1894). Gradually, however, Austen's proposed explanation of head as a cold-climate deposit gained acceptance. Wood (1882)

ascribed the angular drift of nonglaciated areas of England to 'sludging' of frost-riven debris, and in an outstandingly percipient paper Spurrell (1886, p.31) proposed that head may have accumulated by the 'intermittent flowing, under its own weight, of a soil undergoing thaw, that is, in a viscous state'. This description is remarkably consonant with present understanding of the process of *gelifluction*, and indeed with the current interpretation of head deposits. Spurrell also noted the occurrence of up to three units of head separated by wash deposits, and inferred that each represented a distinct cold period. By the end of the nineteenth century, the head of southern England appears to have been regarded by most geologists as a periglacial gelifluction deposit that represents the stratigraphic equivalent of the 'Newer Drift' deposited by the last ice sheet farther north (Geikie, 1894; Edmunds, 1930; Dines *et al.*, 1940).

The accumulation of head on low ground was not the only periglacial effect recognised in southern England by Victorian scientists. Spurrell (1886) suggested that the blind-ended dry valleys or 'coombes' that dissect the South Downs were excavated under periglacial conditions 'by the ice-laden and tumultuous floods of summer' when the ground was frozen and thus impermeable, a process independently advocated by Reid (1887) the following year and still widely accepted as being responsible for at least the final stages of coombe evolution. The rounded 'bowl' shape of the heads of many coombes was subsequently attributed by Bull (1936, 1940) to the operation of *nivation* or snowpatch erosion under periglacial conditions. Nineteenth century researchers also showed a lively awareness of the stratigraphic possibilities of periglacial deposits exposed in coastal cliffs. A fine example was provided by Reid (1892), who noted that Pleistocene deposits on the Sussex coast contain erratics apparently transported by floating ice, and who observed the grounding marks of ice floes in the underlying Tertiary clays. Marine beds overlying these erratic-bearing deposits were found to contain a temperate molluscan fauna, and are themselves overlain by chalky head. From this evidence, Reid inferred that the marine deposits represent a temperate interglacial that separated two distinct periglacial periods.

The interest excited by head deposits and related features in southern England did not extend to most other periglacial phenomena until the present century. Geologists were aware of the importance of frost action, but tended to emphasise its effects in terms of the disintegration of cliffs and the resulting accumulation of scree or *talus* downslope (e.g. Geikie, 1865, pp. 37–38; Avebury, 1902, pp. 220–222). In the evocative language of the day, for example, James Geikie (1877, p. 321) invited his readers to note how '...the hill-tops are buried in their own ruin, and how the flanks are in many places curtained with long sweeps of angular blocks and rubbish, that shoot down from the base of cliff and scaur to the dark glens below. All this,' he added, 'is the work of rain and frost...the havoc affected by frost is yet very considerable'. Early recognition was, however, accorded to *protalus ramparts*, which are ridges of rock debris that accumulate at the foot of perennial snowbeds. Features of this origin were first described by Ward (1873) in the English Lake District, and attributed to former periglacial conditions by Marr (1916, p.202). To Marr also belongs the distinction of identifying *debris cones* (or 'dry deltas' as he termed them) that had formed through the reworking of talus by repeated debris flow activity. The spectacular *tors* or upstanding rock residuals of granite areas and their surrounding *clitter* of frost-weathered boulders also attracted early attention. The importance of periglacial processes in exposing the monolithic tors of Dartmoor was first considered by Jones (1859) and subsequently by Bate (1871) and Albers (1930) in accounts which introduced ideas that were to be hotly debated in later years.

More subtle effects of periglacial activity on British mountains did not entirely escape the attention of early researchers. Small-scale *patterned ground* (the regular arrangement of stones by freezing and thawing of the soil) was first recorded by Ward (1896) on high ground in the Lake District and formed the subject of several articles in the 1930s (e.g. Simpson, 1932; Hay, 1936). Officers of the Geological Survey working in Scotland in the early years of the twentieth century provided succinct descriptions of *mountain-top detritus*, and attributed its formation to frost-weathering of bedrock in Lateglacial times, after the last ice sheet had downwasted below the level of mountain summits (Crampton, 1911; Peach *et al.*, 1912, 1913; Crampton & Carruthers, 1914). These same geologists also gave brief but perceptive accounts of landforms and deposits associated with such detritus, including vegetation patterns, small terraces produced by periglacial mass movement and windblown sand deposits on high plateaux.

This exploratory period was brought to a close by the publication of three important papers that anticipated different strands in the future development of periglacial research in Great Britain. In the earliest of these, Hollingworth (1934) provided an outstanding account of active periglacial processes and landforms in the Lake District. In this he discussed the origin of frost-sorted patterned ground, *solifluction lobes* produced by the downslope movement of soil subject to seasonal freezing and thawing, vegetation patterns, and *turf-banked terraces* that reflect the combined action of wind and periglacial mass movement. The second seminal contribution from this period was made by Paterson (1940), who compared the present effects of frost action in the arctic with structures in bedded gravels near Cambridge. This appears to be the earliest attempt in Britain to interpret relict periglacial phenomena through direct comparison with high-latitude analogues. As a result of his arctic experiences, Paterson was able to demonstrate that large-scale relict polygonal structures on the ground surface are the product of frost-cracking under arctic conditions; that wedge-shaped structures exposed in section in the bedded gravels are the relict equivalents of ice wedges formed in *permafrost* or perennially-frozen ground; and that *involutions* or contorted strata within the Cambridge deposit are similar to those developed in the zone of seasonal thawing (the *active layer*) above permafrost in arctic environments. Others were quick to follow Paterson's lead (e.g. Carruthers & Anderson, 1941). All three types of evidence were to assume great importance in reconstructions of Quaternary palaeoclimates, a major theme of modern periglacial research. Finally, Dines *et al.* (1940) reviewed earlier work on the origin and significance of head deposits and described how such deposits could be mapped and interpreted as climatostratigraphic units within Quaternary successions. This work presaged later research in which the interpretation of periglacial (or periglacially-modified) deposits assumed a vital role in the establishment and correlation of Quaternary chronosequences. By 1940, therefore, some of the main themes

of periglacial research in Great Britain had been outlined. Ensuing decades were to witness their recapitulation, development and enrichment.

1940–65: The widening perspective

The quarter century following the appearance of the three classic accounts summarised above saw the publication of some 50 research papers devoted to periglacial phenomena in Great Britain, though inevitably periglacial features also received frequent mention in Soil Survey Monographs, Memoirs of the Geological Survey and articles devoted to Quaternary stratigraphy. Most of these papers appeared in the final decade of this period. They reveal increasing awareness of the widespread nature of periglacial phenomena in Great Britain, the rich variety of features represented and the importance of periglaciation in Quaternary landscape evolution. Such findings were summarised in a series of contemporary reviews by FitzPatrick (1956a, 1958), Te Punga (1957), Galloway (1958, 1961a) and Waters (1964a,b, 1965).

Amongst the research developments in the British uplands was the recognition of landforms interpreted as *cryoplanation terraces* and *nivation benches*, broad terraces apparently cut across bedrock by the operation of frost-weathering, nivation and solifluction during episodes of Quaternary periglaciation (Guilcher, 1950; Te Punga, 1956; Waters, 1962; Gregory, 1965). Attention was also focused on the contribution of periglacial processes to the formation of tors (Linton, 1955, 1964; Pullan, 1959; Palmer & Radley, 1961; Palmer & Neilson, 1962) and on the role of frost action in the genesis of upland regolith (Tivy, 1962; Waters, 1964a) and talus (Andrews, 1961). Lateglacial solifluction was invoked to account for the development of drift terraces and lobes on the flanks of British hills (Metcalfe, 1950; Galloway, 1961b; Pissart, 1963a, b) and the current importance of wind as a geomorphic agent on high ground emerged strongly from the work of ecologists on the Cairngorm Mountains (Watt & Jones, 1948; Metcalfe, 1950; Burges, 1951). This period also saw the emergence of the earliest studies of the rates and processes of periglacial activity on British mountains. These concerned the operation of frost creep and frost sorting of bare ground on mountains in Snowdonia, the Lake District and the Southern Uplands of Scotland (Miller, Common & Galloway, 1954; Tallis & Kershaw, 1959; Caine, 1963a, b).

Research on periglacial phenomena in lowland Britain during the period 1940–65 not only built on earlier discoveries, but also led to the identification of a number of landforms and soil structures hitherto unrecorded in Britain. The head deposits of southern England continued to attract attention, with studies of their texture, structure, stratigraphy and distribution (e.g. Waters, 1960; Kerney, 1963, 1965; Hodgson, 1964, Gallois, 1965). Attention was also focused on the possible significance of thaw of frozen ground in inducing *cambering* and *valley bulging*, both of which are associated with the deformation of mudrocks overlain by more resistant strata (Hollingworth, Taylor & Kellaway, 1944; Kellaway & Taylor, 1953). In addition, numerous publications during this period demonstrated the widespread occurrence of periglacial involutions and ice-wedge casts in Quaternary drifts (e.g. Arkell, 1947; Dimbleby, 1952; Curtis & James, 1959; Waters, 1961; Watson, 1965a). These were observed to occur not only in areas that remained outside the limits of the last ice sheet, but also well inside these limits, thereby demonstrating the persistence of permafrost conditions after ice-sheet retreat. In Scotland, for example, an appeal by the editor of the *Scottish Geographical Magazine* for information on periglacial phenomena brought reports of ice-wedge casts as far north as Midlothian (Common & Galloway, 1958), Angus (Rice, 1959) and Argyll (Gailey, 1961). The wider distribution of such features in Scotland was ably summarised by Galloway (1961c).

Further evidence of former permafrost in lowland Britain was inferred from the occurrence of *indurated horizons* in soils. Such horizons were first identified by Glentworth (1944, 1954; Glentworth & Dion, 1949) and were attributed to repeated freezing and thawing of the former active layer (FitzPatrick, 1956b; Stewart, 1961; Crampton, 1965). Large-scale patterned ground was also interpreted as indicative of former permafrost conditions in southern and eastern England (Watt, 1955; Shotton, 1960; Williams, 1964). A number of periglacial phenomena previously unrecognised in lowland Britain were first described in the period 1955–65. These included *thermokarst depressions* resulting from the melt-out of massive lenses of ground ice (Cailleux, 1961), circular ramparts resulting from the collapse of former *pingos* or ice-cored hills (Pissart, 1963c), stratified slope deposits or *grèzes litées* (Watson, 1965b), and deposits of *loess* (windblown silt) and *coversand* indicative of powerful wind action across tundra landscapes (e.g. Coombe *et al.*, 1956; Waters, 1960; Straw, 1963).

Many of the papers cited above demonstrate growing awareness of the potential of relict periglacial phenomena for the reconstruction of Late Quaternary palaeoclimates. This awareness was most fully realised by Williams (1965) in a paper devoted to analysis of the former distribution of permafrost in England during the last glacial period. From the distribution of a wide range of relict periglacial phenomena, including ice-wedge casts and polygon structures, involutions and large-scale patterned ground, Williams inferred that continuous permafrost had underlain the Midlands and eastern and southern England, but not ground below 300 m in the south-west of the country. From this evidence he suggested that that under full-glacial conditions the -6 °C (mean annual temperature) isotherm lay along a line from Sussex to the southern edge of Exmoor. This important paper represents the first of several attempts to reconstruct regional palaeoclimates on the basis of evidence provided by relict periglacial phenomena.

Recent research themes

The years since 1965 have seen continued growth in the sophistication and status of periglacial studies in Great Britain. Roughly 200 papers devoted largely or exclusively to British periglacial phenomena were published in this period, and the role of periglacial activity has increasingly formed an intrinsic part of research on such diverse topics as the reconstruction of former glaciers, slope stability, Holocene landscape evolution, Pleistocene stratigraphy and Late Quaternary palaeoclimate. The aims and approaches involved within such research have been equally varied. A major concern of many researchers has been interpretation of relict periglacial deposits and structures in Quaternary stratigraphic sequences. Climatically-diagnostic sediments such as head, coversand and loess have been used along with glacial and fluvioglacial deposits and the evidence provided by peat layers and *palaeosols* (relict soil horizons) to reconstruct the changing nature of the environment through successive stadial (cold) and interstadial episodes (e.g. West, 1980a; Gibbard, 1985; Rose *et al.*, 1985a; Connell & Hall, 1987). Ice-wedge casts in drift deposits have proved particularly valuable as indicators of former permafrost (Worsley, 1987), and involutions in Quaternary sediments have been widely interpreted as indicative of former periglacial conditions (e.g. West, 1980a; Hall & Connell, 1986).

Allied to the use of periglacial deposits as climatostratigraphic indicators within drift sequences has been increasing use of dating techniques to establish ages for such sediments. Radiocarbon (^{14}C) dating of buried or incorporated organic material has been employed to establish the age of head or solifluction deposits dating to the period of the last glacial maximum or earlier (Connell *et al.*, 1982; Sutherland & Walker, 1984; Sutherland, Ballantyne & Walker, 1984; Scourse, 1987), to the Lateglacial period (Dickson, Jardine & Price, 1976; Hall, 1984) and, on Scottish mountains, to the present interglacial (Mottershead, 1978; Ballantyne, 1986d). Radiocarbon dating has also been employed to establish a Lateglacial age for periglacial mudslide deposits (e.g. Chandler, 1976) and postglacial ages for various upland sediments, such as debris-flow deposits (e.g. Harvey *et al.*, 1981; Brazier, Whittington & Ballantyne, 1988) and windblown sand (Ballantyne & Whittington, 1987). Because of its restricted temporal range, however, radiocarbon dating has proved of limited use in dating sediments deposited before *c.* 30 ka BP (i.e. 30 000 'radiocarbon years' before present). For older periglacial sediments, particularly loess and periglacial river deposits, the application of thermoluminescence (TL) dating techniques shows particular promise (e.g. Wintle, 1981, 1990; Seddon & Holyoak, 1985; Wintle & Catt, 1985a, b).

Sedimentological analysis of periglacial deposits has also engaged numerous researchers over the past 25 years, with a wide range of increasingly sophisticated techniques being employed to determine the origin, transport history, geotechnical properties and palaeoenvironmental significance of such deposits. These have included provenance studies of constituent clasts and fines, analyses of clast size and shape, particle-size analyses, measurement of macrofabric and microfabric, scanning-electron microscope studies of grain morphology, micromorphological studies of thin sections, geotechnical analyses, laboratory simulations and stratigraphic differentiation of sedimentary units. Such techniques have been widely applied to head and coombe rock (e.g. Mottershead, 1976, 1982; Harris & Wright, 1980; Harris, 1981a, 1987a; Wilson, 1981; Douglas & Harrison, 1987; Scourse, 1987; Gallop, 1991) as well as to upland regolith (Ragg & Bibby, 1966; Innes, 1986), loess and windblown sand deposits (e.g. Wilson, Bateman & Catt, 1981; Catt & Staines, 1982; Ballantyne & Whittington, 1987) and even coarse upland talus accumulations (Statham, 1976a; Ballantyne & Kirkbride, 1986). A noticeable development has been increased awareness amongst engineering geologists of the geotechnical significance of periglacial deposits (Higginbottom & Fookes, 1971). In particular, much of our understanding of *periglacial mudslides* (translational failures over former permafrost) results from research by engineering geologists on such features and their implications for slope stability (e.g. Hutchinson, 1967, 1974, 1991; Weeks, 1969; Chandler, 1970a, 1976; Hutchinson, Somerville & Petley, 1973; Skempton & Weeks, 1976; Skempton, 1988; Spink, 1991).

Palaeoclimatic reconstructions based partly or exclusively on relict periglacial phenomena have also been attempted. Those for southern Britain refer primarily to the period of the last glacial maximum, c. 18–17 ka BP, and have been based on the distribution of climatically-diagnostic features such as ice-wedge casts, involutions, large-scale patterned ground, pingo ramparts, ground-ice depressions, loess and coversands (e.g. Williams, 1975; Watson, 1977b). Northern Britain lay under the last ice sheet at this time, and palaeoclimatic reconstructions for Scotland have therefore been limited to consideration of the Loch Lomond Stadial of c. 11–10 ka BP, the last period of severe periglaciation to affect Great Britain (e.g. Ballantyne, 1984). In mountain areas, the distribution of relict periglacial phenomena of Loch Lomond Stadial age has also been extensively employed to delimit the extent of contemporaneous glaciers (e.g. Thorp, 1981, 1986; Ballantyne, 1989a), which themselves provide vital evidence for the reconstruction of stadial palaeoclimate.

Data on former rates of periglacial activity, however, remain sparse. Williams (1968) pioneered such investigations by establishing the travel distances of boulders in head deposits, and Ballantyne & Kirkbride (1987) used the volume of debris in Loch Lomond Stadial protalus ramparts to assess contemporaneous rockfall rates, but estimation of former rates of mass transport has otherwise received little attention. More surprisingly, the same is true of current rates of periglacial mass movement on British mountains. Site-specific measurements exist concerning present rates of frost creep, solifluction and rockfall activity, but only one attempt has been made to compare the relative effectiveness of different agencies of periglacial mass movement under present conditions (Ballantyne, 1987a). British geomorphologists have also made curiously little effort to avail themselves of the opportunity provided by periglacial landforms on high ground to study periglacial processes. Conspicuous exceptions include studies of the movement of *ploughing boulders*, which move downslope faster than the surrounding regolith (Tufnell, 1972), nivation processes (Ballantyne, 1985) and the genesis of sorted stripes (Warburton, 1987). A more promising development has been recognition of the importance of bedrock characteristics in determining the nature of frost-weathered regolith, and hence the nature and distribution of periglacial landforms and deposits (e.g. Potts, 1971; Ballantyne, 1984, 1987a, 1991a; Harris, 1987a).

Despite the long history of periglacial research in Britain prior to 1965, appreciation of the scope of periglaciation has continued to widen over the last 25 years. For upland Britain, this has taken the form of the discovery of hitherto unrecorded landforms and deposits, including *protalus rock glaciers* produced by the deformation of ice-rich sediment under talus accumulations (e.g. Dawson, 1977; Sissons, 1979a), valley-floor *solifluction terraces* (e.g. Crampton & Taylor, 1967; Douglas & Harrison, 1987), *earth hummocks* (e.g. Tufnell, 1975; Ballantyne, 1986c), *avalanche impact landforms* (Ballantyne, 1989b) and high-level aeolian or *niveo-aeolian* deposits (e.g. Pye & Paine, 1983; Ballantyne & Whittington, 1987). There

has been increasing awareness, too, of the former role of frost action in shoreline development. Sissons (1974a, 1981a) first proposed that certain raised rock platforms on the west coast of Scotland were cut by a combination of frost weathering and wave action under stadial conditions, a hypothesis subsequently strongly developed by Dawson (1980; Dawson, Matthews & Shakesby, 1987). This mechanism has also been invoked to explain the cutting of erosional shoreline platforms across resistant rock at the margins of former glacier-dammed lakes (Sissons, 1978) and a proglacial lake that occupied the site of Loch Ness in the Scottish Highlands (Firth, 1984). It has also become increasingly recognised that certain abandoned fluvial landforms represent alluviation by braided rivers with flashy *nival* (snowmelt) regimes under periglacial conditions. In lowland areas, periglacial alluviation is represented by floodplain terraces with a complex lithofacies architecture typical of unstable and rapidly-switching gravel bed channels (e.g. Castleden, 1980; Bryant, 1983a,b; Dawson, 1987). In upland areas, fluvioperiglacial sedimentation is represented by relict alluvial fans at the mouths of tributary valleys (e.g. Rowlands & Shotton, 1971; Peacock, 1986).

A number of review papers have attempted to summarise various aspects of the developments outlined above. Some have dealt systematically with topics such as loess and coversands (Catt, 1977), permafrost stratigraphy (Worsley, 1987), solifluction deposits (Harris, 1987a) and ramparted ground-ice depressions (Bryant & Carpenter, 1987). Others have considered periglaciation on a regional basis, for example in northern England (Tufnell, 1969; Boardman, 1985a), southern England (Jones, 1981), upland Britain (Ballantyne, 1987a) and Scotland (Kelletat, 1970a; Sissons, 1976c; Ballantyne, 1984). Others still have provided summaries of the periglacial landforms and deposits within smaller areas, such as Snowdonia (Ball & Goodier, 1970), the Scilly Isles and Cornwall (Scourse, 1987) western Dartmoor (Gerrard, 1988b) and the Isle of Skye (Ballantyne 1991a). Several of the accounts cited above appeared in a single volume, *Periglacial Processes and Landforms in Britain and Ireland* (Boardman, 1987) that represented the outcome of a conference on this topic in 1985. This conference may be seen as marking the coming of age for periglacial research in the British Isles, and recognition of the significant role played by cold-climate nonglacial processes in shaping the evolution of the British landscape. In his introduction to the resulting volume of papers, however, Boardman sounded a cautionary note: progress, particularly over the last quarter century, may have been substantial, but many problems remain to be overcome before the potential of periglacial phenomena for the reconstruction of Quaternary stratigraphy and palaeoenvironments is fully realised. This is probably the major challenge to be confronted by periglacial research in Britain as we enter the new millennium.

Approaches to synthesis: the organisation of this book

Synthesis of current knowledge concerning periglacial phenomena in Great Britain poses three problems of organisation. The first stems from the amazing variety of periglacial phenomena present. These range in scale from microscopic soil structures to entire periglacially-modified landscapes, and often pose difficulties of classification. The second problem concerns spatial variations: many periglacial phenomena occupy specific 'habitats', and are rare or nonexistent outside of these areas; few (if any) occur throughout Great Britain. Finally, periglacial phenomena in Britain have developed at different rates over a very long timescale. Some relict structures exposed along the Norfolk coast, for example, may be over a million years old, whilst certain microforms on the higher parts of Scottish or Welsh mountains may have developed over the last few winters.

These three organisational problems suggest three different modes of treatment: a systematic (topic-by-topic) approach, a regional (area-by-area) approach and an historical (time-slice) approach. In this volume we have attempted to combine these three approaches. Most chapters are organised systematically on a topic-by-topic basis, but are grouped regionally in terms of two major contrasting areas, lowland Britain (Chapters 4–8) and upland Britain (Chapters 9–13). Within these chapters, the phenomena under consideration are placed in temporal context with respect to the Quaternary chronology of Great Britain, and their significance for the reconstruction of particular past environments is assessed. Prefacing this systematic discussion of periglacial phenomena are two further introductory chapters. The first of these (Chapter 2) outlines the temporal pattern of Quaternary environmental change in Great Britain, and thus provides the chronological framework for subsequent discussion. Chapter 3 consists of a brief introduction to the general characteristics of periglacial environments, and is included primarily for the benefit of readers with little prior experience in this field. The final chapter (14) attempts to draw together earlier material in terms of environmental reconstructions relating to three time periods. The book thus falls into four parts: (1) introduction (Chapters 1–3); (2) the periglaciation of lowland Britain (Chapters 4–8); (3) the periglaciation of upland Britain (Chapters 9–13); and (4) reconstruction of periglacial environments (Chapter 14).

Upland and lowland Britain

The rationale behind the division of subject matter in terms of lowland Britain and upland Britain deserves explanation. The prime motive behind this subdivision can be traced to the distribution of periglacial phenomena, in that it is possible to identify distinct 'lowland' and 'upland' assemblages. The 'lowland' assemblage includes a wide range of relict features that are largely or entirely absent on high ground, such as ice-wedge casts, tundra polygons, cryoturbation structures, thermokarst phenomena and pingo scars, together with loess deposits, certain types of mass-wasting deposits and large-scale slope structures. Conversely, upland landscapes support a range of phenomena rarely or never encountered in the British lowlands, such as blockfields, tors, cryoplanation terraces, talus slopes, avalanche landforms, protalus ramparts and rock glaciers. A further fundamental distinction is that whereas the higher mountains of Britain support a number of active periglacial features, the periglacial phenomena of lowland Britain are entirely relict, having been formed under climatic conditions much more severe than those of the present. It would be misleading, however, to present upland and lowland landscapes as mutually exclusive in terms of the range of periglacial phenomena they support, as there is a certain amount of overlap. In particular, patterned ground and solifluction features occur in both upland

and lowland environments, though in general such landforms are better preserved in upland areas, where they have not been obscured by cultivation.

In part, the distinction between upland and lowland periglaciation is a consequence of altitude and relief. The restriction of active periglacial phenomena to high ground reflects the much more severe climate of mountain areas, where ground freezing and snowcover persist throughout much of the winter and strong winds have often stripped away much of the vegetation cover. Similarly, certain landforms such as avalanche boulder tongues, protalus ramparts and rock glaciers occur only at the foot of steep mountain slopes, and hence are absent from the lowland zone. The differences between upland and lowland periglaciation, however, also reflect a number of rather more subtle controls. Foremost amongst these are the effects and limits of glaciation. In upland areas the role of glaciers has been primarily erosive, so that many mountain slopes form rocky cliffs and even gentle slopes support only a thin cover of frost-weathered (and often mobile) regolith. In contrast, the landscape of much of lowland Britain is thickly mantled by glacial, fluvioglacial and periglacial drift deposits. These have formed host materials for the growth of various forms of ground ice and hence the development of a wide range of periglacial features that reflect ground-ice growth and decay, such as ice-wedge casts, ground-ice depressions and cryoturbation structures. The limits of successive Pleistocene glaciations have also exercised a strong control on the *age* of periglacial phenomena in different areas of Britain. The southernmost parts of this island have never experienced glaciation, and here the landscape has adopted its present form during successive episodes of Pleistocene periglaciation. In more northerly lowlands, episodes of periglacial activity have alternated with those of glaciation, so that periglacial processes have modified the deposits laid down by successive ice sheets and their associated meltwater rivers. In the mountains of Britain, though, glacier ice persisted until the very end of the last cold period some 10,000 years ago, and in such areas the imprint of periglacial activity is much more youthful.

The contrasts outlined above have stimulated radically different approaches to the study of periglacial phenomena in upland and lowland Britain. In mountain areas, where periglacial deposits tend to be thin, localised and relatively recent, geomorphologists have concentrated their research within a fairly limited timescale (roughly from the time of the downwastage of the last ice sheet until the present day), as only post-deglaciation landforms and deposits occur in such areas. Periglacial research in the British uplands has therefore concerned mainly Lateglacial and recent landscape evolution, current periglacial processes and rates of activity, and the morphology, distribution, origin and climatic significance of periglacial landforms. In contrast, much research in lowland Britain has been devoted to elucidating the stratigraphic and palaeoenvironmental significance of periglacial soil structures and deposits that developed over a timescale that stretches back a million years or so. Such different research aims imply completely different approaches and techniques, and tend to have attracted different practitioners. Whilst the 'upland' periglacialist clambers across the windswept summits of Scottish or Welsh mountains in pursuit of remote and reclusive landforms, his 'lowland' counterpart is never happier than when ensconced in

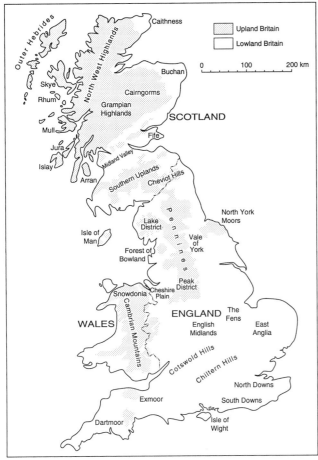

Figure 1.1 Upland and lowland Britain, differentiated principally in terms of relief and altitude. Upland areas are characterised by steep gradients and achieve elevations exceeding about 400 m; lowland areas are generally flat or undulating and lie at lower altitudes.

a deep gravel pit, examining structures created by the melting of ground ice some tens or hundreds of thousands of years ago.

Whilst the terrain characteristics that differentiate 'upland' from 'lowland' Britain are readily identified, it is more difficult to define a precise boundary between the two. In general, the landscapes of upland Britain are characterised by steep slopes, maximum elevations greater than about 400 m, a thin or discontinuous drift cover and absence of arable cultivation. Those of lowland Britain generally lack steep slopes, lie below 400 m, often support a thick cover of drift and for the most part have experienced cultivation. In some areas, of course, these criteria conflict, but there is nonetheless a strong element of spatial correlation amongst them, and it is possible to delimit the two types of terrain in a general way (Figure 1.1). Inevitably, though, areas of intermediate altitude and relief (such as Dartmoor) contain characteristics of both 'upland' and 'lowland' terrain, and indeed 'upland' and 'lowland' periglaciation; the boundary between the two is transitional, not abrupt.

A question of time

One of the most fascinating aspects of the periglaciation of Great Britain is the timespan represented by periglacial phenomena. Some half a million years ago, a soil developed across wide areas of eastern England under a climate fairly similar to that of the present. When climatic deterioration ensued, freezing of the ground caused the development in this soil of numerous periglacial structures, such as ice wedges, frost cracks and involutions, forming a distinct horizon now known as the Barham Structure Soil (Rose et al., 1985a). Sand and silt particles blown across an arid tundra landscape settled on and were incorporated within the Barham Soil, which was then buried by the deposits laid down by an extensive ice sheet. The Barham Soil therefore represents a very ancient periglacial landsurface, yet one that contains remarkably well-preserved periglacial structures that closely resemble similar forms developed beneath the present land surface as little as 11 000–10 000 years ago. At the other end of the periglacial timescale are features that have developed within years or decades rather than millennia. Frost-sorted patterned ground on Scottish and Welsh mountains, for example, is capable of developing over a few winter months, and there is some evidence for a marked increase in mountain-top erosion, solifluction and debris-flow activity within the past two or three centuries (Ballantyne, 1991b). It is important, therefore, to be able to relate the periglacial phenomena of Great Britain to an appropriate timescale of environmental change, such as that outlined in the next chapter.

2
Quaternary environmental change in Great Britain

Periglacial phenomena in Great Britain have developed at a variety of times and under a considerable range of environmental conditions. Some, on high ground, are active at present; others developed during the final millennium of glacial conditions in Britain; many developed when the margin of the last ice sheet lay athwart central England; and other relict features record far earlier episodes of periglaciation. To allow features of differing age to be placed in context, this chapter summarises the chronology and nature of environmental change over the period in which periglacial phenomena are represented in the British stratigraphic record.

Chronologically, evidence for periglacial conditions in Great Britain is restricted to the *Quaternary Period*. This is subdivided into two geological epochs, the *Pleistocene* (2.4 Ma–10 ka BP) and the *Holocene* (after 10 ka BP). The great majority of relict periglacial features identified in Britain are of Middle or Late Pleistocene age (*c.* 810–10 ka BP). The evidence for periglaciation during the Early Pleistocene is largely limited to inferred ice-wedge casts and involutions exposed along the coastline of East Anglia (West, 1980a; West, Funnell & Norton, 1980; Worsley, 1987). For this reason, the summary below focuses on the pattern of major environmental changes during the Middle Pleistocene, the Late Pleistocene and the Holocene. The evidence for earlier glaciations and associated environmental changes is summarised in Bowen *et al.* (1986) and Hey (1991). It must be emphasised, however, that there is at present only limited consensus on the timing and nature of Quaternary environmental changes in Great Britain. A provisional chronostratigraphic framework proposed by Mitchell *et al.* (1973) has since experienced extensive revision as increasingly detailed stratigraphic, palynological, faunal and dating evidence comes to light. Current areas of agreement and contention are outlined in a number of reviews (Bowen *et al.*, 1986; Bowen, 1991; Ehlers & Gibbard, 1991; Ehlers, Gibbard & Rose, 1991a; Gibbard *et al.*, 1991; Worsley, 1991a), and present knowledge concerning the Quaternary glacial stratigraphy of Britain on a region-by-region basis is detailed in a volume edited by Ehlers, Gibbard & Rose (1991b).

The sequence of Quaternary deposits in Britain is subdivided on climatostratigraphic criteria into those representing cold (glacial) and warmer (interglacial) stages (Table 2.1). However, because later glaciations have tended to obliterate the evidence of earlier glacial and interglacial episodes, the stratigraphic record is fragmentary. At present only the most extensive (Anglian) and most recent (Devensian) glaciations are adequately defined in terms of type sections or extent (Figure 2.1). The Devensian glacial stage has been subdivided into *stadials* characterised by a relative deterioration in climate and intervening *interstadials* that represent periods of relative warming.

Within the British succession, however, only the two most recent stadial substages have been formally defined within a geochronological framework: the Dimlington Stadial of 26–13 ka BP, and the Loch Lomond Stadial of 11–10 ka BP (Rose, 1985, 1989a). Several researchers have attempted to relate terrestrial evidence to the continuous oxygen isotope chronology provided by deep-sea cores. Isotope stages are numbered from the most recent, with even numbers being assigned to cold stages and odd numbers to temperate stages. For Britain, there is general agreement that the Anglian Glacial corresponds to Stage 12, the Ipswichian Interglacial to the earliest part of Stage 5 (denoted 5e), the Late Devensian glacial substage to Stage 2 and the present (Flandrian) interglacial to Stage 1. More contentious are proposed correlations of particular isotope stages with the Hoxnian Interglacial, with events between the Hoxnian and Ipswichian Interglacials, and with the alternating stadial and interstadial episodes of the Early and Middle Devensian. The sequence outlined in Table 2.1 is largely based on the views of Rose (1989a) and Ehlers *et al.* (1991a), but a different interpretation has been advocated by Bowen and his co-workers on the basis of amino acid dating (time-dependent isoleucine epimerisation) of molluscs from drift deposits (Bowen & Sykes, 1988; Bowen *et al.*, 1989; Bowen, 1991).

The Anglian glacial stage

The evidence for Anglian glaciation in England is extensive and well defined. The type site for the Anglian is at Corton on the Suffolk coast, where a lower till, the Cromer Till, is successively overlain by outwash sands and an upper till, the Lowestoft Till (Banham, 1971; Pointon, 1978). The Cromer Till has been interpreted as having been deposited by North Sea ice that advanced into East Anglia from the north-east (Perrin, Rose & Davies, 1979), and in places directly overlies Cromerian Interglacial deposits. West (1980a) has shown that the Cromerian–Anglian transition in this area took the form of a progressive change from temperate conditions characterised by grassland with some birch, alder and elm to a sparsely-vegetated periglacial landscape. Late Cromerian and Early Anglian climatic deterioration is also recorded by a well-developed palaeosol, the Barham Soil, which underlies Anglian tills across much of East Anglia (Rose *et al.*, 1985a,b). The presence of ice-wedge casts within this palaeosol demonstrates the development of continuous permafrost before glacier ice reached the area, and associated windblown sands and silts indicate at least local poverty of vegetation cover. From this evidence a picture emerges of arid periglacial conditions outside the limits of the advancing Anglian glaciers. The upper till

Table 2.1: *British Quaternary stratigraphy.*

Stage	Substage	Stadials/Interstadials	Boundary age (BP)	Probable oxygen isotope stages*	
Flandrian Interglacial				1	1
	(Pleistocene/Holocene Boundary)		10 000		
		Loch Lomond Stadial			
			11 000		
	Late Devensian	Windermere Interstadial		2	2
			13 000		
		Dimlington Stadial			
			26 000		
Devensian Glacial	Middle Devensian	Upton Warren Interstadial Complex		3	5a
	Early Devensian	Brimpton Interstadial		4 5a 5b	
		Chelford Interstadial		5c 5d	5c
			122 000		
Ipswichian Interglacial				5e	5e
	(Late/Middle Pleistocene Boundary)		132 000		
				6	
			198 000		
				7	
'Wolstonian' Glacial	(Paviland Glaciation)		252 000	Hiatus? 8	Hiatus? 8
			302 000	Hiatus? 9	
			338 000		
				10	
			352 000		
Hoxnian Interglacial				11 Hiatus?	9
			428 000		
Anglian Glacial				12	12
			480 000		
Cromerian Interglacial Complex				Hiatus	13
	(?Early/Middle Pleistocene Boundary)		810 000		

Based on Rose (1989a) and Ehlers *et al.* (1991a). Elements of the Cromerian Complex and Pre-Cromerian stages are omitted.
* The second set of isotope stages shown are those proposed by Bowen & Sykes (1988), Bowen *et al.* (1989) and Bowen (1991) on the basis of aminostratigraphy.

at Corton, the Lowestoft Till, can be traced across most of eastern England into the English Midlands, and extends as far south as north London (Perrin *et al.,* 1979). According to Rose (1989a), the various facies of Lowestoft Till constitute 'the most extensive recognisable lithostratigraphic unit of the British Pleistocene', and its area of outcrop delimits the extent of Anglian glaciation in England (Figure 2.1). South of this glacial limit the Winter Hill and Westmill Gravels and the Barham Sands and Gravels were deposited as outwash by meltwaters draining the Anglian ice sheet (Rose & Allen, 1977; Gibbard, 1979, 1985).

For many years it was believed that glacial deposits identified in the English Midlands belonged to a more recent glacial stage, named the Wolstonian after a putative type site at Wolston near Coventry, and originally attributed to the period separating the Hoxnian and Ipswichian Interglacials (Shotton,

1953, 1983; Rice & Douglas, 1991). Re-evaluation of the evidence, however, indicates that the till units previously ascribed to the Wolstonian are stratigraphically equivalent to the Lowestoft Till of eastern England, and hence of Anglian age (Rose, 1987, 1989a). This reinterpretation implies that recognition of a 'Wolstonian' Stage should be abandoned, though at present the term persists in informal usage to identify 'glaciations between the Hoxnian and Ipswichian Interglacials' (Bowen *et al.,* 1986; Table 2.1). Farther west, the Anglian Glacial Stage is probably represented by deposits emplaced by the 'Irish Sea' Glaciation. This ice advance extended southwards across Wales and the Bristol Channel, ultimately impinging on the north coast of Somerset, Devon and Cornwall. Bowen *et al.* (1986) argued that this extensive advance of Irish Sea ice antedated Isotope Stage 9 (Table 2.1), and that both the Anglian and Irish Sea Glaciations occurred during Isotope

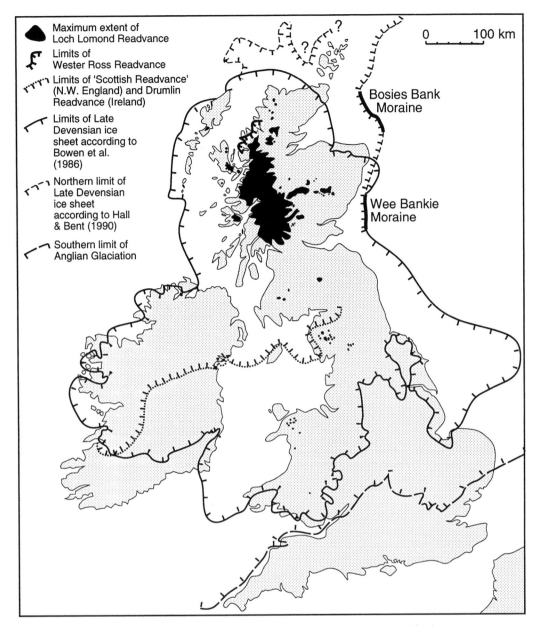

Figure 2.1 Extent of glacial limits in Great Britain and Ireland. Based on a map in Bowen *et al.* (1986), with modifications.

Stage 12. Their map of glacial limits (Figure 2.1) assumes such a correlation.

Although the limits of Anglian and 'Irish Sea' glaciation define the maximum extent achieved by the Pleistocene ice sheets across most of southern Britain, the limit of maximum glaciation is less well established farther north. An extensive submarine erosion surface that cuts across Early and Middle Pleistocene sediments in the North Sea Basin supports tills and glaciomarine deposits, and the latter are in places overlain by freshwater interglacial deposits of inferred Hoxnian age (Cameron, Stoker & Long, 1987). This evidence suggests that the erosion surface was formed during the Anglian glaciation, and indicates confluence of the Scottish and Scandinavian ice sheets at this time. The maximum extent of glacier ice to the

west of Scotland is less well established. At some time prior to the Devensian, glacier ice carried erratics westwards to St Kilda, near the edge of the continental shelf (Sutherland *et al.*, 1984), but the age of this advance has not been established. The upper stratigraphic boundary of the Anglian is marked at several sites in eastern England where the Lowestoft Till is directly overlain by Hoxnian organic sediments that mark a return to temperate interglacial conditions following the waning of the Anglian ice sheet (West, 1980b). There is some evidence, however, for a brief revertance to stadial conditions in the Late Anglian (Bowen *et al.*, 1986). Moreover, a correlation between the Hoxnian and Isotope Stage 9, as advocated by Bowen *et al.* (1989) and Bowen (1991) on the basis of aminostratigraphy, implies a hiatus between the Anglian and Hoxnian.

Glaciation between the Hoxnian and Ipswichian Interglacials ('Wolstonian')

In England, only localised evidence exists for ice-sheet glaciation in the long interval that separates the Hoxnian and Ipswichian Interglacials, unless one accepts that the 'Wolstonian' deposits of the Midlands belong to this period (Straw, 1983; Rice & Douglas, 1991; see above). This dearth of data reflects obliteration or burial of glacial deposits relating to this period by more extensive ice sheets during the Devensian. Bowen *et al.* (1986) concluded that the only convincing glacial evidence relating to this period consists of tills in NE England (the Welton, Basement and Warren House Tills) that locally underlie Devensian till and Ipswichian beach deposits yet overlie gravels attributed to the Hoxnian Interglacial. On the Gower Peninsula of south Wales, however, there is evidence of a north–south movement of glacier ice (the Paviland Glaciation) that postdates the Irish Sea glaciation yet antedates Isotope Stage 7 (Bowen *et al.*, 1989), and hence is demonstrably pre-Ipswichian in age. Bowen *et al.* (1986) and Bowen & Sykes (1988) attributed this event to Isotope Stage 8, and proposed a correlation with the tills of NE England. Similarly, at Kirkhill in NE Scotland, a weathered till that overlies head deposits and a palaeosol containing reworked sediment of inferred Hoxnian age may also be of 'Wolstonian' age (Hall & Connell, 1991) as may a post-Hoxnian erosion surface in the central and northern North Sea Basin (Stoker, Long & Fyfe, 1985; Cameron *et al.*, 1987). However, evidence from sites in Buckinghamshire and Oxfordshire for an additional temperate episode between the Hoxnian and Ipswichian (Green *et al.*, 1984; Briggs, Coope & Gilbertson, 1985) indicates that the pattern of glaciation and environmental change during this time interval was rather more complex than is suggested by assumption of a single glaciation.

The Early and Middle Devensian

Our knowledge of environmental changes during the most recent glacial stage, the Devensian, is rather more detailed than for its predecessors. For the Early and Middle Devensian, though, there are still frustrating gaps and inconsistencies in the palaeoenvironmental record. Amongst these is the lack of an adequate type site for the Ipswichian–Devensian transition and the elusive nature of evidence for Early Devensian glaciation in Great Britain. In the absence of a closely-dated stratotype for the onset of the Devensian, the general pattern of environmental change within the time interval *c.* 120–70 ka BP has been inferred largely from indirect evidence, particularly that provided by deep-sea cores. These suggest that interglacial conditions during Isotope Stage 5e (*c.* 132–122 ka BP) were terminated by cooling around 120 ka BP (Stage 5d), with re-establishment of warmer conditions between *c.* 110 and *c.* 100 ka BP (Stage 5c), renewed cooling around 90 ka BP (Stage 5b) and a final return to warmer conditions at *c.* 85–80 ka BP (Stage 5a). The period *c.* 80–65 ka BP (Stage 4) was marked by further pronounced cooling. Correlation of the Ipswichian Interglacial with Stage 5e is now widely accepted (e.g. Bowen *et al.*, 1986, 1989; Ehlers *et al.*, 1991a), implying that the lower boundary of the Devensian lies at the transition between Stages 5e and 5d.

The difficulty of defining the lower boundary of the Devensian in Britain is compounded by the lack of direct evidence for Early Devensian glaciation. At a few localities in England, such as the Devensian type site at Four Ashes in Staffordshire, intraformational ice-wedge casts indicate permafrost development and thus severe periglacial conditions prior to the Upton Warren Interstadial of *c.* 42 ka BP, but unequivocal evidence of Early Devensian glaciation is lacking in England and Wales (Worsley, 1991a). In Scotland, the evidence is conflicting. The amino acid ratios for molluscs embedded within shelly till in Caithness and on Orkney imply a faunal element of Ipswichian age, prompting Bowen (1991) to suggest that this till was deposited during the Early Devensian, and occupies terrain not subsequently glaciated during the Late Devensian. Sutherland (1991a), however, has sounded a cautionary note, observing that molluscs of Middle Devensian age may also be present in shelly till of Orkney. Moreover, Hall & Whittington (1989) have demonstrated a Late Devensian age for shelly till deposits in SE Caithness, and have pointed out that if the shelly till of Orkney and Caithness represents a single formation, then their findings imply that the tills ascribed by Bowen to the Early Devensian must be of Late Devensian age. If correct, this interpretation also implies that the Late Devensian ice sheet in northern Scotland extended northwards to the Orkney–Shetland channel (Figure 2.1). It has also been suggested that a till unit at Kirkhill in NE Scotland may be of Early Devensian age (Hall & Connell, 1991) but the evidence for this is inconclusive.

Support for Early Devensian glaciation in Britain is indirect and takes four forms. First, Sutherland (1981) inferred Early Devensian glaciation in Scotland from the distribution of high-level shell beds that provide evidence of crustal depression and marine incursion prior to the expansion of the Late Devensian ice sheet. Second, interglacial strata of Ipswichian age are overlain by glaciomarine sediments of inferred Early or Middle Devensian age in the central North Sea Basin (Stoker *et al.*, 1985). Third, ocean core evidence indicates a general global expansion of glacier ice after *c.* 75 ka BP (Ruddiman *et al.*, 1980), and the North Atlantic oceanic polar front appears to have lain well to the south of Great Britain shortly after *c.* 75 ka BP (McIntyre, Ruddiman & Jantzen, 1972). This seems to imply stadial conditions on adjacent land areas, though contemporaneous expansion of glaciers may have been limited by relatively short-lived penetration of polar water into the Atlantic mid-latitudes (Lowe & Walker, 1984). Finally, it seems unlikely that permafrost could have developed in Staffordshire and elsewhere in lowland England at this time without some corresponding accumulation of glacier ice, at least in upland areas.

Although cold conditions appear to have prevailed throughout much of the Early Devensian, the most detailed palaeoenvironmental record we possess for this period relates to warming during the Chelford Interstadial, previously attributed to the period 65–60 ka BP but now believed to have occurred around 100 ka BP during Isotope Stage 5c (e.g. Bowen *et al.*, 1989; Ehlers *et al.*, 1991a; Rendell *et al.*, 1991). Sections in sand pits between Chelford and Congleton in Cheshire contain a layer of peat and organic mud sandwiched between alluvial sands that are overlain by Late Devensian till and locally underlain by a till of indeterminate age (Worsley, 1985, 1991b; Rendell *et al.*, 1991). Structures within the sands indicate that these were deposited under periglacial conditions, but the intervening organic muds contain macrofossil remains of trees, including pine, birch and spruce, that indicate marked warming. Contemporaneous pollen and beetle assemblages from the type site and elsewhere (such as Four Ashes in Staffordshire and Beetley in Norfolk) suggest a more continental climate than that of the present, with slightly lower summer temperatures but markedly lower winter temperatures (Simpson & West, 1958; Coope, 1959; Morgan, 1973; Phillips, 1976; Andrew & West, 1977). At Brimpton in Berkshire, Bryant, Holyoak & Moseley (1983) found evidence for two Early Devensian interstadials of similar boreal forest character to the Chelford Interstadial. Affinities in arboreal pollen suggest that the lower corresponds to the Chelford Interstadial, whilst the upper represents a later warming phase, the Brimpton Interstadial, that has been tentatively assigned to Isotope Stage 5a of c. 80 ka BP (Ehlers et al., 1991a; Table 2.1). The status of other proposed Early Devensian interstadials is less certain. Pollen evidence from organic sediments interbedded with fluvial sands and gravels at a site at Wretton in Norfolk was interpreted by West et al. (1974) as indicating two phases of temperate woodland development, preceded, separated and succeeded by cold open habitat conditions. The upper woodland zone was initially correlated with the Chelford Interstadial on the grounds of similarities in arboreal pollen composition, and the lower woodland phase was attributed to the 'Wretton Interstadial', a hitherto unrecognised warm phase. Subsequent studies of coleopteran (beetle) assemblages and vertebrate remains, however, suggest that the pollen evidence favouring a 'Wretton Interstadial' is misleading (Coope, 1975; Stuart, 1977).

The evidence of permafrost structures in sands overlying interstadial beds at Chelford, Four Ashes and elsewhere indicates that the Chelford Interstadial was succeeded by revertance to a periglacial climatic regime. Apart from the warming implied by the later interstadial sediments at Brimpton (Bryant et al., 1983), cold conditions appear to have persisted into the Middle Devensian (a term conventionally assigned to the period 50–26 ka BP). Insect faunas attributed to the early part of the Middle Devensian indicate a climatic regime of arctic severity (Coope, Morgan & Osborne, 1971; Morgan, 1973; Girling, 1974), though the validity of the radiocarbon determinations on which this interpretation is based is open to question. Lowe & Walker (1984, p. 316) have suggested that at the beginning of the Middle Devensian mean July temperatures were around or below 10 °C, with mean annual temperatures in the range -4 °C to -8 °C. These extreme conditions were interrupted by marked and apparently rapid warming that ushered in a brief interstadial, the Upton Warren Interstadial. At sites in southern and eastern England, contemporaneous beetle assemblages indicate summers marginally warmer than those of the present, with mean July temperatures of perhaps c. 18 °C, but rather more continental conditions (Girling, 1974; Coope & Angus, 1975). Intriguingly though, the associated vegetation appears to have been treeless grassland rather than woodland. This apparent anomaly may reflect continued low winter temperatures, with possible persistence of permafrost (Watson, 1977b; Lockwood, 1979), or slow northwards migration of arboreal species. The age of the Upton Warren event, however, is contentious. Radiocarbon dating suggests a Middle Devensian age of c. 43–40 ka BP (Coope, Shotton & Strachan, 1961; Stringer et al., 1986), but amino acid ratios indicate that the Upton Warren

Interstadial may belong to the Early Devensian, and Bowen et al. (1989) have argued that it may represent Isotope Stage 5a (c. 80 ka BP). The absence of evidence for trees, however, would appear to preclude correlation with the Brimpton Interstadial. On present evidence it seems reasonable to accept that the Upton Warren Interstadial represents Middle Devensian (Isotope Stage 3) climatic warming.

Evidence provided mainly by coleoptera indicates that shortly after c. 40 ka BP conditions of arctic severity had become re-established in Britain (Coope et al., 1971; Coope, 1977b). Coleopteran assemblages relating to this period are dominated by boreal or boreal-montane species that indicate a strongly continental climate characterised by mean annual temperatures as low as -8 °C to -12 °C, with average winter temperatures plunging to between -20 °C and -30 °C and mean annual precipitation reduced to as little as 250–350 mm in lowland England. Contemporaneous pollen assemblages and mammalian faunal remains imply a barren, arid tundra of grasses, sedges and small herbs grazed by reindeer, mammoth, bison, horse and woolly rhinoceros. Many fossiliferous horizons are disrupted by periglacial soil structures, such as ice-wedge casts and involutions. The absence of evidence for glaciation may be attributable to the continentality and aridity of the climate at this time, which itself points to the establishment of a dominantly easterly airflow across Britain and marked reduction of oceanic influence (Lowe & Walker, 1984).

The Late Devensian 1: The Dimlington Stadial

The Late Devensian substage is subdivided into three chronozones: the Dimlington Stadial of c. 26–13 ka BP, the Windermere Interstadial of c. 13–11 ka BP and the Loch Lomond Stadial of c. 11–10 ka BP (Rose, 1985, 1989a). The Windermere Interstadial and Loch Lomond Stadial are often collectively referred to as the Late Devensian Lateglacial, and the Windermere Interstadial has sometimes been called the Lateglacial Interstadial. The dominant feature of the Dimlington Stadial was the growth and subsequent decay of the last British ice sheet, which at its maximum extent covered two-thirds of the present land area of Great Britain (Figure 2.1), the greatest expansion of glacier ice since the Anglian Glacial Stage of the Middle Pleistocene. Lowe & Walker (1984) have suggested that ice-sheet growth in the Late Devensian reflects the re-establishment of an oceanic climatic regime, in contrast to the continental regime that appears to have been dominant in Britain during much of the Early and Middle Devensian. A broad chronology of ice-sheet expansion and contraction during the Dimlington Stadial has been reconstructed from dates relating to sediments buried by or immediately overlying associated glacigenic deposits. Indirect evidence constraining the onset of ice-sheet build up is provided by radiometric dates from a number of sites in Scotland. These relate to organic matter or speleothems that accumulated in a nonglacial environment prior to ice-sheet glaciation, and hence provide limiting ages for glacier expansion. Overall, these dates suggest that much of Scotland remained free of glacier ice until after c. 25 ka BP (Sutherland, 1991b).

A limiting age on the maximum southward expansion of the ice sheet is provided by ^{14}C dates of 18 500 ± 400 yr BP and 18 240 ± 250 yr BP for moss fragments in silts underlying Late

Devensian tills in Holderness near the limit of ice-sheet advance (Penny, Coope & Catt, 1969). These dates are consistent with a thermoluminescence date of 17.5 ± 1.6 ka BP obtained on the loess component of a solifluction deposit that underlies Late Devensian till at Eppleworth near Hull (Wintle & Catt, 1985b), and also with a ^{14}C date of 18 000 +1400/-1200 yr BP for a mammoth bone in a cave sealed by Late Devensian till in the Vale of Clwyd (Rowlands, 1971). Collectively, these dates suggest that the last ice sheet reached its southern limit around 18–17 ka BP. It is unlikely, however, that the limits of the ice sheet are synchronous. There is strong evidence that the culmination of ice-sheet expansion in southern Scotland and England occurred after (possibly long after) ice nourished in the Scottish Highlands reached its maximum extent (Sutherland, 1984, 1991b). This asynchronous behaviour may reflect progressive southwards migration of a zone of increased precipitation associated with movement of the North Atlantic oceanic polar front (Sissons, 1981b).

Although numerous radiocarbon age determinations have been made on organic sediments immediately overlying glacigenic deposits of Dimlington Stadial age, many of these are liable to error (Sutherland, 1980) and at present it is impossible to reconstruct in detail the time-trangressive retreat of the last ice sheet on the basis of radiocarbon dates alone. However, calculations based on the gradients of raised shorelines have led Sutherland (1991b) to suggest that east-central Scotland became ice-free between 16 ka BP and 14 ka BP, and that parts of the Inner Hebrides began to emerge from the ice around 14.5 ka BP. A number of radiocarbon dates suggest that by c. 13 ka BP the remnants of the last ice sheet were confined to the Western Highlands of Scotland (Sutherland, 1991b). This deglacial chronology implies that much of the retreat of the last ice sheet occurred under a severe cold climate regime that reflected the persistence of polar waters around the coasts of Britain until c. 13.5 ka BP or later (Ruddiman & McIntyre, 1973, 1981).

Reconstructions of former directions of ice-sheet movement based on till fabrics, striae, ice-moulded rock, drumlins, carry of erratics and other evidence suggests that the form of the last ice-sheet was complex. Major centres of ice dispersal developed in the mountains of the Northern Highlands and Western Grampian Highlands of Scotland, the Southern Uplands, the Lake District, Snowdonia and south Wales. Minor ice caps or domes formed over the SE Grampians, the Cheviots, the Cairngorms and the northern Pennines, and independent ice caps on the Hebridean islands of Mull and Skye diverted the westward movement of ice from the Scottish mainland. Independent ice caps also developed at this time on the Outer Hebrides and probably Shetland (Clapperton, 1970; Bowen et al., 1986; Thorp, 1987; Catt, 1991a; Peacock, 1991; Sutherland, 1991a,b). The question as to whether higher mountain summits remained above the maximum level of the Late Devensian ice sheet remains contentious. Theoretical models of ice-sheet thickness have suggested that this is unlikely, except perhaps in peripheral areas (Boulton et al., 1977, 1985; Gordon, 1979; Boulton, Peacock & Sutherland, 1991). In NW Scotland, however, there is evidence that a high-level 'periglacial trimline' may delimit the upper surface of the last ice sheet (Ballantyne, Sutherland & Reed, 1987; Reed, 1988; Ballantyne, 1990; Chapter 9), though in the Western Grampians all mountain

summits appear to have been submerged under ice (Thorp, 1987). The status of potential nunataks in other more peripheral mountain areas (e.g. Snowdonia and the Outer Hebrides) remains uncertain.

The southern limits of the Late Devensian ice sheet across England and Wales (Figure 2.1) have been determined by lithostratigraphy, the extent of 'constructional' glacial topography and by the distribution of landforms and sediment sequences indicative of areas that escaped glaciation. In Pembrokeshire and the Gower Peninsula the glacial limit has been defined on the basis of the presence or absence of till overlying raised beach deposits of Ipswichian age (Bowen, 1973a,b, 1974). Farther east, in the Welsh Borderland and Herefordshire, the position of the ice limit has been inferred mainly from morphological evidence, such as the degree of drift dissection (Luckman, 1970). To the west of the Pennines, Late Devensian till has been traced southwards into Shropshire as far as the Wolverhampton area, where delimitation of the ice-sheet maximum has been based on stratigraphic evidence, erratic carry and a general northwards thickening of drift (Morgan, 1973; Worsley, 1991b). Around the southern Pennines the margin is not clearly defined (Burek, 1991) and has been delimited mainly on the extent of till units and 'constructional' drift. Stratigraphic evidence broadly defines the limits of a former outlet glacier in the Vale of York (Gaunt, 1976a; Catt, 1991a,b). The North York Moors remained above the level of the ice sheet, but a low-gradient lobe of ice extended down the coast of Holderness and Lincolnshire and encroached on north Norfolk (Catt & Penny, 1966; Madgett & Catt, 1978; Catt, 1991a,b). The apparently anomalous configuration of this lobe prompted Boulton et al. (1977) to suggest that it may have represented a surge of ice over deformable sediments. South and east of the Late Devensian ice limit (Figure 2.1) the land surface was exposed to severe periglacial conditions throughout the Dimlington Stadial; to the north and west of this line severe periglacial conditions pertained during both ice-sheet advance and retreat.

The question as to whether parts of the Scottish mainland escaped glaciation during the Dimlington Stadial remains open to debate. There is wide consensus that the eastern limit of the ice sheet lies 50–80 km off the coasts of Fife and Angus at a broad belt of hummocky topography known as the Wee Bankie Moraine (Thomson & Eden, 1977; Sutherland, 1984, 1991b; Cameron et al., 1987), but much less agreement concerning the eastern limit of the ice sheet farther north. Sutherland (1984) and Bowen et al. (1986) suggested that the ice sheet terminated in the inner Moray Firth, leaving large parts of Caithness and Buchan ice-free. Hall & Bent (1990), however, have argued that ice from the Moray Firth extended much farther offshore to a submarine moraine complex, the Bosies Bank Moraine (Figure 2.1), and that Buchan, Caithness and the Orkney Islands were all completely overridden by the mainland ice sheet. To the north and west of Scotland the offshore limits of the ice sheet are uncertain, but Sutherland & Walker (1984) have demonstrated that a small area in northern Lewis escaped glaciation during the Dimlington Stadial. The independent ice cap that developed on the Outer Hebrides was probably confluent with the mainland ice sheet to the east (Ballantyne, 1990) but failed to reach St Kilda, 60 km west of the Outer Hebrides on the edge of the continental shelf (Sutherland et al., 1984).

Controversy also surrounds the status of readvances that purportedly interrupted overall retreat of the Late Devensian ice sheet. Increased understanding of the complex nature of glacial depositional environments has led to rejection of several proposed readvance episodes in England and Wales (Worsley, 1985; Bowen *et al.*, 1986) but two continue to attract some support. The first is a putative 'Gwynedd Readvance' that was interpreted by Whittow & Ball (1970) in terms of a south-westwards readvance of Irish Sea ice across Anglesey as far as the north coast of the Lleyn Penisula in NW Wales. McCarroll (1991), however, has demonstrated that the evidence in the latter area is consistent with only a single glacier advance and retreat, and thus casts doubt on the reality of this readvance. Even more hotly disputed is the evidence favouring a 'Scottish Readvance' that may have affected north and west Cumbria (Figure 2.1). The concept of a 'Scottish Readvance' was vigorously attacked by Thomas (1985), who described it as 'devoid of stratigraphic or chronological foundation', but stoutly defended by Huddart (1991) on the basis of a detailed analysis of the stratigraphy and sediment associations of the Cumbrian lowlands. Huddart's evidence is suggestive but not unequivocal, and at present it seems wise to adopt an agnostic stance regarding this event. In Scotland, too, a number of proposed readvances of the last ice sheet have been discarded as new evidence accumulated (Ballantyne & Gray, 1984). In the NW Highlands, however, a chain of moraine ridges demarcates the limits of a readvance, the Wester Ross Readvance, that has been tentatively dated to 13.5–13.0 ka BP (Robinson & Ballantyne, 1979; Sissons & Dawson, 1981; Ballantyne *et al.*, 1987; Figure 2.1). Abrupt drops in the marine limit at sites on the west coast of Scotland also provide evidence for stillstands or minor readvances of the ice sheet, and in SW Scotland one such event has been dated to *c.* 13 ka BP (Sutherland, 1991b). A minor readvance of the ice sheet across Caithness has been proposed on stratigraphic grounds (Hall & Whittington, 1989; Hall & Bent, 1990). The causes of these events are uncertain, but Sissons (1981b) has made the interesting suggestion that some at least may reflect reinvigoration of the wasting ice sheet by an increase in snowfall as the average position of the atmospheric polar front moved northwards across the Highlands at the end of the Dimlington Stadial.

The Late Devensian 2: The Windermere Interstadial

Rapid northwards migration of the North Atlantic oceanic polar front (and hence the average latitude of the associated atmospheric front) is also widely recognised as the main cause of rapid climatic warming in Britain at the end of the Dimlington Stadial. According to Ruddiman & McIntyre (1981), the oceanic polar front stood at the latitude of Portugal or northern Spain during much of the stadial, implying that Britain was surrounded by cold polar waters and probably pack ice. Such conditions help to explain the relatively arid periglacial environment of Britain during this period. By 13 ka BP, or possibly somewhat later (Bard *et al.*, 1987), the oceanic polar front stood well to the north of Britain, possibly at the latitude of Iceland or thereabouts. In consequence, Britain experienced a return to temperate conditions during the period known as the Windermere Interstadial, a term conventionally ascribed to the period 13–11 ka BP.

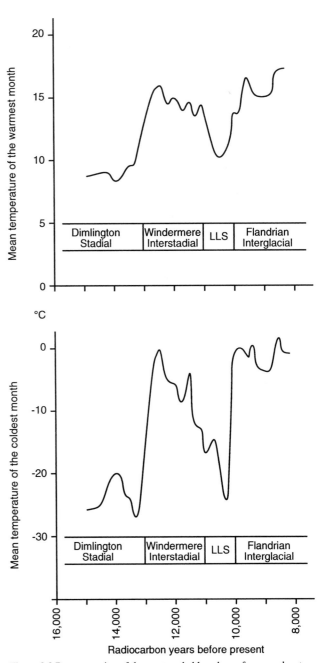

Figure 2.2 Reconstruction of the most probable values of mean palaeotemperatures for the warmest (top) and coldest (bottom) months of the year, based on radiocarbon-dated coleoptera assemblages, for the period 15–8 ka BP. LLS = Loch Lomond Stadial. After Atkinson *et al.* (1987).

Our knowledge of changing environmental conditions during the Late Devensian Lateglacial is much more detailed than for earlier periods of stadial–interstadial transition. Much of our information concerning the changing climate of this period comes from studies of insect (particularly coleoptera) assemblages in Lateglacial sediments at sites in England, Wales and southern Scotland. These have been employed in the construction of a remarkably detailed palaeotemperature record for the period 14–8 ka BP (Atkinson, Briffa & Coope, 1987; Figure 2.2). This record suggests that during the final millennium of the Dimlington Stadial, Britain continued to experience pleniglacial conditions under which the average temperature of the warmest month was no higher than 10 °C and that of the coldest month plunged to between -20 °C and -25 °C. Within the period 13.3–12.5 ka BP, and probably within the period 13.0–12.7 ka BP, there was an astonishingly rapid warming of the British climate: summer temperatures increased by *c.* 7–8 °C, and winter temperatures by *c.* 25 °C. Atkinson *et al.* (1987) estimated that by *c.* 12.5 ka BP the climate of England and Wales was as warm as at present, though more continental in character, with a mean monthly temperature range of perhaps 0–17 °C. In southern Scotland the temperature of the warmest month at this time was probably of the order of 14–15 °C, though it is likely that rather cooler conditions prevailed farther north (Bishop & Coope, 1977). The coleopteran evidence suggests that the thermal zenith of *c.* 12.5 ka BP was succeeded by a period of gradual then more accelerated temperature decline. At first cooling seems to have affected mainly winter temperatures, but pronounced summer cooling was evident from about 11.4 ka BP onwards. This marks the transition to the final short-lived episode of extreme cold in Great Britain, the Loch Lomond Stadial, a period described in more detail below.

Palaeobotanical studies have yielded a wealth of information about the vegetation cover of Britain during the Late Devensian Lateglacial, and collectively suggest strong regional differentiation of vegetation patterns over this period (Pennington, 1977a; Walker, 1984). Early in the Windermere Interstadial the vegetation cover was largely dominated by grasses, sedges and herbs that thrive on open ground and unstable soils, and which represent an early stage in the interstadial succession. Shortly after *c.* 13 ka BP such open-habitat conditions were widely replaced by a shrub vegetation dominated by juniper, willow and, in the west, crowberry (*Empetrum*). This, in turn, was succeeded by the widespread development of birch woodland over much of Britain, though *Empetrum*-dominated heath continued to predominate in central and northern Scotland. In some Lateglacial pollen profiles, however, there is evidence for a short revertance phase during the Windermere Interstadial, characterised by an increase in minerogenic sedimentation in lake basins, a decline in the frequency of birch, juniper and *Empetrum* pollen and a concomitant increase in the representation of open-taxa pollen (e.g. Pennington, 1975, 1977a). The status of this episode is debatable: it is evident only at a limited number of sites, and there is only very slight evidence for a similar fluctuation in the coleopteran record (Coope, 1975).

The Late Devensian 3: The Loch Lomond Stadial

The final period of extreme cold to affect Great Britain was the Loch Lomond Stadial, conventionally assigned to the period 11–10 ka BP. This dramatic conclusion to the Devensian was ushered in by renewed southwards movement of the oceanic polar front, which according to Ruddiman, Sancetta & McIntyre (1977) reached its southernmost position off the coast of SW Ireland around 10.2 ka BP. The cause of this return of cold water to the seas around the British Isles is uncertain, but could be interpreted as a transitory phenomenon caused by rapid meltwater influx into the North Atlantic during the disintegration of the Laurentide and Scandinavian ice sheets (Ruddiman & McIntyre, 1981; Broeker, Peteet & Rind, 1985). Whatever the cause of renewed cooling at this time, the effects on the British landscape

were remarkable. The most obvious consequence was the recrudescence of glacier ice in upland areas, an event referred to as the Loch Lomond Readvance. The limits of this readvance are in many areas defined by end and lateral moraines, drift limits and periglacial trimlines that mark the upslope transition from glacial drift and ice-scoured bedrock to terrain that bears the imprint of contemporaneous frost action (Gray & Coxon, 1991). These features have enabled geomorphologists not only to delimit the lateral extent of the Loch Lomond Readvance with great accuracy in most areas, but also to produce three-dimensional reconstructions of individual ice masses.

During the Loch Lomond Stadial, by far the largest ice mass developed over Western Grampians and the NW Highlands of Scotland. At its maximum development, this great icefield extended southwards from Glen Torridon in NW Scotland to the southern end of Loch Lomond and eastwards from the tidewater lochs of the NW Highlands to the east end of Loch Tay (Figure 2.1). A detailed reconstruction of the dimensions of the icefield in part of the Western Grampians suggests that the ice-shed reached an altitude of around 700 m over Rannoch Moor (Thorp, 1986), and a similar altitude was achieved by the ice mass that occupied much of the NW Highlands. The configuration of this ice mass was strongly influenced by topography, and in most places it took the form of a complex network of transecting glaciers overlooked by nunataks (Figure 2.3). On mountains peripheral to the main ice mass a number of independent icefields and ice caps also developed, for example in the SE Grampians, on the Gaick plateau and on the mountains of Mull and Skye (Gray & Brooks, 1972; Sissons, 1972, 1974b; Ballantyne, 1989a). Much more numerous, however, were the relatively small cirque and valley glaciers that developed at this time on most areas of high ground in Britain from northernmost Scotland to the Brecon Beacons in south Wales. Over 200 Loch Lomond Readvance glaciers have been mapped in the Scottish Highlands and Inner Hebrides (Sutherland, 1984), and the limits of many others have been identified in more southerly uplands, such as the English Lake District (Sissons, 1980a), Snowdonia (Gray, 1982), the Southern Uplands of Scotland (Cornish, 1981), the western Pennines (Mitchell, 1991), Cader Idris and the Brecon Beacons. The upper limits of glaciation during the Loch Lomond Stadial have proved of enormous importance for differentiating upland periglacial phenomena that formed under stadial conditions prior to *c.* 10 ka BP from those that subsequently developed under the milder conditions of the Flandrian (present) Interglacial. Those periglacial features known to occur well within the limits of the Loch Lomond Readvance (such as shallow solifluction sheets and lobes, ploughing boulders, small-scale sorted patterned ground and niveo-aeolian sand deposits) can only have developed during the Flandrian after the decay of the last glaciers. Conversely, periglacial features that occur immediately outside the limits of the readvance, but never within these limits, can reasonably be assumed to have developed under cold stadial conditions prior to *c.* 10 ka BP (Ballantyne, 1984).

Radiocarbon determinations made on marine shells in sediments transported or over-ridden by glacier ice in SW Scotland have yielded a maximal age of *c.* 10 900 yr BP for the culmination of the Loch Lomond Readvance (Sutherland, 1984; Gray & Coxon, 1991). A younger limiting age is provided by a date of 10 560 ± 160 yr BP obtained on plant detritus buried under till

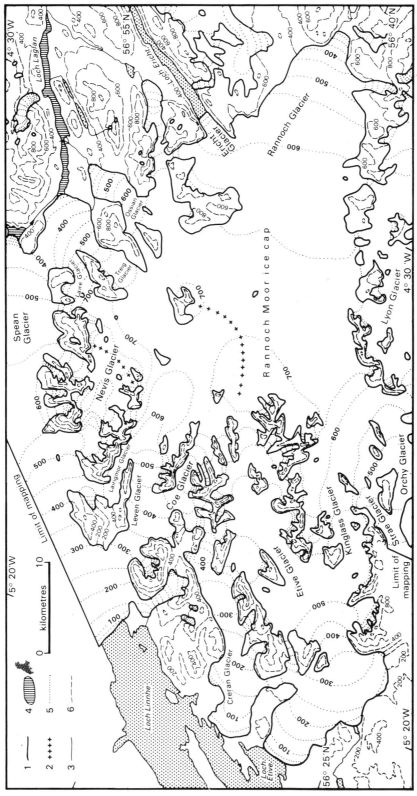

Figure 2.3 Three-dimensional reconstruction of the icefield that occupied the western Grampian Highlands during the Loch Lomond Stadial, as reconstructed by Thorp (1986). (1) Glacier limit. (2) Ice shed. (3) Overflow from glacier-dammed lake. (4) Glacier-dammed lake. (5) Glacier surface contours at 50 m intervals. (6) Land surface contours at 200 m intervals.

inside the limit of the glacier that occupied Loch Lomond. Taking into account the stratigraphic context, this date suggests that this glacier did not achieve its maximum extent until sometime after *c.* 10.5 ka BP (Rose, Lowe & Switsur, 1988). Similarly, radiocarbon ages obtained for the shells of molluscs collected from a glaciomarine bed immediately inside the limit of the Loch Lomond Readvance at the mouth of Loch Creran suggest that the former Creran Glacier reached its maximum extent between 10.5 ka BP and 10.0 ka BP (Peacock *et al.,* 1989). Attempts to reconstruct the time-transgressive retreat of the large icefield in the Western Grampians by dating the basal organic sediments within enclosed lake basins at increasing distances from the margin of the icefield have been frustrated by the uncertainties and errors associated with such age determinations (Lowe & Walker, 1980; Sutherland, 1980). Gray & Coxon (1991, p. 97) have suggested that the large number of radiocarbon determinations made on basal sediments within the area occupied by the readvance 'suggest that widespread stagnation of the ice masses had occurred by 10 200 yr BP'. However, given the large latitudinal range of the Loch Lomond Stadial glaciers, the enormous variation in the sizes of these glaciers and the movement of climatic patterns associated with migration of the oceanic polar front, it seems likely that the initiation, culmination and termination of the Loch Lomond Readvance exhibited considerable spatial diachroneity.

The altitudinal distribution of Loch Lomond Readvance glaciers, and in particular that of reconstructed equilibrium line altitudes (ELAs), has been employed by several researchers in attempts to reconstruct aspects of stadial palaeoclimate. Average 'regional' reconstructed ELAs rise gradually southwards from 495 m for the western Southern Uplands of Scotland to 540 m for the English Lake District and 600 m for Snowdonia (Sissons, 1980a; Cornish, 1981; Gray, 1982). Across the Scottish Highlands and Inner Hebrides, however, a more complex pattern is evident, with a general eastwards rise in reconstructed ELAs from around 300 m in the Inner Hebrides and SW Highlands to 1000 m in the Cairngorms, and a steep northwards rise parallel to the Highland Boundary in the area of the SE Grampians (Figure 2.4). These trends imply marked aridity in the area of the Cairngorms during the stadial, and a very steep decline in precipitation both eastwards and north from the Highland Boundary. This pattern was interpreted by Sissons (1979c, 1980b) in terms of a predominance of southerly snow-bearing winds associated with the eastwards passage of warm or occluded fronts across northern Britain, with depressions following more southerly tracks than at present. Analysis of reconstructed ELAs has also permitted calculation of summer temperatures for the period of glacier growth. These indicate a slight northwards and eastwards decline in mean July stadial sea-level temperatures from 7.5 °C for the English Lake District and 7.0 °C for the Western Grampians (Sissons, 1980a,b) to 6.0 °C for the SE Grampians and Isle of Skye (Sissons & Sutherland, 1976; Ballantyne, 1989a). These estimates are encouragingly consistent with those made on the basis of contemporaneous coleoptera (Coope, 1977b). Atkinson *et al.* (1987) inferred from coleopteran assemblages that at *c.* 10.5 ka BP the mean temperatures of the warmest and coldest months stood at around 10 °C and -17 °C respectively in England and Wales, and that mean annual temperatures declined to perhaps -5 °C during the coldest part of the stadial.

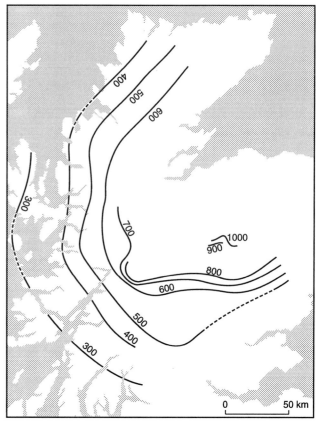

Figure 2.4 Generalised regional pattern of the ELAs of Loch Lomond Stadial glaciers at their maximum extent in the Highlands of Scotland. Based on a map by Sissons (1980b), with modifications.

From a palaeoclimatic standpoint, the nature of deglaciation is of interest. Many upland valleys that supported glaciers during the Loch Lomond Stadial are occupied by 'hummocky moraines', apparently chaotic assemblages of glacially-deposited mounds and hummocks. Interpretation of these features as the product of *in situ* ice stagnation encouraged a belief that glacier advance was abruptly terminated by rapid climatic warming (e.g. Sissons, 1979c, Gray, 1982). It is now accepted, however, that many areas of 'hummocky moraines' reflect sediment deposition at the margins of retreating but still active glaciers (Eyles, 1983; Benn, 1990, 1991; Bennett, 1990). The prolonged survival of active glaciers after the culmination of the Loch Lomond Readvance suggests that glacier retreat may have begun as a result of snowfall diminution under cold conditions, rather than in response to warming (Benn, Lowe & Walker, 1992). If so, permafrost and related periglacial phenomena may have developed immediately inside the readvance limits prior to the rapid thermal amelioration that terminated the stadial (cf. Atkinson *et al.,* 1987).

The vegetation of the Loch Lomond Stadial reflected the prevailing severe climate. Pollen spectra from numerous sites throughout Britain indicate a landscape dominated by grasses, sedges and other herbaceous plants characteristic of bare or disturbed soils (Pennington, 1977b). Particularly well represented in the north and west were *Rumex* and *Artemisia*. Herbaceous assemblages dominated by the latter have been found at sites in eastern Scotland, and have been interpreted as indicative of regional arid-

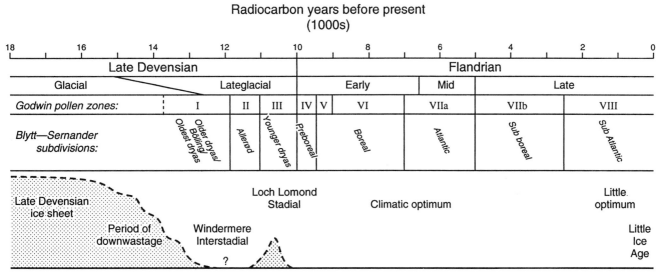

Figure 2.5 Chronological subdivision of the last 18 000 years.

ity (Birks & Mathewes, 1978; Tipping, 1985), an inference consistent with conclusions drawn from the altitudinal trend of contemporaneous equilibrium line altitudes (Sissons, 1979c, 1980b).

The Flandrian Interglacial

The Flandrian Interglacial is conventionally recognised as covering the period 10 ka BP to the present, even though the climatic warming that terminated the Loch Lomond Stadial may have begun rather earlier. The Flandrian stage is thus coincident with the Holocene Epoch. Climatic variation in NW Europe during the Late Devensian Lateglacial and Flandrian has been zoned according to a scheme devised at the beginning of this century by the Scandinavian botanists Blytt and Sernander on the basis of peat stratigraphy, and their zonation corresponds closely to pollen zones established for the British Isles by Godwin (1956). Although the Blytt–Sernander subdivisions are now recognised to be oversimplistic, their terminology (Figure 2.5) is still widely employed and thus forms a useful framework for discussion of Flandrian environmental changes in Britain. In the account below, attention is focused on the effects of such changes in upland Britain, as only at high altitudes (above 500–600 m) is there evidence for periglacial activity under the interglacial conditions of the Holocene.

The change to warmer conditions at the end of the Loch Lomond Stadial was rapid, perhaps as much as 1 °C per decade, and by c. 9.8 ka BP the climate of England appears to have been as warm as at the peak of the Windermere Interstadial (Coope, 1977b; Atkinson et al., 1987; Figure 2.2). This sudden transformation coincided with the return of the warm waters of the North Atlantic Drift to the coasts of Britain. Off the Scottish coast, water temperatures may have risen from near zero to slightly below present-day values within a few decades (Peacock & Harkness, 1990). The effects on the landscape were equally dramatic. Minerogenic sedimentation in upland lakes and depressions was abruptly replaced by organic accumulation as vegetation responded rapidly to climatic amelioration, and by 9.8–9.7 ka BP juniper scrub had replaced the open tundra

vegetation of the preceding stadial throughout most of the Scottish Highlands. Temperatures continued to rise throughout the Boreal period, reaching a maximum in early Atlantic times when mean annual temperatures in England and Wales were 1.3–1.6 °C higher than at present (Lamb, Lewis & Woodroffe, 1966). One consequence of the rapid climatic change that occurred at the Lateglacial–Flandrian transition was the effective cessation of all periglacial activity in lowland Britain; thenceforth, conditions favourable for periglacial activity were restricted to high ground.

After the climatic optimum of the early Atlantic period the British climate experienced several phases of deterioration and amelioration, but at no time in the Flandrian did the amount of change approach that of the Late Devensian Lateglacial. During the final 500 years of the Atlantic period (c. 5.5–5.0 ka BP) there was a revertance to cooler conditions followed by partial recovery and a second (though less pronounced) optimum within the period 4.5–3.5 ka BP. The later part of the Sub-Boreal period was characterised by renewed thermal deterioration, and by the beginning of the Sub-Atlantic chronozone around 2500 yr BP mean annual temperatures in England and Wales were about 2 °C below those of the Atlantic climatic optimum. Throughout NW Europe there was a pronounced increase in wetness and storminess at this time, and a climate of mild winters and cool summers prevailed (Lamb, 1977). The final two millennia of the Sub-Atlantic period witnessed a series of relatively short-term fluctuations in climate, though variations in mean annual temperatures in England and Wales probably did not exceed 1.0 °C in amplitude. Temperatures reached a peak around AD 400 and during the 'little optimum' of AD 1100–1300, then declined to a minimum during the seventeenth and eigthteenth centuries, during a period widely if misleadingly known as the Little Ice Age. Superimposed on the above broad pattern of temperature change were significant variations in precipitation. Deuterium isotope analyses of pine macrofossils in the Cairngorm Mountains suggest that particularly wet conditions prevailed in the Scottish Highlands at around 7300 yr BP, 6200–5800 yr BP, 4200–3940 yr BP and 3300 yr BP

(Dubois & Ferguson, 1985). These episodes have been linked to periods of pine woodland decline and concomitant peat accumulation in the Western Grampians (Bridge, Haggart & Lowe, 1990).

During the Early Flandrian a succession of woodland types developed in upland Britain, and in most areas the treeline achieved its maximum altitude during the Atlantic climatic optimum. In the Cairngorms, Pears (1975a) found tree stumps embedded within blanket peat at nearly 800 m, but the maximum altitudes reached by trees on the more exposed westerly uplands of Britain tended to be at least 200 m lower. The decline of some upland pine forests may have been initiated as early as 7000 yr BP (Pears, 1975a; Birks, 1977) as a consequence of deterioration in the base status of soils under the relatively wet conditions of the Atlantic period, a change that was accompanied by the initial formation of blanket peat in western uplands. Although some revival of pine colonisation occurred during subsequent drier phases (Bridge *et al.,* 1990), the onset of cooler and wetter conditions at the beginning of the Sub-Atlantic chronozone was marked by further extensive peat formation. Peat growth rates at intermediate altitudes in northern England and Scotland were highly variable. Around 600 m in the Pennines, blanket peat reaches a depth of up to 5 m, the result of relatively rapid growth since the beginning of Atlantic times, but in the Cairngorms Pears (1975b) found peat depths of only 0.6–1.5 m at altitudes between 600 m and 800 m. The widespread occurrence of blanket peat at such elevations is important in that peat not only insulates the underlying regolith from freeze–thaw cycles, thus preventing the development of periglacial forms, but also obscures relict features that developed during the Late Devensian Lateglacial.

Above the limits of forest colonisation and peat development, vegetation succession followed a very different course (Pennington, 1974). The climax vegetation in this higher zone was high montane grassland, which Pearsall (1968) considered a natural development from *Rhacomitrium* heath. This grassland consisted of species associated with Lateglacial pollen spectra in upland Britain, such as *Festuca ovina, Deschampsia flexuosa, Agrostis tenuis, Salix herbacea* and species of *Vaccinium* and *Empetrum*, suggesting an upwards migration of such plants in the Early Flandrian, when climatic amelioration allowed their replacement by more successful competitors on low ground. In general, it was only on the very highest ground that true arctic-alpine communities survived Early Flandrian climatic warming. The Flandrian stage therefore saw the development of the present three vegetational zones on high ground: a lowermost zone of peat, often containing the remains of vanished forests; an intermediate zone of montane grassland; and an upper zone of more open vegetation including arctic-alpine species.

It is pertinent to consider in some detail the climate of the Little Ice Age of the sixteenth to nineteenth centuries AD, as it is arguable that this period witnessed a renaissance (or at least an intensification) of periglacial activity on high ground. Mean annual temperatures in England and Wales were at least 0.6 °C lower than recent averages, and the decline in winter temperatures was even more pronounced (Lamb, 1981). Polar water dominated the North Atlantic as far south as latitude 62 °N between *c.* 1600 and 1830 (Lamb, 1979), so it is likely that mean temperatures in Scotland fell even further below present-

day levels, bringing mean annual temperatures on the very highest mountains to below 0 °C. A probable consequence of this depression in temperature was deeper ground freezing, though the amplitude of temperature depression was probably insufficient to re-establish permafrost, even on the highest ground. Another more visible consequence was the widespread establishment of perennial snowbeds on particularly high Scottish mountains such as Ben Nevis and the Cairngorms (Manley, 1949, 1971a; Sugden, 1971). Even Ben Wyvis (1046 m) retained perennial snowpatches. In 1770 Thomas Pennant wrote of it that 'snow lies in the form of a glaciere throughout the year', and in a gazetteer compiled a century later (Wilson, 1873) it is recorded that Ben Wyvis had not been snow-free within living memory, apart from September 1826 after an exceptionally warm summer. The survival of snow on Ben Wyvis during the second half of the nineteenth century was symptomatic of increased snowfall that accompanied thermal improvement after the coldest years of the Little Ice Age. These occurred in the seventeenth and eighteenth centuries, when precipitation in England and Wales was rather less than at present (Lamb *et al.,* 1966; Thom & Ledger, 1976). Conversely, reported snowfalls were most frequent towards the end of the eighteenth century and later, at least in lowland Scotland (Pearson, 1976). There is, however, no convincing evidence for the re-establishment of glaciers in Britain during the Little Ice Age. Sugden (1977a) has presented lichenometric data favouring a Little Ice Age origin for certain high-level corrie moraines in the Cairngorms, but this proposal has been invalidated by radiocarbon dating and biostratigraphic evidence, both of which indicate a Loch Lomond Stadial age for these features (Rapson, 1985).

Although the relatively cold winters of the Little Ice Age almost certainly favoured an intensification of periglacial activity on high ground in Britain, a possibly more significant aspect of the climate of this time was an increase in the frequency of violent storms. The approach of cold polar waters to the northern coasts of Britain resulted in enhanced oceanic (and hence atmospheric) thermal gradients, thus causing a marked increase in both the periodicity and severity of storm events (Lamb, 1977, 1979, 1984, 1985; Whittington, 1985). It has been suggested that particularly violent storms during the Little Ice Age may have induced widespread slope failure and stripped the protective vegetation cover from exposed plateaux, thereby exposing the underlying regolith to frost action and other forms of periglacial activity (Ballantyne, 1991b).

Present-day environmental conditions on British mountains

Present climatic conditions on British mountains lie between the extremes represented by the climatic optimum of Atlantic times and the Little Ice Age of the sixteenth to nineteenth centuries AD, and hence may be regarded as broadly 'average' for the Flandrian stage. The dominant feature of the present mountain climate of Great Britain is its strongly maritime character, which reflects both proximity to the sea and the moderating influence of the North Atlantic Drift. The principal consequences of this maritime influence are steep lapse rates, high humidity, strong winds and limited seasonal variations in cloudiness, precipitation and temperature. There is, however,

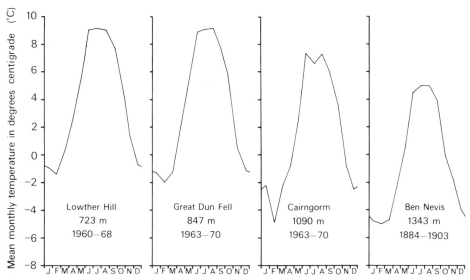

Figure 2.6 Mean monthly screen temperatures at four stations on British mountains.

an unfortunate dearth of data on the upland climate of Britain. The most complete record for high ground consists of the regular meteorological observations made on the summit of Britain's highest mountain (Ben Nevis, 1343 m) over a 20 year period between 1884 and 1903 (McConnell, 1988). A number of shorter records have been obtained for high ground in Wales, the Pennines and the Southern Uplands, however, and an automatic weather station on the summit of Cairngorm (1245 m) has been recording data since 1977. Summaries of the available data are given by Taylor (1976) and Barry (1992), and a lively account of Scotland's mountain weather has been written by Moran (1988).

One striking feature of the the climate of British mountains is the absence of extreme cold. The absolute minimum air temperatures recorded on the summits of Ben Nevis and Cairngorm are -17.3 °C and -16.5 °C respectively, and data from various mountains suggest that screen temperatures below about -10 °C are infrequent. The mean annual temperature on Ben Nevis is fractionally above 0 °C (Manley, 1971b), and mean monthly air temperatures are below zero for only four months of the year at altitudes below 800–900 m and for no more than six months on the highest summits (Figure 2.6). The limited data available suggest that there are on average 30–40 ambient (air) freezing cycles per year, the majority probably related to the alternation of warm and cold airmasses during periods of cyclonic activity rather than diurnal heating and nocturnal cooling. Although permafrost is absent, at higher altitudes there may be in some years an annual ground freezing cycle of 2–6 months duration, though the ground may thaw completely at any time during the winter in response to incursions of warm maritime air. The maximum recorded depth of freezing is c. 0.5 m (Ragg & Bibby, 1966), though freezing to greater depths is theoretically possible (Halstead, 1974). Diurnal freeze–thaw cycles, however, penetrate no more than 50–60 mm into the ground and are ineffective in causing freezing under all but the shallowest cover of snow. Snowcover also mitigates the effect of longer ('cyclonic') freezing cycles, and available data suggest that on average only four or five freezing cycles per year penetrate to

depths exceeding 100 mm at 650–750 m altitude in the Southern Uplands of Scotland (Halstead, 1974; Ballantyne, 1981).

Westerly airstreams and frontal structures rise sharply on meeting the mountain barrier of western Britain, giving rise to increased cloud cover and precipitation. On Ben Nevis, for example, the mean cloudiness exceeds 80%, and during the winter months only about 10% of the possible sunshine hours are recorded. Precipitation gradients are steep, and on westerly mountains such as those of Snowdonia and Wester Ross may exceed 4.5 mm m^{-1} yr^{-1} (Unwin, 1969; Ballantyne 1983), though the average precipitation gradient for Britain is approximately 2.4 mm m^{-1} yr^{-1} (Bleasdale & Chan, 1972; Burt, 1980). In consequence, all but the most easterly uplands receive over 2000 mm yr^{-1} mean annual precipitation, with some western mountains receiving more that 4000 mm yr^{-1}. Cyclonic and (less frequently) summer convectional rainstorms with intensities exceeding 50 mm in 24 hours are common. Duration of winter snow-lie increases approximately linearly with altitude (Figure 2.7). In the Scottish Highlands average snow-lie (>50% cover) generally exceeds 100 days per year at 600 m and ranges from about 150 to 180 days per year at 900 m. Farther south, snow-lie is rather less prolonged: the equivalent figures are 105 days per year for the summit area (880 m) of Cross Fell in the Pennines and about 70 days per year for 900 m on Snowdon (Manley, 1971b). Snowpatches in favoured locations on high Scottish mountains may survive for several years before melting. The most persistent examples occur at 1075 m in a deep corrie on Braeriach, in the Cairngorms, and at 1160 m in Observatory Gully on Ben Nevis.

Perhaps the most notable feature of the upland climate of Great Britain is the strength and persistence of the wind, which reflects concentration and acceleration of airflow as it passes over mountain barriers. Birse & Robertson (1970) estimated that the mean wind velocities on mountains throughout Scotland generally exceed about 30 km h^{-1} at altitudes above 800 m, and Manley (1952) calculated that the average wind speed on Great Dun Fell in the Pennines is about 36 km h^{-1},

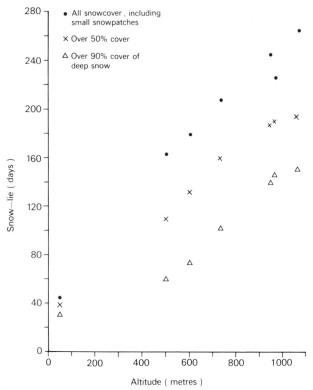

Figure 2.7 Snowcover gradients on An Teallach (NW Scotland), based on data for the period 1976–79.

whilst that on Ben Nevis is about 55 km h^{-1}. Such averages, however, conceal the ferocity of some mountain storms. Measurements on the summit of Cairngorm have shown that during the period 1979–87 the maximum 24-hour average windspeed recorded each year ranged from 93 to 146 km h^{-1}. Even more spectacular are the figures for the strongest gust recorded annually over the same period, which range from 177 to 275 km h^{-1}. The severity of the present climatic environment on British mountains is thus a reflection not of extreme cold, frequent freezing cycles or deeply-frozen ground, but of high humidity, intense precipitation, prolonged snowcover and strong blustery winds. Climatically, the mountains of Great Britain presently experience a distinctive *maritime periglacial* regime that is markedly different from those of arctic, subarctic and alpine areas where there are much greater extremes of cold.

The bracing climate of British mountains exercises a rigorous control on the nature of upland vegetation communities. Trees and shrubs of normal stature cannot grow in exposed areas, and the 'natural' or potential tree-line has been estimated to lie at around 700 m in the Cairngorms and to decline irregularly westwards and northwards. Plants that have colonised exposed summits and ridges are adjusted to survival under conditions of high winds, heavy snowfall, prolonged snow-lie and a growing season of 3–4 months or less. Tansley (1968) estimated that above 900 m roughly half of the cover consists of hemicryptophytes (half-buried plants with persistent buds) and a further 25% consists of chamoiphytes (dwarf woody plants) such as *Vaccinium* or *Empetrum* species. Above the areas of widespread blanket peat cover, however, the nature of the vegetation in the montane grassland and arctic-alpine communities differs from area to area,

depending amongst other factors on the base status of the soil, itself inversely related to the acidity of the underlying rock. On the relatively base-rich soils that have developed on metamorphosed limestone in the southern Highlands, for example, there is considerable species diversity. *Festuca ovina* and *Alchemilla alpina* are common on gentle slopes, and an association of *Nardus stricta* together with various lichens and xerophilous mosses occupies exposed cols and plateaux. Amongst the higher plants, *Silene acaulis*, *Loiseleuria procumbens* and *Carex bigelowii* are common, together with shrubs such as *Empetrum nigrum* and *Vaccinium* species, and various rarer but typically arctic-alpine plants such as *Dryas octopetala* and various species of *Salix* and *Saxifraga*. Elsewhere, however, the diversity of the mountain flora is poorer. *Calluna* and *Vaccinium* communities and grasses such as *Deschampsia flexuosa*, *Juncus triffidus* and *Nardus stricta* occur on slopes covered by thin blanket peat, and *Rhacomitrium* heath (*Rhacomitrium lanuginosum* and *Carex bigelowii*) covers wide tracts of exposed plateau. The soils on which these communities have developed are characteristically thin and acidic. Podsols, peaty podsols, rankers, brown soils and azonal skeletal soils occur on most areas of high ground above the altitude of blanket peat, though on glacially-scoured terrain bedrock outcrops are common, and on some plateaux both the vegetation cover and the soil have been stripped by the wind, leaving sterile, featureless deflation surfaces.

Periglacial environments in space and time

This chapter has outlined the dramatic changes in climate and environment that have affected Great Britain over the last million years or so. Ocean-core evidence suggests that during the greater part of this period glacial or periglacial conditions were the norm in Britain, and that mild interglacials were the exception. During cold stages, glacier ice developed in upland areas and advanced into the lowlands, but the stratigraphic record consistently indicates that such advances were both preceded and succeeded by prolonged periods of severe periglaciation. Inevitably, though, much of the evidence relating to particular periglacial intervals has been destroyed or buried by subsequent glacial advances, and that which survives may be seen only in widely-spaced exposures. In consequence, despite much careful and ingenious research, our knowledge of the chronology of Quaternary periglaciation is still very fragmentary.

Despite the complexity of Quaternary environmental change in Britain, it is possible to differentiate five distinct 'periglacial regions', each delimited by the extent of major glacial advances, and each of which is distinguished from its neighbours in terms of the length of time over which the present land surface has been exposed to periglacial conditions (Figure 2.8). Thus in southern Britain, beyond the limits of glaciation, the landscape has experienced recurrent periglaciation throughout much of the Quaternary period. Farther north, in much of central and eastern England, successive episodes of Middle and Late Pleistocene periglacial activity have modified a predominantly drift-covered landscape since the area was last traversed by a great ice sheet during the Anglian Glacial Stage. The third region comprises Caithness, the Orkney Islands and part of NE Scotland, which may last have experienced glaciation in the Early Devensian and thus been exposed to severe periglacial conditions throughout much of the Late Devensian, though

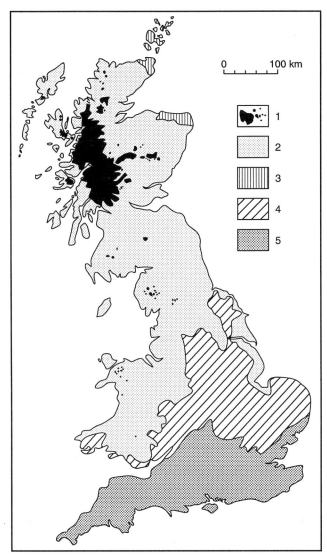

Figure 2.8 The 'periglacial regions' of Great Britain. 1. Areas occupied by glacier ice during the Loch Lomond Stadial: Holocene periglaciation on high ground only. 2. Other areas within the limit of the Late Devensian ice sheet: severe periglacial conditions as the ice sheet retreated (after *c.* 18 ka BP in England and Wales) and again during the Loch Lomond Stadial, together with Holocene periglaciation on high ground. 3. Areas that may have been glaciated during the Early Devensian but not the Late Devensian: severe periglacial conditions throughout the Dimlington and Loch Lomond Stadials. 4. Areas not glaciated since the Anglian: successive periods of prolonged severe periglacial conditions after *c.* 428 ka BP. 5. Unglaciated areas that have experienced successive periods of periglacial conditions throughout the Quaternary.

opinion is divided on this point. The fourth and largest region includes most of Wales, west-central England, northern England and much of Scotland. Here periglacial activity has modified the present land surface only in the latter part of the Dimlington Stadial after the time-transgressive withdrawal of the last ice sheet and again during the Loch Lomond Stadial. Finally, the uplands that were covered by the glaciers of the Loch Lomond Stadial have experienced periglacial conditions only since the the retreat of these glaciers, and hence most evidence for periglaciation in these areas is related to Holocene periglacial activity on high ground.

The above scheme, however, describes only the spatial dimensions of periglaciation across the present land surface of Great Britain, much as the related map of glacier limits (Figure 2.1) concerns only the spatial limits of different glacial episodes. Of equal if not greater importance for our understanding of past environments is reconstruction of the vertical dimensions of glaciation and periglaciation. Under the drifts that mantle much of lowland Britain are preserved complex stratigraphic sequences of glacial, fluvioglacial, periglacial and interglacial deposits, many of which contain structures diagnostic of former periglacial conditions. Sections within these sequences allow us tantalising glimpses of ancient periglacial landscapes, and the structures they contain provide valuable evidence of contemporaneous environmental conditions. At the other extreme, ascent of some of the higher British mountains may take us past the limits of at least two major ice advances, those of the Loch Lomond and Dimlington Stadials, from slopes affected only by Holocene periglaciation to summits that stood proud of the ice sheets for much of the later Pleistocene and experienced modification under a succession of periglacial environments of differing severity.

Finally, it is important to appreciate that the nature and effects of periglaciation in Britain have not been uniform, but have varied enormously throughout the Quaternary in response to climates of differing severity. The periglacial environment of East Anglia immediately prior to the encroachment of the Anglian ice sheet across the area was one of perennially-frozen ground, severe winter cooling and at least local aridity; that of the western Highlands of Scotland during the Loch Lomond Stadial was very cold, wet and stormy; that of British mountains at present is milder, but still very wet and windy. The range of periglacial features that have developed at a particular time inevitably reflects the nature of prevailing climatic and other environmental conditions. If we are to employ periglacial phenomena as palaeoenvironmental indicators, therefore, it is important that we identify appropriate present-day analogues for comparison. Some indication of the range and characteristics of present-day periglacial environments is given in the following chapter.

3
Periglacial environments

Research carried out on periglacial phenomena in present-day cold environments is vital to our understanding of the formation and significance of their Pleistocene counterparts in Great Britain. Observations of periglacial processes in arctic, subarctic and alpine environments not only throw light on how relict features developed in Britain during Pleistocene cold stages, but also provide us with analogues for the conditions under which such relict features formed. For scientists studying the changing environments of the Pleistocene, therefore, relict periglacial landforms, deposits and sedimentary structures represent a valuable source of information concerning past environmental (and particularly climatic) conditions. Such information must be employed with caution, however, for in the past, as at present, periglacial environments spanned a wide range of very different climatic types. Present-day analogues of past environments must therefore be selected with care if valid conclusions are to be drawn concerning the significance of relict periglacial phenomena.

Periglacial climates

Three major locational factors determine the characteristics of global climates in general, and of periglacial climates in particular. These are latitude, altitude and continentality. Individually and in combination, these factors account for much of the range in present-day periglacial climates. At high latitudes, summer daylight and winter darkness are continuous, and even during summer the intensity of solar radiation is low. The annual temperature cycle, comprising cool summers and very cold winters, is therefore more important geomorphologically than short-term diurnal freezing and thawing cycles. Superimposed on such latitudinal climatic variations are the influences of altitude and continentality. Since atmospheric temperatures fall with height (generally at around 0.5–0.6 °C per 100 m rise), marked contrasts in mean annual temperatures exist between lowland and upland regions at similar latitudes, and cold climates occur in mountainous regions even at low latitudes. In the extreme case of the high Andean summits of Peru, for example, there is little seasonal variation in either length of day or the intensity of received solar radiation, so that diurnal freezing and thawing cycles are geomorphologically dominant, and the annual temperature cycle is of little importance. Maritime influences also have a strong mitigating effect on temperature extremes and generally result in a subdued seasonal cycle. Short-term frost cycles are common, however, due largely to the successive dominance of warm and cold airmasses in winter. The moderating influence of the sea is progressively lost with increasing continentality, so that continental interiors experience a far greater seasonal temperature range than maritime locations at

similar latitudes. In continental environments, the depth of winter ground freezing may be tempered by remnant summer warmth in the ground, despite low winter air temperatures.

Classification of climates into a limited number of discrete categories tends to obscure the true diversity of climatic variation, with many intermediate types lying on the boundaries between recognised climatic zones. It is nonetheless convenient to summarise the range of periglacial climates by means of a simple classification such as that presented in Table 3.1. This is based on the work of Troll (1944, 1958), Tricart (1970), Péwé (1974) and French (1976). The last-mentioned author defined four main periglacial climatic groups, namely high arctic, continental, alpine and climates with limited temperature range. A fifth category, polar desert, is added in Table 3.1, and French's 'high arctic' climatic type is modified to include only the arctic tundra biozone that lies between the treeline and the polar desert.

In the arctic, polar desert climates occur at latitudes north of about 77° N and are characterised by a strong seasonal pattern of temperature fluctuation, with cool summers and very cold winters. The mean temperature of the warmest month is less than 10 °C, and annual precipitation is usually less than 250 mm. Ground conditions in summer are generally dry and, since vegetation cover is sparse, wind action is often important. Beneath a shallow *active layer* (the near-surface layer where temperatures rise above zero in summer), ground temperatures remain below zero throughout the year. Such perennially frozen ground is called *permafrost*, and its nature and distribution are discussed in detail later in this chapter. Arctic tundra climates prevail to the south of the polar desert zone and are therefore less severe, though still dominated by the annual temperature cycle. Despite generally low precipitation totals, ground conditions in summer are often wet, especially during the snowmelt period, because the underlying permafrost prevents downward percolation of meltwater. Continental climates, with their greater temperature range, are generally associated with discontinuous permafrost. Although precipitation totals are higher than in the arctic zone, the ground is often drier throughout much of the summer because of higher rates of evaporation. The alpine periglacial climate corresponds to the mountain subtype of Tricart's humid climatic zone. It is characterised by generally higher winter temperatures than the polar, arctic and continental climates, and pronounced seasonal and diurnal temperature fluctuations. Orographic effects result in higher precipitation totals, with most winter precipitation in the form of snow. Permafrost is usually absent or discontinuous. The final group of periglacial climates includes those of maritime areas and those of low latitude high mountains. In the former, seasonal temperature ranges are modest and frontal activity

Table 3.1: *Examples of periglacial climatic types.*

Climatic type	Example	Latitude	Altitude (m)	Mean annual temperature (°C)	Range of mean monthly temperatures (°C)	Mean annual precipitation (mm)
Polar desert	Eureka, Ellesmere Island, Canada	80° N	2	-19.6	43.9	76
Arctic tundra	Sachs Harbour, Banks Island, Canada	72° N	5	-14.5	34.0	93
	Tuktoyaktuk, Mackenzie Delta Canada	69° N	18	-10.7	39.5	130
Continental	Fort Good Hope, NWT, Canada	66° N	53	-7.7	46.9	284
	Surgut W. Siberia	61° N	40	-3.8	40.0	401
Alpine	Tärnaby, Sweden	66° N	447	-0.3	23.3	626
	Niwot Ridge, Colorado, U.S.A.	40° N	3743	-3.1	20.8	854
Low temperature range	Vincocaya, Peru	15° S	4380	+1.9	6.0	263
	Jan Mayen, Norway	71° N	40	0.0	8.0	365

causes high precipitation totals. In the latter, there is little seasonal variation in temperature but a wide diurnal range throughout the year.

Vegetation

The most important ecological boundary associated with periglacial environments is the *treeline*, which was defined by Hustich (1966) as the absolute polar or altitudinal limit of trees regardless of species, though dwarf varieties such as the dwarf birch (*Betula nana*) may occur beyond it. In Scandinavia, birch (*Betula pubescens*) generally represents the treeline in both arctic and montane environments, but in North America and Siberia coniferous trees extend to the treeline with no intervening deciduous zone (Wardle, 1974). A second forest boundary identified by Hustich (1966) is referred to as the *physiognomic forest line*. This is defined as the limit of continuous forest, and between it and the treeline is a transition zone known as the *subarctic forest-tundra* or *taiga* (Larsen, 1974). In this zone vegetation consists of a mosaic of forest and open tundra. In continental areas this subarctic forest-tundra ecotone may be several hundreds of kilometres wide. It corresponds roughly with the continental periglacial climatic type (Table 3.1). There is, however, little altitudinal difference between the treeline and physiognomic forest line in most mountainous regions, the belt of discontinuous tree-cover being of insignificant extent. Here the

treeline provides a convenient lower boundary for the alpine climatic zone. Above the treeline lies the *alpine tundra* ecotone, which is dominated by low-growing shrubs, herbs, grasses, sedges, mosses and lichens (Rune, 1965). In North America, the arctic treeline coincides closely with the southern limit of continuous permafrost. In Siberia, however, continuous permafrost extends well to the south of the treeline. Here, ground temperatures at depth have responded so slowly to Holocene climatic warming that they are still affected by the very cold conditions that prevailed during the last Pleistocene cold stage (Brown, 1970).

Polewards of the treeline lies the arctic tundra. The term *tundra* is widely applied to ecosystems in which the plant cover consists of low herbaceous dwarf shrub or lichen vegetation, and trees are absent. The northward increase in climatic severity in the arctic tundra zone is paralleled by a general decline in plant diversity and plant cover. Webber (1974) defined three major belts of tundra vegetation, namely low arctic, high arctic and polar desert, each corresponding to a progressively more severe climate. Apart from on exposed rock surfaces, plant cover is generally continuous in the low arctic tundra and in better-drained areas is dominated by low shrubs, herbs, grasses and sedges (Figure 3.1). Wetter sites are characterised by grasses, sedges and mosses. In the high arctic tundra, herbs, grasses, mosses and lichens dominate, with occasional prostrate shrubs. Cover is often discontinuous and bare rock is frequently

Figure 3.1 Tundra vegetation near Tuktoyaktuk, Mackenzie Delta area, NWT, Canada. Note the ground ice exposed in the small slump scar.

exposed. Vegetation is sparse in the polar desert zone, and consists largely of lichens. Vascular plants and mosses occur only in more favoured (often moister) locations in this zone (Figure 3.2). Within this general pattern, variations in wetness and exposure to wind lead to considerable local variation in vegetation type, so that exposed habitats in the low arctic zone may support high arctic and even polar desert plant communities. Similarly, habitats with prolonged snowcover often support a sparse vegetation cover due to the very restricted length of growing season following snowmelt. The southern hemisphere polar regions have evolved their own distinctive biota, but these do not offer analogues for Pleistocene Britain, and therefore will not be discussed here (see Sugden (1982) for a review of the characteristics of Antarctic plant communities). In the alpine zone, plant diversity tends to decrease with increasing altitude, but again local variations in exposure, drainage and snow accumulation result in a mosaic of plant communities. Tundra and alpine plants display a range of adaptions to cold winters and short cool summers. These include low or prostrate growth, a dominance of perennial species, cushion forms, preformed flower buds enabling rapid seed production and ripening, vegetative reproduction by underground rhyzomes, bulbs, rooting from stems, and an ability to resist both frost and drought (Billings, 1974).

Soils

Periglacial pedogenesis is dominated by two factors: first, low temperatures and ground freezing; and second, soil drainage.

Plant growth and the decay of organic matter are both retarded by low temperatures, and pedogenic processes such as chemical weathering and the translocation of organic matter and mineral soil also operate more slowly than in warmer environments. Seasonal freezing and thawing of the soil is frequently associated with ice segregation (see below), frost-sorting and the formation of patterned ground (Chapter 6). These processes lead to a churning or stirring of the soil and the breakup and deformation of soil horizons (Rieger, 1974). Such *cryoturbation* also retards pedogenesis. In arctic tundra lowlands, soil drainage is often impeded by the presence of permafrost, and in consequence the ground may be saturated during much of the summer. In the alpine zone, however, topography exerts a stronger control and freely-drained soils are more common.

Most poorly-drained tundra soils have essentially similar profiles (Rieger, 1974), consisting typically of a superficial organic mat, a black or dark brown waterlogged peat horizon and an underlying dark grey or bluish gleyed horizon in which anaerobic conditions lead to reduction of iron compounds to the ferrous state. The thickness of the peat horizon tends to decrease polewards. In low arctic tundra areas the entire active layer may be composed of peat. Better-drained tundra sites are generally characterised by a thin organic litter overlying a dark, acidic, organic-rich A horizon, usually no more than 15 cm thick (Rieger, 1974). The underlying mineral B horizon is paler in colour and often has a platy structure due to the effects of seasonal ice segregation. The B horizon merges downwards into parent material, which may be weathered bedrock, till, aeolian deposits, solifluction deposits, colluvium or alluvium. Such

Figure 3.2 Polar desert, Fosheim Peninsula, Ellesmere Island, NWT, Canada.

soils have been classified as weakly podzolic *arctic brown soils* by Tedrow (1977). Polar desert soils reflect the severity and aridity of the climate. Little organic matter is present and aridity reduces the significance of downward translocation of soil materials. Upward moisture movement often dominates and surface salt efflorescences may occur. In many cases no pedological horizons are discernible. In better-drained alpine areas, leaching is more effective, and thin, acid podzols may develop (Ellis, 1979) as part of a catenary sequence that includes arctic brown soils. Periglacial soils may be preserved as palaeosols within Pleistocene stratigraphic sequences. They are usually associated with cryoturbation and other periglacial structures, and may represent important stratigraphic horizons. They offer the opportunity for detailed palaeoclimatic reconstructions, as we shall see in later chapters.

Fauna

The vertebrate faunas that form part of the tundra ecosystem are characterised by a restricted diversity of species (Sugden, 1982). Only about 70 bird species breed in the arctic, representing less than 0.01% of the world total, and only 23 mammals out of a global total of 3200 occur north of the treeline (Dunbar, 1968). This lack of diversity also extends to the insect population, though as field workers will testify, the abundance of one species, the mosquito, often makes up for the dearth of others! The predator–prey relationship in periglacial ecosystems is also

relatively simple. Herbivores include small rodents such as voles (*Clethrionomys* Spp.) and lemmings (*Lemmus* Spp.), the arctic hare (*Lepus timidus*), reindeer or their North American equivalent the caribou (*Rangifer tarandus*) and musk ox (*Ovibos nivicola*). The smaller herbivores are particularly characterised by cyclic fluctuations in numbers, with a periodicity of several years. Since small mammals provide prey for such species as the weasel (*Mustela* Spp.), arctic fox (*Alopex lagopus*), wolf (*Canis lupus*) and predatory birds, the populations of these predators also fluctuate in response to abundance of food supply. Larger herbivores also provide food for wolves and bears. With a few exceptions the distribution of vertebrates is circumpolar and this wide distribution was probably facilitated by the presence of land bridges during periods of low sea level during Pleistocene cold stages.

Distinctive Quaternary fossil vertebrate assemblages are occasionally preserved in Britain within sediments deposited under periglacial conditions, particularly alluvial gravels. However, these often exhibit greater species diversity than in present-day arctic environments, and sometimes include a number of extinct species such as the woolly mammoth (*Elphas primigenius*), woolly rhinoceros (*Coelodonta antiquitatis*), giant deer (*Megaloceros giganteus*) and cave bear (*Ursus spelaeus*). It appears that in Eurasia a broad periglacial *steppe-tundra* zone developed outside the limits of ice-sheet growth during the Late Quaternary cold stages. This was characterised by a longer growing season and less extreme seasonal varia-

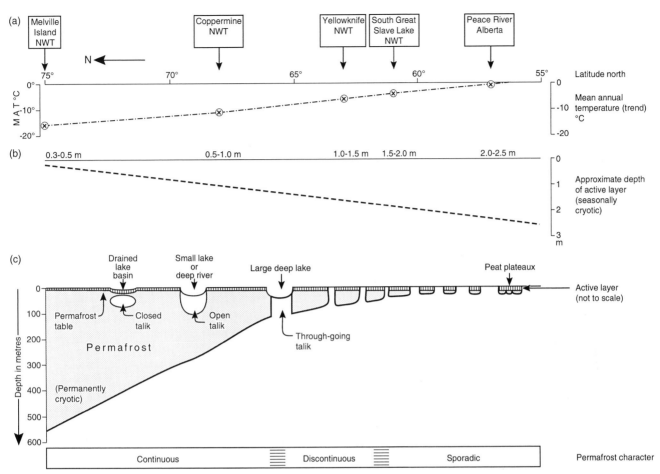

Figure 3.3 Permafrost characteristics along a north–south transect through NW Canada. Modified from Brown (1970) and Lewkowicz (1989).

tions in length of day than is the case in present arctic tundra areas (Hoffmann, 1974). In consequence, vegetation cover included a larger proportion of xerophytic species and grasses. This vast area of open steppe-tundra also supported a much larger range of grazing animals, many of which became extinct at the end of the Pleistocene. Some, such as musk oxen and reindeer, survived in the arctic (though the former did so only in North America), while others such as the wild horse (*Equus przewalskii*) adapted and survived in the southern Eurasian steppes (Hoffmann, 1974). The most useful and ubiquitous faunal palaeoenvironmental indicators, however, are fossil insect assemblages. Most species represented in Quaternary assemblages may still be found today in environments conducive to their success. Since insect populations respond rapidly to climatic change through migration, dated Quaternary assemblages can provide valuable information on terrestrial environmental change (e.g. Coope, 1975, 1977a,b; Atkinson *et al.*, 1987).

Permafrost

Permafrost is generally defined as ground in which the temperature remains below 0 °C over at least two consecutive years (Brown & Kupsch, 1974; Harris *et al.*, 1988). It is therefore a purely thermal condition, independent of the presence of ice in the ground. To overcome the difficulties that arise from applying the terms 'frozen' and 'unfrozen' to ground that may or may

not contain ice or water, two new terms have been introduced. These are *cryotic*, meaning at a temperature of less than 0 °C, and *noncryotic* meaning at a temperature greater than 0 °C (Harris *et al.*, 1988). Thus, permafrost is by definition perennially cryotic ground that may or may not contain ice. Where unconsolidated fine-grained sediments form the substrate, however, permafrost is frequently ice-rich. Although permafrost is not present in all alpine areas and is discontinuous in the subarctic zone, it presently underlies most areas north of the arctic treeline and many high alpine tundra locations. As we shall see in later chapters, there is good evidence that permafrost was widespread in Great Britain during successive Quaternary cold stages.

Permafrost environments

During episodes of climatic cooling, permafrost develops when the mean annual ground temperature falls below 0 °C. Permafrost does not occur at the ground surface (except under perennial ice-cover), as a near-surface zone termed the *active layer* thaws annually in response to summer warming. In the polar desert zone, active layer thicknesses generally do not exceed 0.5 m, but they increase southwards in response to increased heating associated with warmer and longer summers (Figure 3.3). The thickness and temperature regime of permafrost are controlled largely by mean annual air temperature,

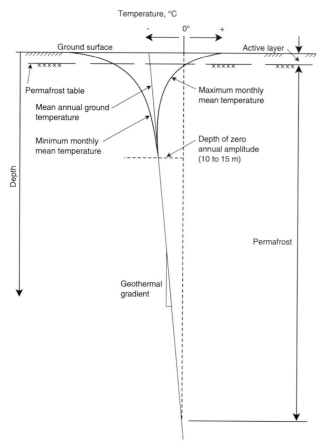

Figure 3.4 Typical temperature regime in permafrost. After Brown (1970).

The annual air temperature cycle responsible for raising active layer temperatures above zero in summer is propagated downwards into the permafrost, but is rapidly attenuated with depth. At a depth of 10–15 m the annual temperature wave is barely detectable (Figure 3.4) and this depth is called the *depth of zero amplitude*. Here the temperature is close to the mean surface temperature, and between 1 °C and 5 °C warmer than the mean annual air temperature (Brown & Péwé, 1973), averaging about 3.5 °C warmer (Haugen, Outcalt & Harle, 1983). Research in Siberia has suggested that the southern limit of continuous permafrost corresponds to a temperature of -5 °C at the depth of zero amplitude, indicating a mean annual air temperature of around -8.5 °C (Brown & Péwé, 1973), though locally continuous permafrost occurs south of this isotherm. Brown *et al.* (1981) utilized the -8 °C isotherm to delimit the boundary between continuous and discontinuous permafrost in North America, and the -1 °C isotherm to define the southern limit of discontinuous permafrost (Figure 3.5). On some maps, the discontinuous permafrost zone has been subdivided into two subzones: one with widespread discontinuous permafrost and the other with sporadic discontinuous permafrost with less than 30% of the surface area underlain by perennially cryotic ground (Harris *et al.*, 1988). Some authors, however, have been reluctant to accept that specific isotherms may delimit the extent of permafrost. Williams & Smith (1989), for instance, stressed the multitude of factors that influence ground thermal regimes and pointed out that the permafrost zonation shown on many maps of the arctic actually represents a gradual southwards transition from continuous through discontinuous to sporadic permafrost, and that local factors lead to wide variations in the mean air and ground surface temperatures associated with this transition.

In mountainous regions, altitudinally-defined zones of continuous and discontinuous alpine permafrost may be present. In southern Norway, for instance, King (1983) showed discontinuous permafrost to be present above 1200 m, where mean annual temperatures are below -1.5 °C. This limit descends northwards to 750 m in northern Sweden. The corresponding continuous permafrost limits are 2060 m in the south and 1600 m in the north, and are approximately defined by the -6 °C isotherm. In North America, discontinuous permafrost is present in the Rocky Mountains where mean annual air temperatures are below about -1 °C (Péwé, 1983), the lower limit rising from just over 1000 m at latitude 60° N to about 4500 m at latitude 20° N. Such alpine permafrost was certainly present in the British uplands during Quaternary stadials when extensive glacier cover failed to develop, the most recent such period being the Loch Lomond Stadial.

The nature and significance of permafrost

Although dry permafrost (perennially cryotic ground in which ice is not present) may have limited geomorphological significance, such situations are rare. The presence of permafrost may also be of limited significance within bedrock if moisture is not readily available. Where water is present, however, it may migrate slowly through the rock and subsequently freeze, causing the growth of ice lenses that prise apart the rock (Hallet, 1983; Walder & Hallet, 1985, 1986; Chapter 9). Where unconsolidated fine-grained sediments form the substrate, permafrost is often ice-rich. Such *ground ice* has been classified by Mackay

the thermal conductivity of the substrate and ground surface conditions (Brown & Péwé, 1973; Brown *et al.*, 1981; Williams & Smith, 1989). In areas marginal for the development of permafrost the thickness of winter snowcover is critical in determining permafrost distribution, as a thick blanket of snow effectively insulates the underlying ground against subzero temperatures (Nicholson & Granberg, 1973). In high arctic areas, however, permafrost is ubiquitous, and continual redistribution of snow by the wind makes snow depth a much less significant factor (Taylor *et al.*, 1982).

In North America and Eurasia two principal permafrost zones are present, the zone of *continuous permafrost* and the zone of *discontinuous permafrost*. Under the very cold climates of high latitudes permafrost is continuous, except beneath large bodies of water, where *taliks* of perennially noncryotic ground may occur (Figure 3.3). Farther south, where climate is less severe, the insulation provided by vegetation, rivers, small ponds and deep snowcover results in breaks in the continuity of the underlying permafrost. Since these insulating factors are spatially highly variable, a complex pattern of discontinuous permafrost results. In Canada, permafrost is between 60 m and 100 m thick at the southern limit of continuous permafrost. Thickness increases northwards to more than 600 m in the northern parts of the Canadian arctic archipelago, and decreases southwards through the discontinuous permafrost zone (Brown *et al.*, 1981; Figure 3.3). In Siberia, permafrost is colder, deeper and extends farther south than is the case in North America.

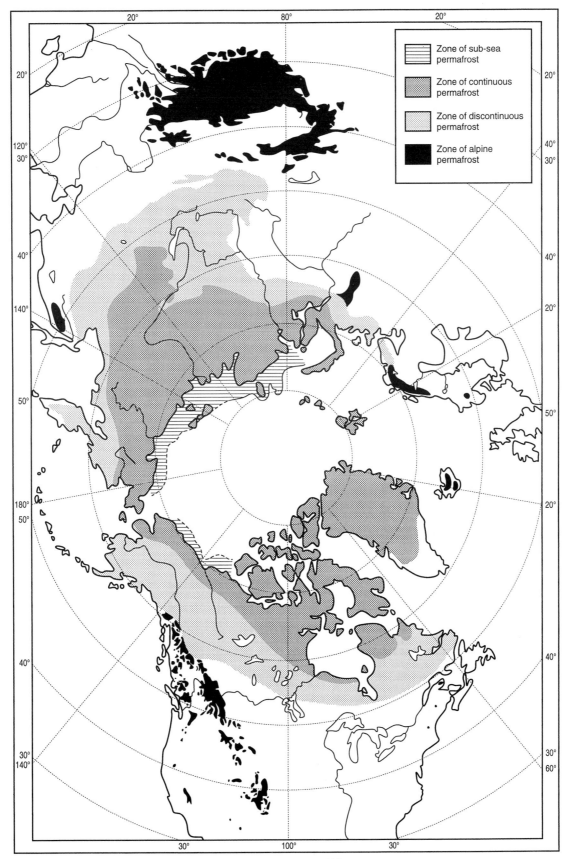

Figure 3.5 Present distribution of permafrost in the northern hemisphere. After Péwé (1983).

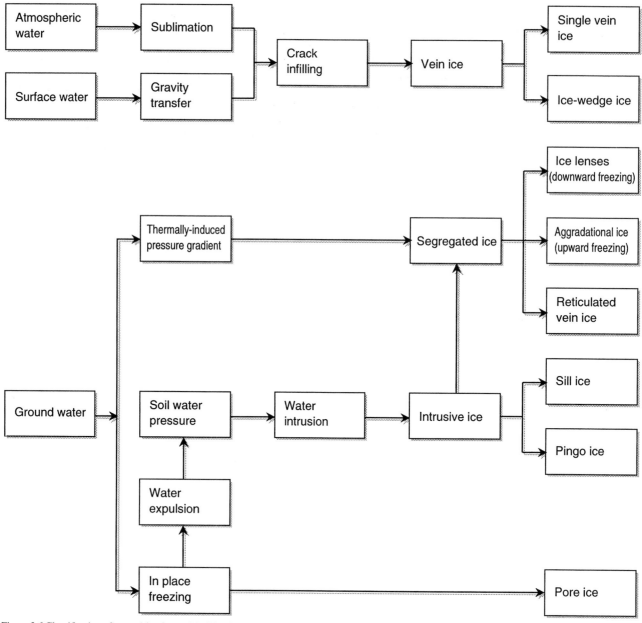

Figure 3.6 Classification of ground-ice forms. Modified from Mackay (1972).

(1972) into four major types; pore ice, segregation ice, vein or wedge ice, and intrusive or injection ice (Figure 3.6). The main effect of pore ice is to cement frozen sediment. Such ice-cemented permafrost is hard and rock-like in character. Water can nevertheless move very slowly through some frozen soils, since thin films of unfrozen water separate ground ice from mineral grains, even at temperatures well below zero (Williams & Smith, 1989). However, such water movement is so slow that for practical purposes permafrost can be considered imperme-able, although over long periods of time water migration within permafrost may enhance its ice content. Despite its hardness and apparent rigidity, ice-rich permafrost deforms in a manner similar to glacier ice when subjected to large and continuous stresses. Deformation of ground ice allows slow permafrost creep to occur above a certain stress threshold, and gravitational

forces on valley sides may be sufficient to cause such creep (Morgenstern, 1985; Chapter 7). Segregation ice develops as a result of soil water migration towards an advancing freezing front, behind which the water freezes to form lenses or thin lay-ers of ice. Vein or wedge ice results from the entry of snow or percolating water into vertical thermal contraction cracks within the permafrost, and takes the form of vertical wedges that taper downwards. Intrusive or injection ice generally devel-ops at sedimentological boundaries during permafrost aggrada-tion as a result of the accumulation and subsequent freezing of pressurised water. It forms horizontal layers or large lenticular bodies often referred to as *massive ice*. Not all massive ground ice results from freezing of water within permafrost, however. In some instances snowbeds or glacier ice may become buried by rapidly-accumulating sediments and be preserved as massive ground ice within aggrading permafrost (e.g. French & Harry, 1988, 1990).

Although segregation ice is widespread in perennially frozen fine-grained sediments, it is also important in the active layer, where it develops during winter freezeback. Many periglacial processes are influenced by ice segregation in the seasonally-frozen near-surface layer. The resulting landforms are not restricted to the permafrost zone, but may occur wherever deep seasonal ground freezing occurs. Their presence in the British landscape as relict features does not, therefore, provide unequivocal evidence for former permafrost. In contrast, wedge ice and massive ground ice develop only in association with permafrost, and relict structures and landforms created by the melting of such ice bodies provide important evidence for the former distribution and extent of perennially frozen ground. In Chapter 4 the morphological and stratigraphical evidence for the former presence of wedge ice is reviewed. This generally constitutes the best available evidence for the distribution of continuous permafrost in Britain during Quaternary cold periods. Chapter 5 considers landforms that developed as a result of the formation and subsequent melting of injection ice phenomena. These provide less certain evidence for continuous permafrost, though discontinuous permafrost at least is implied. The influence of permafrost on periglacial landform development, however, extends far beyond these relict ground ice phenomena and is a recurrent theme of most subsequent chapters of this book.

PART 2
The periglaciation of lowland Britain

Lowland Britain generally lies below 400 m in altitude, supports few steep slopes, is largely underlain by sedimentary rocks and is extensively mantled by drift deposits. South of a line drawn between the Bristol Channel and Essex, lowland Britain never experienced glaciation, and in essence represents a relict Pleistocene periglacial landscape. North of this line both glacial and periglacial influences are evident in the landscape and in the underlying drift stratigraphy. The following five chapters examine the characteristics of lowland periglaciation, and of associated landforms, deposits and soil structures. Amongst the most widespread of these are ice-wedge casts and relict tundra polygons, features that reflect the former cracking of permafrost under conditions of very severe winter chilling (Chapter 4) and thus provide conclusive evidence for periods of extremely cold climate. In some areas there is also evidence for the former development of large bodies of ground ice, which on melting have formed pingo scars, thermokarst depressions and other ground-ice hollows (Chapter 5). The effects of near-surface seasonal freezing and thawing of the ground under periglacial conditions are evident in the form of relict patterned ground features and cryoturbation structures (Chapter 6). On slopes, such seasonal ground freezing resulted in mass movement of the former active layer, which in some areas resulted from solifluction and in others from the sliding of masses of intact soil over the former permafrost table (Chapter 7), whilst more deep-seated slope movements associated with the development or degradation of permafrost affected some areas of argillaceous bedrock. Chapter 8 summarises the effects of fluvial and aeolian activity on lowland Britain under the severe periglacial conditions of the past.

4
Ice-wedge casts and relict tundra polygons

One of the most striking features of permafrost terrain, particularly when viewed from the air, is the presence of extensive polygonal networks formed by shallow troughs in the ground surface (Figure 4.1). These *tundra polygons* range from a few metres to tens of metres in diameter. The surface troughs mark the locations of ice wedges that extend downwards into the permafrost from the base of the active layer. These wedges result from the repeated opening of tension cracks caused by thermal contraction during winter (see below). Subsequent thawing of the permafrost often fails to destroy the wedge structures, since as the ice wedges melt, the resulting voids become filled by sediment slumping in from above. The former presence of ice wedges is thus indicated by ice wedge casts within the host sediments. Such casts probably constitute the most valuable evidence of former permafrost available to the Quaternary geologist (Worsley, 1987). In this chapter the formation of ice wedges and tundra polygons will first be outlined, before consideration is given to their occurrence and significance as relict features.

Ice wedges, sand wedges and tundra polygons

Ice wedges

Ice wedges consist of vein ice, and take the form of vertical wedge-shaped dykes penetrating downwards into permafrost (Figure 4.2). They may be between 10 mm and 3 m wide and penetrate up to 10 m into the frozen ground (Péwé, 1974), and presently constitute the most widely distributed form of ground ice in the North American arctic (Mackay, 1972). Lachenbruch (1962) has provided a detailed analysis of the mechanism of ice-wedge formation, based on a contraction theory first proposed by Leffingwell (1915). This analysis centres on the thermal contraction of near-surface permafrost that results from severe winter cooling. Tensile stresses set up by such contraction are relieved by cracking, resulting in a polygonal network of vertical cracks. Snow and hoarfrost may enter these open cracks, and meltwater may run into them and refreeze, preventing the cracks from closing fully as the permafrost warms and expands in summer. At this initial stage, cracks are marked by thin veins of ice penetrating down into the permafrost from the base of the active layer (Figure 4.3). Since the tensile strength of ice is less than that of frozen sediment, thermal contraction of the permafrost in subsequent winters reopens contraction cracks along the pre-existing network of ice veins. Ice wedges thereby grow incrementally as each successive phase of contraction cracking is accompanied by filling of the open crack with more ice (Figure 4.3).

Mackay (1974) measured winter thermal contraction crack widths of around 10 mm in established ice wedges on Garry Island in the Canadian arctic, but spring warming and expansion of the permafrost caused compression of the infilling ice crystals and reduced the widths of the resulting ice veins to only a few millimetres at most, and often to less than 1 mm. Over a six-year period Mackay observed that on average only 40% of established ice wedges on Garry Island cracked in any one year, this percentage tending to decline as snow depths increased. On Richards Island, Mackay (1986a) monitored the initiation of ice wedges in ground exposed by the drainage of a thaw lake. In their first few years some wedges widened by as much as 35 mm yr[1], but growth rates depended on site-specific factors such as vegetation cover, snow depth and the ground thermal gradient. In many cases open cracks penetrating the permafrost did not become filled with ice because plugs of mud from the active layer prevented water and snow entering the crack (Mackay, 1988a). Such cracks probably closed completely as the permafrost warmed during the following summer. The growth rate of ice wedges is therefore very variable and appears to decrease with time. It is likely that a 1 m wide ice wedge takes many hundreds or even thousands of years to form, but variability in growth rates makes estimation of age on the basis of wedge width very uncertain.

The simplest model for thermal contraction cracking assumes that permafrost behaves as an elastic material (Lachenbruch, 1962; Romanovskij, 1977). However, the predictive power of this model is limited, as it implies progressive cracking from the surface downwards, whereas it is widely reported that cracking is instantaneous, and sometimes accompanied by an audible report. Lachenbruch (1962) outlined an alternative viscoelastic model. This predicts that cracking is most likely during periods of rapid cooling at temperatures below -20 °C, when the frozen ground is relatively brittle. Observations by Mackay (1974) have confirmed that on Garry Island cracking begins at *c.* -20 °C. The viscoelastic model suggests that crack spacing (and hence the diameter of tundra polygons) is related to the depth of cracking and the mechanical properties of the frozen ground. Thus large-diameter polygons are likely to be associated with deep permafrost cooling. Lachenbruch suggested that crack spacing may be less in fine-grained sediment than in coarse, due to the greater plasticity of frozen fine-grained sediments.

The incremental growth of wedge ice produces foliation parallel to the sides of the wedge (Figure 4.2). Such foliation is due to aligned ice crystal fabrics and the incorporation of air bubbles and silt layers (Black, 1973, 1978). Compressional stresses developed in the permafrost during summer warming are usually accommodated by upward deformation of the frozen sediments adjacent to ice wedges (Figure 4.2). Mature ice wedges are typically 1.0–1.5 m wide and penetrate up to about 4 m below the permafrost table (Harry & Gozdzik, 1988). Since the

Figure 4.1 (a) Orthogonal low-centre polygons 7–15 m in diameter near Barrow, Alaska. Photograph by R.I. Lewellen. (b) Hexagonal high-centre polygons 10–20 m in diameter, Ellesmere Island, Canada.

Figure 4.2 (b) Ice wedge in organic-rich silts, Wilber Creek, near Livengood, Alaska. Photograph by T.L. Péwé.

Figure 4.2 (a) Ice wedge exposed by river erosion, Ellesmere Island, Canada.

active layer thaws each summer, ice wedges can only develop below the permafrost table. However, if climatic cooling occurs, the active layer will become thinner and upward aggradation of permafrost will take place. This may be associated with upward extension of ice wedge development to produce complex forms (Mackay, 1974; Harry, French & Pollard, 1985; Romanovskij, 1985; Figure 4.4). Ice wedges that develop some time after the accumulation of their host sediments are called *epigenetic ice wedges*. Where sediment deposition occurs over permafrost, however, upward aggradation of the permafrost may result in ice wedges extending upwards to keep pace with sediment accumulation. The resulting complex structures reflect the rate and periodicity of sediment accumulation and are called *syngenetic ice wedges* (Figure 4.4).

Tundra polygons

The location of an ice wedge beneath the active layer is generally marked at the ground surface by a linear trough that links up with other troughs to form a polygonal network. Such tundra polygons occur in two main forms, termed *low-centre polygons* and *high-centre polygons* (Figure 4.1). The former tend to

develop in flat, wet areas, where polygon troughs are bounded by distinct ridges caused by the upward deformation of sediment adjacent to ice wedges. High-centre polygons have gently domed centres bounded by deeper troughs. Differential thawing of the troughs may contribute to the development of high-centred forms. Both low-centre and high-centre tundra polygons vary considerably in size, but generally range from 15 m to 50 m in diameter (Harry & Gozdzik, 1988). Larger polygons may be subdivided by smaller ones, possibly indicating increased severity of winter cooling (French, 1976). The plan form may be roughly hexagonal with angles between sides tending to approximate 120°, or orthogonal with a tendency towards 90° junctions (Figure 4.1). Hexagonal patterns are considered to indicate a more or less synchronous development of cracks, whilst orthogonal patterns appear to result from the initial formation of random primary cracks, followed by the development of secondary cracks that progressively subdivide the area (Lachenbruch, 1962). Excellent examples of a wide range of tundra polygon forms in the Colville Delta area of Alaska have been illustrated by Walker (1983).

Sand wedges and sand-wedge polygons

Although tundra polygons are most commonly associated with ice wedges, under circumstances where wind action provides abundant aeolian sediment, thermal contraction cracks may become sediment-filled rather than ice-filled. Sand is the most

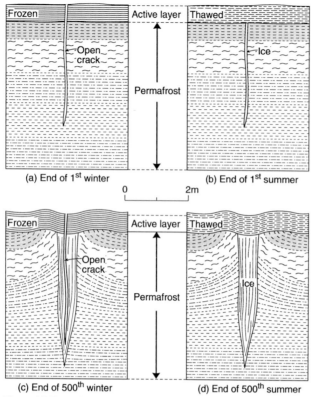

Figure 4.3 Formation of ice wedges by thermal contraction cracking. After Lachenbruch (1962).

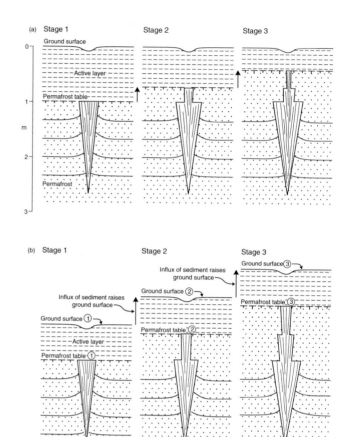

Figure 4.4 Upward extension of an ice wedge: (a) due to thinning of the active layer (epigenetic ice wedge); (b) due to sediment accumulation (syngenetic ice wedge).

common fill material for sediment wedges, and the term *sand wedges* has been applied to the resulting structures (Péwé, 1959). Although arid conditions may both increase the supply of aeolian sediment and reduce the supply of snow, sand wedges are not necessarily indicative of regional aridity. On Banks Island in the Canadian arctic, for example, ice wedge and sand-wedge polygons occur within 2 km of each other (Worsley, 1984; Seddon, 1984), but the latter are confined to sandy substrates where deflation supplies plentiful windblown sand to fill thermal contraction cracks. Here the sand-wedge fill material displays vertical laminations, within which are included very thin laminae of ice (Worsley, personal communication, 1991). A continuum between ice wedge polygons and sand-wedge polygons also occurs in Victoria Land in Antarctica, where ice wedge polygons form in moister coastal areas and sand wedge polygons in the dry interior (Black, 1973). In the McMurdo Sound area of Antarctica, active sand wedges are 0.25–1.00 m wide at the surface and penetrate downwards to a maximum depth of 3 m (Figure 4.5). They are filled with mainly structureless fine to medium sand, and at the ground surface they form orthogonal patterns 10–30 m wide (Figure 4.6; Péwé, 1974). Sand-wedges often extend to the surface, since the mineral fill is not destroyed by thawing of the active layer.

Soil wedges and related phenomena

In permafrost areas with continental climates, severe ground chilling in winter is followed by deep thawing in summer, giving a thick active layer. Winter ground cooling is often associated with thermal contraction cracking of the active layer, the cracks subsequently filling with soil materials (Romanovskij, 1977, 1985; Jahn, 1983). This leads to the formation of *soil wedges*. Where thermal contraction cracking also extends into underlying permafrost, ice wedges may develop beneath such active layer soil wedges. In Siberia, soil wedges also occur in seasonally-frozen ground south of the permafrost limit. Similarly, in northern Fennoscandia, sand-filled soil-wedges have developed, forming polygonal nets up to 12 m in diameter with wedges penetrating downwards to a depth of 0.7 m (Seppälä, 1982, 1987). Thermally-induced cracking of the ground surface may even occur during periods of particularly severe winter cooling in temperate environments (Washburn, Smith & Goddard, 1963; Aartolahti, 1970). If sediment enters these *frost cracks* they may be preserved within the host sediments as thin veins no more than a few millimetres thick and around 0.5 m deep.

Clearly, then, sediment-filled wedge structures may develop as a result of thermal contraction cracking under a wide range of cold climates, and may or may not be associated with permafrost. Where deep seasonal freezing occurs, but not permafrost, soil wedges may approach the dimensions of permafrost sand wedges and the casts of true ice wedges. Great care is therefore necessary in interpreting relict wedge structures, especially where they are small. As Worsley (1987) has warned, a tendency for geologists and geomorphologists to

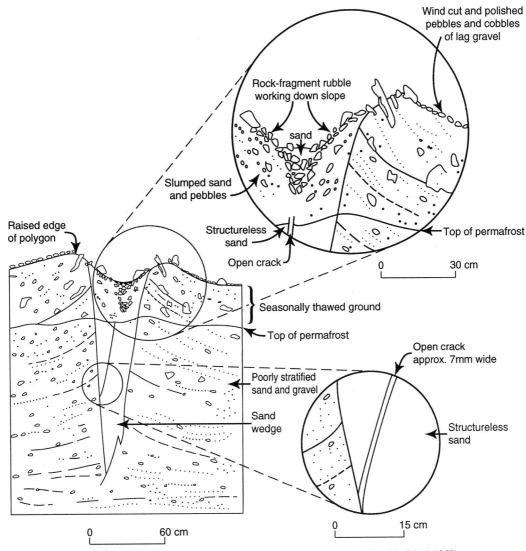

Figure 4.5 Sketch of an active sand wedge from the Taylor Dry Valley, McMurdo Sound, Antarctica. After T.L. Péwé (1959).

interpret almost all wedge structures as ice wedge casts has in some cases led to unjustifiable conclusions regarding the occurrence of permafrost within the British Quaternary stratigraphic record.

Ice-wedge formation and climate

Péwé (1966a,b, 1969, 1974) has discussed the environmental conditions necessary for ice wedge formation in Alaska. He concluded that continuous permafrost that periodically cools rapidly to below -15 °C to -20 °C in winter is necessary for wedge growth, and that this corresponds approximately with mean annual air temperatures below -6 °C to -8 °C. Snow depth in winter must be limited, certainly less than 1.4 m, to allow sufficient ground cooling. In the comparatively continental climate of Siberia, however, ice wedges have formed under rather higher mean annual temperatures. Romanovskij (1985) reported ice wedge formation in frozen loams where mean annual air temperatures are -2.5 °C or lower, and in sands and gravels where mean annual air temperatures are below -6 °C. In Alaska, ice wedges

are now known to have developed in perennially-frozen peaty sediments in areas where the mean annual air temperature is around -3.5 °C (Hamilton, Ager & Robinson, 1983), and wedges in peaty sediments occur in northern Québec where the mean annual air temperature is -5 °C (Payette, Gauthier & Grenière, 1986). The preservation potential of ice wedges formed in a peaty substrate is low, however (Worsley, 1986), and ice wedges have not been found in minerogenic sediments at these sites. Soil wedges are apparently capable of forming in seasonally-frozen loams where mean annual temperatures are slightly above 0 °C, and in sand and gravels where mean annual temperatures are -1 °C or less, but in both cases severe winter cooling of the ground is necessary (Romanovskij, 1985).

It is apparent, then, that many local factors may influence the incidence of thermal contraction cracking in permafrost regions. This inevitably reduces the precision of climatic inferences based on the presence of relict wedge features. The development of ice wedges, however, invariably indicates the presence of permafrost, and is generally associated with rapid cooling of the ground during winter.

Figure 4.6 Sand-wedge polygons 10–30 m in diameter in the Taylor Dry Valley, McMurdo Sound, Antarctica. Photograph by Professor T.L. Péwé.

Mechanism of wedge casting and characteristics of casts

Thawing of permafrost, whether in response to regional climatic amelioration or to more localised changes in vegetation cover, gradually lowers the permafrost table and leads to thawing of ice wedges. The troughs defining tundra polygons become deeper (French & Egginton, 1973) and adjacent or overlying sediments slump into the void left by the melting ice, thus preserving its cast within the host sediments. If the host material is ice-rich, as is often the case in silty sediments, then thawing releases excess meltwater and causes considerable deformation of ice wedge casts as the surrounding ground consolidates (Harry & Gozdzik, 1988). The resulting cast may then have a bulbous contorted form, lacking a sharp apex at the base (Péwé, 1974; Van Vliet-Lanoë, 1985). Where tundra polygons are developed in river gravels, however, little thaw deformation is likely and excellent casts are frequently preserved (Seddon & Holyoak, 1985). Infilling of the void left by a melting ice wedge often produces downwarping of adjacent host sediments (Figure 4.7(a)), and where these are sandy, microfaulting may be preserved (Vandenberghe, 1983a). Conversely, upwards compressional structures resulting from wedge growth may also occur in sediments adjacent to wedge casts (Figure 4.7(b)).

Wedge casts occur in a variety of host materials, most frequently in tills, outwash sands and gravels, periglacial river sediments and periglacial slope deposits. They sometimes penetrate pre-Quaternary bedrock, such as clays, mudstones and shales (Paterson, 1940; Allen, 1984; McManus, 1966). Strati-

(a)

(b)

Figure 4.7 (a) Ice wedge cast at Stanton Harcourt, Oxfordshire, showing downward deformation of adjacent sediment due to slumping. Note the small stratigraphically superimposed cast penetrating the main cast from higher in the sedimentary sequence. Scale divisions 0.5 m. Photograph by Dr M.B. Seddon. (b) Detail of an ice-wedge cast at Ponterwyd, 20 km east of Aberystwyth, showing upward deformation of adjacent host sediment. The wedge is filled with steeply-dipping gravel. Photograph by the late Dr E. Watson, by permission of Mrs S. Watson.

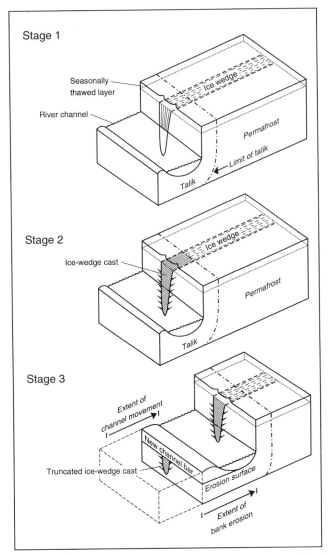

Figure 4.8 Mechanism of casting of interformational and intraformational ice wedges due to lateral migration of braided periglacial river channels. After Seddon & Holyoak (1985).

Figure 4.9 Aerial photograph of polygonal crop marks underlain by ice-wedge casts, near Adwick-Le-Street, Yorkshire. The polygons are developed in sand and gravel and have a random orthogonal pattern. Larger polygons, defined by primary wedge casts, appear to be subdivided into smaller polygons by narrower, less continuous secondary wedges. University of Cambridge Collection of Aerial Photographs.

graphically, ice-wedge casts and sand wedges may be classified as intraformational, interformational or supraformational (Worsley, 1966a). Intraformational casts indicate episodic sedimentation separated by prolonged periods of permafrost conditions under which tundra polygons are formed. These are then buried by the next sediment pulse. Such intraformational wedge casting appears to occur most frequently in fluvial sediments. In an important study of polygon development within active braided river systems in arctic Canada and Spitsbergen, Seddon & Holyoak (1985) described localised melting of permafrost and associated ice-wedge casting caused by channel migration, rather than regional climatic amelioration. They pointed out that where water bodies are more than about 3 m deep they do not freeze completely in winter, and therefore promote melting of the underlying permafrost to form a talik (Figure 4.8). Channel migration across an area of active ice-wedge polygons may cause erosion of the upper part of the wedges and, due to exposure in the channel bank and migration of the subchannel talik, thawing of the remaining lower portions of the wedges. Slumping in of sediment may then preserve a truncated wedge cast, which is subsequently buried by fluvial gravels. If channel migration continues away from the area, renewed permafrost development and thermal contraction cracking may occur, producing a new generation of ice-wedge polygons. Thus intraformational wedge casts may be formed throughout the sediment body, with wedges associated with a series of discontinuous erosion surfaces.

Interformational wedge casts often occur at sedimentary boundaries marking significant changes in sediment character and depositional environment. Such interformational structures may be of considerable climatostratigraphic significance. For instance, if the upper surface of a periglacial gravel formation is penetrated by ice wedge casts, and overlain by till, one might infer that continuous permafrost was present prior to the advance of glacier ice over the site. If till is incorporated as wedge-filling, one might further deduce that permafrost thawing and wedge casting had not taken place until *after* the site had become ice-covered, or until after ice retreat from the site, depending on ground thermal conditions. Supraformational ice wedge casts and sand wedges are also of particular geomorphological interest. These penetrate downwards from the present ground surface and offer the potential for tracing polygonal networks in plan. Because the infilling sediment often differs from the surrounding host sediments, soil drainage along the outcrop of the wedge casts often contrasts with that of the adjacent sediments. Vegetation growth is in consequence affected, producing polygonal 'crop marks' that reflect the pattern of the former tundra polygons (Figure 4.9).

Figure 4.10 Ice-wedge cast, Ponterwyd, Mid Wales. Note slumping of host sediments into the upper parts of the cast. Scale 1 m. Photograph by the late Dr E. Watson, by permission of Mrs S. Watson.

Recognition of permafrost wedge casts

The recognition of ice-wedge casts within Quaternary sedimentary sequences clearly offers valuable potential for palaeoclimatic reconstructions. However, there is a danger that wedge structures not genetically related to thermal contraction cracking in permafrost may be mistaken for true ice-wedge casts. Johnsson (1959) listed numerous possible origins for wedge-like structures, including solution subsidence, sediment disturbance by tree roots, infilling of steep-sided gullies, glaciotectonic faulting, water escape and density deformation, cryoturbation and the inclusion of wedge-shaped sediment bodies during solifluction or landsliding. Certain characteristics of true ice-wedge casts may, however, be taken as diagnostic (Johnsson, 1959; Black, 1976). First, it is of prime importance to establish that a wedge structure exposed in vertical section continues in a horizontal direction, to form part of a polygonal net. This allows recognition of nonlinear structures such as solution pipes and tree root disturbances, and nonpolygonal linear structures such as water escape features. Wedge casts normally taper downwards to a point, and are approximately vertical. If a cast is irregular, possesses a bulbous base or is obliquely inclined, other wedge structures with more characteristic forms should be sought in the vicinity. In any case, wedge structures that occur in isolation provide an insecure basis for palaeoenvironmental interpretation. A third diagnostic characteristic of true ice-wedge casts is the preservation of upturned host sediments against the sides of a cast. Although not always preserved, especially where wedge structures have been partly filled by slumping from the sides, such upturning provides important evidence that the structures represent true ice-wedge casts. The infill of wedge structures may also provide useful evidence. As the infill of wedge casts is usually derived from overlying sediment, it is often characterised by steeply-inclined or vertical slump fabrics. Where the fill material consists of gravel, stones within the cast are generally aligned vertically (Figure 4.10). Stratification within the filling is sometimes arcuate downwards across the cast, forming 'sag structures' (Watson, 1981). Finally, where surface erosion has not caused truncation of wedge casts, wedge structures often extend downwards from the base of a cryoturbated horizon that corresponds to the former active layer.

Since the material filling sand wedges is emplaced as the wedges develop, permafrost thaw often causes very little disturbance of such wedges and their surrounding host sediment (Vandenberghe, 1983b). Relict sand wedges therefore closely resemble their active counterparts, retaining a sharp wedge boundary, upturning of adjacent host sediments, and even verti-

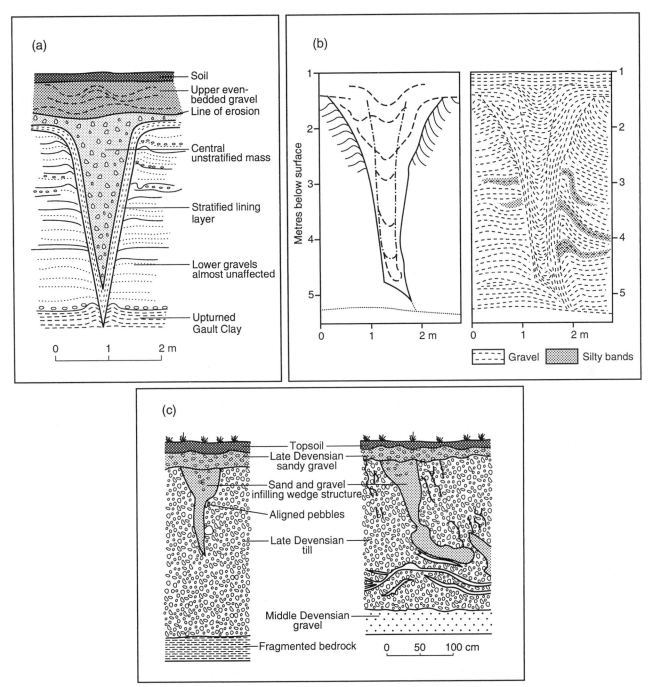

Figure 4.11 Examples of Devensian ice-wedge casts in lowland Britain. (a) Wedge cast located north-west of Cambridge. Host material: fluvial sand and gravel. After Paterson (1940). (b) Wedge cast near Ponterwyd, mid Wales. Host material: angular slope gravels. After Watson (1981). (c) Wedge casts located north-west of Wolverhampton. Host material: Late Devensian till. After Morgan (1971a).

cal lamination of the sand fill. Not all sand wedges taper downwards to an apex, however. Many terminate in a broad feather-edge of interdigitating host material and sand infill. This develops during sand wedge formation if the winter contraction crack does not always occupy the central axial position. Successive cracks produce fingers of sand that fan out at the base of the wedge.

Permafrost wedge casts and tundra polygons in Great Britain

Just as ice wedges represent the most widely distributed form of ground ice in many arctic environments, so ice wedge casts are the most common indicators of former permafrost in Britain. Relict sand wedges have also been reported, but these are much less common, being restricted to areas where circumstances favoured filling of thermal contraction cracks with sand. Also occasionally

recognised are small-scale thin fissures, often less than 1 m long
and only a few millimetres wide. Such structures may result from
desiccation rather than thermal contraction. If, however, they
occur in association with other cryogenic features such as involutions, a periglacial origin may be considered, though they are not
necessarily diagnostic of former permafrost. Examples of well-
preserved thermal contraction structures are described below to
illustrate their nature and mode of occurrence. The stratigraphy of
British thermal contraction structures is then reviewed.

Ice-wedge casts

A perceptive description of ice-wedge casts in Britain appears
in a pioneering paper by Paterson (1940), who investigated
wedge structures of probable Devensian age in a gravel pit near
Cambridge. Paterson's account is particularly useful in illustrating characteristics diagnostic of true ice-wedge casts. He
described casts up to 4 m deep and 2 m wide that penetrated fluvial sands and gravels, and in places the underlying Gault Clay.
Upward bending of the Gault Clay against the sides of casts was
clearly developed and well preserved (Figure 4.11(a)).
Slumping of sand and gravel into the wedges during melting
probably accounts for the formation of a layer of gravel lining
each cast, with pebbles aligned parallel to the sides. In the outer
parts of the wedges, sand and gravel derived from particular
levels in the host sediments could be traced downwards as
streaks extending to the bases of the casts. The inner parts of the
wedges were filled with an unstratified mass of sandy loam
derived from surficial sediments. This pattern of infill was interpreted by Paterson in terms of the melting of wedge ice both
inwards from the sides and downwards from the surface.

A second detailed study of Devensian ice-wedge structures
was provided by Watson (1981), who described 41 individual

Figure 4.12 Late Devensian ice-wedge casts in fluvial gravels at Baston, in
Lincolnshire. Two cycles of ice-wedge development are indicated by truncation of the smaller cast by the larger. Photograph by Professor P. Worsley.

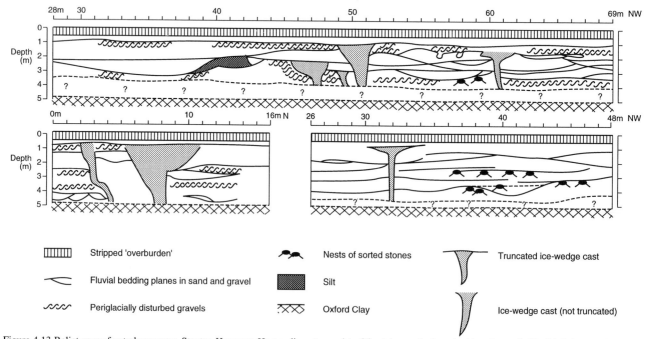

Figure 4.13 Relict permafrost phenomena, Stanton Harcourt. Host sediments consist of fluvial gravels deposited in a low-relief braided river environment.
After Seddon & Holyoak (1985).

Figure 4.14 Ice-wedge cast in fluvial gravels of the River Bain, Tattershall Thorp, Lincolnshire. Photograph by Dr D.T. Holyoak.

casts developed in slope deposits and fluvioglacial gravels in west-central Wales. Watson recognised various diagnostic features, including deformation of adjacent host sediments, thin linings of silt derived from adjacent host materials, sag structures in cast infills and clasts aligned parallel to the sides of casts (Figures 4.10 and 4.11(b)). He also noted that wedge casts generally penetrated more deeply in angular slope gravels than in waterlain gravels (up to 4.6 m in the former, 3.5 m in the latter), and were generally wider (up to 2.4 m in the former compared with 1.4 m in the latter). Watson concluded that thermal contraction cracking and the formation of tundra polygons occurred immediately after the retreat of the last ice sheet, under a climate sufficiently severe to cause the development of continuous permafrost.

Periglacial structures are often well preserved in sands and gravels deposited by braided rivers. Excellent examples of large ice-wedge casts within fluvial sediments occur, for instance, at Baston in Lincolnshire (Worsley, 1987; Figure 4.12). The nature and origin of intraformational and supraformational ice-wedge casts developed in fluvial gravels have been studied in detail by Seddon & Holyoak (1985) at Stanton Harcourt in Oxfordshire, near the confluence of the River Windrush with the Thames, and at Tattershall Thorpe in Lincolnshire, where

the River Bain enters the Fenland Basin. Their interpretations were based on analogy with processes observed in actively-accumulating braided river sediments on Banks Island and Spitsbergen. At Stanton Harcourt, a large gravel quarry called Dix's Pit revealed 4–6 m of fluvial gravels overlying Oxford Clay. The largest wedge casts at this site are over 2 m wide at the top and penetrate the entire gravel sequence. Smaller intraformational casts originate at various levels. The upper parts of these are sometimes truncated by laterally-extensive erosional surfaces (Figure 4.13). The untruncated wedge casts extend downwards from disturbed horizons that are generally less than 1 m thick. The presence of cryoturbation structures and nests of stones within these disturbed horizons indicates that they represent former active layers, and hence mark former land surfaces preserved within the sedimentary sequence. It appears that lateral migration of braided river channels at this site led to pulses of sedimentation separated by prolonged periods of subaerial exposure, during which tundra polygons developed across exposed gravel surfaces. The most clearly-defined cryoturbated layer lies close to the present surface, immediately below the sandy topsoil, and is associated with large supraformational wedge casts.

Complex intraformational ice-wedge casts are also common at this site. These include two types of superimposed cast. The first comprises a cast within a cast, formed as a result of renewed thermal contraction cracking and ice-wedge growth along the line of an earlier ice-wedge cast. The second, referred to as *stratigraphically superimposed*, consists of later wedge casts that originated higher in the sedimentary sequence and penetrated earlier, lower features (Figure 4.7(a)). Abundant intraformational ice-wedge casts are also present at Tattershall Thorpe, in the periglacial fan gravels of the River Bain (Figure 4.14). At both sites it appears that truncation, casting and superimposition of wedge casts resulted from lateral migration of river channels and their underlying taliks. Worsley (1987), however, has proposed an alternative possible mechanism for ice-wedge casting at these sites, based on his observations on Banks Island in Canada. He suggested that the surface troughs overlying ice wedges may have become occupied by standing or flowing water. This may have caused thermal erosion and induced thawing of the wedge ice. The accumulation of sediment in the resulting voids would then have created casts. Irrespective of the relative validity of the two mechanisms outlined above, however, wedge casting in these fluvial sediments appears to have occurred without significant climatic changes.

As mentioned earlier in this chapter, ice-wedge casts not only occur in unlithified Pleistocene deposits, but also penetrate certain sedimentary rocks, such as Tertiary clays, Jurassic and Triassic mudstones and Carboniferous shales. A particularly extensive exposure of bedrock wedge structures occurs near Oldbury on Severn, Gloucestershire (Allen, 1984). At this site a polygonal network of wedge casts is developed in soft Triassic mudstone on an intertidal rock platform called High Heron Rock. Here the wedge casts are up to 0.65 m wide, up to 1.3 m deep and filled with pebbly sands. Their original dimensions were almost certainly greater, however, as erosion of the rock surface has beheaded the wedge structures so that their original depths are unknown. The strata on either side of most casts are upturned, forming zones of disturbance up to 3.4 m wide that define the polygonal pattern of the wedges across the rock plat-

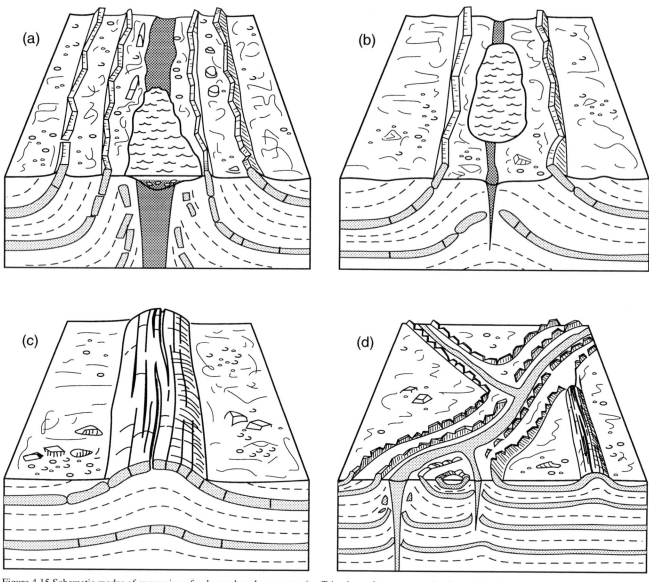

Figure 4.15 Schematic modes of expression of polygonal wedges penetrating Triassic mudstone as seen in plan and (partly speculative) vertical section: (a) deep or little-eroded wedges; (b) moderately deep or moderately-eroded wedges; (c) shallow or much-eroded wedges; (d) the full assemblage of structures commonly found at High Heron Rock, Gloucestershire. After Allen, 1984.

form (Figure 4.15). Both orthogonal and nonorthogonal junctions occur, and a moderately strong preferred alignment is evident (Figure 4.16). This is probably related to strike and dominant joint orientation.

On both banks of the Severn Estuary, in the vicinity of High Heron Rock, Allen (1987) recognised similar ice-wedge casts exposed in vertical sections within Triassic and Jurassic mudstones. Pebbles in the sandy or gravelly fills are aligned parallel to the wedge sides. Allen argued on lithological grounds that the fills were derived from the Main Terrace gravels of the Lower Severn. As these are compositionally related to the Late Devensian till of the Midlands, Allen concluded that the Severn wedges developed during the Dimlington Stadial. Bradshaw & Smith (1963) and Harris (1989) have shown that structures resembling those described by Allen also occur in Triassic mudstones on the south Glamorgan coast of south Wales. Strata

are upturned on either side of vertical fissures, some of which are filled with sandy gravel (Figure 4.17). Harris concluded that these structures represent truncated ice-wedge casts that probably developed contemporaneously with those farther up the Severn Estuary. Ice-wedge casts that penetrate bedrock also occur in Yorkshire (Dimbleby, 1952) and Scotland (McManus, 1966).

Most British ice-wedge casts are 2–4 m long, though at many sites they extend downwards from erosional unconformities and hence are probably truncated. Length:width ratios range from about 10:1 to about 4:1. Structures with length:width ratios greater than 10:1 have been classified as frost fissures or frost cracks (Galloway, 1961c; John, 1973). The possibility that wedge-shaped structures genetically unrelated to ice-wedge casts may be misinterpreted as permafrost phenomena makes it important that all other possible mechanisms of formation be

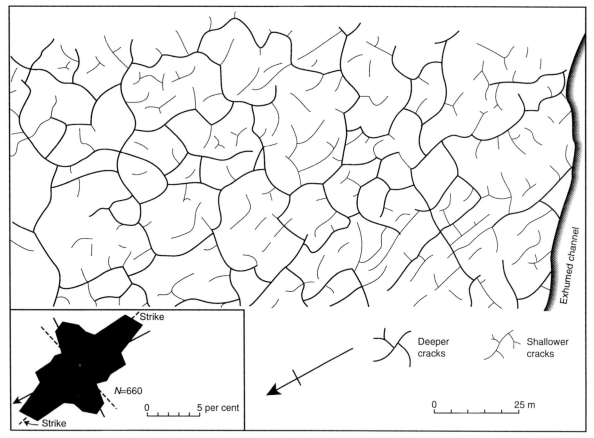

Figure 4.16 Plan and orientations of ice-wedge cast outcrops, High Heron Rock, Gloucestershire. After Allen (1984).

Figure 4.17 Possible ice-wedge cast penetrating Triassic mudstone at Sully, South Glamorgan.

considered and rejected before such structures are accepted as true ice-wedge casts.

Relict sand wedges

Relict sand wedges are much less widespread than ice-wedge casts. Their distribution in lowland Britain appears to have been determined mainly by the availability of aeolian sand to form the wedge infill. Probably the finest examples so far discovered

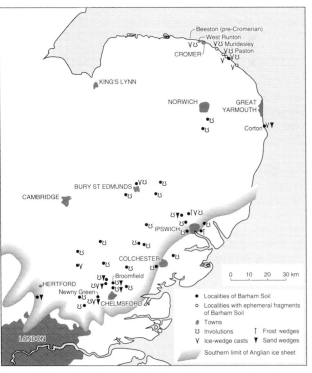

Figure 4.18 Location of pre-Hoxnian ice-wedge casts, sand-wedge casts and involutions in East Anglia. After Rose et al. (1985a).

Figure 4.19 Sand-wedge polygons associated with the Barham Soil at Newney Green, near Chelmsford, Essex. The polygons are picked out by the wedge filling of lighter-coloured sand. These polygons were overlain by Anglian till, and were exposed during quarry working. Photograph by J. Rose.

in Britain are those associated with an extensive palaeosurface known as the Barham Soil, which underlies Anglian till in East Anglia (Rose *et al.* 1985a,b). This palaeosol is preserved in fluvial sands and gravels (the Kesgrave Sands and Gravels; Chapter 8), and incorporates not only relict sand wedges, but also ice-wedge casts, frost cracks and other periglacial structures. The relict sand wedges associated with the Barham Soil occur in Essex and SE Suffolk (Rose *et al.* 1985a; Figure 4.18). At Newney Green, near Chelmsford, a network of sand-wedge polygons was discovered by Rose *et al.* (1978). The pattern is complicated by the presence of sand-filled involutions, but includes polygons up to 10 m in diameter (Figure 4.19). The wedges at Newney Green are infilled by well-sorted medium and fine windblown sand, with 2–5 mm thick laminae arranged in a mirror-image pattern on either side of the cast axis (Figures 4.20 and 4.21). At Broomfield, cross-cutting groups of laminae are present in wedge casts, indicating changing thermal contraction crack locations during sand-wedge growth. The casts themselves are 0.5–2.0 m deep and 5–75 cm wide at the top. In some cases thaw consolidation of ice-rich host sediments has caused deformation of cast margins, but generally the original wedge forms are preserved. The formation of sand wedges

rather than ice wedges in the Barham Soil may indicate formation under cold arid conditions. However, Rose *et al.* (1978) suggested that local topographic factors may have influenced the nature of thermal contraction crack filling, with sand wedges forming in well-drained sites where an abundant supply of aeolian sand was available, and ice wedges forming elsewhere.

In general, the best evidence that sand-filled wedge structures originated as sand wedges (and are not simply ice-wedge casts filled with sand) is provided by the presence of a laminated fill in which increments of wedge growth can be identified. Unfortunately, active sand wedges are often filled with structureless sand and the presence of such an infill makes differentiation of many relict forms difficult. Evidence for downward slumping of the cast infill may suggest melting of ice. However, since thermal contraction cracks may sometimes be filled partly by ice and partly by sand to form composite wedges, it is likely that a range of intermediate cast forms is preserved in the stratigraphic record. Identification of true sand-wedge casts may well be restricted to a limited number of ideal cases, leaving the origin of a significant proportion of wedge structures open to debate.

Figure 4.20 Sand wedge in the Barham Soil, Broomfield near Chelmsford. Photograph by Professor J. Rose

Relict tundra polygons

Relict supraformational tundra polygons can sometimes be seen as crop marks on aerial photographs. For example, nonorthogonal polygonal crop marks, mostly 20–30 m in diameter, are preserved on the surface of the Number 4 Terrace of the Worcestershire Avon near Abbot's Salford (Shotton, 1960). Polygonal crop markings on a similar scale, but with a random orthogonal pattern, are developed in sand and gravel near Adwick-Le-Street, Yorkshire (Figure 4.9). These polygons consist of a primary network with diameters of 20–50 m subdivided by secondary fissures that define smaller polygons 5–10 m in diameter. The geometric pattern and scale of such crop marks generally forms the basis for their interpretation as relict thermal contraction polygons. Where vertical sections through the polygon borders expose the underlying wedge casts, however, much firmer conclusions may be drawn concerning their nature and origin.

Morgan (1971a,b; 1973) identified 17 areas with polygonal crop markings to the north and west of Wolverhampton, in the English Midlands, and a trench excavated for a gas pipeline enabled him to investigate the nature of the underlying wedge structures. The polygons have 4–6 sides, but tend to be nonorthogonal. The surface pattern is defined by lines 0.5–1.5 m wide at the ground surface and the polygons themselves range from 1.5 m to 12.3 m in diameter (Figure 4.22), with larger 'master polygons' up to 19.2 m wide in two areas. The polygons investigated by Morgan are developed in a silty clay till of Late Devensian age, and therefore must postdate deglaciation of the area. Their associated ice-wedge casts taper downwards, are filled with sand and gravel, rarely exceed 1.9 m in depth and have an average length:width ratio of about 4.5:1 (Figure 4.11(c)). Deformation of wedge casts and the development of flame structures along their edges suggest slumping and flow of ice-rich till during wedge melt-out. Morgan concluded that the fairly regular nonorthogonal polygons in this area indicate a single phase of thermal contraction cracking of a uniform substrate. He also inferred that the shallow penetration of wedges and correspondingly small size of polygons indicate relatively shallow winter cooling of the permafrost, and hence a climatic regime of only moderate severity. As this example demonstrates, at sites where the three-dimensional geometry of wedge casts can be determined, and detailed analysis of the cast filling and the host sediments is possible, revealing inferences may be made concerning the environment of tundra polygon formation and the processes associated with wedge casting.

The stratigraphy of wedge casts and relict tundra polygons

A summary of the stratigraphy of wedge structures indicative of former permafrost in Britain is given in Figure 4.23, where a 'conventional' chronological framework is adopted. Uncertainties relating to the status of the 'Wolstonian' Glacial Stage (Bowen et al., 1986; Rose, 1987; Ehlers et al., 1991a) and the number of glacial–interglacial cycles between the Hoxnian and Ipswichian periods (Chapter 2) should, however, be borne in mind. As might be anticipated, by far the greatest number of recorded wedge casts date from the last glacial stage, the Devensian. However, earlier episodes of permafrost are also indicated by wedge casts at certain sites in lowland Britain.

Pre-Cromerian

The earliest evidence for ice-wedge development and hence former permafrost in lowland Britain occurs in sediments underlying Cromerian Interglacial deposits on the north Norfolk coast at Beeston, West Runton and the Paston–Mundesley area (Figure 4.18). These wedge casts have been ascribed to the Beestonian cold stage. At West Runton, West (1968) observed a system of relict ice-wedge polygons exposed on the foreshore. Here the ice-wedge casts penetrate silts ascribed to the preceding Pastonian Interglacial. At Beeston, a complex series of ice-wedge casts is present within a sequence of Beestonian fluvial muds, sands and gravels. Six stratigraphic groups were identified by West (1980a), possibly indicating distinct episodes of permafrost development separated by periods of fluvial activity. Alternatively, however, some at least of the different groups of wedge structures may reflect casting as a result of migration of river channels across a braided floodplain during a single episode of permafrost (cf. Seddon & Holyoak, 1985). Involutions also provide evidence for Early Pleistocene periglacial conditions at these sites (Chapter 6).

Figure 4.21 Close up of laminations in the sand wedge shown in Figure 4.20. Photograph by Professor J. Rose.

The Anglian Glacial Stage

West (1980a) and Rose *et al.* (1985a,b) have argued that permafrost became established in England in response to climatic cooling in the last part of the Cromerian Interglacial, prior to the onset of the Anglian Glacial Stage. Worsley (1987), however, was sceptical of the evidence for this, suggesting that the truncated wedge structures at West Runton on which their argument is based could equally well be pre-Cromerian in age. Worsley added a further note of caution concerning the status of the Cromerian–Anglian stratigraphic transition, raising the possibility that it might bridge at least one glacial–interglacial cycle that is not currently recognised in the 'standard' British succession. There is abundant evidence, however, that permafrost was widespread in East Anglia in the early part of the Anglian Glacial Stage prior to the advance of glacier ice across the area. Tundra polygons, frost cracks and a cryoturbated active layer characterised the contemporary ground surface, which is now preserved beneath Anglian till in the form of the widespread horizon known as the Barham Soil (Rose *et al.*, 1985a,b). Relict sand-wedge polygons associated with the Barham Soil were described earlier in this chapter. Ice-wedge casts and polygons have also been recorded, the wedge casts being around 4 m deep, but reaching a maximum of more than 8 m (Rose *et al.*,

1985a). At Corton in Suffolk (Figure 4.18) a polygonal system of ice-wedge casts penetrates Cromerian Interglacial deposits (Gardner & West, 1975). The wedge casts here were apparently truncated, probably as a result of erosion by the Anglian ice sheet, then buried under a deposit of Anglian till. Also at Corton, Ransom (1968) observed ice-wedge casts in the Corton Beds, which apparently represent outwash sands that overlie an early Anglian till, the Cromer Till. The Corton Beds and their associated wedge structures are, in turn, overlain by a later Anglian till, the Lowestoft Till. This suggests that permafrost developed not only prior to glaciation in the early Anglian, but also in the interlude between two successive advances of the Anglian ice sheet. It is probable that permafrost affected a large area of lowland Britain during the build up and expansion of the Anglian ice sheet, but conclusive evidence for this outside the area of East Anglia and its vicinity has not yet emerged. Wedge structures that formed at this time in other areas either have not yet been identified, or may have been destroyed during later erosional episodes.

The post-Hoxnian, pre-Ipswichian time interval

The uncertainties surrounding this part of British Quaternary stratigraphy have been discussed in Chapter 2. In the absence of

Figure 4.22 Polygonal crop marks 1 km south of Four Ashes, Staffordshire. The average polygon diameter is around 5 m. Photograph by A.V. Morgan.

absolute dates, correlations between sites largely rely on lithostratigraphic and biostratigraphic evidence. Since the sediments, periglacial structures, floras and faunas of one cold stage are likely to be similar to those of another, such correlations are often equivocal. It appears highly probable that more than one cold stage within the time interval separating the Hoxnian and Ipswichian Interglacial Stages is represented in the terrestrial sedimentary record.

At Hoxne in Suffolk, fluvial sediments directly overlying Hoxnian Interglacial deposits are penetrated by two sets of wedge structures (Gladfelter, 1975). Those of the lowest group, however, are only 0.45 m in length, and their origin as true ice-wedge casts is open to question. However, wedge casts of the second series, some 0.8 m higher in the sequence, are well developed and penetrate downwards more than 1.5 m. Although the underlying Hoxnian Interglacial deposits provide a maximum age for these casts, the absence of overlying datable sediment precludes their assignation to any particular cold stage. The uppermost sedimentary unit at this site consists of unstratified pebbly sands and silts, and postdates periglacial disturbance of the underlying fluviatile sediments. This surface deposit mantles large areas of East Anglia. It probably includes an aeolian component in its upper horizons, and may well be of Devensian age (Gladfelter, 1975).

Lower Palaeolithic hand-axes of the Acheulian type are associated with Middle and Late Hoxnian Interglacial deposits, and at the Hoxne type site Acheulian artefacts have also been recovered from the lower part of the fluvial sediments that over-

lie the interglacial deposits. The wedge structures described above occur within these fluvial sediments, suggesting that Acheulian artefacts may have been produced over a timespan that includes the cold stage immediately following the Hoxnian Interglacial (Gladfelter, 1975; Wymer, 1977). Acheulian hand-axes have also been recovered at other sites from sediments that probably date to this cold stage. The presence of Acheulian artefacts within ice-wedge casts developed in fluvial gravels adjacent to the Rivers Roach and Crouch, in SE Essex, led Gruhn & Bryan (1969) to conclude that wedge development occurred here during a post-Hoxnian but pre-Ipswichian cold stage. The possibility that the archaeological material was reworked by later river action before incorporation into the wedge casts must, however, be borne in mind. Polygonal crop marks associated with the casts in SE Essex indicate a network of contemporaneous tundra polygons, and in some places such polygons are unconformably overlain by Devensian loess. Polygon development and wedge casting at this site must therefore predate this phase of loess accumulation.

A section near Marsworth in Buckinghamshire has revealed a sequence of periglacial deposits that overlie organic muds and gravels containing a temperate fossil assemblage and underlie further organic muds that also contain temperate faunal remains. Both sets of organic deposits infill channels. The lower channel is cut in chalk bedrock and the upper series of channels is cut into the surface of the periglacial deposits (Figure 4.24). These consist of colluvial chalky muds overlain by frost-weathered and

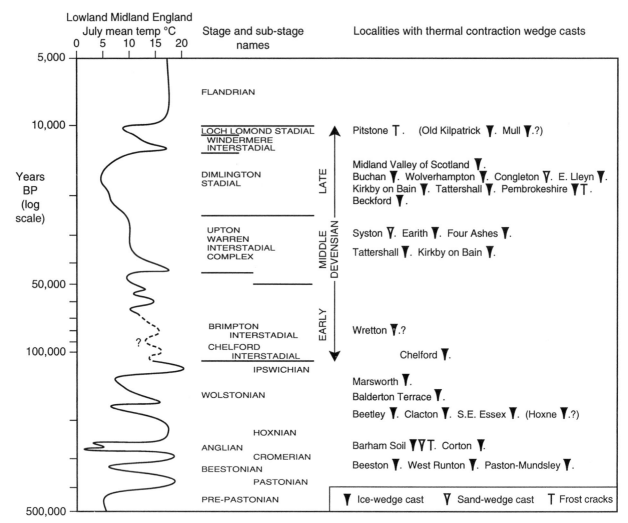

Figure 4.23 A summary of the stratigraphy of thermal contraction crack structures in Britain. After Rose et al. (1985b).

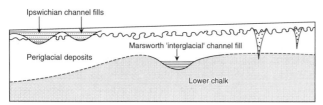

Figure 4.24 Diagrammatic representation of the stratigraphy of ice-wedge casts at Marsworth, Buckinghamshire. After Worsley (1987).

soliflucted chalk (coombe rock). Worsley (1987) has reported the presence of a series of ice-wedge casts near the top of the coombe rock deposit. These wedge structures link up to form a large-scale polygonal network in plan. The upper channel fill incorporates a mammalian fossil assemblage similar to Ipswichian Interglacial assemblages found elsewhere in Britain (Green *et al.*, 1984). In the lower channel deposits, travertine (precipitated calcite) occurs and uranium series dating of this travertine indicated an age range of 140–170 ka BP. Collectively, this evidence suggests that a severe periglacial climate prevailed during the interval between precipitation of travertine in the lower channel and the

onset of the Ipswichian Interglacial. During this interval, deposition of the coombe rock occurred, followed by formation of large-scale tundra polygons. Continuous permafrost may not necessarily have been present as the coombe rock accumulated, but is implicit in the development of the ice-wedge polygons.

Probably the best documented evidence for the presence of continuous permafrost immediately prior to the Ipswichian Interglacial comes from the Balderton Terrace, a sinuous low-lying tract 1.5–3.0 km wide that separates the Rivers Witham and Trent between Newark and Lincoln (Brandon & Sumbler, 1991; Figure 4.24). This terrace consists of sand and gravel (the Balderton Sand and Gravel) that infills a broad shallow channel cut in Lower Jurassic mudstone bedrock. This channel marks a former more easterly course of the River Trent. The Balderton Sand and Gravel contains planar gravel beds, cross-bedded sands and gravels and lenticular interbeds of trough cross-bedded sand. Extensive subplanar erosion surfaces occur throughout the sequence and truncate syndepositional intraformational ice-wedge casts. Sediment deposition is thought to have occurred on the floodplain of a braided river. Truncation of wedges and ice-wedge casting probably resulted from lateral

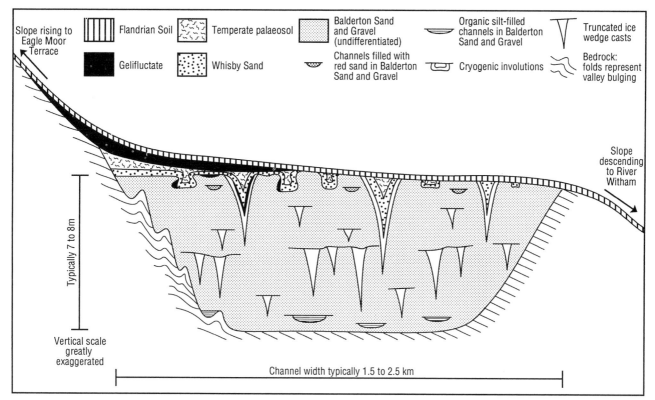

Figure 4.25 Schematic stratigraphic cross-section of the Balderton Terrace, between Newark and Lincoln, Lincolnshire. After Brandon & Sumbler (1991).

migration of river channels and their associated subchannel taliks . Complete wedge casts up to 5 m deep and 1 m wide are present at the top of the sequence, where they extend downwards from a cryoturbated layer. Overlying the Balderton Sand and Gravel, and forming the fill of the uppermost wedge casts, is a medium-grained sand (the Whisby Sand), which is interpreted as being of fluvio-aeolian origin. This sand contains evidence of postdepositional weathering and clay translocation, possibly due to pedogenesis under temperate climatic conditions (Brandon & Sumbler, 1991). In places the Whisby Sand is overlain by coversand and gelifluctate (Figure 4.25).

Dating of the Balderton Sand and Gravel and its contained ice-wedge casts rests on several independent lines of evidence. First, the geomorphological and topographical relationships between the Balderton Terrace and other terrace deposits in the area indicate formation during a post-Hoxnian, pre-Ipswichian cold stage. Secondly, a mammalian fauna found in the lower parts of the gravel suggests accumulation during a pre-Devensian cold stage, since two of the taxa present, straight-tusked elephant (*Palaeoloxodon antiquus)* and rhinoceros (*Dicerorhinus*), were extinct in Britain before the end of the Ipswichian Interglacial. Limb bones of horse (*Equus caballus*) and bison (*Bison* cf. *priscus*) are also larger than those from Devensian deposits, but comparable with pre-Devensian examples. The fauna includes both temperate and arctic species and suggests a rapidly deteriorating climate during the early stages of gravel accumulation. A third line of evidence relates to electron spin resonance dating of mammoth and straight-tusked elephant teeth. This indicates ages in the range 130–190 ka BP and is supported by amino acid analysis of molluscs from the

Balderton Sand and Gravel that indicate correlation with Isotope Stage 6 (the cold stage immediately preceding the Ipswichian). Finally, the occurrence of a slightly-worn Acheulian hand-axe in the Balderton Sand and Gravel and the absence of any later types suggests deposition during a pre-Ipswichian cold stage. Brandon & Sumbler (1991) concluded that two generations of ice-wedge casts are present in the Balderton Terrace. The intraformational casts developed first, during sand and gravel aggradation in a pre-Ipswichian cold stage. The uppermost interformational casts and associated cryoturbation structures appear to be much younger, probably of Devensian age, since they incorporate the Whisby Sand and its associated palaeosol, both of which are thought to date to the Ipswichian Interglacial.

Worsley (1987) has argued that the ice-wedge casts within fluvial gravels at Stanton Harcourt (described earlier in this chapter) may also belong to an immediately pre-Ipswichian permafrost event. His argument is based on the correspondence in height between the gravel surface at Stanton Harcourt and adjacent terrace fragments that form the Summertown-Radley terrace system of the upper Thames Valley. In places this terrace system incorporates Ipswichian deposits that infill channels on its upper surface, evidence that led Briggs *et al.* (1985) to conclude that the Summertown–Radley terrace gravels accumulated during a cold stage immediately predating the Ipswichian Interglacial. However, both radiocarbon assay and thermoluminescence dating suggest that accumulation of the Stanton Harcourt gravels occurred during the Devensian, though the former technique indicates a Middle Devensian age and the latter an Early Devensian age for the deposits (Seddon & Holyoak, 1985). In the face of this conflicting evidence, it is prudent to conclude that the

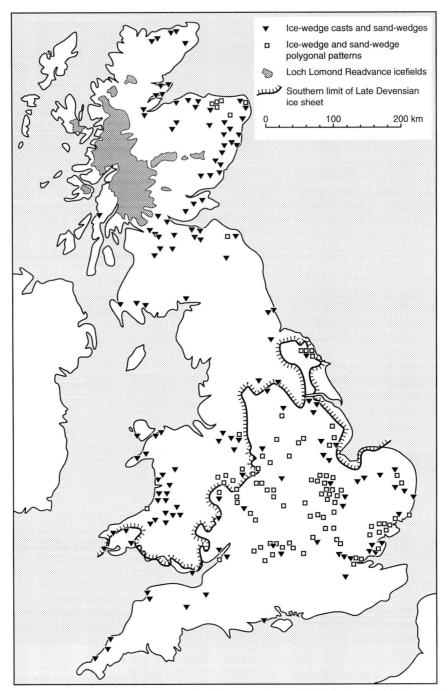

Figure 4.26 Distribution of wedge casts and polygonal crop marks of probable Devensian age in Britain. Based on Galloway (1961c), Worsley (1966a,b), Williams (1968); Watson (1977b), West (1977), Clayton (1979), Straw (1979), and Rose *et al*. (1985a), with additions by the authors.

age of ice-wedge formation at Stanton Harcourt remains unresolved.

Despite such problems of dating, the examples cited above indicate that at least one episode of permafrost development occurred in southern England within the long and enigmatic time interval that separated the Hoxnian and Ipswichian Interglacials. In Britain, however, the status of Middle Pleistocene cold stages remains in the realm of conjecture and contention. Resolution of the detailed permafrost stratigraphy of this period represents a major challenge for Quaternary geologists.

The Devensian Glacial Stage

Much of East Anglia, like most of southern Britain, lay beyond the limits of the Late Devensian ice sheet (Figure 4.26). Early Devensian sediments and structures in these areas have therefore escaped erosion by later glacial activity. At Wretton in Norfolk, for example, fluvial sands and gravels deposited by a braided river system under a predominantly periglacial climate directly overlie Ipswichian Interglacial deposits (West *et al.*, 1974). These fluvial sands and gravels constitute the lower ter-

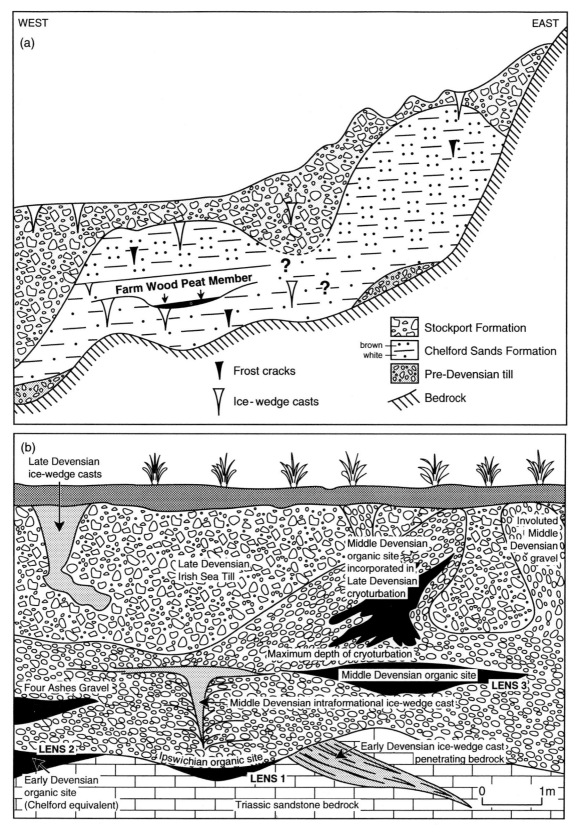

Figure 4.27 (a) Schematic cross-section through the Pleistocene succession of east Cheshire, showing the stratigraphy of ice-wedge casts of probable Devensian age. After Worsley (1987). (b) Schematic cross-section through the Pleistocene succession exposed at the Four Ashes pit in Staffordshire between 1967 and 1970. After Morgan (1973).

Table 4.1: *Summary of the stratigraphy at Chelford.*

- - - - - - - - - - suprafomational ice-wedge casts and sand wedges
Stockport Formation Till (Dimlington Stadial, c. 26–13 ka BP)
- - - - - - - - - - - - - - - - interformational ice-wedge casts- - - - -
Chelford Sands (brown) intraformational frost cracks
- - - - - - - - - - - - - - - - intraformational ice-wedge casts- - - - - -
Chelford Sands (white)
Farm Wood Peat Member (Chelford Interstadial, c. 100 ka BP)
- - - - - - - - - - - - - - - - intraformational ice-wedge casts- - - - - -
Chelford Sands (white) intraformational frost cracks
- -

race of the River Wissey and and incorporate layers of wind-blown coversand together with various periglacial structures including ground-ice depressions (Chapter 5), involutions (Chapter 6) and ice-wedge casts. Pollen analysis of organic muds preserved within the ground-ice depressions and in channels located beneath, within and above the terrace sands and gravels indicated a generally cold continental grassland environment interrupted by two temperate woodland substages (West *et al.*, 1974), but later studies of insect faunas and vertebrate remains have cast doubt on the reality of these temperate episodes (Coope, 1975; Stuart, 1977). Ice-wedge casts are present in the fluvial sediments deposited after both of the supposed 'woodland' phases at Wretton. These casts are often truncated by erosional surfaces within the fluvial sequence, and it is likely that beheading and casting of wedges resulted from the migration of braided river channels and their underlying taliks.

At the Chelford type site on the Cheshire Plain there is a more secure stratigraphic framework (Table 4.1; Figure 4.27(a)). Intraformational wedge casts occur within the fluvial Chelford Sands Formation, interformational casts occur at the boundary between this and the overlying Stockport Formation till and suprafomational casts penetrate the surface of the Stockport Formation till (Worsley, 1966a; 1987). An important stratigraphic marker at this site is provided by the Farm Wood Peat Member, which is correlated with the Chelford Interstadial (Worsley, 1985). This interstadial organic deposit occupies a channel within a sequence of white sands. Associated with it is a contemporaneous landsurface that is represented in the succession by a thin palaeosol. Wedge structures occur immediately below the organic channel fill and extend down from the palaeosol (Figure 4.27(a)). The former landsurface represented by the palaeosol was apparently stable throughout the Chelford Interstadial and possibly longer. It is possible that the wedge casts below the channel fill and associated palaeosol represent ice-wedge formation in the Early Devensian immediately prior to the onset of interstadial conditions. However, Rendell *et al.* (1991) have argued that the Chelford Sands underlying the palaeosol may antedate the Ipswichian Interglacial, in which case the associated wedge casts may be of much greater antiquity. No ice-wedge casts have been observed immediately above the interstadial deposit, though deep seasonal freezing after the end of the interstadial may be indicated by structures thought to have been formed by thermoerosional processes (Worsley, 1987).

Some distance above the Chelford Interstadial horizon, a

major unconformity terminates the white sands. Above this boundary the Chelford Sands are brown in colour. The unconformity truncates a series of ice-wedge casts at the top of the white sand sequence. A phase of tundra polygon formation and decay therefore preceded erosion of the white sands and subsequent accumulation of the brown sands. The brown sands contain vertical crack-like structures marking incipient ice wedges (ice veins) or frost cracks. In places the original upper surface of the brown sands appears to have been preserved, and further ice-wedge casts descend from this surface. These interformational casts are filled with sediment derived from the overlying (Late Devensian) Stockport Formation till, leading Worsley to conclude that casting postdated the advance of the last ice sheet over this site. Finally, suprafomational ice-wedge casts within the Stockport Formation till mark the development of continuous permafrost within the glacial deposits exposed by retreat of the last ice sheet. This probably occurred immediately following ice-sheet deglaciation during the later stages of the Dimlington Stadial, between *c.* 18 ka BP and *c.* 13 ka BP.

The Chelford sequence demonstrates the presence of permafrost prior to the Chelford Interstadial, either during the Early Devensian or prior to the Ipswichian Interglacial, and again at some time following the Chelford Interstadial. It also proves that permafrost was present at this site both immediately before and some time after the encroachment of the Late Devensian ice sheet across Cheshire. A similar palaeoclimatic interpretation is indicated by the sequence of deposits exposed in a gravel pit at Four Ashes in Staffordshire (Morgan, 1973). Here several organic lenses are interstratified with fluvial gravels (Figure 4.27(b)). At the base of the sequence is an organic deposit considered, on the basis of palaeobotanical evidence, to be of Ipswichian Interglacial age. A second organic lens, higher in the sequence, has been correlated on biostratigraphic grounds with the Chelford Interstadial. A third organic unit, higher again in the gravel sequence, has been attributed to the Upton Warren Interstadial on the basis of both palaeobiology and radiocarbon dating. Two phases of ice-wedge formation and decay apparently preceded the deposition of Late Devensian till at Four Ashes: an Early Devensian phase immediately postdating the Chelford Interstadial, and represented by ice-wedge casts that penetrate through the gravels into the underlying bedrock; and a later phase that followed the Upton Warren Interstadial but predated the arrival of the Late Devensian ice sheet. As at Chelford, suprafomational ice-wedge casts penetrate the surface of the Late Devensian till, indicating the re-establishment of continuous permafrost following ice-sheet deglaciation (Morgan, 1971a, 1973).

Evidence for the presence of continuous permafrost in NE Scotland during part of the Early Devensian has been presented by Connell & Hall (1987). A quarry section at Kirkhill in Buchan revealed a complex sedimentary sequence that includes two till units separated by periglacial slope deposits. At the top of the lower till unit, beneath the periglacial sediments, is a palaeosol of probable Ipswichian Interglacial age (Connell, *et al.* 1982; Connell & Hall, 1987). The lower till is therefore considered to be pre-Ipswichian in age. The upper till has been assigned to an Early Devensian glaciation on the basis of regional lithostratigraphy. Ice-wedge casts are developed in the intervening periglacial slope deposits. If the age of the upper till unit has been correctly interpreted, these wedge casts imply

Table 4.2: *Characteristics of ice-wedge casts of assumed Devensian age.*

Location	Dimensions	Host sediment	Cast fill	Source
Sites outside the limits of the Late Devensian ice sheet				
NE Yorkshire	Ice-wedge casts 3 m long; polygons 30 m in diameter.	Jurassic grits	Sand and till	Dimbleby (1952)
Highcliffe, Hampshire	Ice-wedge casts 3 m long and 0.9–1.2 m wide.	Gravel	Gravel	Lewin (1966)
Tregunna, Cornwall	Two ice-wedge casts 0.5 m wide.	Head	Gravel	Scourse (1987)
Sites inside the limits of the Late Devensian ice sheet				
Congleton, Cheshire	Sand-wedges 0.7–2.9 m long.	Late Devensian till	Sand, with stones in top 0.5 m	Worsley (1966b)
Dyfed, Wales	Ice-wedge casts 3m long. Length : width ratios < 10:1.	Late Devensian outwash	Sand and gravel	John (1973)
	Frost fissures up to 4 m long. Length:width ratios >10:1.			
	Frost cracks up to 0.7 m long. Length:width ratios > 30:1.			
Gwynedd, N Wales	Ice-wedge casts up to 2.7 m long.	Late Devensian outwash	Sand and gravel	Saunders (1973)
Scotland: 31 sites outside Loch Lomond Readvance	Ice-wedge casts up to 5 m long, mainly 1.5–3.0 m. Length: width ratios 4.3:1 to 7.5:1.	Late Devensian outwash	Sand and gravel	Galloway (1961c)
	Frost cracks with length: width ratios of 10:1 to 20:1			
Fife, Scotland	Ice-wedge cast 2.5 m long, 0.85 m wide.	Carboniferous shale	Sand and till	McManus (1966)

establishment of permafrost prior to the encroachment of an Early Devensian glacier across Buchan.

Although supraformational ice-wedge structures found immediately below the present ground surface in areas to the south of the Late Devensian ice limit in England and Wales may have developed during any of the Devensian permafrost phases discussed above (or indeed in some cases during pre-Devensian cold stages), it is often assumed that most developed or were last active during the Dimlington Stadial. At most sites datable material is absent, but at Syston in Leicestershire and Beckford in Worcestershire radiocarbon dating of organic silts has shown that terrace gravels containing large numbers of ice-wedge casts accumulated sometime after the Upton Warren Interstadial. At Syston, organic lenses dating to around 37 ka BP occur within the gravel terrace deposits of the Soar Valley. The gravels that overlie the organic lenses contain ice-wedge casts and cryoturbation structures (Bell *et al.*, 1972). Pollen and insect assemblages indicate terrace accumulation in a cold treeless environment, and the presence of ice-wedge casts near the top of the sequence suggests subsequent climatic cooling, the development of permafrost, and the formation of tundra polygons. At Beckford it appears that prior to about 27 ka BP solifluction was the dominant process, but later climatic cooling was accompanied by fluvial gravel aggradation, ice-wedge growth, aeolian activity, and finally by river incision (Briggs, Coope & Gilbertson, 1975).

The distribution of structures that have been interpreted as Devensian ice-wedge casts, sand-wedge casts and tundra polygons is shown in Figure 4.26. A summary of the characteristics of wedge structures that have not been discussed in detail earlier in this chapter is given in Table 4.2. Sites are subdivided into those to the south of the Late Devensian ice limit, where wedge casts may date from one of several Devensian permafrost phases, and those to the north of the ice limit, where supraformational wedge formation must have followed ice-sheet deglaciation after *c.* 18 ka BP. There remains, however, considerable uncertainty concerning the presence of widespread continuous permafrost during the subsequent Loch Lomond Stadial of *c.* 11–10 ka BP. The nature of the evidence for ice-wedge formation at this time is discussed below.

Evidence for ice-wedge formation during the Loch Lomond Stadial

Wedge casts have been reported to occur in sediments of Loch Lomond Stadial age at low altitudes on the Island of Mull and in Strath na Sealga in NW Scotland (Sissons, 1976c, 1977b), but the status of these casts is questionable. Descriptions of those

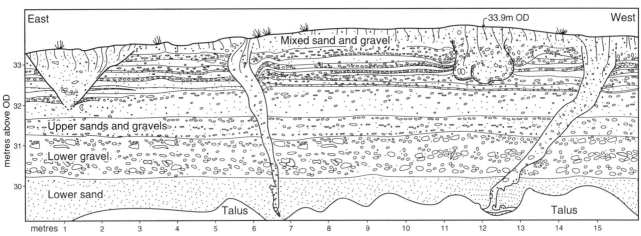

Figure 4.28 Supposed ice-wedge casts and periglacial involutions at Old Kilpatrick, on the north bank of the River Clyde near Glasgow. After Rose (1975).

on Mull have never been published and the supposed 'wedge cast' mapped by Sissons (1977b) in Strath na Sealga is extremely dubious (Ballantyne, unpublished data). Wedge casts in raised beach gravels below the Lateglacial marine limit at two sites in western Scotland may imply that continuous permafrost developed at sea level during the stadial (Gailey, 1961; Rose, 1975) but the evidence at both sites is equivocal. Rose has described two deep, thin wedge structures that penetrate raised beach deposits and underlying outwash sands and gravels at Old Kilpatrick on the Firth of Clyde. The wedges are 0.9 m wide, and extend downwards for a distance of 5.5 m from immediately below the ground surface. Their lower parts are somewhat irregular in outline and one wedge apparently grades downwards into a thin horizontal layer of gravelly sand (Figure 4.28). Gravels adjacent to the casts show both upward and downward deformation against the sides of the wedges. Associated with these wedge structures are several near-surface bowl-shaped structures and a broad V-shaped deformation. Rose interpreted the wedges as epigenetic ice-wedge casts, but offered no explanation for the associated shallower structures. Worsley (1987) has cast doubt on Rose's interpretation, suggesting that both the near-surface structures and the wedges may have resulted from processes of soft rock deformation associated with dewatering. He suggested that the wedge structures may have been injected upwards from the basal sands through the overlying beach deposits. The age of these structures is also contentious. On the basis of evidence provided by radiocarbon dates from marine shells and the altitude of the Old Kilpatrick site, Rose argued that the raised beach deposits are of Windermere Interstadial age (*c.* 13–11 ka BP) and that ice-wedge formation took place at this site during the succeeding Loch Lomond Stadial. Worsley (1987), however, pointed out that the shoreline chronology is ambiguous, though it does not preclude a Loch Lomond Stadial age for these structures. Taken overall, it appears that the evidence presented by Rose (1975) for low-level continuous permafrost at Old Kilpatrick during the Loch Lomond Stadial must be treated with caution.

Gemmell & Ralston (1984) have described eight sites in NE Scotland with polygonal crop marks, most of which occur on the gently-sloping surfaces of outwash terraces or raised beach deposits. They argued that in at least two cases (Pugeston and

Redcloak, near Montrose and Stonehaven respectively), ice-wedge polygons date from the Loch Lomond Stadial. Armstrong & Patterson (1985), however, presented evidence that retreat of the Late Devensian ice sheet in this area took place under cold climatic conditions, and that all the sites described by Gemmell & Ralston may reflect ice-wedge formation during the later stages of the Dimlington Stadial. It must be concluded, therefore, that no examples have yet been discovered of wedge casts of undisputed Loch Lomond Stadial age. Although other geomorphological evidence points to the presence of at least *discontinuous* permafrost in Britain during the Loch Lomond Stadial, the case for the development of widespread *continuous* permafrost at this time remains equivocal (Chapter 14).

The significance of wedge casts and polygonal crop marks

The above discussion highlights the widespread occurrence of wedge casts and polygonal crop marks throughout Great Britain. Though the ages and dimensions of these features vary considerably, most imply the former presence of continuous permafrost and winter temperatures sufficiently low to induce thermal contraction cracking to depths in excess of 2 m. Smaller wedge structures and narrow cracks a few millimetres wide and less than a metre long, may, however, represent thermal contraction cracking of seasonally frozen ground, in which case a continental climate with a large annual temperature range is indicated. Although such soil-wedge casts may sometimes be distinguished by their stratigraphic position (at, rather than below the contemporaneous ground surface) or by the nature of their infill, they add to the uncertainty of identifying 'true' wedge casts indicative of permafrost. Small-scale polygonal cracking may also result from desiccation rather than thermal contraction, so that small crack structures must be interpreted with caution. Indeed, some larger wedge-like structures formed by a range of processes not related to permafrost may resemble 'true' ice-wedge casts or sand wedges when seen in section, and it is likely that some geomorphologists have been overenthusiastic in interpreting them as such. Despite this difficulty, wedge casts provide the most valuable indicators of former permafrost in the British Pleistocene succession. They demonstrate that continuous permafrost underlay extensive areas during the

Beestonian, Anglian and Devensian cold stages, as well as during the ill-defined cold stage(s) that separated the Hoxnian and Ipswichian Interglacials. Ice-wedge casts and sand wedges have also provided vital evidence in terms of the detailed climatostratigraphy now established for the Devensian Glacial.

It is of interest to consider briefly the implications of areas in which firm evidence for polygonal marking and wedge casts have not been found. In SW England, for instance, despite the abundant evidence for other manifestations of former periglacial activity (particularly frost-weathered bedrock and thick deposits of head), there are no reports of polygonal markings and wedge structures are generally too small to be considered true permafrost wedge casts. This suggests that though this area certainly experienced periglacial conditions during several Pleistocene cold stages, these were insufficiently severe to induce the development of *continuous* permafrost. Alternatively, winter temperatures may have been insufficiently cold to cause widespread thermal contraction cracking of frozen ground. Even more conspicuous is the absence of polygonal markings and wedge casts within areas occupied by Loch Lomond Stadial glaciers in the mountainous parts of Britain. The doubtful nature of the few sites where possible Loch Lomond Stadial ice-wedge casts have been recognised was noted above. The broader question of whether widespread continuous permafrost existed in Britain during the Loch Lomond Stadial is explored in greater detail in Chapter 14.

Detailed analyses of wedge cast form and stratigraphy, together with associated host sediment deformation structures, have provided valuable information concerning local palaeoenvironmental circumstances. Degradation of ice-rich permafrost is indicated by the deformed nature of wedge casts and by slump structures in adjacent sediments. Truncation and superimposition of casts within fluvial gravel sequences implies the presence of permafrost during sediment aggradation in a braided river system with unstable migratory channels. The implications of sand wedges are less certain. Although sand wedges are particularly characteristic of cold arid environments, they also develop locally under moister conditions, particularly where there is a nearby source of windblown sandy fines. Caution must therefore be exercised in inferring former regional aridity from relict sand wedges.

The pages of this chapter have demonstrated the importance of wedge structures and relict tundra polygons in Pleistocene palaeoenvironmental reconstructions. These features are, however, by no means the only periglacial structures or landforms indicative of former permafrost conditions, as the following chapters will show.

5
Pingos and related ground-ice phenomena

This chapter considers the geomorphological consequences of the development and subsequent melting of massive ice bodies within near-surface sediments. Such *intrusive ice* or *injection ice* forms in two main ways. First, water expelled ahead of the freezing front during freezing of coarse sediments may be intruded into adjacent unfrozen sediments, where it eventually freezes to form a massive body of ground ice. Second, subpermafrost groundwater in areas of discontinuous permafrost may be injected under artesian pressure into unfrozen sediment. Subsequent freezing of the injected water again leads to the development of massive ground ice. In either case the resulting ice body may cause bulging of the surface and the growth of a blister-like hill, called a *pingo* (Worsley, 1986; Pissart, 1988). Alternatively, sill-like intrusions of massive ground ice may develop, and these, unlike pingos, may have little or no surface expression. Pingos are generally circular or oval in plan, and may achieve diameters of up to 600 m and heights of up to 50 m (Figure 5.1). Markedly elongate forms have also been reported. On Prince Patrick Island in the Canadian arctic, for instance, Pissart (1967) observed a pingo 1300 m long and 8.75 m high that formed an esker-like ridge. Similar features occur on Banks Island (Pissart & French, 1976; French & Dutkiewicz, 1976) and Spitsbergen (Liestøl, 1977). Pingos develop in a variety of unconsolidated sediments, including till, slope deposits and alluvial silts, sands and gravels.

Melting of the ice core of a pingo leaves a more or less circular depression surrounded by a low ridge or *pingo rampart* (Figure 5.2). Such depressions have been termed *pingo scars* and have been found at various locations in Europe, including England and Wales. Not all circular depressions represent pingo scars, however, and identification of relict pingos requires careful consideration of site characteristics and comparisons with modern analogues. Moreover, pingo scars are not the only manifestation of the melt-out of massive ground ice bodies. Thawing of ice-rich permafrost in arctic lowlands causes differential ground subsidence and widespread slope instability (Harry, 1988). The resulting *thermokarst topography* is irregular and is often dominated by enclosed hollows that mark sites formerly underlain by massive ice beds. In the North American arctic, thermokarst terrain is dominated by shallow *thaw lakes*. In many parts of Siberia, however, the characteristic thermokarst forms are wide, shallow depressions termed *alas*, which often do not contain lakes. Research in eastern England has suggested that some areas of the Fenlands may partly owe their surface configuration to thermokarst processes associated with permafrost degradation towards the end of the Late Devensian glacial stage.

Characteristics of active pingos

Classification of pingos is based on the mechanism by which water is supplied to the growing ice core. Pingos resulting from water expulsion during permafrost aggradation are termed *closed-system pingos* (Mackay, 1972, 1979a), while those fed by subpermafrost groundwater percolation are known as *open-system pingos* (Müller, 1959; Holmes, Hopkins & Foster, 1968). Closed-system pingos are particularly well developed in the Mackenzie Delta area of Canada, where the formation of thaw lakes has led to the development of taliks in the underlying permafrost. Should drainage of a lake occur, due for instance to fluvial erosion, refreezing of the talik results partly from upward and inward advance of the permafrost. Expulsion of porewater during refreezing causes water to collect in unfrozen sediment below the centre of the former lake. This water eventually freezes to form the pingo ice core (Figure 5.3). Mackay (1977b, 1979a, 1983b) has demonstrated the development of a lens of water beneath growing closed-system pingos. Such sub-pingo water progressively freezes onto the base of the ice core, causing an increase in pingo height. This water is under sufficient pressure to support the overburden weight, but loss of water through lateral faults in pingos may lead to surface subsidence. The shape of closed-system pingos reflects the planform of the taliks from which they have developed. Circular to elliptical taliks beneath thaw ponds produce dome-shaped pingos, while refreezing of the long narrow taliks that underlie the channels of large rivers results in the formation of elongated ridge-like pingos (Pissart & French, 1976; French & Dutkiewicz, 1976). Mackay (1979a) stressed the continuum that exists between pingo mounds and areas underlain by tabular sill-like layers of injection ice. In addition, some pingo cores may contain a significant proportion of segregation ice.

Measurements by Mackay (1981a) in the Mackenzie Delta area suggest that closed-system pingos grow rapidly at first, but that their rate of growth decreases with time. One large pingo was shown to be currently growing at about 2.3 cm yr^{-1}, and is estimated to be between 1100 and 1500 years old (Mackay, 1986b). Others, on Banks Island, are probably more than 3500 years old (French & Dutkiewicz, 1976; Pissart & French, 1976). As growth proceeds, the sides of pingos become steeper and tensional stresses develop near their summits. These eventually cause opening of radial dilation cracks that expose the ice core to melting (Figure 5.4), and the pingo begins to subside. Complete melting of the ice core leaves a crater-like depression surrounded by a rampart (Figure 5.2).

Open-system pingos result from groundwater movement beneath shallow discontinuous permafrost. Water that infiltrates into the ground on hillsides percolates downslope through permeable slope sediments or fractured bedrock, causing the gener-

Figure 5.1 A large closed-system pingo near Tuktoyaktuk in the Mackenzie Delta area, NWT, Canada.

Figure 5.2 A collapsed pingo near Tuktoyaktuk in the Mackenzie Delta area. Photograph by Professor T.L. Péwé.

(a)

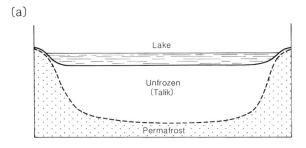

(b)

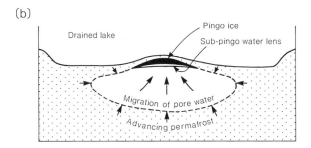

(c)

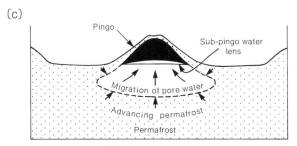

Figure 5.3 Formation of closed-system pingos (after Mackay, 1983b). (a) Talik develops beneath a thaw lake. (b) Lake drainage leads to refreezing of the talik as the permafrost table advances inwards; this is associated with porewater expulsion. (c) Progressive refreezing of the talik leads to pingo growth.

Figure 5.4 Radial dilation cracks in the surface of a small closed-system pingo in the Mackenzie Delta area. The pingo is approximately 125 m in diameter and 12 m high. Photograph by Professor J.R. Mackay.

ation of artesian groundwater pressures under permafrost at the foot of the slope or on the valley floor. Injection of this subpermafrost groundwater into the weakest portion of the permafrost is followed by freezing to form a pingo core (Figures 5.5 and 5.6). As with closed-system pingos, continued growth eventually leads to the opening of dilation cracks at the surface, exposure of the ice core to melting and collapse of the pingo. In central and northern Alaska, for example, open-system pingos are more frequent below south-facing slopes (Holmes *et al.*, 1968; Hamilton & Obi, 1982). On the higher parts of such slopes, larger and more frequent openings in the permafrost allow infiltration of surface water and resultant build up of artesian pressures beneath the shallow permafrost downslope. In central Alaska, open-system pingos occur in clusters, with old pingo scars often disturbed by the growth of later pingos, indicating a cycle of growth and decay. These pingos often occur at or near the transition between slope sediments and valley fill deposits. In the Brooks Range, most pingos are associated with thick deposits of colluvium or alluvium, but in some cases ice cores have developed at shallow depth beneath bedrock, at sites where fracture systems have concentrated groundwater flow (Hamilton

& Obi, 1982). Open-system pingos are usually round or elliptical in plan and are generally smaller than the largest closed-system forms. In Central Alaska, for instance, they reach maximum widths of less than 500 m and heights of up to 35 m (Holmes *et al.*, 1968). Dating of organic silts exposed in the sides of a 27 m high open-system pingo in the Brooks Range indicated a maximum age of 1015 ± 80 yr BP (Hamilton & Obi, 1982), and most open-sytem pingos in North America are estimated to be hundreds if not thousands of years old (Holmes *et al.*, 1968).

A particular relationship between glaciers and open-system pingos has been observed in Spitsbergen by Liestøl (1977). Here permafrost is largely continuous, but the glaciers have a subpolar thermal regime, so that permafrost extends only beneath their marginal zones. Beneath the warm-based ice in the central parts of glaciers, meltwater percolates into the unfrozen substrate then migrates under hydrostatic pressure beneath the permafrost that underlies the glacier margins (Figure 5.7). In consequence, artesian pressures develop in the sub-permafrost groundwater beyond the glacier margins, allowing the growth of open-system pingos in areas of thinner or weaker permafrost.

Degradation of open-system pingos in the continental discontinuous permafrost zone of Mongolia has been described by Babinski (1982). Here, at a latitude of approximately 47° N, summer temperatures are much higher than in the arctic (up to + 41 °C) and summer insolation is much stronger. A range of thaw processes occur, including the effect of summer insolation on the south-facing sides of pingos. This causes instability in the active layer, resultant exposure of the pingo core and lateral migration of a thaw pond across the pingo. Rapid spring thawing may also lead to slumping of sediment down the sides of pingos, which in turn exposes their cores to melting. In arctic areas, solifluction and permafrost creep also cause movement of sediment down the flanks of pingos. As a result, peripheral sediments often show downslope distortion and overturning (Figure 5.8). Since pingo sides may be as steep as 45–50 °, rates of mass movement may be high. Transported sediment is deposited

(a)

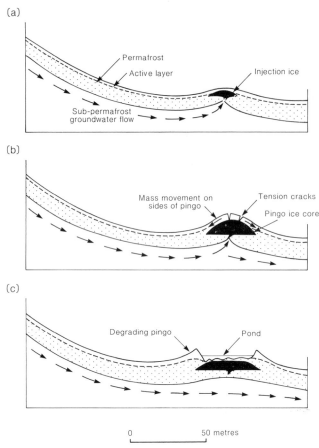

(b)

(c)

Figure 5.5 Cycle of formation and decay of open-system pingos. After Holmes *et al.* (1968).

around the pingo margins, so that when the ice core melts, the pond marking its former location is surrounded by an annular ridge or rampart (Figure 5.2). Thickening of near-surface sediments at pingo margins may also result from lateral compression during pingo growth. Where this has occurred, marginal sediments may display compressional deformation structures.

Other ice-cored mounds

Cryogenic mounds somewhat smaller than pingos are common within the discontinuous permafrost zone and also occur in areas of sporadic permafrost. Among the most common such forms are *palsas* and *mineral palsas*. Palsas are small mounds of peat, generally a few metres high, with perennially frozen cores containing high concentrations of segregation ice (White, Clark & Rapp, 1969; Seppälä, 1972, 1986, 1988). They occur in bogs in the discontinuous permafrost zone. Localised deflation of snowcover is thought to create areas of more severe winter cooling where perennially-frozen ground may survive. Ice segregation causes updoming of the surface and the creation of a palsa mound. Like pingos, palsas evolve through a cycle of growth and decay, and melting of core ice may leave small shallow ponds marking their former locations. However, since palsas are developed in peat bogs, such features are unlikely to survive for long following climatic amelioration, and relict Pleistocene forms have not been found.

Of greater palaeoenvironmental significance are mineral palsas (Dionne, 1978; Pissart, 1983; Gangloff & Pissart, 1983), also referred to as *cryogenic mineral mounds* (Payette, Samson & Lagarec, 1976; Lagarec, 1982) or *permafrost mounds* (Åkerman & Malmström, 1986). Mineral palsas are mounds developed in mineral soils rather than peat, and are differenti-

Figure 5.6 Ice core of a small open-system pingo exposed by placer mining near Fairbanks in Alaska.

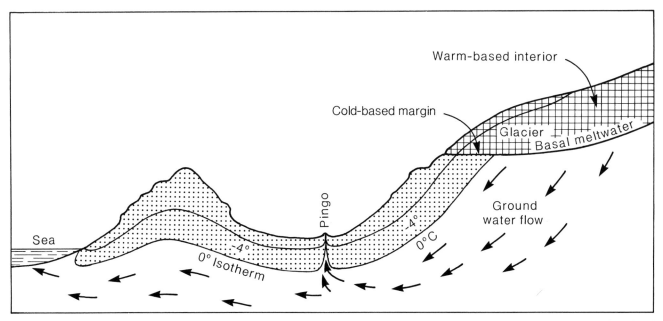

Figure 5.7 Formation of open-system pingos in Spitsbergen, due to meltwater migration from subpolar glaciers under adjacent permafrost. After Liestøl (1977).

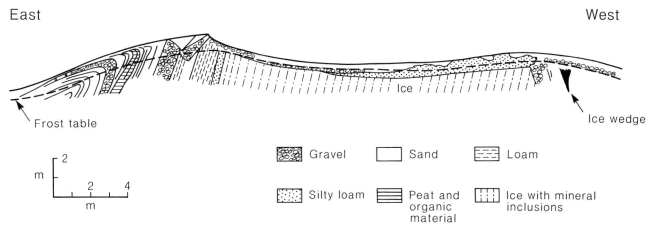

Figure 5.8 Section through the flanks of a pingo on Banks Island in the Canadian arctic. After Pissart & French (1976).

ated from pingos on the grounds that they reflect the growth of segregation ice rather than injection ice (Wrammer, 1973; Pissart, 1983). However, as many pingos also contain significant amounts of segregation ice (Mackay, 1979a), the distinction between the two types of landform may be less clear-cut than this criterion suggests. A detailed description of mineral palsas developed in marine silts and clays in subarctic northern Québec has been provided by Pissart & Gangloff (1984). The mounds in this area are up to 40 m in diameter and 6 m in height, and excavation has revealed updoming of clays and overlying gravels above a core of ice-rich frozen silt. Thawing of these mineral palsas leaves a circular ridge surrounding a thermokarst pond, the ridge consisting of sediment layers dipping parallel to the former surface slope of the mineral palsa. Such ridges were considered by Pissart & Gangloff (1984) to result partly from lateral compression and tilting of sediments during mound growth, and partly from mass wasting of sediment down the flanks of the mounds.

The absence of a clear distinction between cryogenic mounds with cores of injection ice and those with cores of frozen soil and segregation ice has been emphasised by Åkerman & Malmström (1986). They showed that in the discontinuous permafrost zone of northern Sweden, permafrost mounds 20–50 m in diameter and 3–8 m high contain both injection ice and segregation ice. The mechanism of mound formation they proposed resembles that for open-system pingos, but they suggested that water migration and injection-ice formation take place initially beneath seasonally-frozen ground rather than permafrost (Figure 5.9). They inferred that selective deflation of snow from the resulting small injection-ice mounds promotes severe localised winter cooling, segregation ice growth and survival of the ice-rich core through subsequent summers. The development of tension cracks as a result of mound growth may eventually lead to exposure and thawing of the icy cores of such mounds, and ultimately to the development of thermokarst ponds encircled by low ramparts. Small circular open-system pingos and pingo-like mounds in northern Scandinavia have also been described by Lagerbäck & Rodhe

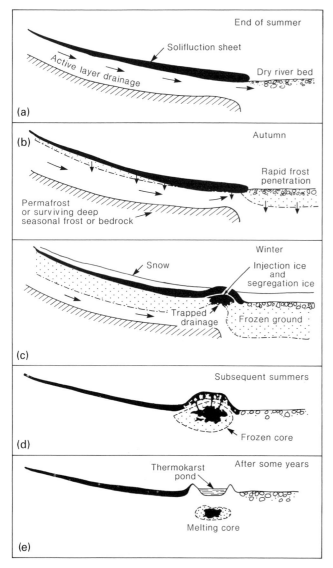

Figure 5.9 Suggested mechanism for the growth and decay of pingo-like ground-ice mounds in the Abisko area of northern Sweden. After Åkerman & Malmström (1986).

(1985, 1986). The pingos they observed are 20–80 m in diameter, 2–7 m high and occur at the bottoms of slopes. A borehole through one such pingo revealed a mantle of till and a core comprising 4.5 m of intercalated layers of ice, silt and sand underlain by over 3 m of pure ice. The upper ice layers may represent segregation ice, but the underlying massive ice may reflect injection. Other ground-ice mounds in the area were interpreted as being equivalent to mineral palsas. Climatically-induced decay of both the pingo-like forms and palsa-like forms has resulted in the development of circular thaw ponds up to 60 m in diameter, often surrounded by low ramparts.

Small *seasonal frost mounds* may also form on permafrost terrain. These reflect updoming of the ground surface at sites where artesian pressures develop in groundwater trapped between the permafrost table and an overlying layer of seasonally-frozen ground (e.g. Pollard & French, 1984, 1985; Pollard, 1988), but are usually ephemeral features and are unlikely to survive in relict form. Surface ice accumulations called *icings*

or *aufeis* are also common in permafrost environments (Washburn, 1979). These consist of sheets of ice that accumulate at the surface during the autumn freezeback. Some are associated with rivers, and occur when hydrostatic pressures build up in water that is trapped under a layer of thickening surface ice. If the overlying ice cover ruptures, the trapped water escapes, overflowing river banks and freezing as it does so. Icings also develop at the base of slopes, where artesian pressures may be generated in water trapped between the permafrost table and the overlying freezing soil. If such water escapes to the surface, it again spreads out over the ground and freezes. Icings may be several metres thick and cover hundreds of hectares, and burial by alluvium may lead to their incorporation into the permafrost as ground-ice layers.

The climatic significance of pingos and related ground-ice features

As pingos form only in permafrost evironments, pingo scars in temperate environments such as Great Britain provide evidence of former permafrost. In particular, scars inferred to represent open-system pingos imply former discontinuous permafrost, whilst those attributed to closed-system pingos imply former continuous permafrost. The implications of such features in terms of limiting mean annual air temperatures (MAAT) are much more difficult to establish. Estimates of limiting MAAT for pingo formation have been derived in part by correlating the present MAAT in North America with the distribution of active pingos. On this basis the maximum or 'warm-side' MAAT has been estimated to lie between -1 °C and -3 °C for open-system pingos, and between -4 °C and -6 °C for closed-system pingos (cf. Washburn, 1979, 1980). Mackay (1988b), however, has advanced several trenchant arguments against this procedure. He pointed out that as many pingos form over hundreds or thousands of years, the present pingo distribution must have evolved over a long period of changing climate and is not necessarily related to present MAAT. He also noted the uncertainties implicit in attempting to relate MAAT to mean annual ground temperature (MAGT). Although the latter is *on average* 3–4 °C warmer than the former in North American permafrost environments, local factors such as snowcover and variations in thermal offset (the difference in mean annual temperature between the top and base of the active layer) imply that this relationship is not consistent. Moreover, rate of pingo growth is both time- and temperature-dependent; at MAGTs close to the threshold for pingo development, much longer time periods are required for pingos to grow to a size sufficient to produce a rampart upon collapse. Mackay calculated that for MAGTs of around -1 °C, hundreds of years would be required for small pingos to grow to full size, and large pingos would require periods in excess of a thousand years to achieve maturity. If climatic fluctuations during pingo growth are taken into account, Mackay argued, the limiting MAGT for small pingos is unlikely to exceed -2 °C, whilst the limiting MAGT for large pingos to begin growing would be about -3 °C to -4 °C. Assuming a difference of 3–4 °C between MAGT and MAAT, these figures correspond to a limiting 'warm-side' MAAT of -5 °C to -6 °C for small pingos and -6 °C to -8 °C for large pingos. Other problems relate to the fact that pingo distribution reflects availability of favourable sites as well as climatic limitations.

Because of the uncertainties associated with estimating ther-

Figure 5.10 Thaw lakes in the Mackenzie Delta area.

mal constraints on pingo growth, Mackay restricted his conclusions to the cautious statement that warm-side limiting MAATs previously inferred from pingo scars 'are several degrees celsius too warm'. Bearing in mind the caveats of Mackay's arguments, it is useful to consider the climatic implications of pingos that have developed under favourable conditions at the limits of their climatic tolerance. Such a situation occurs in northern Scandinavia, where small open-system pingos show signs of gradual decay (Lagerbäck & Rodhe, 1985). The present MAAT at the pingo sites is -3.5 °C, but the current degradation of these forms suggests that this is too high to sustain pingo growth, which was probably initiated under the colder conditions of the eighteenth and nineteenth centuries. Similarly, the small hybrid pingo-like mounds described by Åkerman & Malmström (1986) also appear to be near their climatic limits at sites where the MAAT is estimated to be around -4 °C. Such evidence suggests that *sustained* MAATs no higher than -4 °C to -5 °C are necessary to permit the growth of small open-system pingos. The growth of large open-system pingos would appear to imply a more rigorous regime, with mean annual air temperatures no higher than perhaps -6 °C to -8 °C. As Mackay (1988b, p. 508) has pointed out, specification of a limiting MAGT is much more difficult for closed-system pingos. It is therefore imprudent to infer precise limiting temperatures for formation of these landforms.

Thermokarst depressions

Disturbance of the thermal equilibrium of permafrost may be triggered either by climatic changes or by changes in ground surface conditions. If such disturbances cause a rise in ground surface temperature, *permafrost degradation* results, and is manifest in a lowering of the permafrost table. Degradation of ice-rich permafrost is accompanied by subsidence as ground ice thaws. The depth of subsidence depends partly on the depth of thaw, and partly on the ice content of the upper layers of permafrost prior to thaw. This may be considerable. In alluvial silts in the Mackenzie Delta area, for example, Pollard & French (1980) recorded ice contents of 50–70% by volume in the uppermost 5 m of permafrost. Similarly, in the thick loess (windblown silt) deposits of northern Alaska, ice accounts for 60–70% by volume of the upper permafrost to depths of at least 20 m (Carter, 1988). Regional climatic warming is therefore likely to lead to widespread disruption of the ground surface as zones of ice-rich permafrost subside (French & Egginton, 1973). Similar disruption may also accompany loss of vegetation cover due, for instance, to fire (Mackay, 1977c; Burn & Smith, 1988). Another cause of permafrost degradation is the development of small ponds that persist through to the autumn freezeback. Such ponds often form over low-centre tundra polygons, or as a result of widening and deepening of the troughs above large ice wedges (Harry & French, 1983; Chapter 4). Release of latent heat during freezing of pond waters locally reduces the severity of winter ground cooling and thus enhances degradation of the underlying permafrost. This may cause subsidence that further increases the depth of the pond. If pond water depth increases to more than one or two metres, then winter freezing may not reach the pond floor, and a talik develops in subjacent sediments. This in turn causes further subsidence as ground ice melts, and the pond grows into a shallow *thaw lake* (Figure 5.10).

Figure 5.11 Part of an alas developed in loess, 50 km north-west of Yakutsk, Siberia. Photograph by Professor T.L. Péwé.

Figure 5.12 Retrogressive thaw flow slide, Ellesmere Island, NWT, Canada. Note the ground ice exposed in the arcuate headwall.

Where a high excess ice content is restricted to the upper few metres of permafrost, as is the case in much of arctic Canada and Alaska, thaw lakes are only a few metres deep. However, in deposits where ice-rich permafrost persists to greater depths, such as the loess of northern Alaska, larger and deeper depressions may develop (Carter, 1988). The widespread *alas* depressions of Siberia also appear to reflect deep subsidence caused by the degradation of permafrost that supports or formerly supported abundant excess ice at depth (Czudek & Demek, 1970a; Figure 5.11). Bosikov (1988) has stressed the importance of small ponds in initiating localised permafrost degradation and hence alas development, and has shown that in Siberia such lakes are most likely to form during summers with higher than average precipitation.

Lateral enlargement of thaw lakes takes place mainly by thermoerosion of ice-rich permafrost exposed along lake shores. Small slumps occur as the permafrost thaws, and the sediment that is released spreads across the lake floor to form laminated deposits (French & Harry, 1983). Shore retreat may be rapid, particularly at sites where thick ground ice is exposed in the sides of larger depressions. A near-vertical scarp develops and an arcuate embayment may form as the exposed ground ice melts back. Near the base of the slope, water-saturated sediment released by scarp retreat forms mudflows (Figure 5.12). Such thaw-induced slope failures have been termed *retrogressive thaw flow slides* or *ground-ice slumps* (McRoberts & Morgenstern, 1974; Lewkowicz, 1987). They commonly have retreat rates well in excess of 10 m yr^{-1} (French & Egginton, 1973; Lewkowicz, 1987; Chapter 7). Many thaw lakes eventually drain, often through being tapped by migrating river channels or as a result of shoreline recession in coastal areas. Large alas depressions may simply dry up when summer evaporation rates exceed the supply of water from precipitation and ground-ice thawing. Subsequent refreezing of taliks underlying thermokarst depressions may lead to the growth of closed-system pingos (Figure 5.3).

Morphological evidence of Pleistocene pingos and related cryogenic mounds

The former presence of pingos or related ground-ice features has been inferred at sites where enclosed depressions, often surrounded by low ramparts, are preserved in the landscape. Most reported examples of relict pingo scars take the form of circular or oval depressions with diameters in the range 25–250 m. The majority occur in clusters that may contain hundreds of individual depressions. On slopes they tend to be elongated downslope. Pingo scars occur in a variety of sediment types and are often located on plains, valley floors and lower valley sides where groundwater seepage takes place (Flemal, 1976; De Gans, 1988). The depth of the basin left by pingo collapse depends on the size of the ice core and whether this contained a component of mineral-rich segregation ice as opposed to soil-free injection ice. As a result of sediment slumping into the thaw pond during decay, pingo basins are often broadly conical in form. In the Netherlands, pingo depressions range from 2 m to 17 m in depth, though depths of 4–5 m are characteristic (De Gans, 1988). Here, however, the pingo depressions are 100–350 m wide, rather greater than most reported examples.

Flemal (1976) has suggested that many Pleistocene pingo scars represent the cyclical growth and decay of open-system rather than closed-system pingos. This inference is supported by the common occurrence of interfering and superimposed forms. These suggest that several 'generations' of pingos developed at the same locality, a characteristic of open- but not closed-system forms (Müller, 1959). Most ramparted pingo scars therefore indicate the former presence of discontinuous permafrost. However, larger-scale depressions in the Netherlands are developed above impermeable clay, with the floors of the depressions rooted in sand layers. From this evidence, De Gans (1981, 1983, 1988) postulated that such scars represent the sites of closed-system pingos that developed under conditions of continuous permafrost. Radiocarbon dating of organic material buried under ramparts suggests that these Dutch pingos were active during the coldest part of the Late Weichselian (Late Devensian) cold stage.

The following diagnostic criteria have been proposed by De Gans (1988) for recognising true pingo scars.

(1) Minimum depth of depression 1.5 m and minimum diameter 25 m.
(2) Bottom of depression lies below the level of the surrounding ground and is floored by sediment that is sufficiently permeable to allow migration of groundwater.
(3) At least part of a rampart is present, and contains sediment derived from the depression, often as outward-dipping strata.
(4) Pingo scars occur on flat ground or slopes of up to 5 °.
(5) Pingo scars are accompanied by other permafrost phenomena.

Four rather different criteria, however, have been advocated by Mackay (1988b).

(1) The volume of the rampart should approximate that of the depression.
(2) The presence of peripheral deposits associated with mass wasting, stream deposition and debris flow.
(3) Casts of dilation crack ice trending across the ramparts.
(4) The presence of peripheral normal faults in ramparts.

Not all ramparted ground-ice depressions necessarily represent the sites of former pingos. Pissart (1956, 1965, 1974) and Pissart & Juvigne (1980) have described in detail an extensive area of ramparted depressions in the Hautes Fagnes area of Belgium (Figure 5.13(a)). Here circular depressions average around 80 m in diameter and ramparts are 1.0–7.5 m high. A section through one rampart provided clear evidence that mass movement of sediment down the sides of a former cryogenic mound was responsible for rampart formation (Figure 5.13(b)). A peat layer buried during rampart accumulation was shown to be of Allerød (i.e. Windermere Interstadial) age, suggesting that the ramparts and hence the associated mounds developed during the succeeding Younger Dryas (i.e. Loch Lomond Stadial) cold stage. However, the ramparted depressions of the Hautes Fagnes are developed not on a footslope or valley floor, but in weathered shales and quartzites on a plateau surface. The topographic location of these features therefore suggests that the ancestral mound forms are unlikely to have developed above a core of injection ice. Pissart concluded that the Hautes Fagnes depressions mark the former presence of mineral palsas rather than open-system pingos, and that these features developed when the ground was underlain by discontinuous permafrost.

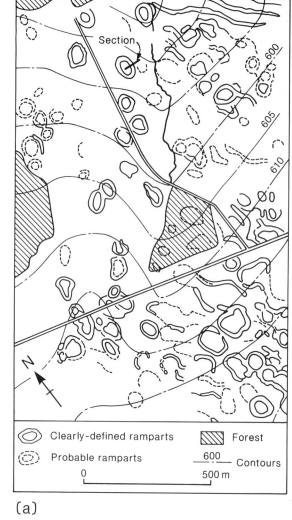

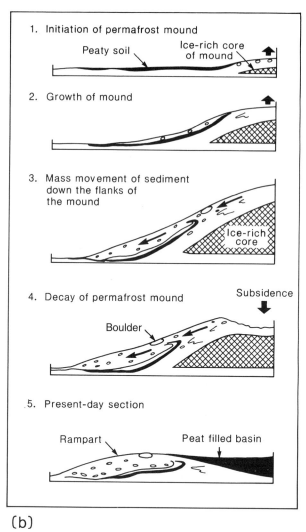

(a)   (b)

Figure 5.13 (a) Map of ramparted depressions in the Hautes Fagnes area, Belgium. (b) Section through a rampart and suggested sequence leading to its formation. After Pissart & Juvigne (1980).

Evidence for Pleistocene pingos and related cryogenic mounds in Britain

Features interpreted as pingo scars or 'ramparted ground-ice depressions' have been identified at various sites in southern Britain, particularly in Wales, East Anglia and the Thames basin. Examples have also been found as far north as the Whicham Valley in Cumbria and the Isle of Man (Figure 5.14). Such landforms therefore occur both south and north of the Late Devensian ice-sheet maximum in England and Wales, though none have yet been reported in NE England or Scotland. However, as Worsley (1977) has suggested, the distribution of known pingo scars and related features may reflect the spatial scope of investigations rather than their true distribution. A useful survey of known examples of ramparted ground-ice depressions in Britain by Bryant & Carpenter (1987) has shown that such features occur almost exclusively on gently-sloping ground close to valley floors. A similar distribution is evident for features interpreted as pingo remnants in Ireland, where many are associated with spring-lines at the foot of south-facing slopes (Coxon & O'Callaghan, 1987). In England and Wales, ramparted ground-

ice depressions occur on a variety of substrates, including fluvial and fluvioglacial sands and gravels, soliflueted chalk (coombe rock) and diamictons of glacial or periglacial origin.

Some of the most spectacular ramparted depressions in Britain are those investigated by Sparks, Williams & Bell (1972) in East Anglia (Figure 5.15). In their account of these features, however, Sparks and his fellow researchers cautiously eschewed the term 'pingo scar', preferring to refer to the East Anglian features by the broader generic term 'ground-ice depressions'. These depressions lie within the Anglian glacial limits, but beyond the limits of the Late Devensian ice sheet. They occupy almost driftless areas of low-lying chalk, between drift-mantled hills to the east and the Fens to the west. The depressions in this area are 10–120 m in diameter and up to 3 m deep. They tend to occur in clusters, in concentrations of 30–50 per square kilometre. The frequent occurrence of superimposed forms suggests that they represent several cycles of ground-ice growth and decay. Particularly striking examples of ramparted depressions occur at Walton Common in Cambridgeshire. At this site the bulk of the ramparts consists of a mixture of fine chalk rubble and sand, but their outer margins overlie sands with thin organic

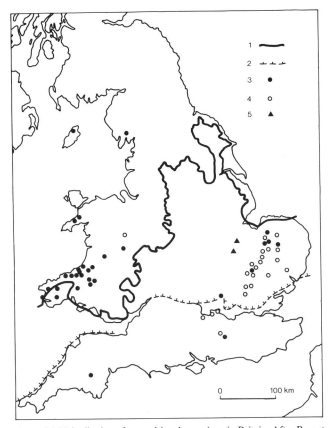

Figure 5.14 Distribution of ground-ice depressions in Britain. After Bryant & Carpenter (1987) and Hutchinson (1991). 1: Maximum extent of the Late Devensian Ice sheet. 2: Maximum extent of Pleistocene glaciation. 3: Relict open-system pingos or ground-ice mounds identified by surface relief. 4: Crop marks indicating possible relict open-system pingos or ground-ice mounds. 5: Sites of possible closed-system pingos.

Figure 5.15 Ramparted depressions, possibly representing open-system pingo scars, Walton Common, Norfolk. Irregular ridges result from the superposition of successive generations of ground-ice mounds. Cambridge University Collection of Air Photographs.

seams, suggesting that the ridges resulted primarily from mass movement down the sides of dome-shaped mounds.

Sparks *et al.* (1972) differentiated two generations of ground ice depressions on the basis of their morphology, namely fresh features with clearly-defined perimeter ramparts and older features with subdued ramparts that show up mainly as crop marks. Excavation of a pipeline trench to the south of Walton Common provided confirmatory evidence for two generations of depressions at this site. At one location Sparks and his co-workers observed a small depression nested within a larger depression, and separated from it by periglacial chalk rubble (Figure 5.16(a)). Pollen analysis of organic muds within the lower depression indicated that filling began at the end of the Dimlington Stadial, while peat occupying the upper depression yielded pollen assemblages of Flandrian age. From this evidence, Sparks *et al.* (1972) concluded that the older generation of depressions represents ground-ice mounds that developed during the Dimlington Stadial of *c.* 26–13 ka BP, whilst the fresher forms represent renewed formation of ground-ice mounds during the Loch Lomond Stadial of *c.* 11–10 ka BP. In general, features belonging to the older generation of depressions are more widely distributed than those of the younger generation. This suggests that the former may have developed under a more severe climatic regime with widespread permafrost, though it is equally possible that their wider distribu-

tion reflects more prolonged permafrost conditions during the Dimlington Stadial than during the short-lived Loch Lomond Stadial. It is noteworthy, however, that depressions of the younger generation tend to be restricted to zones of groundwater seepage in footslope locations (Figure 5.16(b)). Given the permeability of the underlying chalk, such sites must have been particularly favourable for ground-ice growth under conditions of shallow discontinuous permafrost (Worsley, 1977).

In southern England, a cluster of ramparted depressions has been found on gently-sloping ground on the southern valley side of the River Wey near Elstead in Surrey (Figure 5.14), a site underlain by uncemented well-sorted Cretaceous Greensands. One of these depressions has been investigated in detail by Carpenter & Woodcock (1981). The basin they studied is around 100 m long, 60 m wide and at least 3.5 m deep, and is surrounded by a low, discontinuous rampart of apparently structureless sand. This is least conspicuous on the upslope side, but stands up to 1 m above the valley-floor bog surface on the northern and south-western margins of the depression. Inner rampart slopes generally exceed 20 °, with a maximum gradient of 35 °, but distal slopes seldom exceed 5 °. The basin contains a few tens of centimetres of reworked sand and laminated silts, overlain by up to 3 m of detrital muds containing occasional lenses of clay and silt. Pollen analysis indicates that the oldest infilling deposits are of Loch Lomond Stadial age, but no dateable buried organic material has been found beneath the rampart. Carpenter & Woodcock interpreted the depression as the site of an open-system pingo of Dimlington Stadial age, noting that its location (on a spring line over permeable sands) would have been particularly propitious for the accumulation of injection ice beneath thin permafrost.

In west and mid-Wales, localities with ramparted ground-ice depressions lie both north and south of the Late Devensian ice-sheet limit (Figure 5.14). East of Llangurig, for example, circular or oval ramparts enclosing peat-filled depressions occur on fan gravels that overlie the floor of a large glacial valley inside

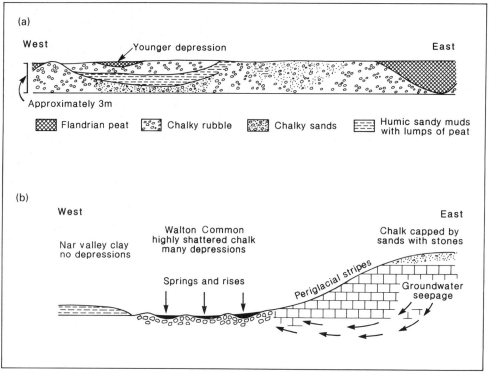

Figure 5.16 (a) Section through the southern part of Walton Common. (b) Hydraulic conditions at Walton Common. After Sparks *et al.* (1972).

the ice-sheet limit. On sloping ground, the ramparts are open-ended upslope and semicircular or horseshoe-shaped in plan, and many superimposed forms are present (Figure 5.17(c)). The depressions range from 42 m to 120 m in diameter, and augering of one site indicated a maximum depth of 6.5 m. A section through the lowest part of one semicircular rampart revealed imbricate gravel units flexed into a gentle anticlinal structure and overlain by sandy solifluction deposits. Pissart (1963a) interpreted this folding as indicative of sediment compression during pingo growth, and he inferred that the overlying deposits represent solifluction of sediment down the flanks of the former pingo. As with the Wey Valley pingo described above, the Llangurig site is characterised by a permeable substrate at a slope-foot location, a situation particularly favourable for the development of open-system pingos under conditions of shallow discontinuous permafrost.

Ramparted depressions also occur in many of the valleys of SW Wales (Figure 5.14). Those in the valleys of the Rivers Cledlyn and Cletwr have been described in detail by Watson (1971, 1972, 1976, 1977b) and Watson & Watson (1972, 1974). On gentle slopes the depressions and surrounding ramparts are roughly circular in plan, but on steeper ground ramparts tend to be horseshoe-shaped, with the upslope section missing (Figures 5.17 and 5.18). The depressions are mainly developed in poorly-sorted diamictons, probably soliflucted tills, that contain around 20% clay. In the Cledlyn Valley the external diameters of circular and oval depressions range from around 60 m to 180 m, though the long axes of elongate features attain *c.* 250 m. The ramparts are up to 5 m high. Augering in the centre of one roughly circular depression (W in Figure 5.17(a)) revealed 3.35 m of peat resting on a clay-silt fill more than 6.9 m thick (Figure 5.19). The basin beneath the fill appeared to be roughly sym-

metrical with very steep (52–55°) sides. On slightly steeper ground, a depression with a rampart on its downslope side and flanks (U in Figure 5.17(a)), proved to be underlain by a 9 m deep asymmetrical basin flanked by slopes of up to 66° on the downslope side and up to 60° on the upslope side (Figure 5.19). Immediately upslope of its deepest part the floor was irregular and hummocky. This was considered by Watson & Watson (1972) to represent upslope rampart material which had collapsed into the basin. In other similar depressions on ground with gradients of up to 6°, basins were shallow, suggesting that collapse of the upslope ramparts had largely infilled the underlying basins.

The ramparted depressions in the Cletwr Valley resemble the Cledlyn group but are in places more markedly elongate downslope, with complex superimposed forms (Watson & Watson, 1974; Figure 5.17(b)). Excavation of one rampart revealed steeply-inclined beds below the rampart crest, which were interpreted by Watson (1977a) as indicating outward compression associated with the growth of an ice core (Figure 5.20). Dipping clasts in the outer rampart slope suggest mass movement down the flanks of a former pingo-like mound. On the basis of their form and location on lower valley slopes and valley bottoms, Watson & Watson (1974) concluded that the Cletwr Valley depressions represent the remains of open-system pingos (Figure 5.21).

The age of the ground-ice mounds or pingos represented by the Cledlyn Valley and Cletwr Valley depressions is uncertain. In the Cledlyn Valley, radiocarbon dating of peat immediately overlying clay-silt basin fills yielded ^{14}C ages ranging from 10 080 ± 320 yr BP to 9360 ± 340 yr BP (Watson, 1977b). The validity of these dates was confirmed by pollen analyses that indicate commencement of peat accumulation at the beginning

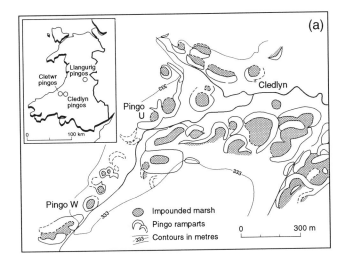

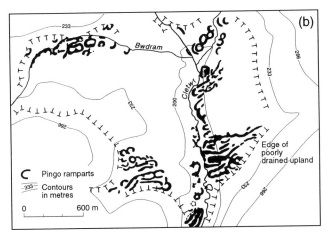

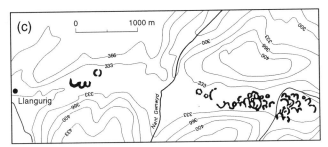

Figure 5.17 Possible open-system pingo scars: (a) Cledlyn Valley, Dyfed (after Watson & Watson, 1972); (b) Cletwr Valley, Dyfed (after Watson & Watson, 1974); (c) Llangurig, mid-Wales (after Pissart, 1963a).

Figure 5.18 (a) Ramparted depressions in the Gwili Valley near Llanpumsaint, Dyfed, Wales. Note the large U-shaped feature near the road and the small circular depressions in the right centre of the photograph and beyond the road. Cambridge University Collection of Air Photographs. (b) Part of a circular rampart and depression in the Cledlyn Valley, Dyfed, Wales.

of the Flandrian (present) Interglacial (Handa & Moore, 1976). The dates suggest that ground-ice mounds last developed at these sites under conditions of discontinuous permafrost during the Loch Lomond Stadial of 11–10 ka BP. Watson (1977b) has even suggested that pingo growth may have begun with a change from continuous to discontinuous permafrost at the beginning of the Windermere Interstadial of 13–11 ka BP, and continued during the Loch Lomond Stadial. It is, however, equally possible that pingos last developed at these sites during the Dimlington Stadial (26–13 ka BP). If so, the apparent absence of organic sediments of Windermere Interstadial age in the basins underlying the ramparted depressions may be explained in three possible ways: (1) survival of ground ice through the Windermere Interstadial; (2) nonaccumulation of Windermere organic deposits; and (3) burial of Windermere Interstadial deposits under minerogenic sediments during the Loch Lomond Stadial. As many kettle holes and other depressions received a substantial infill of minerogenic sediment during the Loch Lomond Stadial, the final explanation may be the most plausible.

The interpretation of these features as the remnants of open-system pingos has been questioned by Pissart & Gangloff (1984) on the grounds that the substrate may have been insufficiently permeable to permit the formation of injection ice. They argued that the ramparted depressions in Wales represent collapsed mineral palsas rather than pingos. However, given the similarity between the scars of open-system pingos and those formed by the collapse of mineral palsas in terms of both morphology and location, this argument cannot be resolved on present evidence. The two groups of features, moreover, may actually form the end-members of a continuum of ground-ice forms,

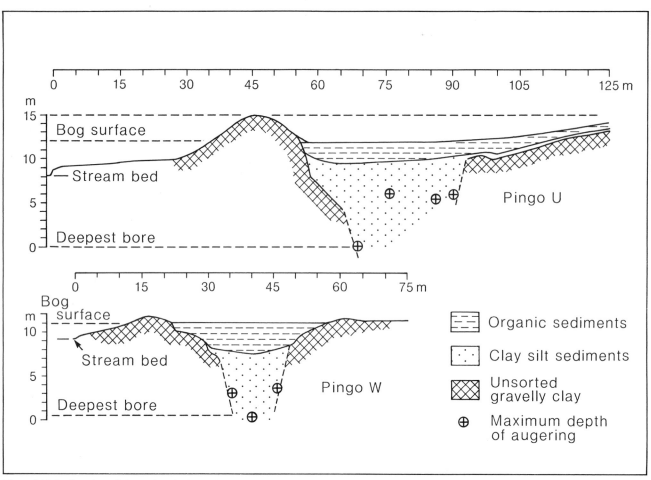

Figure 5.19 Sections through depressions in the Cledlyn Valley (see Figure 5.17(a)). After Watson & Watson (1972).

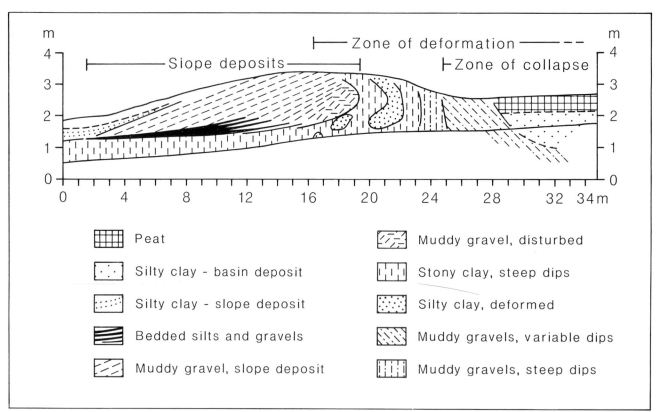

Figure 5.20 Excavated rampart in the Cletwr Valley. After Watson (1976).

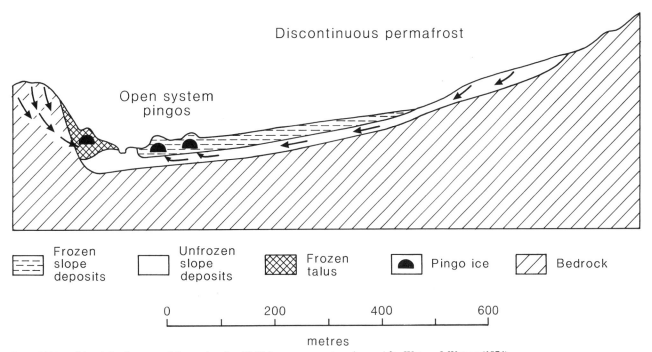

Discontinuous permafrost

Open system pingos

| | Frozen slope deposits | | Unfrozen slope deposits | | Frozen talus | | Pingo ice | | Bedrock |

```
0          200         400         600
```
metres

Figure 5.21 Possible origin of ramparted depressions in mid-Wales as open-system pingos. After Watson & Watson (1974).

and may have rather similar palaeoenvironmental implications (Åkerman & Malmström, 1986).

The size and location of some ramparted depressions in other localities, however, certainly favour their interpretation as the scars of mineral palsas rather than pingos. Into this category falls a group of enclosed depressions mapped by Miller (1990) near Brent Tor on Dartmoor, since these are rather small and occur in a gentle col rather than at the foot of a valley-side slope. The Brent Tor depressions are mainly circular in plan and have rim diameters of up to 43 m, though one elongate basin attains a length of 60 m. The depressions are only about 1 m deep and are developed in silty head derived from the underlying shales. Augering of one depression showed that the underlying basin reaches a depth of only 1.75 m below the ground surface. The basin infill was found to consist of a basal clay layer overlain by 1.34 m of organic silt and peat. The age of these presumed ground-ice hollows is uncertain, but as they occur in Devensian head deposits, a Late Devensian (Dimlington Stadial or Loch Lomond Stadial) age appears likely. Peat-filled depressions even smaller than those near Brent Tor are developed in terrace sands and gravels of Devensian age in the valley of the River Wissey at Wretton in Norfolk (West et al., 1974). These are only 5–10 m wide and up to 3 m deep. Their limited size suggests that they represent either seasonal or short-lived perennial ground-ice mounds. With reference to a section exposed in one such depression, West (1987) proposed that differential freezing of coarse terrace deposits and fine-grained channel fill would have favoured the development of injection ice or segregation ice in the latter, raising a ground-ice mound. The features are associated with (but not necessarily contemporaneous with) ice-wedge casts and cryoturbation structures, and evidence from molluscan fauna and pollen assemblages indicates a dry continental climatic regime at the time of basin formation.

The northernmost ramparted depressions of which we have knowledge occur in the Whicham Valley in Cumbria and at Ballaugh on the Isle of Man. Both groups have been interpreted as the scars of open-system pingos. The Whicham Valley features are up to 65 m wide and occur in valley-floor lacustrine clays that overlie outwash gravels. Detailed reconstruction of the stratigraphy of one depression by Bryant, Carpenter & Ridge (1985) revealed that the basin infill consists of a thin layer of gyttja, overlain in turn by peat, dumped material and mineral soil. Intriguingly, the underlying sands and gravels were observed to form a dome within the lacustrine clays under the deepest part of the depression. Although the origin of this diapiric structure is uncertain, its formation may have been associated with upward water migration during the growth or melting of the former ice core. At Ballaugh the depressions are up to 140 m in diameter and overlie gravels, being developed on the gently-sloping surface of an alluvial fan (Watson, 1971). Both sites lie within the limits of the Late Devensian ice sheet, and hence represent the development of pingos or similar ground-ice mounds either in the last few millennia of the Dimlington Stadial or as a result of renewed permafrost development during the Loch Lomond Stadial.

As the examples discussed above illustrate, most evidence for the former development of pingos and related ground-ice mounds in Britain takes the form of surface depressions that are often (but not always) at least partly encircled by ramparts. In the central London area, however, Hutchinson (1980, 1991) has located 25 subsurface depressions that appear to represent buried pingo scars. These basins are developed in London Clay of Eocene age, and are both infilled by and buried under sands and sandy gravels of probable Devensian age (Figure 5.22). Basin widths are between 20 m and 150 m, with width:length ratios of 0.4:0.9. The basins are funnel-shaped and extend up to 25 m below the general surface level of the London Clay, which

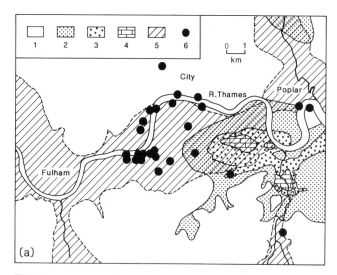

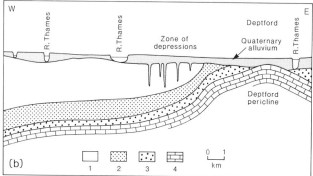

Figure 5.22 Distribution of enclosed depressions, possibly marking the locations of former open-system pingos, in the Lower Thames Valley. 1: London Clay 2: Woolwich and Reading Beds. 3: Thanet Beds. 4: Chalk. 5: Former naturally-flowing artesian areas. 6: Locations of known depressions. After Hutchinson (1980).

nowhere exceeds 30 m in thickness in the area of basin development. In at least five cases the basins penetrate the base of the London Clay, and the underlying Woolwich and Reading Beds are intruded 6–12 m upwards into the basins. Even where the base of the London Clay has not been penetrated, tongues of underlying sediment have been injected upwards into the floors of the basins. Hutchinson (1980) demonstrated that the subsurface basins are located where the London Clay wedges out against the Deptford pericline, which brings the Chalk and Lower Tertiary aquifers to the surface, to form sites of artesian groundwater flow (Figure 5.22). He suggested that the basins were enlarged, if not initiated, by the growth of open-system pingos at sites where groundwater seepage would have favoured injection-ice formation. A Devensian age appears most likely for pingo formation, but it is possible that they are older.

Hutchinson (1991) has also attempted to reconstruct the probable hydraulic conditions that prevailed during formation of these buried depressions. He proposed that they were probably initiated by localised river scour, and suggested that artesian pressure beneath the London Clay then caused cracks to form, allowing groundwater to be injected upwards and feed the growth of pingo ice cores. He also suggested that the diapiric intrusion of the underlying Woolwich and Reading Beds into

the London Clay beneath the basins reflected zones of particularly high artesian pressures and possible weakening of the Woolwich and Reading Beds by freezing and thawing. The relationship between pingo growth and diapiric intrusion of underlying beds is not clear, but it appears likely that diapirism was associated with pingo decay. In this context it is noteworthy that similar upward intrusion of sediments beneath pingos has been observed by various other authors, including Bik (1969), Flemal (1972) and Mackay (1978), and, as noted earlier, is evident in a section cut through a ramparted depression in the Whicham Valley in Cumbria (Bryant et al., 1985).

Large thermokarst depressions

Although ramparted depressions constitute the most conspicuous expression of the decay of massive ground ice in lowland Britain, more subtle evidence of ground-ice melting is also to be found, notably in the Fenland of eastern England. In this area some near-circular depressions about a kilometre wide have been attributed to thermokarst processes initiated by the thawing of ice-rich permafrost. The southern part of the Fenland Basin is bounded partly by Jurassic and Cretaceous sediments and partly by Pleistocene fluvial terrace gravels, the latter marking the former courses of rivers draining into the Fenland Basin. Within the Fenlands there occur slightly raised 'islands' of terrace gravels that in places display polygonal crop marks indicative of former ice wedges. Radiocarbon age determinations indicate that much of the terrace gravel is younger than 42 ka BP (West, 1991). The Fenland depressions are developed in these gravels and in places penetrate through to the underlying Jurassic and Cretaceous clays. Large embayments filled with Flandrian deposits are also cut directly into the Jurassic and Cretaceous clays around the Fenland edge (Figure 5.23(a)).

The general characteristics of the larger Fenland depressions have been summarised by Burton (1987). These generally exceed 1000 m in major diameter and are completely enclosed by higher ground except for a single outlet, sometimes only 60 m wide, towards which the entire depression drains. No constructional ramparts are evident around the margins of the depressions. The edges of the depressions are mantled by coarse gelifluctate overlain by finer solifluction deposits with a loessic (aeolian silt) component. At Conington Fen and Farcet Fen near Peterborough, for example, peat-filled oval and circular flat-floored depressions reach a depth of 6.0–7.5 m (Figure 5.23(b)). Most have a shallow outlet channel or are linked by such channels. These features were attributed by Burton (1987) to degradation of ice-rich permafrost and thermal erosion, in a manner akin to alas formation in Siberia. Polygonal patterns in adjacent terrace surfaces attest to the former presence of ice wedges, and Burton suggested that these, together with pore ice, could have provided sufficient ground ice to permit thermokarst degradation, without the need to postulate the former existence of massive ice beds. At Mepal Fen, 45 km south-east of Peterborough, a depression 2.2 km in diameter occurs within terrace gravels of probable Middle Devensian age (Figure 5.23(c)). This depression is bounded by a series of arcuate scarps and floored by clay that probably represents mudflow deposits. Burton inferred from this evidence that thermokarst subsidence had triggered rapid lateral thermal erosion, in the form of retrogressive thaw slumps, as ground ice became exposed in the sides of the

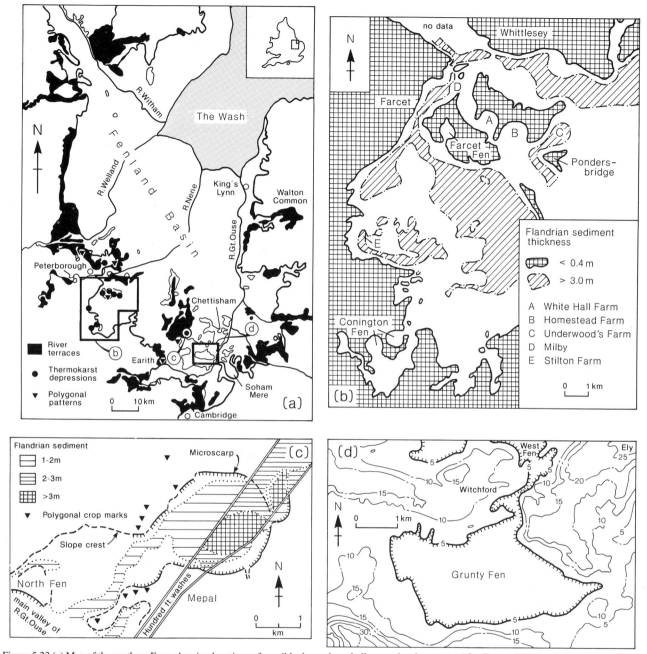

Figure 5.23 (a) Map of the southern Fens, showing locations of possible thermokarst hollows and embayments. After Burton (1987) and West (1991). (b) The western margin of the Fens, south of Whittlesey in Cambridgeshire. After Burton (1987). (c) Mepal Fen, Cambridgeshire. After Burton (1987). (d) Map of Grunty Fen, Cambridgeshire. After West (1991).

depression. Although the thermokarst depressions of the Fenland cannot be dated directly, their occurrence within terrace gravels of presumed Middle Devensian age suggests that they developed during the Late Devensian. The loessic component within the gelifluctate that mantles their margins is probably of Dimlington Stadial age (Chapter 8). These relationships led Burton to conclude that depression formation also occurred during the Dimlington Stadial, though he rightly points out that thermokarst degradation is not necessarily an indication of climatic warming, and may simply reflect the evolution of ice-rich permafrost terrain under a stable cold climate. Interestingly, he also identified a possible earlier thermokarst depression near

Ely that may have developed in Early or Middle Devensian times.

Other thermokarst features in the Fenland have been identified by West (1991), who suggested that shallow depressions such as Grunty Fen and Chettisham Fen may represent former thaw lakes, similar in form and origin to those in the Canadian arctic. In arctic areas, wave action sometimes causes elongation of such lakes into D-shapes, with their long axes transverse to the dominant wind direction and the straight shoreline on the upwind side (Harry & French, 1983). The form and orientation of Grunty and Chettisham Fens (Figure 5.23(d)) led West to speculate that their development was controlled by winds blowing from the north-east. Such winds may have reflected either the development of a blocking anticyclone, or may have taken the form of katabatic airflow associated with the Late Devensian ice sheet to the north. West also suggested that the marked embayments cut into the southern edge of the Fenland Basin represent a further expression of periglacial activity. Lacustrine sediments dated to c. 18 ka BP imply damming of The Wash by the lobe of ice that extended down the east coast of England at the time of the Late Devensian ice-sheet maximum and consequent ponding of water in the Fenland Basin. West proposed that thermoerosion and wave action along the shores of this shallow glacier-dammed lake may have caused retrogressive thaw slumping, the effects of which are now marked by embayments in the Fenland margin.

Although the Fenland is clearly rich in periglacial phenomena that reflect the degradation of ice-rich permafrost, it does not have a monopoly of such forms. At West Moor in the southern Vale of York, for example, a 2.5 km wide circular depression in fluvioglacial and deltaic deposits has been interpreted by Gaunt (1976b) as a relict alas depression, and other smaller depressions in this area may have a similar origin. The floor of the depression is occupied by drift relating to the period of the Late Devensian ice-sheet maximum, which implies that the depression formed no later than the first half of the Dimlington Stadial. A rather different thermokarst feature is represented by an elongated hollow in the flood plain of the River Ouse in Suffolk. This hollow is some 350 m long, 50–60 m wide, and up to 1.8 m deep, though augering has suggested that it was originally slightly larger. A spring provides a constant input of water at the western end of the basin. Coxon (1978) argued that freezing of springwaters under periglacial conditions during the Late Devensian may have caused the formation of a surface icing or naled that became buried by aggrading river gravels. He proposed that subsequent thaw of this buried ice body led to subsidence of the overlying gravels and consequent formation of the present hollow. These examples suggest that thermokarst phenomena may be of much more widespread occurrence in lowland Britain than the present documentation of such features would suggest.

The significance of ground-ice phenomena in Britain: an evaluation

The known distribution of landforms indicative of the melt-out of massive ice bodies or ice-rich permafrost in Britain must be regarded as no more than provisional. Postformational modification, burial under Flandrian peat and the effects of agriculture are all likely to have obscured or obliterated such features. Perhaps even more important is a lack of appreciation of the diagnostic characteristics of thermokarst depressions, pingo scars and related ramparted depressions amongst Quaternary geologists and geomorphologists working in Britain. Absence of evidence is not evidence of absence, and it is not unlikely that enclosed depressions cursorily dismissed as kettle holes may have an entirely different origin and palaeoenvironmental significance. Although there is no reason to doubt the explanation of wide, shallow, rimless depressions in the Fenland and elsewhere as thermokarst features, most supposed 'pingo scars' that have been identified on the British landscape pose problems of interpretation. These difficulties stem, as Bryant & Carpenter

(1987) have pointed out, from equifinality: on decay, different types of ground-ice mound may produce similar landforms. This difficulty prompted Bryant & Carpenter to favour description of such landforms as 'ramparted depressions', rather than 'pingo scars', which has genetic implications and which they believed should be employed only when there is reasonable evidence that the ancestral mounds formed above a core of injection ice. Such evidence, however, has often proved elusive or equivocal. Despite the diagnostic criteria proposed by De Gans (1988), neither size nor morphology offers a simple means of differentiating pingos from other ice-cored mounds, and the presence of compressional structures in ramparts indicates only the growth of an ice core, not the origin of the ice. The more rigorous criteria proposed by Mackay (1988b) are of limited use unless extensive sections occur through and beyond the encircling ramparts, which is rarely the case. In essence, interpretatation of the palaeoenvironmental implications of ground-ice depressions in Britain raises three critical problems. The first, alluded to above, is the interpretation of such depressions: do they represent open-system pingos, closed-system pingos, seasonal frost mounds or mineral palsas formed by the growth of segregation ice? The second relates to the implications of such features in terms of former climatic regime. The third concerns uncertainties associated with the dating of these landforms.

Despite the difficulties inherent in the interpretation of ramparted (or indeed rampart-less) ground-ice depressions, a number of conclusions can be reached. The first is that the forms hitherto reported are unlikely to represent closed-system pingos analogous to those of the Mackenzie Delta area. Most supposed 'pingo scars' in England and Wales occur in clusters at the bottom of valley-side slopes and overlie permeable beds, characteristics that strongly favour an open-system origin with subpermafrost groundwater feeding a growing ice core. This conclusion does not deny the possible former existence of closed-system pingos in Britain. Structures resembling large-scale pingo scars have been detected by seismic and sonar surveys of offshore areas formerly exposed at the surface by glacio-eustatic lowering of sea level, for example west of Anglesey (Wingfield, 1987) and in the northern North Sea (Long, 1991). If these do indeed represent pingo scars they may well be of the closed-system type. Others may lie buried under Flandrian peat, for example in the Fenland. Equally, the great majority of ground-ice depressions are simply too large (and certainly too deep) to represent former seasonal ice mounds. The depth of infill in most reported examples exceeds 3 m and in some cases exceeds 6 m (Bryant & Carpenter, 1987), and such depths are clearly incompatible with the development of shallow seasonal mounds.

It is more difficult, however, to differentiate the remnants of true open-system pingos from those of mineral palsas, particularly as information concerning the characteristics of the latter is scarce. The locational circumstances that favour the growth of open-system pingos (slope-foot situations and a permeable substrate) would also appear to favour the growth of thick bodies of segregation ice, and hence the growth of mineral palsas. Interestingly, though, the case for interpreting ramparted depressions as collapsed mineral palsas is generally made for rather small features (e.g. those at Brent Tor on Dartmoor), those that occur in locations apparently inappropriate for pingo growth (e.g. the Hautes Fagnes plateau in Belgium) or those developed where the substrate is clay-rich and hence possibly of insufficient permeability to permit the formation of injection ice (Pissart & Gangloff, 1984). These criteria, however, apply to only a minority of the supposed 'pingo scars' hitherto identified in Britain; most are certainly large enough to represent the sites of former open-system pingos, and occur at sites favourable for pingo development. Moreover, palsas – mineral or otherwise – are often rather subdued domes underlain by at most a few metres of ice. Pingos frequently take the form of steep-sided hills with a much deeper ice core. For a rampart and depression to form, sufficient sediment must be lost from the flanks of a ground-ice mound not only to create the ramparts but also to cause a marked loss of overburden so that a distinctive depression is formed when the ice core melts. On these grounds it appears reasonable to argue that pingo degradation is much more likely to cause the formation of substantial ramparted depressions than the collapse of mineral palsas. Deep depressions (e.g. those of Llangurig and SW Wales, and the buried depressions under central London) seem particularly likely to represent collapsed pingos rather than the remnants of mineral palsas. Whilst numerous authors have rightly advocated a critical approach to the identification of enclosed depressions as pingo scars (e.g. Bryant & Carpenter, 1987; West, 1987; De Gans, 1988; Pissart, 1988), it appears probable that most of the ramparted depressions hitherto documented in Britain do indeed reflect the growth and decay of open-system pingos. Whilst this interpretation can rarely be established with confidence, at most sites it appears to represent the most *plausible* explanation of the evidence available.

The second problem identified above relates to the palaeoclimatic implications of pingo scars and related ramparted depressions. Whilst there is general agreement that the development of both open-system pingos and mineral palsas reflects the occurrence of thin discontinuous permafrost, it is much more difficult to infer limiting climatic parameters such as MAAT. Although discontinuous permafrost may under favourable circumstances occur where the MAAT is as high as -1 °C to -3 °C, Mackay's (1988b) arguments convincingly demonstrate that the development of even small (<100 m diameter) open-system pingos is likely to imply a limiting 'warm-side' MAAT of no higher than -5 °C to -6 °C. Current degradation of small pingos and hybrid 'palsa-pingos' in northern Scandinavia at sites where present MAAT is -3.5 °C to -4.0 °C lend support to this interpretation (Lagerbäck & Rodhe, 1985; Åkerman & Malmström, 1986). It is noteworthy that these observations also suggest that small open-system pingos and mineral palsas may have similar climatic thesholds. If so, the problem of discriminating between 'true' pingo scars and those resulting from the decay of mineral palsas may be less important than has hitherto been perceived.

The final problem concerns the dating of pingo scars and related features. With the exception of the buried depressions identified by Hutchinson (1980) and the possible 'older' thermokarst depression observed by Burton (1987) near Ely, all documented examples are probably of Late Devensian age. Discrimination of those that developed during the Dimlington Stadial (26–13 ka BP) from those that reflect renewed permafrost development during the Loch Lomond Stadial (11–10 ka BP) is often more difficult. There is reasonable stratigraphic evidence to suggest that the Wey Valley depressions, the 'older generation' of subdued features in East Anglia, the small depressions of Brent Tor and the large thermokarst depressions of the Fenland and Yorkshire all represent the development of ground-ice mounds (or, in the case of the thermokarst depressions, ice-rich permafrost) during the Dimlington Stadial. Only in the case of the 'younger generation' ramparted depressions of East Anglia is there definite evidence for the growth of ground-ice mounds during the Loch Lomond Stadial. The ramparted depressions of SW Wales, mid-Wales, the Isle of Man and Cumbria may reflect the development of ground-ice mounds either during the Dimlington Stadial or the Loch Lomond Stadial, or possibly (especially at sites supporting superimposed forms) both time periods. Those sites that lie inside the limit of the Late Devensian ice sheet (e.g. Llangurig, Ballaugh and the Whicham Valley) all imply the development of permafrost in the interval between the ice-sheet maximum (c. 18 ka BP) and the onset of warmer conditions at the beginning of the Flandrian at c. 10 ka BP.

In sum, the evidence outlined in this chapter indicates that there is reasonable evidence for the formation in southern Britain during the Dimlington Stadial of both open-system pingos and smaller ground-ice mounds. These features imply conditions of thin discontinuous permafrost and MAATs no higher than about -5 °C. The Dimlington Stadial also witnessed the development of thermokarst depressions within ice-rich permafrost in the Fenland and Yorkshire. In Wales, Cumbria and the Isle of Man, open-system pingos developed at favourable slope-foot locations at some time after the retreat of the last ice sheet from these areas. Evidence from East Anglia indicates renewed development of ground-ice mounds (and hence discontinuous permafrost) during the Loch Lomond Stadial. These suggest that MAATs sank to below c. -5 °C at this time, at least in eastern England. The regional significance of these patterns is difficult to assess, however, as the distribution of pingo scars, thermokarst depressions and related forms is incompletely documented.

6
Active layer processes: cryoturbation and patterned ground

In all periglacial areas there exists a surface layer of ground which freezes in winter and thaws in summer. If it overlies permafrost, this zone of annual freezing and thawing is called the *active layer*. Repeated freezing and thawing within the active layer often results in the formation of distinctive small-scale landforms and sedimentary structures. Such features developed throughout Great Britain under permafrost conditions during Quaternary stadials, and many are now preserved in relict form. In periglacial areas lacking permafrost, however, seasonal freezing and thawing of the ground may produce similar landforms and structures. These are also preserved in both upland and lowland Britain, and are often difficult to distinguish from features associated with an active layer over permafrost. Indeed, some frost-action landforms are actively developing under conditions of fairly shallow annual ground freezing on the higher parts of British mountains at present (Chapters 10 and 11). In lowland Britain, two main categories of such features may be distinguished: first, relict patterned ground and cryoturbation phenomena, and second, periglacial slope deposits. This chapter considers the nature and significance of patterned ground and cryoturbation features. Periglacial slope processes and deposits are discussed in Chapter 7.

Active layer processes

The active layer is generally shallow in the high arctic (often extending to a depth of no more than 0.5 m), but gradually thickens southwards in response to increases in the duration and intensity of summer thawing. The depth of thaw depends largely on the air thawing index (the cumulative number of degree-days above 0 °C), and tends to be greater in more continental climates with warmer summers. However, factors such as the nature of the substrate, its ice content, slope aspect and vegetation cover may also have a considerable influence on thaw depth. For instance, where ground-ice contents are high the active layer is likely to be thinner since more heat is required to cause thawing to a given depth. In Spitsbergen, for example, Bakkahøi & Bandis (1988) measured an active layer thickness of 0.7–1.05 m in an ice-rich clay soil, compared with between 1.2 m and 2.1 m in gravel fill material. Thawing of the active layer takes place from the surface downwards in response to above-zero air temperatures. Refreezing during winter, however, occurs not only from the surface downwards as heat is conducted out of the ground by subzero air temperatures, but also from the permafrost table upwards, as heat is lost to the colder permafrost below. During winter freezeback of the active layer, therefore, an unfrozen zone of progressively diminishing thickness separates a freezing front advancing upwards from the permafrost table from one advancing more

rapidly downwards from the surface. In the examples depicted in Figure 6.1, this unfrozen layer persisted from mid-September to mid-January near Fairbanks, Alaska, in 1970–1, and for a period of 35 days from mid-September 1986 at Hornsund in Spitsbergen.

The geomorphological significance of active layer freezing and thawing depends largely on the nature of the substrate. Where bedrock is exposed at the surface, mechanical weathering is likely to be enhanced (Chapter 9). Where unlithified sediments form the substrate, much depends on their granulometry. In fine-grained soils (fine sands, silts and clays), porewater does not freeze at 0 °C, but at some temperature slightly below zero. Moreover, freezing of water in fine-grained soils does not take place instantaneously at a specific temperature, but occurs over a temperature range, the soil water progressively freezing as temperatures fall. Some soil water therefore remains unfrozen even at temperatures well below zero (Burt & Williams, 1976; Williams, 1988; Williams & Smith, 1989). A direct consequence of this is the occurrence of *ice segregation* during freezing of fine-grained soils: water is drawn to the freezing front, leading to the development of lenses of clear ice within the frozen soil (Figure 6.2).

The process of ice segregation

Cooling of the ground surface leads to the initiation of soil freezing as soil temperatures fall below 0 °C. The 0 °C isotherm moves slowly downwards into the soil from the surface, and behind it soil water undergoes progressive freezing. Latent heat, released as the phase change from water to ice takes place, is conducted upwards to the surface. The rate of heat flow is controlled by the thermal conductivity of the frozen soil and the thermal gradient, which itself depends on the surface temperature and the thickness of the frozen layer. Laboratory experiments have shown that ice segregation does not take place at the freezing front, but some distance behind it (Loch & Kay, 1978; Miller, 1980; Williams & Smith, 1989). Between the level at which freezing commences and the level at which ice segregation takes place is a 'frozen fringe', where unfrozen soil water occupies smaller pores and surrounds ice in larger pores, separating it from adjacent mineral grains. Water is drawn to the frozen fringe from the unfrozen soil below, and migrates through it to feed ice lenses growing along its upper boundary. Ice lenses continue to grow as long as freezing of water migrating from below releases sufficient latent heat to maintain the heat flow dictated by the thermal gradient in the frozen soil above. If the supply of water is reduced, cooling of the frozen fringe occurs. This in turn results in the formation of more ice in soil pores, which reduces the permeability of the fringe (Konrad

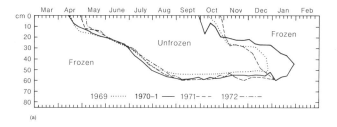

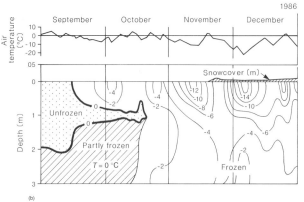

Figure 6.1 (a) Active layer thermal regime, near Fairbanks, Alaska, 1969–72. After Viereck & Lev (1983). (b) Active layer refreezing, 1986, Hornsund, Spitsbergen. After Chmal, Klementowski & Migala (1988).

& Morgenstern, 1982) and effectively shuts off the supply of water to the ice lenses above. The frozen fringe then advances downwards through the soil until equilibrium conditions are re-established between water entering the fringe at its base, ice formation at its upper boundary, and latent heat flow to the ground surface above. When such conditions are satisfied, ice lens growth recommences at a lower level. In this way, successive ice lenses develop parallel to the advancing freezing plane as moist, fine-textured soils undergo downward freezing (Figure 6.2).

The amount of ice segregation in the active layer (or seasonally-frozen layer in nonpermafrost areas) depends largely on soil granulometry, moisture supply and the rate of freezing. Coarse-grained soils have relatively large pores; as a result, depression of the freezing point is negligible and water shows little tendency to migrate towards the freezing front. Fine-grained soils, however, have smaller pores, and are associated with considerable freezing point depression. This causes strong migration of water towards the freezing front (Williams & Smith, 1989). If sufficient water is available, ice segregation will occur in such fine-grained soils, which are therefore termed *frost susceptible*. However, though clay-rich soils are frost susceptible in terms of granulometry, their low permeability may restrict water migration and hence ice lens growth. Significant ice segregation is therefore only likely in clay soils where the rate of freezing is extremely slow.

Figure 6.2 Ice lenses in alluvial silt, Fosheim Peninsula, Ellesmere Island, Canada.

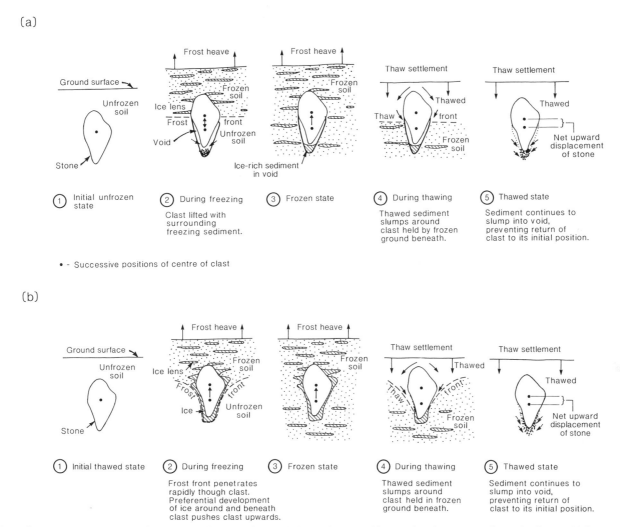

Figure 6.3 (a) Upward frost sorting of stones according to the 'frost-pull' theory. (b) Upward frost sorting of stones according to the 'frost-push' theory.

Ice segregation within seasonally-frozen soil increases its volume, causing upward expansion of the ground surface. Such *frost heave* progressively raises the level of the ground surface by an amount equal to the total thickness of ice lenses, plus any increase in soil volume due to the formation of *pore ice* within intergranular voids. If groundwater seepage in the lower part of the seasonally-frozen layer maintains a water supply during the entire period of freezing, segregation ice may develop throughout this layer, and frost heaving of the surface will be considerable. Where no such external water supply is available, ice lenses are generally concentrated near the ground surface and, in permafrost areas, near the base of the active layer. In its frozen state therefore, the seasonally-frozen layer (or parts of it) may contain a considerable volume of clear ice lenses. This ice represents a volume of frozen water in excess of the natural saturated water content of the soil in its unfrozen state. The presence of such ice lenses has important consequences during thaw, when they release water that cannot be accommodated in the available soil pore space. This excess water must be expelled before resettlement of the soil can occur. In ice-rich frozen soils, therefore, thawing is associated with very high water contents and resettlement (or *thaw consolidation*), and is

accompanied by gradual seepage away of excess water. Moreover, in permafrost environments ice segregation may not be confined to the period of winter freezeback. Mackay (1983a) observed that renewed ice segregation may take place at the base of the active layer during the summer thaw period as a result of downward water migration from the thawing soil above. This water is able to enter the still-frozen lower part of the active layer and cause ice lens growth, because of the presence of unfrozen water films within the frozen soil. The ice content near the base of the active layer is therefore likely to be high, causing supersaturation when it eventually thaws towards the end of the summer.

Ice segregation during winter freezing and the release of excess water during summer thaw consolidation are responsible for the formation of a variety of periglacial phenomena, as well as cryogenic deformation of near-surface sediments (*cryoturbation*). Such processes are not restricted to the active layer in permafrost environments, but also occur in seasonally-frozen ground in nonpermafrost areas. In the latter case, of course, upwards freezing and preferential ice segregation near the base of the frozen layer do not occur, as the temperature of underlying soil remains above the freezing point throughout the year.

Figure 6.4 Examples of periglacial patterned ground. (a) Sorted circles, formed in the zone of continuous permafrost near Resolute Bay on Cornwallis Island, NWT, Canada. The circle centres are *c.* 2 m in diameter. (b) Sorted stripes in the discontinuous permafrost zone, Niwot Ridge, Front Range, Colorado, USA Stripe spacing 1.5–2.0 m. (c) Nonsorted circles, zone of discontinuous permafrost, Mount Albion, Front Range, Colorado. USA. (d) Nonsorted stripes, zone of continuous permafrost, Ellesmere Island, NWT, Canada. Stripe spacing *c.* 2 m.

Frost-sorting processes

Laboratory studies in the 1950s and 1960s, particularly by Corte (1961, 1962a,b, 1963, 1966), showed that repeated freezing and thawing of poorly-sorted sediments in the presence of moisture leads eventually to segregation of fines and clasts. Corte demonstrated that repeated freezing and thawing from the surface downwards causes fines (sand, silt and clay) to migrate slowly down whilst stones tend to move upwards towards the surface. Two mechanisms have been proposed to explain the upward sorting of clasts; these are referred to as the 'frost pull' and 'frost push' hypotheses (cf. Washburn, 1979, pp. 86–91). Upwards sorting of stones by frost pull is believed to occur when the top of the stone is reached by a descending freezing front. Further ice segregation as the freezing front continues to descend causes heave of the overlying frozen soil, and the attached stone is pulled upwards as a result (Figure 6.3(a)). Thawing from the surface downwards leads to resettlement of the soil around the stone, which remains supported by still-frozen soil at its base. In this way, repeated cycles of freezing and thawing cause stones to migrate slowly upwards to the surface. In contrast, the frost push hypothesis is based on the higher thermal conductivity and lower heat capacity of stones in comparison to adjacent moist fines. Since the freezing of moist

fines releases latent heat, the freezing front will advance more rapidly through a stone than through surrounding soil matrix, allowing ice segregation at the base of the stone. This is believed to result in stones being pushed upwards through the adjacent soil (Figure 6.3(b)). One drawback of this hypothesis is that it requires stones to be pushed upwards through *frozen* soil, but Washburn (1979) has suggested that this may be plausible in view of the fact that fine-textured soils remain incompletely frozen, even at temperatures several degrees below 0 °C.

Corte (1962a, 1966) showed experimentally that a laterally-advancing freezing front tends to push both fines and stones ahead of it, but that fines are able to migrate faster than stones. The net result is a tendency for lateral sorting of fines in the direction of advance of the freezing front. The effect of upward freezing from the base of the active layer has been investigated by Mackay (1984), who found that if clasts in the active layer are porous, water may percolate through them from their unfrozen (upper) sides, to feed ice lens growth on their frozen (lower) sides, thus inducing upward frost push. This mechanism is, of course, restricted to permafrost areas, but in other respects the processes of both vertical and lateral frost sorting are identical in the active layer above permafrost and in seasonally-frozen ground where permafrost is absent.

Figure 6.5 (a) Earth hummocks (nonsorted circles), Ellesmere Island, NWT, Canada. (b) Thufur, Okstindan Mountains, Norway. (c) Section through a thufur, showing updoming of mineral and organic layers.

Patterned ground

The term *patterned ground* is employed to refer to terrain that exhibits regular or irregular surface patterning, most commonly in the form of circles, polygons, irregular networks or stripes (Figure 6.4). Step-like, oval, lobate and garland patterns are also sometimes referred to as patterned ground. Such patterns are commonly defined by microrelief (alternating mounds and depressions, ridges and furrows), by vegetation cover (alternating vegetated and unvegetated ground) or by the regular alternation of fine and coarse debris at the ground surface. Although patterned ground may be produced by a variety of mechanisms under nonperiglacial conditions, for example in hot arid environments (Hunt & Washburn, 1966; Cooke & Warren, 1973) on intertidal platforms (Kostyaev, 1973; Ball, 1976) and on debris-covered glaciers (Washburn, 1956a; Ballantyne, 1979a), landforms of this type are most frequently associated with cold environments where ground freezing and thawing is the dominant formative mechanism. Following Washburn (1956b), most researchers accept a distinction between *sorted patterned ground*, where the pattern is defined by the alternation

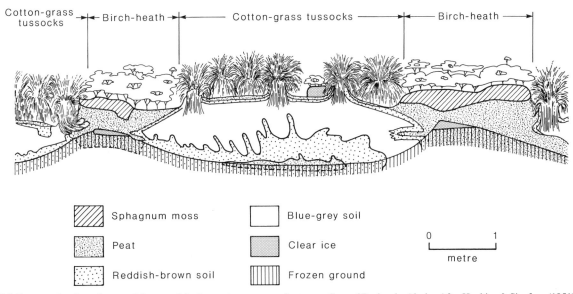

Figure 6.6 Cross-section through an earth hummock in the continuous permafrost zone, Seward Peninsula, Alaska. After Hopkins & Sigafoos (1951).

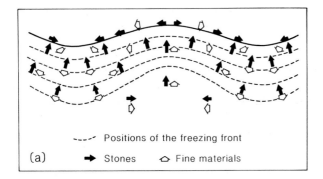

(a)

--- Positions of the freezing front
➤ Stones   ⬠ Fine materials

(Figure 6.5(a)). Tarnocai & Zoltai (1978), for example, described hummocks with average heights of 0.4–0.6 m and diameters of 1.1–1.6 m in the Mackenzie Delta area of Canada but found that in the central Canadian arctic hummocks tend to be smaller, averaging 0.32 m in height with diameters of 0.8–1.4 m. They emphasised the variability of hummock size even at a single site. Similar but larger hummocks occur on the Seward Peninsula, Alaska, with diameters of 6–9 m (Hopkins & Sigafoos, 1951; Nicholson, 1976), but here the sorting of coarser stones into the interhummock troughs blurs the distinction between nonsorted and sorted patterned ground. Smaller well-vegetated hummocks are widespread in permafrost-free tundra areas, particularly on poorly drained ground, and are often referred to by the collective

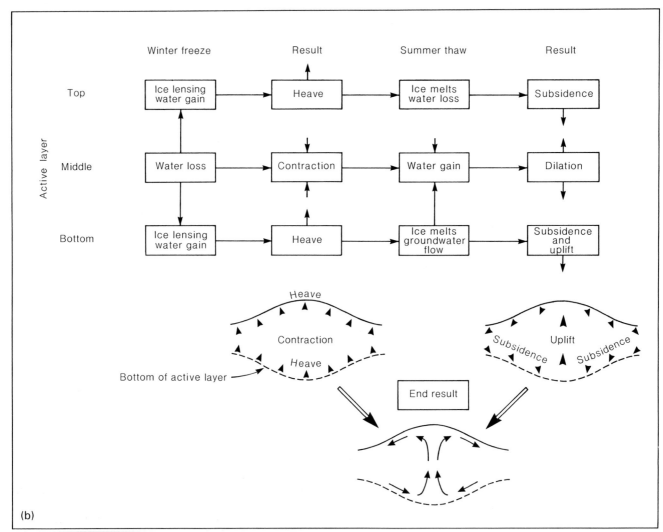

(b)

Figure 6.7 (a) Displacement of fine material within an earth hummock associated with movement of an undulating freezing front, as proposed by Nicholson (1976). (b) Model of earth hummock development as proposed by Mackay (1979b, 1980).

of fine and coarse debris, and *nonsorted patterned ground*, where the pattern is formed by microrelief and/or vegetation cover.

Nonsorted patterned ground

A widepread phenomenon in arctic and subarctic lowlands is patterned ground consisting of roughly circular dome-shaped hummocks that form a cellular pattern over the ground surface

term *thufur*. These tend to be more knob-like than those developed over permafrost, averaging c. 0.5 m in height and c. 1.0 m in diameter (Figure 6.5(b)). In Iceland, thufur are known to be capable of forming within a few years or decades under present conditions (Thorarinsson, 1951; Schunke, 1975, 1977; Scotter & Zoltai, 1982; Schunke & Zoltai, 1988).

On gently-sloping sites hummocks are sometimes elongate downslope. Where the gradient exceeds about 6 ° such elongate hummocks may grade into nonsorted stripes that form a ridge-and-furrow pattern following the line of maximum slope (Lundqvist, 1962; Nicholson, 1976; Figure 6.4(d)). In general, hummocks develop in fine-grained, frost-susceptible sediments, and are associated with tundra vegetation or open subarctic forest (Schunke & Zoltai, 1988). The thickness of vegetation and organic litter is often greater within the inter-hummock troughs. This has the effect of insulating the troughs from high summer air temperatures and thus reduces the thickness of the underlying active layer. The permafrost table is therefore bowl-shaped beneath the hummock, like a mirror image of the ground surface (Figure 6.6). Excavation of earth hummocks usually reveals disruption of underlying sediment, often with intrusions of organic-rich soil into the hummock centres. In permafrost areas, tongues of sediment frequently extend upwards from the lower part of the active layer (Figure 6.6). Trenches cut through hummocks in nonpermafrost areas generally reveal updoming of sediments and organic horizons, but in such cases disruption is confined to the hummock, and tends not to extend below the depth of inter-hummock troughs (Figure 6.5(c)).

In permafrost environments, the formation of earth hummocks is generally attributed to mass displacement of active layer soils during freezing and/or thawing. Such mass displacement has frequently been explained by the establishment of *cryostatic pressures* within the active layer during winter freezeback. According to this hypothesis, such pressures are generated when an unfrozen layer is squeezed between freezing fronts advancing downwards from the surface and upwards from the permafrost table (French, 1988). Nicholson (1976), for instance, argued that slight variations in vegetation cover, snow-cover, soil texture or soil moisture are sufficient to cause local variations in the rate of downward freezing. He suggested that ice segregation along inclined sections of the advancing freezing front sets up lateral heaving pressures, which in turn result in lateral displacement of unfrozen fine-grained sediment towards zones of impeded frost penetration. Frost sorting of fines ahead of the undulating freezing front may also induce lateral migration of sediment in the manner described by Corte (1962a, 1966). Nicholson suggested that a slow, cell-like circulation of sediment becomes established above the permafrost table, with updoming of the surface above areas of net upward sediment displacement (Figure 6.7(a)). He believed that once hummocky microrelief is established, thawing of ice-rich soil close to the surface leads to creep and flow of soil down the flanks of hummocks, thus completing the cell-like circulation of sediment.

This model of hummock development has been challenged by Mackay & Mackay (1976), who pointed out that in frost-susceptible soils migration of water from unfrozen soil to the freezing front may result in desiccation of the former. Such desiccation induces tension in the unfrozen soil water rather than cryostatic pressure, and leads to compaction of the desiccated soil. Mackay & Mackay also observed that pressures in clay-rich soils are greater during thaw than during freezing, which led them to propose that wetting and expansion of the desiccated sediment during thaw results in upward mass displacement as the surrounding frost-heaved ground melts. Mackay (1979b, 1980) subsequently developed an equilibrium model of hummock development in permafrost areas (Figure 6.7(b)). He considered that two-sided freezing of the active layer leads to ice segregation in its upper parts (due to downward freezing from the surface) and near the base of the active layer (due to upward freezing from the permafrost table), with desiccation of the intervening soil. Thaw of the ice-rich upper part of the active layer leads to settlement and radial displacement of sediment down the flanks of hummocks. Subsequent melting of ice lenses at the base of the active layer causes slow sediment movement down the sides of the bowl-shaped depressions in the permafrost table beneath the hummocks. Meltwater also percolates down the sides of such depressions and refreezes, forming ice lenses that cause further upward heave of the hummock centres. Mackay envisaged that thawing of the ice-rich sediments at the base of the hummocks towards the end of each summer produces a semi-liquid mud that is less dense than the overlying soil, and hence rises upwards as a diapiric tongue into the centre of each hummock.

Mackay supported his model with experimental evidence and grain-size measurements that favour the notion of diapiric injection of mud into hummock centres. He estimated that the rate of soil 'turnover' within hummocks in the western Canadian arctic is of the order of 350 years. Mackay's model, however, applies only to extant hummocks, and does not explain their initial formation. He suggested that they may have their origins in bare mud patches on otherwise vegetated surfaces. Such bare areas would allow deeper thawing in summer, causing the development of the bowl-shaped depressions in the permafrost table that are thought to influence mass displacement of soil. Though Mackay's model may well apply to the western Canadian arctic, where his experiments were carried out, it seems unlikely to offer a general explanation for hummock initiation, particularly in areas where vegetation cover is complete. Moreover, his model cannot explain the origin and dynamics of thufur in nonpermafrost environments.

The transition in some areas from hummocks to nonsorted stripes with increasing gradient implies the addition of a component of downslope mass movement to the freezing and thawing processes responsible for hummock formation. As outlined in more detail in the next chapter, thawing of ice-rich soil leads to downslope resettlement (creep) and slow viscous flow (gelifluction) under the influence of gravity. Such downslope movement would appear to explain the progressive elongation of hummocks as the gradient increases, and their eventual replacement by nonsorted stripes on some slopes.

As noted earlier, the term 'nonsorted patterned ground' also encompasses patterns that reflect the alternation of vegetated and unvegetated terrain. *Nonsorted circles* are unvegetated circular areas of fine-grained soil amid otherwise vegetated ground (Figure 6.4(c)) and on slopes commonly merge into nonsorted stripes (Figure 6.4(d)). Such circles typically range from 0.5 m to 5 m in diameter, and their surfaces are usually slightly domed above the level of the surrounding ground. In many cases they appear to be genetically related to earth hummocks. Similar features have been referred to as mud boils (Shilts, 1978). As with earth hummocks, nonsorted circles are underlain by bowl-shaped hollows in the permafrost table. They may show evidence of diapiric extrusion of sediment over the vegetated surface, particularly where the soil comprising the active layer has a low liquid limit (Shilts, 1978), and liquefies readily if overcharged with meltwater during thaw. Of somewhat different origin, but similar form, are *stony earth circles*,

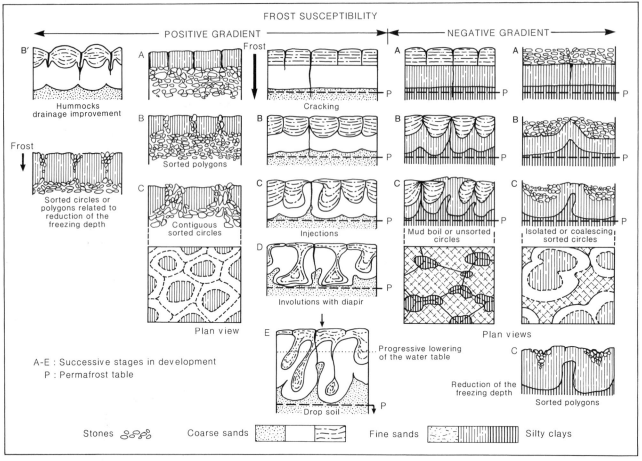

Figure 6.8 Evolution of periglacial patterned ground in relation to contrasts in frost susceptibility, differential frost heave, thermal gradient and drainage. After Van Vliet-Lanoë (1988a).

often found in exposed, snow-free sites in the subarctic and on alpine tundra. Breaks in the vegetation cover are caused by a combination of wind erosion, frost heave and disruption by needle ice, and take the form of circular or elliptical patches of bare soil. Absence of vegetation cover promotes deflation, accelerated seasonal freezing and thawing and more frequent short-term freeze–thaw cycles, all of which prevent vegetation recolonisation. Increased frost sorting brings small stones to the surface, to produce the characteristic stony cover (Williams, 1958; Thorn, 1976b; Williams & Smith, 1989).

Sorted patterned ground

In contrast to the nonsorted forms described above, sorted patterns develop on unvegetated or sparsely-vegetated soils that contain abundant clasts embedded in fine sediment. On level or gently-sloping ground, sorted patterns consist of 'cells' of fine soil surrounded by stony borders that may be slightly raised or depressed relative to the fine centres (Figure 6.4(a)). These concentrations of clasts sometimes take the form of an irregular *sorted net* surrounding fine cells of varying configuration, but elsewhere form regular polygons or circles. Such patterns tend to become elongated downslope as the gradient increases, and are eventually replaced by *sorted stripes*, which consist of alternating bands of coarse and fine debris that follow the direction

of maximum slope (Figure 6.4(b)). Such stripes reflect the simultaneous operation of lateral sorting and downslope mass movement. It is widely reported that rates of downslope movement are greater for fine stripes than for adjacent stony stripes (e.g. Benedict, 1970; Mackay & Mathews, 1974; Coutard, Gabert & Ozouf, 1988). This appears to reflect the greater frost susceptibility of fine stripes, which encourages gelifluction as well as frost creep. However, surficial creep due to the growth and collapse of needle ice may contribute to accelerated movement of surface clasts in small-scale stripe patterns (Higashi & Corte, 1972; Walton & Heilbronn, 1983; Lewkowicz, 1988).

Pattern widths for sorted nets, polygons, circles and stripes vary from a few centimetres to several metres, and generally increase as margin clast size increases (Goldthwait, 1976). Although sorted patterns are known to develop in seasonally-frozen ground where permafrost is absent, larger forms appear to be diagnostic of deep annual freezing, and often (but not always) permafrost. Williams (1975, p.96) considered that relict sorted patterns incorporating boulders in their borders 'strongly suggest the former presence of permafrost', while according to Goldthwait (1976, p.34) 'extensive ice-rich permafrost is implicit in every regional occurrence of large patterned soils (>2 m)'. Washburn (1985) also regarded sorted patterns greater than 2 m wide as indicative of permafrost, though

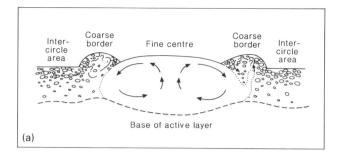

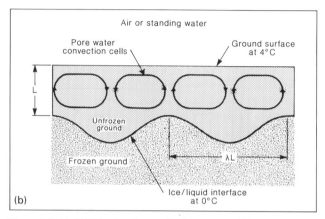

Figure 6.9 (a) Model of sediment displacements within sorted circles according to Hallet & Prestrud (1986). (b) Model of Rayleigh free convection of porewater during active layer thawing, with resulting uneven thawing of the permafrost table, according to Ray *et al.* (1983).

smaller forms may develop in non-permafrost environments as a result of seasonal ground freezing and thawing. There are records, however, of sorted circles and nets exceeding 1–2 m in width in nonpermafrost areas where edaphic or hydrological conditions are particularly favourable (e.g. Troll, 1944, pp. 592–600; Conrad, 1946; Lundqvist, 1962, pp. 78–82). Ballantyne & Matthews (1982) have shown that sorted circles up to 3.5 m in diameter may form at the margins of retreating glaciers even in the absence of permafrost.

Most researchers have envisaged the development of sorted patterns in terms of a circulatory model involving upward movement of sediment in cells, migration of surface and near-surface sediment from the cells to adjacent borders and compensatory movement of material towards the cell centres at depth. Sorting occurs because sediment moving outward towards the margins near the surface is coarser than that moving inwards towards the cell centres at depth. There is less consensus, however, regarding the causes of such circulatory movement of sediment. One school of thought attributes lateral displacement of clasts towards pattern margins and fines towards pattern centres to the establishment of inclined freezing planes as the ground freezes in winter. Following the principles established by Corte (1962a, 1966), this should result not only in vertical (upwards) sorting of clasts, but also in their migration away from centres of impeded frost penetration (Nicholson, 1976). Once established, this effect should be self-perpetuating, as the freezing plane will tend to descend more rapidly below the zones of clast concentration than below the intervening cells of fine sediment. There have been numerous

suggestions as to how inclined freezing planes may be initiated (Washburn, 1979), but two possibilities have been particularly favoured. One involves contraction cracking of the ground as a result of cooling or desiccation. The resulting polygonal cracks act as freezing centres by allowing the penetration of cold air, so that clasts become progressively concentrated in the cracks, forming sorted polygons (Ballantyne & Matthews, 1983). An alternative, termed 'primary frost sorting' by Goldthwait (1976), considers that randomly-spaced concentrations of fines act as centres of slower frost penetration, and thereby generate inclined freezing planes within the soil. Due to their greater frost susceptibility, these finer centres are also liable to greater frost heaving. Frost sorting causes clasts to move towards the surface, and to migrate away from the finer-textured areas of slow frost penetration. At the onset of thaw the surface of the finer centres is updomed due to greater frost heaving, so that surface clasts experience lateral gravity-induced sorting during thaw consolidation. Sorted circles may initially be formed, but ultimately the interaction of adjacent circles may result in the formation of polygons.

The above ideas have been further developed by Van Vliet-Lanoë (1988a,b,c; 1991), who investigated the characteristics of sorted and nonsorted patterns in coastal areas of Spitsbergen. She observed that in all cases these are associated with textural contrasts within the active layer sediments, and suggested that a combination of differential frost susceptibility and ground cracking causes mass displacement of sediment during freezing (Figure 6.8). The model rests on the existence of sediment layers with different frost susceptibility and the assumption of ground cracking as the fundamental control of network size; it may therefore apply specifically to the Spitsbergen sites, rather than having more general applicability.

The second school of thought attributes the convection-like circulation of soil thought to be responsible for the formation of sorted patterns to an unstable density configuration within thawing ground. Chambers (1967) pointed out that frost sorting of clasts to the surface may itself produce density inversion during thaw, at which time underlying fines have low densities and are very wet following melting of ice lenses. He envisaged that the low density fine soil at depth may undergo diapiric intrusion into the overlying layer of clasts, with sinking of clasts around the margins of the resulting sorted circles. He also suggested that annual upwelling of fines and sinking of adjacent clasts may continue until the borders extend down to the permafrost table, when they become 'anchored', and the sorted circle form is stabilised.

This concept was developed by Palm & Tveitereid (1977), Hallet & Prestrud (1986), Hallet *et al.* (1988), Harris & Cook (1988), and Cook (1989). Hallet & Prestrud observed liquefaction of fines in circle centres, and measured updoming of centres and radial displacement of near-surface sediment during thaw. Organic material on the surface of the circle centres was found to extend downwards into the active layer adjacent to stony borders, indicating subduction of soil material around the margins of the cells. Their field data showed that such movements resulted from density inversion caused by thaw consolidation of near-surface sediment above loosely-consolidated recently-thawed sediment at depth. Liquefaction of this low-density saturated soil in cell centres is followed by upward intrusion into the denser soil above, with concomitant sinking

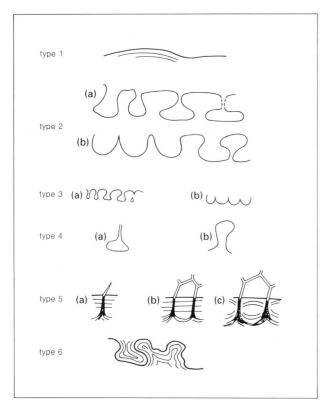

Figure 6.10 Morphological classification of cryoturbation structures in terms of symmetry, amplitude:wavelength ratio and pattern of occurrence. After Vandenberghe (1988).

of denser sediment at the cell margins (Figure 6.9(a)). Theory predicts that density-driven convection cells ('Rayleigh free convection') should have a width:depth ratio of *c.* 3.6. Such ratios were observed in Spitsbergen by Hallet & Prestrud, and in the Jotunheimen mountains of Norway by Cook (1989). If it is assumed that sorting extends to the base of the active layer in permafrost environments, then the dimensions of relict sorted patterns may be used to estimate former active layer thickness; sorted circles with diameters of 2 m, for example, would imply formation in an active layer 0.5–0.6 m deep.

Finally, the question of how regularly-spaced sorted patterns are first initiated has been addressed by Ray *et al.* (1983) and Gleason *et al.* (1986). They pointed out that during thaw of frozen ground, meltwater at the thaw plane (at 0 °C) is less dense than overlying soil porewater, provided the temperature of the latter does not rise above 4 °C. These authors proposed that this density inversion may drive convection of water through the soil pores, with downflow of warmer water creating loci of preferential melting at the base of the active layer (Figure 6.9(b)). In consequence, a pattern of regularly-spaced peaks and troughs is formed in the permafrost table, the spacing being controlled by the size of the convection cells. Advocates of this model suggest that somehow the underlying pattern of thaw in the permafrost table is transferred to the surface, and this controls the dimensions of sorted patterns. How this link is effected, however, is unclear. In addition, the water density differences are small, and would appear incapable of driving significant convectional movement of porewater except in very permeable soils.

Cryoturbation

Sediment strata and soil horizons are often severely disturbed by repeated freezing and thawing of the ground. We have seen that the processes responsible for the formation of patterned ground cause lateral and vertical displacement of the soil, so that formerly horizontal beds become distorted. Similar disturbance of near-surface sediments may also occur without any surface expression. The general term for distortions resulting from mass displacement of soil is *involutions*, and where such involutions are produced by freezing or thawing of the ground they are termed *periglacial involutions*, *cryoturbation structures* or simply *cryoturbations*. Not all involutions are of periglacial origin. Very similar structures may form as a result of depositionally-induced density inversion in rapidly-accumulating sediments (e.g. Dzulynski, 1963; Anketell, Cegla & Dzulynski, 1970). Care is therefore necessary in interpreting involutions in terms of former periglacial environments.

Attempts to classify and interpret cryoturbation structures in Pleistocene sediments are beset by difficulties arising from the diversity of forms produced and uncertainty as to the conditions under which these may have developed. A useful classification has recently been proposed by Vandenberghe (1988), who recognised six fundamental types (Figure 6.10): (1) individual folds of small amplitude (depth) but large wavelength; (2) fairly regular, symmetrical and intensely convoluted forms with amplitudes of 0.6–2.0 m; (3) forms similar to (2) but with much smaller amplitudes; (4) solitary 'teardrop' or diapiric forms; (5) upwards injected sediment in cracks; and (6) irregular deformation structures. It is widely recognised that such cryoturbation structures are polygenetic, and three main modes of formation have been proposed. First, these features have been widely interpreted as load structures that result from density inversion during thaw of frozen ground. Melting of ice-rich frozen ground may leave a loose, saturated sediment that is liable to liquefy as it consolidates. As a result, this low density sediment at depth is injected upwards into overlying horizons (often producing flame-like structures; Figure 6.11), whilst the denser sediment above sinks downwards into the liquefied layer (e.g. Gullentops & Paulisen, 1978; Vandenberghe & Van den Broek, 1982). This process is probably responsible for most cryoturbations of types 2–4 in Figure 6.10. A second possible mechanism involves liquefaction of near-surface sediment due to the generation of high porewater pressures during autumn freezeback (*cryohydrostatic pressures*). It is argued that unfrozen soil confined between a descending freezing front and an underlying permafrost table may experience raised porewater pressures and consequent deformation, though Mackay & Mackay (1976) have shown that ice segregation is likely to reduce water pressure in the confined zone, rather than raise it. Gullentops & Paulisen (1978) argued that cryoturbation due to cryohydrostatic pressures is most likely where a hydraulic head is present due to lateral seepage of groundwater between the permafrost table and a downward-advancing freezing front. Finally, different rates of frost penetration and ice segregation in freezing soil, together with the fact that water in finer sediments freezes at lower temperatures than water in coarser sediments, may generate differential frost heaving pressures, causing mass displacement of sediment (Corte, 1971). In this way repeated freezing and thawing may lead to the gradual deformation of sediments into cryoturbation structures (Pissart, 1985; French, 1986;

Figure 6.11 Examples of cryoturbation structures: (a) active layer, Ellesmere Island, Canada (type 3(b) of Vandenberghe (1988)); (b) cryoturbation of fluvial sands, near Mons, Belgium (type 2(a)).

Figure 6.12 Chalkland patterned ground near Elveden, Suffolk. Note transition from regular polygons through elongates to stripes as the gradient increases. Cambridge University Collection of Air Photographs.

Figure 6.13 Chalkland patterned ground, Thetford Heath, Suffolk. Cambridge University Collection of Air Photographs.

Van Vliet-Lanoë, 1988a,b,c). Williams & Smith (1989) and Van Vliet-Lanoë (1991) concluded that many cryoturbation structures form by this mechanism rather than as a result of cryohydrostatic pressures or thaw-induced density displacements.

Cryoturbation is favoured in flat, poorly-drained areas, but is not restricted to the permafrost zone. Although differentiation between cryoturbations and nonperiglacial involutions may be problematic, structures constrained by a former permafrost table may sometimes be distinguished by termination at a constant depth, by flat-bottomed 'teardrop' forms, and in places by

Figure 6.14 Chalkland patterned ground developed on a flat, nearly horizontal surface, Brettenham, Norfolk. Cambridge University Collection of Air Photographs.

the presence of ice-wedge casts in subjacent sediments. Small cryoturbations may reflect differential frost heaving or unstable density stratification during thaw in areas with shallow active layers (low mean annual temperatures), or in much warmer areas where permafrost is absent but where seasonal ground freezing occurs. On the other hand, widespread, large-scale structures (type 2) appear likely to develop in areas with a thick active layer, or as a result of regional degradation of ice-rich permafrost during periods of climatic amelioration. Clearly, cryoturbations in themselves are somewhat ambivalent palaeoenvironmental indicators.

Quaternary patterned ground in lowland Britain

Nonsorted patterned ground

Despite their small size, Pleistocene periglacial patterned ground phenomena are still visible as surface markings in certain areas of lowland Britain. Indeed, some of the most dramatic periglacial features in Britain, at least when viewed from the air, are the nonsorted patterned ground phenomena of the East Anglian chalklands (Figures 6.12, 6.13 and 6.14). Patterns in this area include regular nets with diameters of 5–10 m on low-gradient hill crests and plateau areas (Williams, 1964; Watt, Perrin & West, 1966; Evans, 1976; Nicholson, 1976). These grade into elongated doughnut- or garland-shaped patterns on gentle slopes, and into stripes where the gradient is between 2 ° and 6 °. On steeper slopes, gelifluction was apparently too active for patterns to develop (Evans, 1976). Stripe widths are commonly between 5 m and 7 m. Such patterned ground occurs where frost-weathered chalk is overlain by a thin layer of aeolian sand. Sometimes, as at Thetford Heath in Norfolk, chalky till rather than weathered chalk underlies this sand. Excavation reveals an updoming of the chalk regolith beneath the centres of nets, with only a thin cover of sand (Figure 6.15(a)). Here the sand is mixed with fine chalky loam and is consequently calcareous. In contrast, beneath the net margins the sand is up to 1.5 m thick, noncalcareous and often of coarser texture. Chalkland grasses tend to cover the calcare-

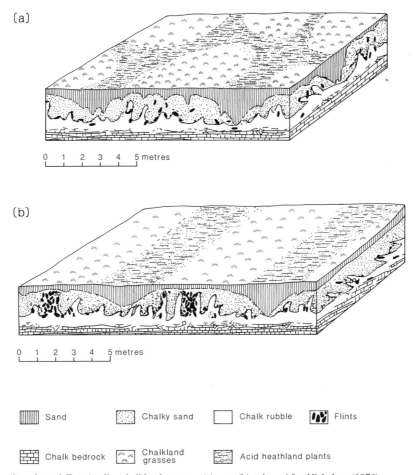

Figure 6.15 Typical cross-sections through East Anglian chalkland patterns: (a) nets; (b) stripes. After Nicholson (1976).

ous centres of the nets, while heather (*Calluna vulgaris*) occupies the noncalcareous sandy margins. Sections excavated across stripe patterns show them to be structurally similar to the nonsorted nets, with the somewhat broader grass-covered stripes corresponding to zones with a thin surficial calcareous sandy layer, and the alternating *Calluna* stripes corresponding to zones with a thicker noncalcareous sand layer (Figure 6.15(b)).

The weathered chalk or chalky till beneath the patterned ground is cryoturbated, with tongues extending upwards into the sandier sediment above and pockets of sand penetrating downwards into the chalky substrate (Figure 6.15). The sand incorporated in such involutions commonly exhibits a laminar structure, with individual laminae up to 4 mm thick, a feature attributable to former ice segregation. Within stripe patterns, involutions also show downslope distortion indicative of frost creep and gelifluction during formation, and vertical concentrations of steeply-dipping clasts beneath the centres of the grass-covered nets and stripes (Figure 6.15) probably indicate upward injection of sediment during pattern formation. Nicholson (1976) noted the similarities in scale, plan and cross-sectional form between chalkland patterns and active nonsorted circles and stripes developed over permafrost in Alaska. Morphological similarities are also evident between the chalkland patterns and nonsorted patterned ground on Ellesmere

Island in the Canadian high arctic (Figure 6.16). If such features are valid analogues for the chalkland patterns, it seems likely that the latter developed when continuous permafrost was present in eastern England, probably during the Devensian (Williams, 1965). The depth of cryoturbation beneath the chalkland patterns suggests a former active layer thickness of 1.5–2.5 m. The dimensions of nonsorted patterns in the arctic tend to be related to active layer depth: those observed by Nicholson in Alaska are 6–9 m wide and overlie an active layer *c.* 2 m deep, whereas those on Ellesmere Island (Figure 6.16) have widths of only 2.5–3.5 m, and overlie an active layer only 0.6 m deep. Over 500 patterned ground locations were mapped in England by Williams (1965; Figure 6.17), who noted the absence of patterns on chalklands farther west, for instance in the Chiltern Hills, the Weald and Hampshire. This may be due to a lack of coversand in these areas, as such sand appears to be important in preserving patterned ground and in generating corresponding surface vegetation patterns. Evans (1976), however, found similar patterns on the Yorkshire Wolds, and in Lincolnshire, Berkshire and Wiltshire, though they cover only a small proportion of the chalkland surface in these areas. The altitude of chalkland patterns tends to increase north and west of the Breckland area of East Anglia, with altitudinal ranges of 12–46 m above sea level in Breckland, increasing to 85–122 m in Yorkshire.

Figure 6.16 Active nonsorted circles (nets) and stripes, Ellesmere Island, NWT, Canada. Diameters of circles and spacing of stripes 2–3 m; active layer depth 0.6 m.

Sorted patterned ground

Sorted stripes as striking in appearance as the chalkland patterns described above are well exposed on the quartzite ridge known as the Stiperstones, in Shropshire (Figure 6.18; Goudie & Piggott, 1981). This ridge rises to an altitude of about 525 m and consists of steeply-dipping beds of resistant siliceous sandstone and conglomerate that form a series of angular tors on its crest (Chapter 9). The slopes flanking the ridge have gradients of between 3° and 15°, and are covered with alternating stripes of boulders and vegetated finer soil. The unvegetated stone stripes range from 0.8 m to 9.8 m in width (mean 3.2 m), and the vegetated finer stripes from 0.8 m to 9.5 m (mean 3.0 m). Clasts are angular, up to 2.5 m long, and consist of quartzite derived from the tors above. Many elongate clasts are aligned in a downslope direction. The stripes at this site occur on slopes of 6–15° (rather steeper than the chalkland nonsorted stripes), while polygons are present on the more gently-sloping ridge crest (Figure 6.18). The scale of sorting within these patterns suggests that they developed in association with permafrost. The ridge stood beyond, but close to, the Late Devensian ice limits. Frost sorting, downslope mass movement and frost weathering may therefore have been active throughout a large part of the Devensian. The same is probably true of similar sorted stripes developed downslope of granite tors on Dartmoor (Te Punga, 1957), though most large-scale sorted patterns on

British mountains probably postdate the last ice-sheet maximum (Chapter 10).

Periglacial patterned ground was probably widespread in the British lowlands beyond the ice limits during the Devensian and earlier glacial periods. These patterns, however, have been largely obscured or destroyed by Holocene slope processes, and by the effects of cultivation. In the Yorkshire Wolds, for instance, archaeological excavations at the deserted medieval village of Wharram Percy revealed Late Devensian sorted stripes that are not now visible on the ground surface (Ellis, 1981). The stripes at this site consist of alternating chalk rubble and silty fines aligned at an angle slightly oblique to the slope direction. The gradient is only 3°, and stripe spacing is about 1 m. The coarser stripes form ridges, separated by silt-filled troughs around 0.2 m deep. Clasts in the ridges are steeply inclined, and tend to dip into the slope. The silt-rich material infilling the troughs and burying the pattern is of loessic (i.e. windblown) origin, and accumulated during the Late Devensian. The stripes were probably active during the period of initial loess accumulation. As this example illustrates, even though the surface expression of active layer processes has disappeared across much of the British lowlands, the subsurface effects of freezing and thawing are often still evident in Pleistocene deposits. Cryoturbation structures (some of which

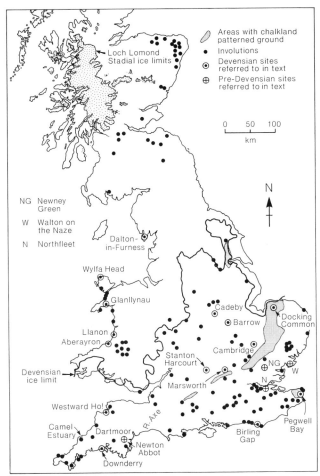

Figure 6.17 Sites of periglacial patterned ground and involutions. Chalkland patterns according to Williams (1965) and Evans (1976).

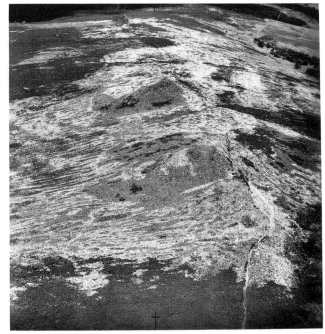

Figure 6.18 Sorted stripes, Stiperstones, Shropshire. On the ridge crest (centre right) is an area of irregular sorted polygons that grade into stripes. Cambridge University Collection of Air Photographs

must have developed under now-vanished patterned ground) are probably the most common periglacial structures in unconsolidated sediments. Many, but not all, relate to Devensian periglaciation.

Cryoturbation structures in lowland Britain

Periglacial involutions are not only common, but also show great diversity of size and configuration. Not all necessarily indicate permafrost, but those that directly overlie ice-wedge casts are generally assumed to have developed in the active layer above a former permafrost table. In the absence of subsequent erosion, the amplitude of cryoturbation may also indicate the thickness of the former active layer. Periglacial involutions exposed in Quaternary sediments usually have one or more of the following characteristics: (1) a wide range of fold styles, from gentle undulations to sharp irregular piercement structures; (2) restriction to a layer no more than 3 m thick in which the degree of disturbance may decrease with depth; (3) inclusion of sediments with contrasting textural properties; and (4) the presence of erect clasts (Worsley, 1977).

Pre-Devensian cryoturbation phenomena

As cryoturbations develop only within the near-surface zone of annual freezing and thawing, their presence in sections cut through Quaternary deposits effectively identifies the location of former landsurfaces that were subject to cold-climate conditions. They are therefore important stratigraphic horizons, and often provide valuable palaeoclimatic evidence. A good example of a complex sequence of cryoturbations, indicating at least three phases of periglaciation, occurs at Walton-on-the-Naze in Essex (Hails & White, 1970). Here the sedimentary succession consists of Red Crag (an Early Pleistocene shelly sand), successively overlain by sands, organic silty clay, stratified sands and silts, outwash sands and gravels, and finally loessic silt (Figure 6.19(a)). The earliest phase of cryoturbation affected the Red Crag and overlying sands, but not the succeeding organic silts (Figure 6.19(b)). A layer 1.5–2.0 m thick is cryoturbated, and in places ice-wedge casts penetrate the Red Crag and underlying London Clay. Accumulation of the organic silty clays and stratified sands and silts was followed by a minor phase of cryoturbation (or possibly load casting) prior to accumulation of the overlying outwash sands and gravels (Figure 6.19(c)). During and after outwash deposition intense cryoturbation partly destroyed current bedding, creating involutions that penetrate downwards into the underlying unit of bedded sands and silts (Figure 6.19(d)). A final phase of permafrost development is indicated by an ice-wedge cast that is partly filled with loessic silt, and which penetrates through the outwash deposit from its upper surface. This sequence suggests that an initial phase of permafrost development immediately followed deposition of the Red Crag and overlying sands. Cryoturbation appears to have occurred in a relatively thick (1.5–2.0 m) active layer, or during permafrost degradation. Accumulation of the organic silts and bedded sands and silts was then followed by minor deformation, possibly due to seasonal freezing and thawing in the absence of permafrost, or alternatively attributable to loading and dewatering. Periglacial conditions, possibly with permafrost, were again present during deposition of the overlying

(a)

G	BRICKEARTH	Thin loam on higher ground, locally occupying open wedges in gravels.
F	GRAVELS	Fluvioglacial gravels, generally sandy, locally stratified.
E	ALTERNATING SANDS AND SILTS	Alternating beds of silty clay and fine sand. Beds often contorted and sometimes totally disrupted by involutions.
D	CARBON- ACEOUS SILTY CLAY	Bed of silty clay with pebbles of flint and quartzite.
C	SANDS	Grey-green silty sand with scattered pebbles above a coarse red-brown sand.
B	RED CRAG	Mixed sand and comminuted shell debris, strongly current bedded.
		UNCONFORMITY
A	LONDON CLAY	Strong blue-grey clay.

(b)

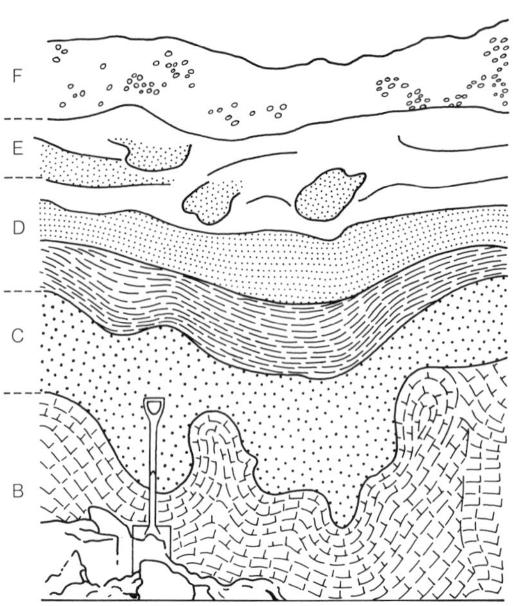

(c)

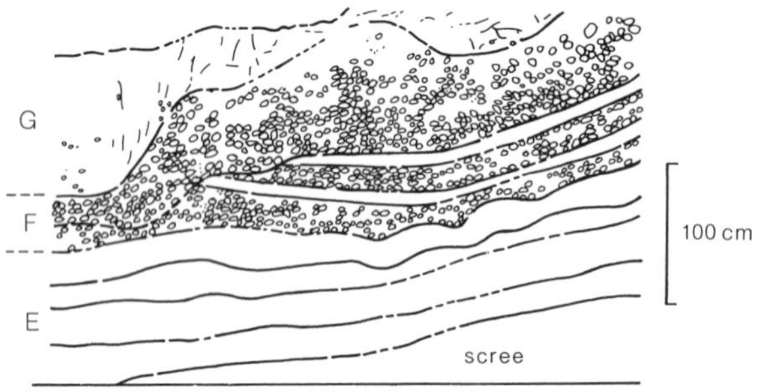

(d)

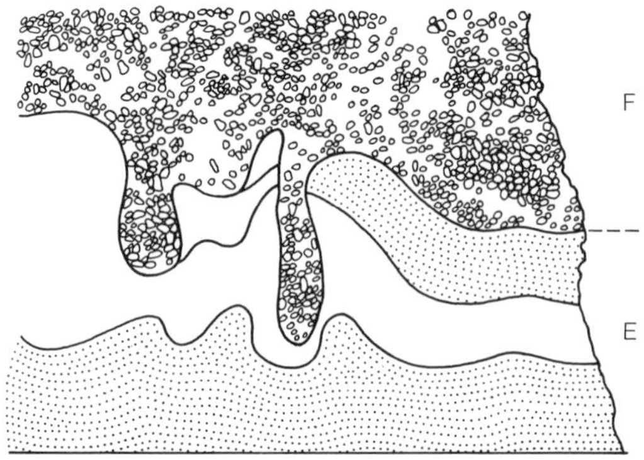

Figure 6.19 Involutions at Walton-on-the-Naze, Essex (after Hails & White (1970)): (a) generalised succession; (b) first and second phases of cryoturbation, affecting the Red Crag (Bed B), and the alternating sands and silts (Bed E) respectively; (c) small involutions in the alternating sands and silts, with undisturbed and disturbed outwash sands and gravels above; (d) involutions in base of outwash gravels.

outwash sediments, since these contain both severely cryoturbated and relatively undisturbed bedding. It is likely that this phase of proglacial sedimentation corresponds to the Anglian glacial maximum. If so, a pronounced unconformity exists between the outwash deposits and overlying loess, as the latter is probably of Late Devensian age (Chapter 8). The loess-filled ice-wedge cast within the outwash presumably reflects a return to permafrost conditions during the Devensian cold stage.

The possibility of establishing a detailed cryostratigraphy from sequences of periglacially-disturbed sediments is well illustrated by the example above. Elsewhere, other complex sequences occur, such as in the fluviatile silts and sands ascribed to the Beestonian cold stage in East Anglia. At the Beeston type site, for instance, West (1980a) observed three horizons with cryoturbation structures, each associated with ice-wedge casts. Elsewhere in East Anglia, the widespread

palaeosol known as the Barham Soil (Figure 4.18) also exhibits widespread evidence for cryoturbation and ice-wedge development. This soil underlies till deposited by the Anglian ice sheet, and is associated with wind-blown sand and silt deposits (Rose & Allen, 1977; Allen, 1983). The soil is developed on fluvial gravels of the Kesgrave and 'Pebble Gravel' Formations, deposited when the lower Thames flowed through East Anglia (Rose et al., 1985a; Chapter 8). The periglacial structures themselves, however, are superimposed on an interglacial soil (the Valley Farm Soil), so that clay contents in the cryoturbated horizons are significantly higher than those of the underlying gravels. The Barham Soil involutions generally take the form of festoons (Figure 6.20), with intervening symmetrical lobe-shaped bowls of either coarser or finer material. Those with gravel cores show a mean width of 0.98 m and a mean depth of 0.53 m, while those with sand cores have a mean width of

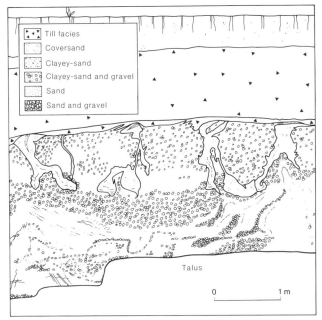

Figure 6.20 Involutions at Newney Green, near Chelmsford. After Rose *et al.* (1985b).

0.58 m and a mean depth of 0.35 m (Allen, 1983). Rose and his co-workers have emphasised the variability in form of the involutions, and noted the coexistence of structures of contrasting size, with smaller involutions either truncating or nested within larger structures. More complex forms, resembling type 4 of Vandenberghe's classification (Figure 6.10), are also present.

At Newney Green, near Chelmsford, the Barham Soil involutions form a polygonal pattern in plan, with diameters around 1.5 m, while at the Barham type site involutions of similar dimensions have a circular plan form. Also at Barham, cryoturbation of Anglian till and coversand immediately below the present ground surface indicates a second, post-Anglian, periglacial phase. Although reconstruction of the thickness of the former active layer on the basis of the amplitude of involutions in the Barham Soil is difficult (since superimposed forms probably reflect more than one phase of activity), maximum depths of cryoturbation of around 0.8 m at Barham and Badwell Ash, 1.5 m at Broomfield and 1.79 m at Newney Green give some indication of former maximum active layer depths. Sites in NE Essex where there are cryoturbation structures of equivalent age to those in the Barham Soil have been described by Bridgland (1988).

Possible evidence for periglaciation in southern England dating to a post-Hoxnian, pre-Ipswichian cold stage is found at Northfleet, on the southern side of the Thames Estuary. The Northfleet valley contains coombe rock into which overlying solifluction deposits are cryoturbated. These cryoturbated sediments are buried by aeolian silts that are themselves overlain by temperate freshwater silts. The solifluction gravels contain Palaeolithic hand-axes of the Levalloisian type, which is associated with 'Wolstonian' cold stage gravels at some sites in southern England. However, Levalloisian hand-axes also occur in deposits of Ipswichian and Devensian age. Although the soliflucted gravels at Northfleet and their associated cryoturbation structures have been provisionally assigned to a post-Hoxnian, pre-Ipswichian cold stage, the possibility remains

that they are of Devensian age, the overlying temperate silts having been deposited during a Devensian interstadial rather than during the Ipswichian Interglacial (Wymer, 1988). The age of large dome-like structures in the upper 3 m of the Oligocene Bovey Beds near Newton Abbot in Devon is also uncertain. These affect clay and lignite beds, are up to 6 m wide and have been attributed by Dineley (1963) to cryoturbation of unspecified but probably pre-Devensian age. Dineley suggested that they developed during downward advance of thickening permafrost, but they could represent load structures unrelated to frost action.

Devensian cryoturbation phenomena

The majority of cryoturbation structures hitherto observed in Britain relate to the Devensian cold stage. Many authors have reported a consistent maximum depth of disturbance that may correspond to the position of a former permafrost table or to the former depth of seasonal freezing. Unfortunately, cryoturbation structures are generally exposed only in vertical section, so that their plan form often remains unknown. As will be seen later, some idea of their three-dimensional character enables firmer conclusions to be drawn regarding processes of formation, and, in particular, any genetic relationship there might have been with surface patterned ground. Most cryoturbation structures have been attributed to the Late Devensian, though cryoturbations and associated ice-wedge casts of inferred Early Devensian age were observed by West *et al.* (1974) at Wretton in Norfolk. At this site cryoturbations occur in three horizons separated by deposits interpreted by West and his co-workers as representing two warmer substages, though the validity of this interpretation now appears questionable (Chapter 2). The involutions reflect upward injection of fine muds into overlying sands, together with irregular or amorphous forms involving muddy layers within the gravels. These structures have vertical amplitudes of less than 1 m, and form polygonal patterns in plan. The presence of ice-wedge casts indicates that permafrost was present during formation of the involutions.

In southern England, cryoturbation structures often occur in weathered chalk (Figure 6.21). On the Sussex coast and on the Isle of Thanet in Kent the boundary between the weathered chalk and overlying aeolian silts is commonly intensely cryoturbated (Kerney, 1965; Williams, 1971), and about 40% of the coastal cliffs in which the chalk of the South Downs is exposed show near-surface involutions. These consist of rounded pockets of chalky sand or loess that penetrate downwards into weathered chalk (Williams, 1980, 1987). On gradients of less than 4° they form polygons in plan, with fine-grained centres and margins formed of vertically-oriented chalk rubble. They become elongated into short sinuous stripes on gradients of 4–9°, but on steeper slopes were apparently prevented from forming by mass movement processes. Similar cryoturbations underlie much of the East Anglian chalklands, including areas with large-scale patterned ground. Flame structures of chalk rubble rise upwards from *in situ* chalk, separating pockets of aeolian sand (Watt *et al.*, 1966; Williams, 1973; Figure 6.22). These pockets appear to have resulted from diapiric displacement of supersaturated sediments. Larger-scale cryoturbation phenomena occur on the upper valley sides of the River Yare in Norfolk, in the form of narrow, wedge-shaped

Figure 6.21 Involutions developed in colluvial silt and coombe rock, Marsworth, Buckinghamshire. Photograph by Dr J.G. Evans.

(a)

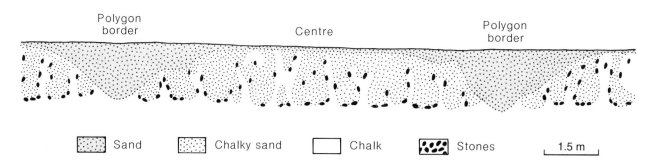

Sand   Chalky sand   Chalk   Stones   1.5 m

(b)

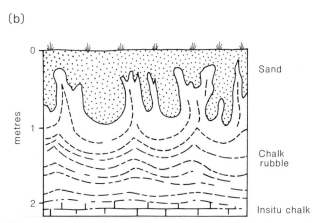

Figure 6.22 Cryoturbation structures associated with chalkland patterned ground in East Anglia: (a) from Williams (1973); (b) redrawn from a photograph taken at Lakenheath, Suffolk, in Watt *et al.* (1966).

mafrost table, and disturbance of the sensitive clays probably resulted from freezing and thawing. Similar diapiric intrusions of weathered chalk into overlying sediments have been described by Hawkins (1952) near Newbury in Berkshire, Kerney (1965) in the Isle of Thanet in Kent, and Paterson (1971) at Gore Hill in Berkshire. Tertiary clays have also experienced cryogenic injection into overlying sands at Claygate in Surrey (Te Punga, 1957) and into overlying gravels at Stanstead Abbots in Hertfordshire (Shilston, 1986). Exceptionally large-scale involutions with vertical amplitudes of up to 6 m are present in the dry valley of the Birling Gap in Sussex (Williams, 1971). Here, rounded pockets of flinty weathered chalk extend downwards into underlying chalk regolith. These large-scale structures are probably too deep to have developed in a former active layer, but may have formed during the thaw of underlying permafrost at the end of the Dimlington Stadial.

At Pegwell Bay, Broadstairs and other locations in Kent there are irregular pocket-like involutions that penetrate 1.5–2.0 m into the underlying weathered chalk. These underlie loess containing a palaeosol of Windermere Interstadial age that also shows signs of cryoturbation (Kerney, 1965). The intense cryoturbation of the weathered chalk probably dates to the Dimlington Stadial. The later, less severe phase of cryoturbation affecting the interstadial palaeosol almost certainly occurred during the subsequent Loch Lomond Stadial. Similar cryoturbated palaeosols of Windermere Interstadial age occur under colluvium in the chalk downs of Buckinghamshire and Berkshire (Evans, 1966; Paterson, 1971). Farther north, inside the Late Devensian ice limits, periglacial involutions of Loch Lomond Stadial age are widespread. These are described in detail later in this section.

Cryoturbation structures in fluvial gravels of apparent Devensian age have also been observed in Cambridgeshire (Paterson, 1940), Norfolk (Straw, 1980) and the upper Thames valley west of Oxford (Seddon & Holyoak, 1985), in all cases at sites where ice-wedge casts extend downwards from the cryoturbated horizons. The cryoturbations themselves characteristically take the form of gravel flame structures separated by spheroidal or festoon-like bulbs of the overlying finer-grained sediment. Gravel clasts within the flame structures are generally steeply inclined. On the floor of a dry valley at Docking Common in Norfolk, two cryoturbated sandy gravel units are separated by a well-developed palaeosol that is itself intensely cryoturbated (Figure 6.23). Organic material from the palaeosol yielded radiocarbon ages of 19 300 ± 300 yr BP and 24 000 ±

ridges of weathered chalk (coombe rock) that penetrate approximately 1 m upwards into overlying sands (Leeder, 1972). It appears that diapiric intrusion of the clayey weathered chalk into the overlying sands resulted from thixotropic liquefaction under saturated conditions within a former active layer. Abundant groundwater was apparently held above the per-

Figure 6.23 Cryoturbated palaeosol beneath upper sandy gravels at Docking Common, Norfolk. Photograph by Professor A. Straw.

550 yr BP (Straw, 1980, 1991). These dates suggest that the lower gravels experienced cryoturbation during the Early or Middle Devensian, and that accumulation and cryoturbation of the upper gravels occurred during the Dimlington Stadial.

In his classic study of periglacial structures in a gravel exposure near Cambridge, Paterson (1940) stressed the role of differential ice segregation in the formation of cryoturbations and argued that this reflected uneven frost penetration from the surface downwards. He drew an analogy between the Cambridgeshire involutions and sorted circles he had observed on Baffin Island and in West Greenland, suggesting that frost heaving was responsible for the presence of steeply-inclined clasts. Similarly, Seddon & Holyoak (1985) described clusters or nests of clasts within cryoturbation structures at Stanton Harcourt (Figure 6.24). The nests are up to 1.5 m apart, form an irregular polygonal network in plan, and were interpreted as representing the sites of sorted circles or nets. At Stanton Harcourt the cryoturbated layers apparently mark surfaces exposed during the accumulation of gravels in a braided river system (Chapter 8). The most extensively cryoturbated zone lies 0.5–1.2 m below the present ground surface, but two lower cryoturbated horizons are also present, each underlying erosion surfaces within the gravels. Ice-wedge casts indicate that continuous permafrost conditions were present during gravel deposition, probably in the Early or Middle Devensian. The maximum thickness of cryoturbation in the most extensive layer is 0.6 m (Seddon, 1984). This may indicate the former active layer

depth, though truncation of this layer as a result of channel migration could have occurred prior to the accumulation of overlying sediment. Rather larger cryoturbations occur in periglacial fluvial gravels in the Axe Valley south of Axminster, and involve deformation of the upper 2–3 m of gravels (Shakesby & Stephens, 1984). Although the gravels at this site are pre-Ipswichian in age, this area experienced prolonged periglaciation during the Devensian, and it is likely that ground freezing and associated cryoturbation would last have been active at this time.

Elsewhere in the south and west of England, beyond the limits of successive Pleistocene ice sheets, periglacial involutions are widespread in frost-weathered regolith, gelifluction deposits and aeolian deposits (Figure 6.25). These structures are frequently exposed in coastal sections. Scourse (1987) noted that involutions in this area are particularly common in head developed over slate bedrock, and concluded that the highly frost-susceptible silt-rich matrix of the head facilitated their formation, while the platy clasts allow the structures to be easily distinguished. At Westward Ho! in North Devon (Figure 6.17), raised beach gravels have been reworked by Devensian gelifluction, and experienced inwashing of silty fines. Cryoturbation structures developed in the resulting diamicton are now exposed in plan on the present-day beach platform and form sorted nets 2–3 m wide, with borders of vertical cobbles. These structures indicate a significant period of frost sorting following deposition of the parent gelifluctate. Cryoturbations

Figure 6.24 Involutions in the form of 'stone nests', associated with sorted nets, Stanton Harcourt, Oxfordshire. Photograph by Dr M.B. Seddon.

also occur on Dartmoor, where they take the form of festoon-like structures up to 1.8 m deep resulting from downward deformation of poorly-sorted gravelly head into the underlying weathered granite (Waters, 1961). The latter consists of a poorly-sorted gruss of sand-sized quartz grains, mica flakes and kaolinite clay, and it seems likely that a thaw-induced unstable density configuration led to the overlying gravelly head sinking downwards into the finer-grained substrate.

To the north of the Devensian ice limit, cryoturbation structures are common in near-surface layers of glacial and fluvioglacial sediments. Watson (1965a, 1977a), Watson & Watson (1971) and Harris & McCarroll (1990) have described a variety of such cryoturbation forms in Wales. Most commonly these involve irregular downward penetration of pockets or festoons of silty or sandy sediment into underlying gravel or till. The diapiric structures of gravel and till that separate the festoons contain vertically-erected clasts. At Llanon, on the coast of Cardigan Bay (Figure 6.17), highly irregular involutions involve deformation of loessic silts, waterlain gravels and underlying till (Figure 6.26(a)). The horizontal dimensions of these involutions vary from less than 1 m to more than 3 m, but their vertical amplitude is rarely greater than 1.5 m. The irregularity of these forms suggests mass displacement of sediment in response to cryogenically-induced density contrasts (Watson, 1965a). Less commonly encountered are regularly-spaced involutions in which fine-grained bulbous structures are separated

by flame-like upward intrusions of gravel. Such features were exposed in plan on the beach at Aberayron in October 1961 (Watson, 1965a). At this site they took the form of circular patterns of erect gravel with finer-grained centres and resembled sorted circles, particularly those described by Van Vliet-Lanoë (1988a,b,c) as having developed under a positive frost susceptibility gradient (Figure 6.8). Frost sorting that may have occurred in the absence of other cryoturbation processes may be indicated by vertically-aligned clasts within diamictons (Watson & Watson, 1971, Potts, 1971). Frost sorting of clasts towards the surface tends to cause vertical alignment of elongated stones, and it has been argued that vertical stones within the upper parts of Quaternary sediments therefore provide evidence for a former period of intense frost heaving. Watson & Watson (1971) showed that on sloping ground clasts often have their long axes aligned downslope, but are set on edge, with steeply-dipping intermediate axes.

In north Wales, cryoturbation is particularly well developed and well exposed at Glanllynau, near Pwllheli (Figure 6.17). The sedimentary sequence here is complex, but is dominated by Devensian fluvioglacial sands and gravels. Kettle holes within the sands and gravels contain silts of Windermere Interstadial age (Coope & Brophy, 1972). The main zone of cryoturbation predates the Windermere Interstadial sediments and developed immediately following ice-sheet deglaciation (Harris & McCarroll, 1990; Figure 6.26(b)). These cryoturbations have a

Figure 6.25 Examples of involutions in SW England: (a) Downderry, south Cornwall: upper layer fine head, lower layer coarse head; (b) Camel Estuary, north Cornwall: slaty head deposits.

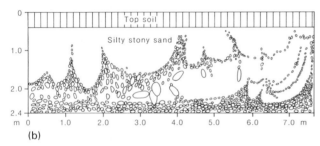

(b)

Figure 6.26 (a) Large-scale irregular involutions at Llanon, Dyfed, west Wales. (b) Dimlington Stadial involutions, Glanllynnau, Gwynedd, north Wales.

maximum vertical amplitude of 1.9 m. A second phase of cryoturbation at Glanllynau involves sands and gravels that postdate the Windermere Interstadial deposits, and apparently dates to the Loch Lomond Stadial. These upper gravels are deformed into underlying sands to form a series of lobate festoons, with the gravel in each festoon arranged parallel to its lower curving boundary. The depth of disturbance is up to 1.5 m. The older involutions that predate the Windermere Interstadial are associated with ice-wedge casts that imply the presence of contemporaneous permafrost. Small wedge-like structures, possibly frost cracks, occur in places within the post-Windermere Interstadial sediments but do not provide conclusive evidence for the presence of permafrost during the Loch Lomond Stadial. Evidence for Late Devensian permafrost in the coastal lowlands of north Wales also occurs at Wylfa Head in Anglesey, where small pocket-like involutions occur in aeolian silty sand. These extend from 0.55 m to 1.1 m below the ground surface, are cir-

cular in plan, with diameters up to 0.35 m (width:depth ratio 0.63), and are picked out by contrasts in colour rather than granulometry. They are associated with a dense, indurated layer below about 0.9 m, which is interpreted as marking a former permafrost table (Harris, 1991).

Various authors have suggested that the Loch Lomond Stadial reintroduced to much of the English Midlands, Wales, northern England and Scotland a periglacial climate of severity sufficient to promote the development of at least discontinuous permafrost. Cryoturbated aeolian coversands in Leicestershire were considered by Douglas (1982) to correspond in age to coversands in Lincolnshire for which radiocarbon evidence indicates accumulation during the Loch Lomond Stadial. At Cadeby, for example, till is intruded upwards into coversand, in places forming club-shaped diapirs up to 1.5 m in amplitude that separate pockets of sand (Figure 6.27(a)). The sands are delicately stratified subparallel to lower edges of the pockets, some of which have flattened bases, and the overall depth of disturbance is around 2 m. At Aston Flamville, between Leicester and Coventry, laminated clays form the underlying beds. Here pockets of coversand are completely enclosed in clay, forming spheroidal sand balls. On the basis of the assumed age of the coversands, Douglas concluded that cryoturbation occurred during the Loch Lomond Stadial, and was probably associated with discontinuous permafrost. Near Barrow in Northamptonshire, surface sands also occupy pockets within underlying clays (Curtis & James 1959; Figure 6.27(b)). These structures appear to have developed as a result of sinking of the sands into less dense and very wet clays, and correspond to type 2(a) of Vandenberghe's classification (Figure 6.10). Further cryoturbations of possible Loch Lomond Stadial age occur in

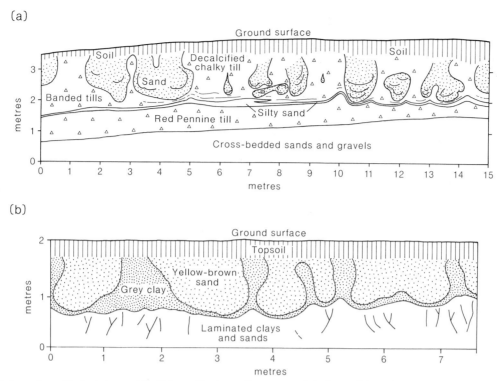

Figure 6.27 Involutions of possible Loch Lomond Stadial age affecting aeolian coversands in the English Midlands: (a) near Cadeby, Leicestershire (after Douglas (1982)); (b) near Barrow, Northamptonshire (after Curtis & James (1959)).

head deposits at Dalton-in-Furness, near the north shore of Morecambe Bay (R. H. Johnson, 1975; Figure 6.17). Here a matrix-dominated upper head forms pockets that penetrate downwards into an underlying gravelly head, the depth of disturbance being up to 2 m.

The occurrence of cryoturbation structures in Scotland was reviewed by Galloway (1961c). He noted that in some cases these had formed where fine sediments overlie coarser ones, in others where coarse sediments overlie finer ones. Four characteristic situations were evident: (1) gravels penetrating downwards into underlying till; (2) sand penetrating downwards into gravel; (3) coarse sand penetrating downwards into fine sand; and (4) gravel penetrating downwards into sand. At one site where gravels penetrate underlying sands, the latter form diapiric upward intrusions with uniformly flattened tops, possibly corresponding to a frozen surface layer. In general, the Scottish features indicate former active layer depths of 0.8–1.8 m. Galloway reported some cryoturbation structures overlain by till in Central Scotland, and attributed these to a Devensian interstadial. The majority of involutions in Scotland, however, appear to have developed in the interval between the retreat of the Late Devensian ice sheet and the end of the Loch Lomond Stadial.

The environmental significance of cryoturbation structures in Britain

Scrutiny of the pertinent literature suggests that three main forms of cryoturbation structure occur in the drift deposits of Britain. These correspond to the following categories in the classification by Vandenberghe (1988; Figure 6.10): (1) types

2(a) and 3(a) (in which downsinking sediments form pocket-like or bulbous intrusions in underlying deposits); (2) types 2(b) and 3(b) (in which larger downward-penetrating festoon-like forms are separated by diapiric injections of underlying sediment); and (3) type 6 (irregular structures). Festoon-like structures in which the width of downward-penetrating structures exceeds their depth (types 2(b) and 3(b)) appear to constitute the majority of cryoturbations in Britain. At some sites, particularly where relatively coarse sediment is intruded upwards into finer deposits, such cryoturbations form a regular network in plan, with fine-grained circular or polygonal centres surrounded by borders of coarser sediment. These forms are probably genetically related to sorted patterned ground. Pocket-like involutions with width:depth ratios less than 1, however, apparently reflect downsinking of relatively dense sediment into underlying soft material of lower density. Where such cryoturbations occur in association with patterned ground, as in the Chalkland patterns of East Anglia, they are often of much smaller scale than the patterned ground itself.

The significance of the patterned ground and cryoturbation structures described in this chapter lies chiefly in the insights they offer regarding near-surface processes operating during Quaternary periglacial periods. Cryoturbated horizons within Quaternary sedimentary sequences are, however, also of considerable stratigraphic importance. Although cryoturbation and patterned ground structures alone need not necessarily imply the former presence of permafrost, those that occur in association with ice-wedge casts (or other indications of former permafrost) provide important evidence for the thickness of the former active layer. This is particularly so where structures terminate downwards at a consistent depth that probably marks the position of the

Table 6.1: *Amplitude (depth) of cryoturbation structures in Britain.*

Site	Depth of cryoturbation	Source
1. Cryoturbations of inferred Anglian age		
Barham Soil, Eastern England	1.0–1.8 m	Rose *et al.*, 1985a
2. Cryoturbations of inferred Early Devensian age		
Wretton, Norfolk	> 1.0 m	West *et al.*, 1974
3. Cryoturbations of Late Devensian age		
3(a). Late Devensian cryoturbations attributed to the Dimlington Stadial		
Scotland	0.8–1.8 m	Galloway, 1961c
Anglesey	1.1 m	Harris, 1991
Dyfed, Wales	1.5–2.5 m	Watson, 1965a
Dartmoor, SW England	1.8 m	Waters, 1961
Cambridgeshire	0.6 m	Paterson, 1940
Stanton Harcourt, Oxfordshire	0.6 m	Seddon & Holyoak, 1985
Pegwell Bay, Kent	2.0 m	Kerney, 1965
Birling Gap, Sussex	6.0 m	Williams, 1971
Gwynedd, Wales	1.9 m	Harris & McCarroll, 1990
East Anglian chalklands	1.0–2.0 m	Williams, 1964; Watt *et al.*, 1966
3(b). Late Devensian cryoturbations attributed to the Loch Lomond Stadial		
Dalton-in-Furness, Cumbria	2.0 m	Johnson, 1975
Gwynedd, Wales	1.5 m	Harris & McCarroll, 1990
Cadeby, Leicestershire	1.5 m	Douglas, 1982
Near Barrow, Northamptonshire	1.0 m	Curtis & James, 1959
Upton, Berkshire	0.6 m	Paterson, 1971

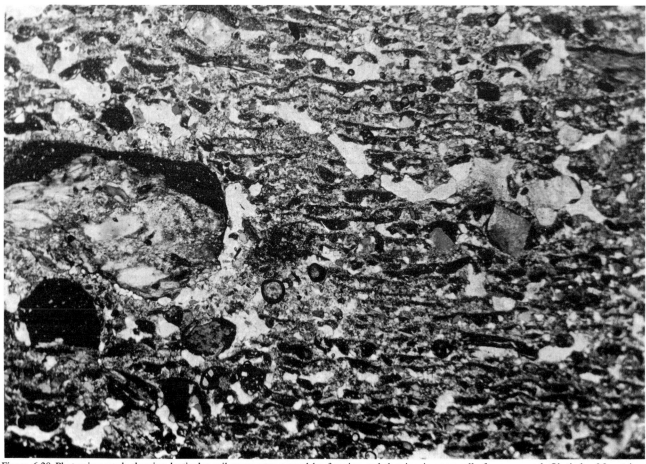

Figure 6.28 Photomicrograph showing lenticular soil aggregates caused by freezing and thawing in seasonally-frozen ground, Okstindan Mountains, Norway. Frame length 4 mm.

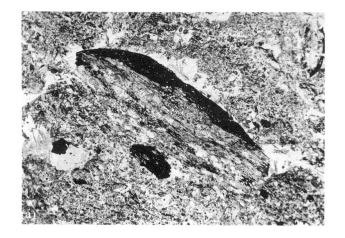

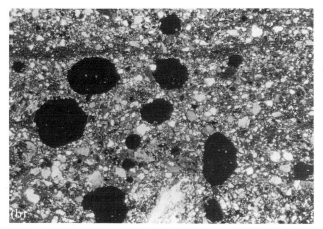

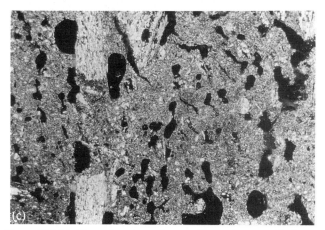

Figure 6.29 (a) Silt cappings (cutans) on sand grains in seasonally frozen soil, Okstindan Mountains, Norway. Frame length 8 mm. (b) Vesicles in the fine-grained centre of a sorted circle, Jotunheimen, Norway. Frame length 14 mm. (c) Large vesicles and vertical frost-sorted granules from the active layer, Klondike Plateau, Alaska–Yukon boundary. Frame length 14 mm.

former permafrost table. In some cases, however, interpretation of the dimensions of cryoturbations in terms of former active layer depth may be misleading. For instance, where structures exceed 3 m in depth it is likely that they relate to phases of permafrost degradation, or some nonperiglacial mode of formation, rather than to cryoturbation in a stable active layer. Where permafrost degradation was not accompanied by widespread instability and

cryoturbation structures developed as a result of seasonal active layer freezing and thawing, their vertical extent may reflect active layer thicknesses during the *waning* of the cold stage, rather than during the period of greatest climatic severity when active layer thickness would have been less. Galloway (1961c) noted that pedological evidence suggests an active layer of no more than 0.6 m in Scotland during the Late Devensian, but contemporaneous cryoturbation structures extend to depths of 0.8–1.8 m.

Conversely, cryoturbation structures of limited vertical extent may reflect either shallow frost penetration in a nonpermafrost environment or a shallow active layer associated with a very severe permafrost climate. Erosion may also truncate a cryoturbated horizon, giving the impression that the active layer was shallower than was actually the case, though such erosion is likely to be evident in the 'beheading' of cryoturbation structures. In addition, Williams (1975) has pointed out that involutions are often initiated at structural boundaries within the active layer, particularly those between sediments with contrasting hydraulic and geocryological properties. The depth at which cryoturbation structures are initiated may therefore be controlled in part by the depth of such sedimentary discontinuities. A further problem is that if the degree of disruption and the vertical extent of the cryoturbation structures increases with time during their formation, immature features whose growth was for some reason prematurely terminated may not truly reflect former active layer depths. Finally, as pointed out earlier in this chapter, not all involutions owe their origins to cryoturbation. Dewatering and load structures are common in waterlain sediments, and may easily be confused with periglacially-induced involutions. Caution is therefore necessary when employing cryoturbation structures and related active layer phenomena as evidence for former environmental conditions. The depths of cryoturbation observed at sites discussed in this chapter are summarised in Table 6.1. The wide range of values serves to emphasise the difficulty of making general statements regarding regional active layer depths on the basis of such data.

Pedological evidence for soil freezing and permafrost

It is widely accepted that the development of segregation ice within freezing soils causes the formation of distinctive and persistent microstructures (Harris, 1985; Van Vliet-Lanoë, 1985). In frost-susceptible soils, ice lenses develop parallel to the advancing freezing front. The resulting frost heave causes compaction of the soil between ice lenses, and, as water migrates towards the growing ice lenses, desiccation further increases soil compaction. The result is a platy or lenticular soil structure, with the thickness of individual aggregates reflecting the spacing of ice lenses. Repeated freezing and thawing causes fine lens-shaped or blade-shaped aggregates to develop. These are usually up to a few millimetres thick and up to a few tens of millimetres wide. Soil freezing from the ground surface also causes downward frost sorting of silt and clay. Thawing of ice-rich soil releases meltwater, and downward percolation leads to further translocation of fine particles. These fines are intercepted by the upper surfaces of the platy aggregates, producing a distinctive fining-upwards pattern through each aggregate, and increasing its density (Figure 6.28). Thus a distinctive and very stable freeze–thaw structure results (Van Vliet-Lanoë, 1982, 1985).

In permafrost environments, concentrations of translocated fines tend to accumulate immediately above the permafrost table (Corte, 1963). In many soils the base of the active layer and the top of the underlying permafrost also constitute a zone of intense ice segregation as a result of refreezing of percolating meltwater late in the summer thaw period (Mackay, 1983a; Guodong, 1983). An important consequence of the above processes is the formation of a hard, dense indurated layer with a platy or lenticular structure at the base of the active layer. Permafrost aggradation (thinning of the active layer, usually in response to lower summer temperatures) is particularly effective in generating such an indurated horizon (FitzPatrick, 1956b; Van Vliet-Lanoë & Valadas, 1983). European soil scientists refer to this indurated layer as a *fragipan* or *consolidated horizon*. The depth of such fragipans in Belgium and northern France has been used to reconstruct Weichselian (Devensian) active layer thicknesses in these areas (Van Vliet-Lanoë & Langhor, 1981). With increasing depth below the permafrost table, ice lenses become more widely spaced, and a blocky or prismatic soil structure develops following thaw.

Downward translocation of fines within the active layer also leads to coatings of clay, silt or fine sand on the upper surfaces of clasts or even coarse sand grains. Such coatings are termed *cutans* (Figure 6.29(a)). FitzPatrick (1956b) observed that in soils on Spitsbergen coatings developed on clasts as surrounding sheaths of ice melted, and Brewer & Pawluk (1975) attributed coatings of fines around stones to liquefaction of active layer soils during thaw. Distinctive smooth-walled bubble-like pores (*vesicles*) are also common in frost-susceptible soils of the arctic and subarctic (Figure 6.29(b),(c)). FitzPatrick proposed that these vesicles reflect the expulsion of air from soil water during freezing, but Harris (1983, 1985) and Van Vliet-Lanoë, Coutard & Pissart (1984) have shown that soil collapse during thaw consolidation leads to liquefaction, causing soil air to form bubbles.

Pedological effects of soil freezing in Great Britain

All of the cryogenic structures described above have been detected in the soils of Great Britain. Lenticular structures indicative of former soil freezing and indurated horizons marking the top of the former permafrost table appear to be particularly common. At Belever Quarry on Dartmoor, for example, head derived from weathered granite exhibits a fine lenticular structure to a depth of approximately 1.5 m below the surface (Figure 6.30(a)). Vesicular pores and silt cappings on skeletal grains provide further evidence for repeated freezing and thawing of the soil at this site. In areas of SW England where thicker fine-grained head deposits occur, such as Lannacombe on the south Devon coast, a strongly-developed platy or blocky structure with aggregates a few centimetres thick extends several metres below the surface (Figure 6.30(b)). These soil structures probably record the former presence of permafrost.

Several studies have also revealed the presence of cryogenic structures in Welsh soils. Well-developed indurated layers with a platy structure and low porosity are common in the South Wales Coalfield, the Brecon Beacons and west Wales (Stewart, 1961; Crampton, 1965). In the coalfield area, at altitudes of 300–450 m, Crampton found indurated horizons at depths of 0.5–0.6 m. In the Brecon Beacons he observed similar indura-tion at depths of 0.4–0.6 m in large valley-floor solifluction ter-races at around 400 m altitude. The base of the indurated layer ranges from 1.6 m to 2 m below the surface in the coalfield area to more than 10 m in the Brecon Beacons. Crampton suggested that induration had developed within permafrost, and inter-preted the top of the indurated layer as representing the former permafrost table. Since these areas were occupied by the last (Late Devensian) ice sheet, induration must have occurred after ice-sheet withdrawal. If such induration indeed represents the former presence of permafrost, it therefore implies permafrost development either in the wake of ice-sheet retreat at the end of the Dimlington Stadial, and/or during the Loch Lomond Stadial of *c.* 11–10 ka BP. Further evidence for soil freezing in these areas was provided by Harris (1981a,b), who analysed silt and fine sand coatings on the upper surfaces of clasts within solifluction deposits in both the Brecon Beacons and the South Wales Coalfield, and showed that these closely resemble coat-ings associated with active solifluction sediments in arctic Norway (Figure 6.30(c)). Similar cryogenic structures occur in North Wales, at Wylfa Head in Anglesey. At this site an indurated layer is present in soliflucted aeolian silty sands at a depth of about 0.8 m below the surface, extending downwards to about 1.5 m (Harris, 1991). The indurated soil has a strong lenticular structure, with silt cappings on skeletal grains (coarse sand and gravel), and steeply-dipping to vertical clasts (Figure 6.31). Above the indurated zone are small cryoturbation struc-tures. As the aeolian silty sands at Wylfa Head accumulated after Late Devensian ice-sheet deglaciation, their subsequent induration and cryoturbation must again date to the final part of the Dimlington Stadial and/or to the Loch Lomond Stadial.

In Scotland, indurated layers occur in many freely-drained soils (Glentworth, 1944, 1954; Glentworth & Dion, 1949; FitzPatrick, 1956b, 1969, 1987; Romans, Stevens & Robertson, 1966). Such layers generally occur 0.4–0.6 m below the sur-face, have sharply-defined upper boundaries but diffuse lower ones, and sometimes extend downwards for several metres (FitzPatrick, 1987). A platy structure is present near the top, but the indurated soil becomes more massive with depth. Vesicular pores and silt cappings on stones are also common. Fine-grained sediments such as certain tills and lacustrine deposits (e.g. around Tipperty and Cruden Bay in NE Scotland; FitzPatrick, 1987) show a distinct subcuboidal structure rather than the more common lenticular or platy structure. FitzPatrick (1956b) considered that induration resulted from a gradual rise in the permafrost table following deglaciation, the depth of the top of the indurated layer marking the base of the active layer at its minimum thickness. As in Wales, soil induration in Scotland may date from the period immediately following ice-sheet deglaciation at the end of the Dimlington Stadial, but it is possi-ble that the last period of permafrost development and fragipan formation was during the Loch Lomond Stadial (Rose *et al.* 1985b).

Romans *et al.* (1966) found horizontal lenticular patches of silt enrichment above the indurated layer in the alpine soils of the Scottish Highlands. These 'silt droplets' were considered to have formed as a result of ice segregation and silt translocation during freezing and thawing, and were observed to incorporate what were interpreted as fragments of an older generation of droplets. Also observed were sand grains with silt cappings, but with disturbed orientations, so that cappings covered the sides

Figure 6.30 (a) Lenticular frost-induced structures in head derived from weathered granite, Belever Quarry, Dartmoor, Devon. (b) Platy to blocky structure probably developed below a former permafrost table, Lannacombe, south Devon. (c) Coating of silt and clay (cutan) on a gravel-sized clast in head deposits derived from Devonian sandstone, Brecon Beacons, south Wales. Frame length 18 mm.

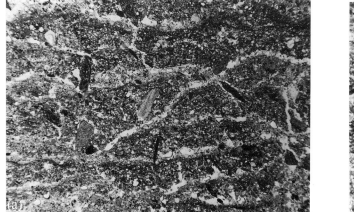
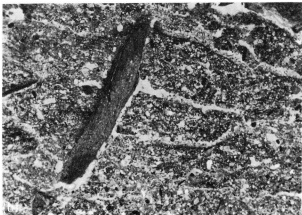

Figure 6.31 Micromorphology of indurated layer formed in silty sand, Wylfa Head, Anglesey, North Wales. (a) Lenticular aggregates caused by ice lensing. Frame length 18 mm. (b) Vertical frost-sorted clasts. Frame length 20 mm.

of the grains. The underlying indurated layer was considered to have formed as a result of permafrost development immediately following Late Devensian ice-sheet deglaciation. Romans and his co-workers speculated that the earlier generation of silt droplets formed during the Windermere Interstadial but were disturbed by frost during the Loch Lomond Stadial, and that renewed silt droplet formation occurred in a seasonally-frozen soil (but without permafrost) during the earliest part of the

Flandrian. There is, however, no independent dating evidence to support this postulated sequence of events. Romans & Robertson (1974) subsequently mapped the distribution of silt droplets in soils at upland sites in Scotland, the Lake District and Wales, and used this evidence in a putative reconstruction of the distribution of discontinuous permafrost in the earliest part of the Flandrian. They argued that discontinuous permafrost was present at this time above 595 m in Wales, 550 m in the Lake District, and 390 m in Scotland, reaching sea level at latitude 60° N. However, as silt droplets are known to develop in seasonally-frozen soils in parts of Norway where permafrost is absent (Harris & Ellis, 1980), this conclusion must be regarded as highly speculative.

The cryogenic soil structures described above all apparently relate to Devensian periglacial environments. However, similar micromorphological phenomena occur in the Early Anglian Barham palaeosol (Rose *et al.*, 1985a). In addition to the ice wedge casts and involutions associated with this palaeosol, indurated platy horizons have been observed at several sites (e.g. Great Waltham, Stebbing, Beazley End, Barham, Fornham Park and Caistor St Edmunds), together with banded fabrics resulting from silt and clay accumulations over the upper surfaces of lenticular peds, and silt-clay cappings on the upper surfaces of sand grains. Such features provide further evidence for the periglacial origin of the Barham Soil, and suggest that analysis of the texture and micromorphology of buried palaeosols has an important role to play in the identification of periglacial horizons in Pleistocene stratigraphic sequences.

Summary

Although patterned ground and cryoturbation are considered under separate headings in this chapter, it appears that in many cases the latter is simply the subsurface expression of the former. However, since relict cryoturbations are generally observed only in vertical sections, the genetic links between them and patterned ground have only rarely been explicitly recognised. It can be argued that of all periglacial features, pat-terned ground and cryoturbations are the least understood. The rich variety of hypotheses that have been proposed to explain patterned ground formation were summarised nearly four decades ago by Washburn (1956b), but progress since then in determining precisely how these features form has been slow. In particular, there remains a major conflict of opinion between researchers who consider that processes associated with the freezing phase dominate and those who consider that thaw-induced processes are of greater importance. As with many geomorphological phenomena, periglacial patterned ground and involutions may well be polygenetic, and include similar forms that have developed in different ways.

The palaeoenvironmental significance of relict patterned ground and cryoturbation structures relates mainly to the insights these phenomena provide concerning former ground temperatures and thermal regimes. Since they result from ground freezing *and thawing*, they do not necessarily provide evidence for permafrost. However, when found in association with diagnostic phenomena such as ice-wedge casts and soil induration, patterned ground and involutions may permit reconstruction of active layer processes and thicknesses. At sites where undisturbed samples of sediment can be retrieved, micromorphological investigations may provide additional detailed information concerning the cryogenic processes affecting near-surface layers. In lowland Britain, periglacial palaeosurfaces exhibiting cryoturbation may form widespread unconformities in Pleistocene sedimentary sequences, and thus provide important stratigraphic horizons. They also furnish valuable evidence concerning contemporaneous climatic regimes and the geomorphological evolution of those areas in which they occur. The outstanding example of this is the Barham Soil of eastern England (Rose *et al.*, 1985a). Moreover, the widespread preservation of delicate, near-surface periglacial structures in the soils of lowland Britain implies remarkably limited geomorphological evolution during the Holocene, and serves to emphasise the extent to which the present landscape owes its form to processes that operated during successive Pleistocene cold stages.

7

Periglacial mass wasting and slope evolution in lowland Britain

Mass wasting was defined by Washburn (1979, p. 192) as 'the movement of regolith downslope by gravity, without the aid of a stream, a glacier, or wind'. Although mass wasting occurs in all climatic regions due to the over-riding influence of gravity, it is in periglacial areas that mass wasting processes often achieve their greatest intensity and efficacy. This is largely because of the abundance of water (particularly soil water released by melting ground ice) during the spring and summer thaw. Those areas of Britain that lay outside the margins of the ice sheets during Pleistocene glacial stages experienced widespread periglacial mass wasting that often resulted in significant modification of slope form. This is particularly evident in southern England, which escaped glaciation altogether (Figure 2.1). In consequence, the landscape of this area is often dominated by the smoothing and rounding effects of mass wasting under former periglacial conditions. Even in glaciated areas, however, unconsolidated sediments have often been extensively reworked by periglacial mass wasting to give smooth convexo-concave valley-side profiles such as those of the Brecon Beacons and the Cheviots. Indeed, in the uplands, periglacial mass-movement processes are still active today (Chapter 11). This chapter begins with a discussion of the nature of periglacial mass wasting processes, then continues by considering the effects of such processes on the landscape of lowland Britain. This involves three main issues: first, the characteristics and implications of periglacial slope deposits; second, the nature of hillslope evolution under periglacial conditions; and third, the origin of large periglacially-induced bedrock structures.

The nature of thawing soils

Thaw-induced instability is primarily responsible for periglacial mass wasting. As we saw in Chapter 6, ice segregation is likely to accompany freezing of wet fine-grained soils. During freezing, water is drawn to the freezing front, maintaining a moisture supply to growing ice lenses. Ice segregation therefore leads to an increase in soil volume proportional to the influx of water during the freezing process. The result is upward frost heaving of the surface. Because thawing takes place from the surface downwards, in response to above-zero air temperatures, downward drainage of meltwater is impeded by the underlying still-frozen soil (Figure 7.1). Since thawing of ice-rich frozen ground releases more meltwater than can be accommodated by the normal soil pore space, the voids left by melting ice lenses can only be closed by the expulsion of excess meltwater. The rate of water expulsion is limited by the permeability of the thawed soil, so that for a given thickness of soil there will be a finite period of time when drainage is taking place. During this time, part or all of the overburden weight is supported by pore-water pressures in excess of hydrostatic (Taber, 1943; Morgenstern & Nixon, 1971). This transfer of stress from inter-granular contacts to porewater reduces the friction between soil grains, and hence the frictional strength of the soil. If downslope gravitational forces exceed the strength of the thawing soil, the soil will respond by failing, either as a flow, or as a slide.

The generation of porewater pressures during soil thaw depends largely on the ratio between the rate of thaw (which governs the rate of meltwater release) and the rate of consolidation (which is controlled by the rate of expulsion of excess water). This ratio, the *thaw-consolidation ratio*, forms the basis of thaw-consolidation theory, on which most geotechnical analyses of slope stability during thaw are based (e.g. McRoberts & Morgenstern, 1974; Nixon & Ladanyi, 1978; Morgenstern, 1981; Harris, 1981b). The thaw-consolidation ratio (R) is defined as:

$$R = \tfrac{1}{2}\alpha\, C_v^{-1/2} \tag{7.1}$$

where C_v is the coefficient of consolidation and α is the rate of thaw, defined by the Newman equation:

$$\alpha = Xt^{-1/2} \tag{7.2}$$

where X is the depth of thaw in time t. Thaw-consolidation theory has been applied to the prediction of slope stability by McRoberts & Morgenstern (1974), who provided a graphic solution for a soil with an assumed friction angle of 25°. For example, assuming soil moisture contents greater than 30% and $R = 0.5$, instability is predicted at gradients in excess of 8.5°; if $R = 1$, gradients as low as 4.5° are likely to be unstable.

The texture of a thawing soil, in particular its clay content, largely controls its geotechnical properties, and therefore its response to thaw-induced high water contents. Key parameters in this respect are the *plastic limit*, which is defined as the moisture content at which the soil ceases to behave as a brittle solid and becomes a plastic or ductile solid; and the *liquid limit*, which is the moisture content at which disturbance of the soil transforms it into a viscous slurry (Terzaghi & Peck, 1967). The *plasticity index* is the difference between the liquid limit and the plastic limit, that is the range of moisture contents over which the soil behaves as a plastic solid. In general, sandy and silty soils have low plastic and liquid limits and low plasticity, while clay-rich soils have higher plastic and liquid limits and a much higher plasticity index. These differences in part reflect the fact that sandy and silty soils derive most of their strength from intergranular friction, and have little cohesion. Clays, on the other hand, have lower frictional strength but are highly cohesive. In addition, sands and silts have much lower void ratios

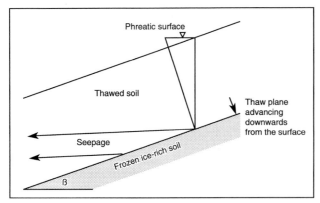

Figure 7.1 Model of a thawing ice-rich slope. Thaw consolidation is only possible when excess water is expelled. After McRoberts & Morgenstern (1974).

and hence lower saturation moisture contents than clays. Sandy and silty soils therefore require less water to reach saturation, and lose their strength more rapidly if the moisture content slightly exceeds the normal saturation level. With their low plastic and liquid limits, sandy and silty soils are particularly sensitive to changes in water content, and are susceptible to flowage if the water content exceeds the liquid limit. Clay soils often respond to the high pore pressures induced by thaw consolidation by failing along clearly-defined slip surfaces, the soil sliding *en masse*, held together by cohesion. In permafrost environments, the failure plane frequently lies immediately above the permafrost table, where the thawing soil is often ice-rich, and above which the active layer is sufficiently thick to generate shear stresses large enough to initiate sliding.

Processes of periglacial mass wasting

Mass wasting in periglacial environments may be subdivided into: (1) the gradual downslope displacement of slope sediments due to repeated freezing and thawing; and (2) more localised rapid slope failures that occur sporadically during thaw of the active layer or underlying permafrost. The former results from the combined effects of *frost creep* and *gelifluction*, as outlined below. However, since both frost creep and gelifluction are caused by frost heaving of the soil and subsequent resettlement during thaw, in practice they are often impossible to differentiate in field measurements (Harris, 1987b; Lewkowicz, 1988), and the term *solifluction* is often used to describe their combined effects.

Slow periglacial mass movements

Frost creep

In permafrost areas the active layer freezes every winter, and where it is composed of fine sediments, ice segregation commonly causes considerable frost heaving. Seasonal ground freezing in nonpermafrost periglacial areas may also be associated with frost heaving, the amount of heave depending on the moisture supply, the frost susceptibility of the soil and the depth and rate of freezing. Subsequent thawing of the ground in spring and summer leads to resettlement as ice lenses melt and the meltwater thus released percolates away. However, where

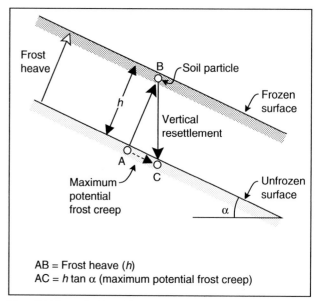

AB = Frost heave (*h*)
AC = *h* tan α (maximum potential frost creep)

Figure 7.2 The mechanism of frost creep.

the active layer is underlain by 'cold' permafrost in which temperatures are below *c.* -5 °C, meltwater may refreeze as it percolates downwards. Such refreezing causes renewed frost heave (Mackay, 1980, 1981b, 1983a). Thaw-induced resettlement takes place under the influence of gravity, so that soil particles displaced by frost heave in a direction perpendicular to the ground surface tend to settle vertically (Figure 7.2). In reality, cohesion between soil grains generally prevents truly vertical resettlement, so that the angle of settling generally lies between vertical and perpendicular to the surface. The difference between the heaving and settling angles is nevertheless sufficient to cause net downslope displacement.

On 2.5–14.0° slopes in Greenland, for example, Washburn (1967) recorded surface frost creep rates of between 4.4 mm yr[-1] and 26.2 mm yr[-1] in a silty diamicton over permafrost. At drier sites frost creep exceeded gelifluction by as much as 5.3:1, but at wetter sites gelifluction generally exceeded frost creep. Measurements of downslope frost creep in the Colorado Front Range by Benedict (1970) showed surface displacements of up to 7.2 mm yr[-1]. Like Washburn, Benedict concluded that frost creep dominates at drier sites where mass wasting is relatively slow, whereas gelifluction dominates at wetter sites where mass wasting is much faster. In relatively coarse-grained regolith, high rates of superficial frost creep may result from the formation of *needle ice*. This consists of long, stem-like ice crystals that grow in bundles perpendicularly outwards from the soil surface. During growth, needle ice crystals may lift surface soil particles and stones by 50 mm or more (Lewkowicz, 1988). Under such circumstances, surface markers may be observed to move downslope at rates of several hundred millimetres per year, but movement generally affects only the uppermost few centimetres of soil (Mackay & Mathews, 1974; Walton & Heilbronn, 1983).

Gelifluction

Annual thawing of seasonally-frozen ground is often associated with a loss in soil strength during thaw consolidation. Where

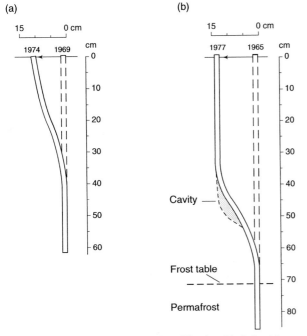

Figure 7.3(a) Profile of soil movement by gelifluction, Okstindan, Norway. (b) Profile of soil movement resulting from plug-like flow, Garry Island, NWT, Canada. After Mackay (1981b).

Table 7.1: *Measured volumetric rates of solifluction.*

Source	Location	Gradient (degrees)	Downslope discharge of soil ($cm^3\ cm^{-1}\ yr^{-1}$)	
			maximum	mean
Rudberg (1964)	Norra Storfjäll, Sweden	10–35	101	39
Benedict (1970)	Colorado Front Range	24		59
Harris (1981b)	Okstindan, Norway	4–14	33	26
Price (1973)	Ruby Range, Yukon	14–22	71	32
Williams (1966)	Schefferville, Québec	4–16	134	114
Mackay (1981b)	Garry Island, NWT, Canada	3–7	52	15
Smith (1988)	Canadian Rocky Mountains	8–22	21	11
French & Lewkowicz (1981)	Banks Island, NWT, Canada	2–4		22–30

the regolith is silty or sandy in texture and the thawing soil is ice-rich, moisture contents frequently exceed the liquid limit during thaw. Under these circumstances, disturbance of the soil structure during consolidation leads to thixotropic liquefaction. As we saw in Chapter 6, on level ground such temporarily fluidised soil is liable to density-driven diapiric displacements, and may form cryoturbation structures and patterned ground. On slopes, however, the fluidised soil is liable to slow downslope flow, or *gelifluction*, a term that refers specifically to slow soil flow associated with thawing ground (Harris, 1981b; Smith 1989). Since it is the release of meltwater from thawing ice lenses that initiates gelifluction, the distribution of excess segregation ice in the soil is a major factor affecting the vertical profile of soil movement. In those areas with seasonally-frozen ground but no permafrost and in those with discontinuous or 'warm' permafrost (permafrost temperatures warmer than *c.* -5 °C), the seasonally-frozen layer freezes only from the surface downwards. Ice segregation draws water to the advancing freezing front and the highest ice contents usually occur near the ground surface. Under these circumstances most gelifluction occurs in spring and early summer as the near-surface soil layers thaw. Movement rates generally decrease with depth (Harris, 1981b; 1987b), with displacement commonly extending down to depths of *c.* 0.5 m (Figure 7.3(a)). Where permafrost temperatures are lower, however, 'two-sided' freezing of the active layer may occur in early winter. Under such circumstances heat is not only conducted out of the ground surface by low air temperatures, but also out of the base of the active layer by the cold permafrost below. Since soil water tends to percolate to the base of the active layer in summer, freezing from the permafrost table upwards results in segregation ice being concentrated in the basal zone. During thaw therefore, little gelifluction occurs until late in the summer when the

ice-rich basal layers melt. When this happens, virtually the entire active layer moves downslope in a plug-like fashion over the deforming basal soil layers (Mackay, 1981b; Egginton & French, 1985; Figure 7.3(b)). If thawing is rapid and ice contents high, rapid displacements may occur, resulting in *active layer detachment slides*. Such rapid slope failures are discussed later in this chapter.

Field measurements indicate that solifluction generally causes surface rates of movement of a few centimetres per year, though rates of 150–350 mm yr⁻¹ have been recorded (Williams, 1959; Rapp, 1962; Holdgate, Allen & Chambers, 1967), and movement may occur on slopes as gentle as 1° (Washburn, 1979). Values for the volume of soil transported downslope by periglacial solifluction, calculated from measured profiles of soil movement, are presented in Table 7.1. Such values tend to be at least an order of magnitude greater than those of soil creep in humid temperate regions (Young & Saunders, 1986). This suggests that in Britain mass wasting was much more effective during Quaternary periglacial periods than during intervening interglacials.

Rates of movement on many alpine slopes are highly variable (e.g. Gamper, 1981), so that sediment tends to build up in areas of slower movement to form step-like landforms with straight or lobate risers. These landforms are usually referred to as solifluction terraces, solifluction sheets or solifluction lobes (Figure 7.4; Chapter 11). Frost sorting and differential rates of downslope displacement may lead to a concentration of clasts at lobe and terrace risers, and when this is so the features are often described as 'stone-banked'. In contrast, 'turf-banked' lobes and terraces lack such sorting, and display vegetation-covered frontal banks. Solifluction lobes have risers up to about 1.5 m high, while those of solifluction terraces may reach 3 m, though terrace risers up to 6.5 m high have been reported (Harris, 1981b). In high-latitude permafrost areas movement rates are

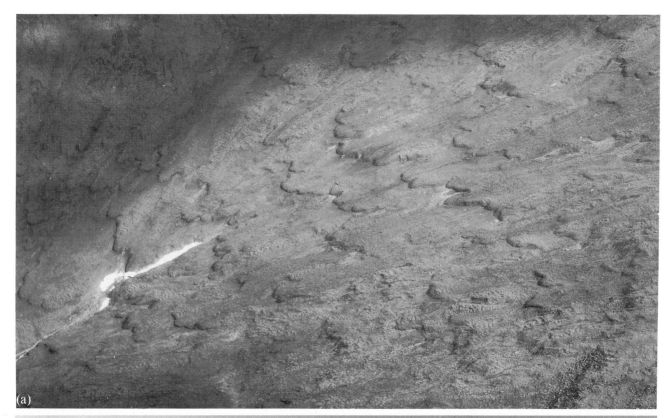

Figure 7.4 (a) Turf-banked solifluction lobes, Nome Creek, Alaska. The lobe risers are 2–3 m high, and the treads are 30–100 m long. Photograph by T.L. Péwé. (b) Stone-banked solifluction lobes, Niwot Ridge, Colorado Front Range, U.S.A..

Figure 7.5 Smooth convexo-concave solifluction slopes underlain by permafrost on Ellesmere Island, NWT, Canada.

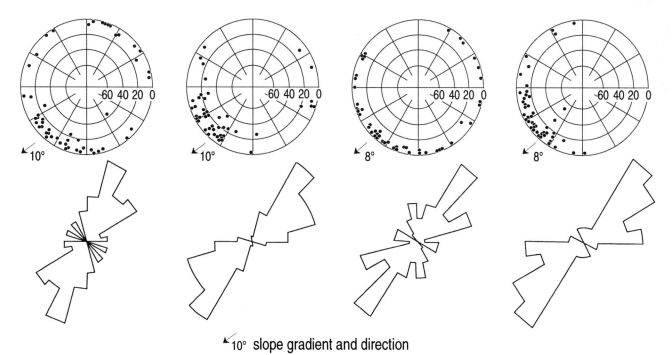

Figure 7.6 Macrofabrics from active solifluction slopes, Jotunheimen, Norway.

often more uniform so that solifluction of the active layer results in smooth convexo-concave slope profiles (Figure 7.5). For example, Holdgate *et al.* (1967, p.60) provided a graphic account of solifluction on Signy Island (Antarctica) occurring 'on such a vast scale that it is often difficult to realize that it is happening. Whole slopes are in motion, and it is not until an obstacle is encountered which causes buckling of the surface material that the extent of the motion is apparent'. Many of the sediments affected by solifluction are diamictons. During gelifluction, clasts are carried downslope embedded within a flowing fine-grained matrix. As a result, such clasts tend to become oriented parallel to the direction of flow, and to dip either parallel to the ground surface, or at an angle slightly less than the slope angle in an imbricate fashion (Washburn, 1979, p.217; Harris, 1981b; Nelson, 1985; Figure 7.6). Microscopic analysis of impregnated undisturbed samples from such deposits shows that elongate sand grains also tend to become oriented downslope, and to dip parallel to the surface (Harris & Ellis, 1980; Harris, 1981a).

Rapid mass movements

Skinflows and active-layer detachment failures

Rapid slope failures affecting the active layer in permafrost areas are necessarily shallow, since displacements take place over the permafrost table. Terminology is confused, since some authors have coined specific terms for thaw-induced rapid mass movements while others have applied landslide terminology developed in non periglacial environments. Process-based classifications distinguish between flow-dominated and slide-dominated failures (Harris 1981b, 1987b; Lewkowicz, 1988). The former are commonly referred to as *skinflows* (McRoberts & Morgenstern, 1974) and the latter, *active-layer detachment slides* (Lewkowicz, 1990) or *active-layer glides* (Mackay & Mathews, 1973). The terms *mudflow* and *earth flow* have been used for skinflow type failures, while active layer detachment slides have also been referred to as *slab slides* and *mudslides*. Both types of failure are generally triggered by rapid thawing of the active layer due to summer rainstorms, unusually warm weather or fires that destroy the insulating vegetation cover. The landforms resulting from skinflows comprise ribbon-like flow tracks, which in Alaska are 10–50 m wide and up to 450 m long (Carter & Galloway, 1981). Active-layer detachment slides range from long narrow slides to broader failures. On Ellesmere Island in arctic Canada, for example, they are up to 750 m long and 150 m wide (Figure 7.7). Movement is fairly rapid, and may last only a few hours or a few days. Failures may be initiated on slopes as gentle as 5° and may continue across gradients of only 2° (Lewkowicz, 1990). Failures of the skinflow type result from soil moisture content in excess of the liquid limit during thaw consolidation leading to soil flowage. They commonly occur in cohesionless silty soils that are highly frost-susceptible and have low liquid limits. Active-layer detachment failures, however, slide over a distinct basal slip surface (Figure 7.8) and suffer little internal disturbance other than compression in the toe zone. Failure is by sliding rather than flow either because soils are clay-rich and therefore cohesive, or because upward freezing from the permafrost table during autumn leads to desiccation and hardening of the middle levels of the active layer. In such cases failure is initiated in late summer when the depth of active layer thaw is close to its maximum. Melting of ice-rich soil immediately above the permafrost table leads to loss of strength and localised shearing, the active layer above sliding as a relatively dry, rigid block.

Mudflows and Debris Flows

Periglacial mudflows are often rather smaller features than the skinflows described above. On Banks Island in Arctic Canada, Egginton & French (1985) observed episodic mudflows that originated as mud bursts, then flowed downslope for a distance of 1–10 m before losing moisture and stabilising. The mud-bursts were initiated immediately below melting snowbanks that probably enhance the water content of the active layer. The mudflow sediments at this site have accumulated in successive sheets or lobes, each up to 0.3 m thick. Similar small-scale surface mudflows in Vestspitsbergen were referred to as 'silt seeps' by Chandler (1972a). In mountainous terrain where slopes are steep, shallow mudflows are often very rapid. They are common in both permafrost and nonpermafrost areas. When water contents are high, movement rates are rapid and the term *debris flow* is then generally applied (Innes, 1983b; Costa, 1984). Debris flows are usually initiated by very heavy rainfall (Caine, 1980). In permafrost areas such as Spitsbergen, the active layer may become saturated during rainstorms and lateral seepage then initiates debris flows over the permafrost table (Larsson, 1982). In nonpermafrost areas such as the Scandinavian mountains, bedrock may form an impervious substrate, so that the overlying soil becomes saturated by heavy summer rainfall events (Rapp, 1960a,b). The flow track is ribbon-like, flanked by bouldery levées and terminates in a lobate front. Clasts are aligned parallel to the levées, but show transverse orientations within the terminal lobe. In many parts of upland Britain debris flows have been active during much of the Holocene, and in some mountain areas represent the main agent of mass transport operating under present conditions (Chapter 12).

Ground-ice slumps

Unlike skinflows and active layer slides, *ground-ice slumps* are not restricted to the active layer. They are initiated by rapid thawing of ice-rich permafrost, often on river valley sides, but also on coastal slopes and on the flanks of thermokarst depressions (Harris, 1987b; Lewkowicz, 1988). Alternative terms include thaw slumps (Washburn, 1979), bimodal flows (McRoberts, 1973), mudslumps (Kerfoot, 1969; Kerfoot & Mackay, 1972), retrogressive thaw-flow slides (Hughes, 1972), and thermoerosion cirques (Czudek & Demek, 1970a). Active ground-ice slumps have steep, arcuate headscarps within which ground ice is exposed. Ablation of the exposed ground ice releases sediment that falls, slides or flows down to a gently-inclined mudflow lobe below (Figure 7.9). Slumps are up to 900 m across in the Mackenzie Delta area, with headwalls 6–12 m high, though larger features have been recorded. French (1976) reported widths of up to 300 m on Banks Island, where headwalls rarely exceed 2–3 m. It appears that ground-ice slumps tend to decrease in size with increasing latitude (Lewkowicz, 1988). Headscarp retreat rates of up to 10 m yr^{-1} have been recorded on Banks Island, where slumps generally stabilize in 30–50 years (French & Egginton 1973). Polycyclic flows are common, in which erosion of stabilized flows exposes further ground ice and new headscarps begin to retreat across the floor of the former slump.

Figure 7.7 Active-layer detachment slides, Ellesmere Island, NWT, Canada.

Figure 7.8 Slickensided slip surface exposed in the track of an active-layer detachment slide, Fosheim Peninsula, Ellesmere Island, NWT, Canada.

Figure 7.9 Ground-ice slump, Fosheim Peninsula, Ellesmere Island, NWT, Canada. The backwall of the slump scar is approximately 150 m across and exposes massive ground ice.

Permafrost creep

The dominant factor affecting the strength of frozen soils is the bonding of constituent particles by ice. This is complicated, however, by the presence of unfrozen water films that surround soil particles and separate them from pore ice (Johnston *et al.*, 1981). It is well known that ice under stress exhibits creep behaviour, and field studies in the Mackenzie Valley, Canada, have revealed that ice-rich permafrost is also liable to creep

deformation (Savigny, 1980; Morgenstern, 1981). On a slope of 15–24° composed of ice-rich frozen clay, creep rates of 2.5–3.0 mm yr^{-1} have been recorded, with deformation extending to depths of nearly 40 m. Huang, Aughenbaugh & Wu (1986) found an exponential relationship between increasing temperature and increasing creep of permafrost, probably reflecting a corresponding increase in unfrozen water content.

Measurements by Bennett & French (1988, 1990) on Melville Island in the Canadian high arctic showed average permafrost creep rates of 1.1 mm yr[-1] at a depth of 0.35 m, and 0.4 mm yr[-1] at a depth of 0.65 m, the decrease reflecting decline in summer temperatures with depth. Creep rates at this site were much less than those recorded in the Mackenzie Delta area. Such differences largely reflect the much lower permafrost temperatures on Melville Island compared with the those in the Mackenzie Delta area (MAGTs -16.5 °C and -3.0 °C respectively). Morgenstern (1981) has suggested that permafrost creep may account for the deformation of mudrocks that is frequently observed beneath large valleys in Britain and elsewhere. Such deformation is known as *valley bulging*, and will be discussed in more detail later in this chapter.

Slopewash

There have been few quantitative studies of the rates of sediment removal from periglacial slopes by surface wash. As in other environments, the efficacy of overland flow in eroding and transporting sediment depends largely on the protection afforded by vegetation cover. Lewkowicz (1988) suggested that in the Arctic a slight increase in denudation due to slopewash occurs with increasing latitude, since the decline in vegetation cover leads to an increase in soil exposure that outweighs the northward decline in runoff. Maximum erosion by surface wash probably occurs in the high-latitude polar desert zone, though studies in Spitsbergen (Jahn, 1961) and Devon Island in the Canadian high arctic (Wilkinson & Bunting, 1975) indicate that even in polar deserts denudation rates due to wash are lower than in many non-permafrost regions. In arctic and alpine tundra zones, erosion due to surface wash is largely restricted to areas where late-lying snow inhibits vegetation cover. In a non-permafrost area of northern Sweden, where slopes are well vegetated, Strömqvist (1983) found that surface wash is much less important than solifluction as an agent of denudation. Similarly, French & Lewkowicz (1981) showed that on Banks Island sediment transport due to solifluction is at least two orders of magnitude greater than that due to surface wash even at sites immediately downslope of late-lying snow patches. It appears, therefore, that mass wasting is the dominant form of denudation in most periglacial environments, with surface wash playing a relatively minor role. Where vegetation is limited and snowmelt generates large volumes of overland flow, however, slopewash processes may assume greater significance. Such conditions may have prevailed during periods of rapidly changing Pleistocene climate, particularly on recently-deglaciated terrain.

Sediments deposited by slopewash are typically fine-grained, sorted and occasionally stratified, and are referred to as *colluvium*. Such deposits are sometimes intercalated with solifluction deposits (*gelifluctates*). For example, a study of sediments within a solifluction lobe in Norway revealed a 0.4 m thick colluvial layer in addition to the anticipated poorly-sorted solifluction deposit (Matthews, Harris & Ballantyne, 1986). This layer consisted of a massive, relatively well-sorted silty sand. Observations on other solifluction lobes in the area showed that periodic collapse of lobe fronts releases fine sediment, which is then redeposited downslope by surface wash and incorporated within solifluction lobes. This example shows that

Figure 7.10 (a) Apron of head mantling raised beach platform, Lannacombe, south Devon. (b) Apron of head mantling raised beach platform, Oxwich Point, Gower Peninsula, south Wales. Cambridge University Air Photograph Collection.

the interdigitation of colluvium and solifluction deposits does not necessarily indicate regional changes in climate, runoff regime or vegetation cover, but may result from localised lobe collapse during thaw.

Periglacial slope evolution in lowland Britain

Data summarised by Saunders & Young (1983) and Young & Saunders (1986) strongly suggest that periglacial solifluction is a far more potent agent of denudation than the various processes that contribute to soil creep under humid temperate conditions. An important consequence of the effectiveness of periglacial slope processes is that many hillslopes in mid-latitude lowlands are essentially relict periglacial landforms that were largely fashioned during successive Pleistocene glacial stages. Such slopes have apparently experienced only limited modification under temperate interglacial conditions. Evidence for periglacial modification of slopes in lowland Britain takes three forms: first, the nature of hillslope and valley-floor deposits; second, the morphology of the slopes themselves; and third, deep-seated deformation of bedrock. In the pages that follow, we consider in turn each of these three aspects of periglacial slope activity.

Periglacial slope deposits

Periglacial slope deposits in lowland Britain are generally referred to as *head*, a term introduced by De la Beche (1839) to describe the unsorted, unconsolidated 'rubble drift' of SW England. Head deposits are commonly assumed to have accumulated primarily as a result of solifluction, but Harris (1987a, p. 209) preferred to define head more broadly as 'slope deposits accumulated through various periglacial mass movement processes', on the grounds that clay-rich head in particular may have accumulated by shallow landsliding over permafrost rather than through the agency of gelifluction or frost creep. This definition also distinguishes between head and colluvium, as the latter reflects the operation of slopewash rather than mass movement. Others have used the term 'breccia' to describe head deposits (e.g. Henry, 1984a,b; Scourse, 1987) but as most head deposits are diamictons (see below), this term may be misleading, and also fails to differentiate between head deposits and scree. For these reasons the term *head* is retained here to refer to slope deposits produced by mass movement under periglacial conditions. The general effect of periglacial mass wasting in lowland Britain has involved the downslope movement of regolith and its accumulation as head on lower slopes and valley floors. These head deposits often infill valley bottoms and form aprons across raised beach platforms in coastal locations (Figure 7.10). As a result of solifluction, hill crests are often rounded, though more resistant strata may crop out to form tors and crags (Chapter 9). Similarly, valley sides are commonly smoothly convexo-concave in profile, though the underlying rockhead geometry may be far less regular.

Head deposits are commonly matrix-supported *diamictons*, that is poorly sorted mixtures of fine matrix material (silty or sandy fines) in which are embedded clasts of various sizes. Jones & Derbyshire (1983) described head as resembling till, but having a higher void ratio, looser fabric, higher permeability and lower mechanical strength. The material incorporated in head deposits is often derived from weathering of the local bedrock, and when this is the case the lithology of underlying rocks has a considerable influence on the sedimentological and geotechnical properties of the head. In some areas, however, unconsolidated superficial sediments (particularly glacial deposits) have also been widely reworked by mass wasting under periglacial conditions. Clasts released by frost weathering of bedrock tend to be angular, but those derived from till may exhibit edge rounding due to glacial abrasion. The importance of parent material in determining the sedimentological and geotechnical properties of head has been emphasised by Harris (1987a), who recognised three categories: (1) head derived from nonargillaceous bedrock; (2) head derived from argillaceous bedrock; and (3) head derived from till. The major subdivision in terms of mass-movement processes is between categories (1) and (2), that is between poorly-sorted diamictons

Figure 7.11 Examples of head deposits formed over different bedrock lithologies: (a) Devonian slate, Whitesands Bay, Cornwall; (b) Granite, Lamorna Cove, Cornwall; (c) Carboniferous Limestone, Heatherslade, Gower, south Wales.

with low clay content and clay-rich heads largely derived from Mesozoic and Tertiary mudstones and clays. In the discussion below, therefore, coarser-grained poorly-sorted head deposits are considered separately from clay-rich deposits since it is likely that different mass-movement processes were responsible for their accumulation. In a review of the engineering significance of periglacial slope deposits in Britain, Hutchinson (1991) also identified two main types of periglacial mass-move-

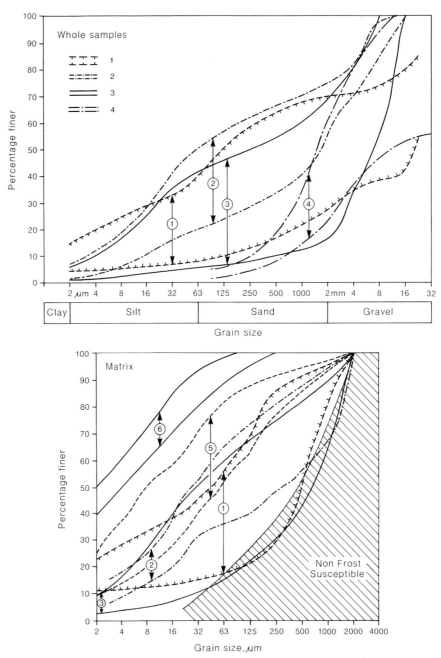

Figure 7.12 Grain-size curves for head deposits derived from different lithologies. 1. Vale of Edale: grits and shales. 2. North Cornwall: sandstones and shales. 3. Woolhope Hills, Herefordshire: limestones, shales and marls. 4. Hampshire Downs: chalk. 5. Morfa-bychan, Dyfed, Wales: mudstones. 6. North Cornwall: weathered mudstones.

ment deposit. These he termed *relic granular solifluction deposits* and *relic clayey solifluction deposits*. He suggested that the first group probably accumulated primarily through gelifluction and frost creep, while the second resulted from shallow periglacial landsliding above a former permafrost table.

Granular head deposits derived from nonargillaceous rocks

Although highly variable in terms of matrix granulometry and clast content, nonargillaceous head deposits form a distinctive category, with several common characteristics (Harris, 1981b,

1987a; Hutchinson, 1991). Such deposits usually consist of angular clasts set in a silty or sandy matrix (Figure 7.11), and the latter is generally sufficiently fine-grained to be frost susceptible according to Beskow's (1935) criteria (Figure 7.12). In most cases, clast content increases with proximity to underlying bedrock (Mottershead, 1971, 1976, 1982; Scourse, 1987). Layers of better-sorted fine-grained material containing relatively few gravel- or pebble-sized clasts have been interpreted either as colluvial bands deposited during periods of increased slopewash (Kidson, 1971; Harris, 1981b; Henry, 1984a), or as the result of redeposition of interglacial soils by solifluction early in a periglacial cycle (Waters, 1964a; Scourse, 1987).

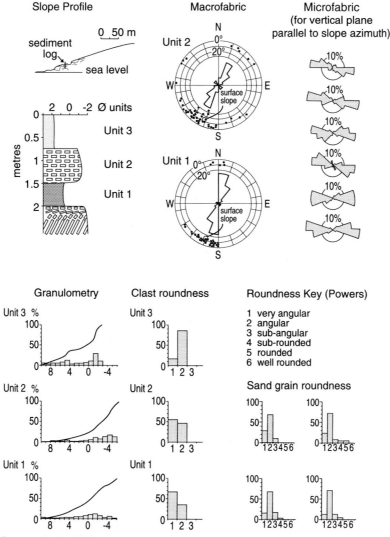

Figure 7.13 Sedimentology of granular head at Whitesands Bay, Cornwall. After Harris (1987a).

A typical example of nonargillaceous head occurs at Whitesands Bay in SE Cornwall (Harris,1987a; Figure 7.11(a)). At this site, three texturally-defined units overlie frost-weathered slate in which the uppermost few decimetres exhibit creep deformation. The three units of head all consist of angular or very angular platy clasts set in a silty sand matrix. Unit 1 has a matrix-to-clast ratio of 1.32:1 by weight, unit 2, 0.47:1 and unit 3, 1.78:1 (Figure 7.13). Clay constitutes less than 5% of the total weight in units 1 and 3, and less than 1% in unit 2; these values correspond to 7% of the < 2 mm fraction in unit 1, 8% in unit 3 and 2% in unit 2. Elongate platy clasts display a strong preferred downslope orientation and a low angle of dip, with most stones dipping at an angle less than the surface slope in an imbricate fashion (Figures 7.11(a) and 7.13). Such macrofabrics are typical of poorly-sorted head deposits and are generally interpreted as indicating accumulation by gelifluction (Stephens, 1961; Kirby, 1967; Watson & Watson, 1967; Kidson, 1971; Mottershead, 1971; Potts, 1971; Harris & Wright, 1980; Wilson, 1981; Cresswell, 1983; Douglas & Harrison, 1984, 1987; Henry, 1984a,b; Scourse, 1987; Gallop, 1991). This interpretation is discussed in more detail below.

Vertical thin sections of undisturbed samples from units 1 and 2 show sand grains with a consistent low angle of dip parallel to that of the clasts. Harris & Ellis (1980) have observed similar imbricate low angle dips of sand grains in active gelifluction lobes and terraces in north Norway.

On the south coast of Devon, between Hope Cove and Start Point, an apron of head extends seawards from an abandoned cliffline, burying the underlying raised shore platform (Figure 7.10(a)). The head deposits form a smooth concave bench that is typical of coastal head terraces throughout SW England and south Wales (Figure 7.10(b)). These deposits again consist of angular clasts set in a silty or sandy matrix, and show typical downslope clast orientations and low angles of dip (Kirby, 1967; Mottershead, 1971). Gallop (1991) discovered a particularly strong preferred orientation of clasts in deposits close to the buried cliff, with clasts aligned downslope dipping roughly parallel to the ground surface. He also found that dips become progressively less steep away from the cliffline, and finally reverse (i.e. clasts dip into the slope) at the distal end of the terrace (Figure 7.14). He concluded that the head accumulated by slow saturated flow due to the thawing of ice-rich sediments,

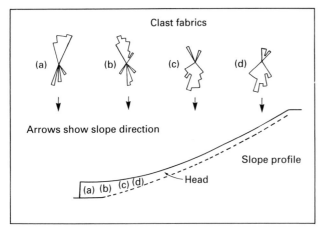

Figure 7.14 Macrofabrics from coarse granular head, Lannacombe, Devon, plotted to indicate dip direction. Redrawn from Gallop (1991).

and that the reversal in clast dips resulted from compression of sediment as movement slowed prior to immobilisation. Gallop envisaged that this caused clast-to-clast collisions and resultant stacking of clasts in an imbricate fashion. At many coastal sites in SW England the main unit of head is mantled by up to 1.5 m of finer-textured 'upper head' (Waters, 1964a; Stephens, 1970; Mottershead, 1976). This upper head is matrix-dominated and often contains only scattered clasts that nonetheless display a preferred downslope orientation. The matrix is silt-rich, and Mottershead (1976) suggested that this uppermost head facies contains a significant component of loess.

Although developed on different bedrock (Carboniferous Limestone), the head deposits of the south coast of the Gower Peninsula in south Wales bear a close sedimentological resemblence to those of Devon and Cornwall. Henry (1984a,b) referred to the main periglacial slope deposit in south Gower as the 'Hunts Breccia'. This unit is 1–5 m thick and overlies a raised beach of Ipswichian age, forming an apron in front of a scree-covered fossil interglacial cliff. Close to the cliff the deposit is almost devoid of matrix and merges into scree, but generally it is matrix-supported. The matrix consists of red silt or silty fine sand, which Henry interpreted as representing reworked interglacial soils. The angular clasts are strongly oriented parallel to the direction of maximum slope and have low-angle dips, giving the deposit a crudely stratified appearance. Soliflucted pre-Devensian till is in places mixed with the head and is the source of rounded erratics in certain facies. Locally underlying the Hunts Breccia is a colluvial silt layer that apparently represents slopewash prior to the onset of significant gelifluction activity. Like the south Gower coast, the head-covered slopes of the Woolhope Hills in Herefordshire lie just outside the limit of Devensian glaciation. Here, at the foot of the eastern side of the Wye Valley, head deposits reach a thickness of over 9 m and consist of a poorly-sorted silty or sandy diamicton containing mainly angular clasts with thin layers and lenses of silty sand (Figure 7.12). A strong preferred downslope orientation of clasts suggests accumulation by gelifluction, but the presence of better-sorted and more rounded gravels on the floors of nearby small tributary valleys indicates reworking of the head by nival streams (Shakesby, 1981).

Scourse (1987) identified five facies of granitic head in west Cornwall. The basal facies (A) he termed 'deformation brec-

cia'. It includes two subfacies. In one of these, the underlying unconsolidated sediments (usually raised beach sands in coastal situations) are incorporated as flame- or tongue-like structures into the base of the head; in the other, downslope overturning of frost-heaved bedrock forms a merging boundary between undisturbed bedrock below and totally reorganised head material above. Facies B occurs close to bedrock headlands in coastal sections. It is clast-supported and consists mainly of large boulders. The matrix is poorly-sorted, and in places almost absent. Scourse considered that facies B represents frost-weathered and heaved blocks that have experienced only limited transport. In contrast, facies C is matrix-dominated, crudely-stratified, contains only occasional clasts, and is composed mainly of sand and granules. It generally forms the basal unit in coastal exposures well removed from bedrock headlands and was interpreted by Scourse as representing solifluction of fine-grained topsoils at an early stage of periglaciation. Facies D is the most commonly observed variant of granitic head and usually overlies facies A, B or C. It consists of a matrix-supported diamicton, with angular clasts set in a silty sand matrix. Lobate structures may be present at the surface, with concentrations of clasts marking the lobe fronts. Clasts display a strong preferred downslope orientation and dip at low angles into the slope. Some larger clasts are shattered and sheared out into 'clast-trails', aligned parallel to the fabric. Loessic silt lenses also occur in this facies. Finally, facies E, the uppermost unit generally observed in these granitic head sequences, comprises extremely large granite blocks set in a granular matrix. Load structures beneath some blocks suggest they were rafted along on the surface by solifluction, but in other cases an erosional contact between the boulders and underlying fines suggests that they may have moved through the underlying finer sediments as *ploughing boulders* (Chapter 11). Facies E probably corresponds to the boulder stripes and boulder streams present downslope of many Dartmoor tors (Williams, 1968; Miller, 1990). Similar features on Bodmin Moor in Cornwall have been interpreted as being of Loch Lomond Stadial age (Brown, 1977).

The most common facies association observed by Scourse in west Cornwall was the coarsening-upward sequence C–D–E, a pattern similar to that previously recognised on lower slopes in Dartmoor by Waters (1964a). Waters argued that the first material to be removed from upper slopes by solifluction was the finer-grained weathered regolith from near the top of the pre-existing soil, and that this was redeposited on lower slopes to form the basal head facies. He envisaged that further solifluction upslope then removed coarser, less weathered rock debris from lower in the pre-existing soil profile, and that this coarser debris was then transported downslope and spread over the top of the earlier finer-grained gelifluctate. In this way, he argued, the coarsening-upwards sequence represents an inversion of the original weathering sequence. Though Green & Eden (1973) did not detect such a coarsening-upward trend in Dartmoor head deposits, Scourse (1987) maintained that inversion of the weathering sequence is generally present in head deposited near the foot of slopes and on valley floors. This assertion is supported by Douglas & Harrison (1985, 1987), who found evidence for a similar inversion of the weathering profile within valley-floor solifluction sheets in the Cheviot Hills (Chapter 11).

The time interval over which these gelifluctates accumulated

Figure 7.15 Overturning of Devonian slate bedrock beneath granular head, Wembury, south Devon.

is often impossible to determine, since they generally lack organic material or palaeosols marking interstadials or interglacials. However, organic lenses within head deposits on the Isles of Scilly have been dated by Scourse (1985) to between 34 and 21 ka BP, dates that represent maxima for the accumulation of overlying head deposits. Moreover, many coastal aprons of head in SW England and south Wales overlie Ipswichian raised beach deposits, also indicating accumulation during the Devensian. Just north of the Late Devensian ice margin in Wales, at coastal sites such as Langland Bay in the Gower Peninsula, and Druidstone and Abermawr in Pembrokeshire, Devensian till overlies locally-derived angular head. At Langland the head itself overlies an Ipswichian raised beach and hence must have accumulated during the Devensian prior to the invasion of south Wales by the Late Devensian ice sheet.

Periglacial deformation of rockhead

A gradational boundary several decimetres thick often separates underlying intact bedrock and overlying head deposits. Within this zone bedrock becomes increasingly shattered in an upward direction, and merges into the base of the overlying slope deposit. This progressive disruption of bedrock may be accompanied by downslope overturning of strata (Figure 7.15). Such bedrock overturning appears to be widespread (Mottershead,

1971; Green & Eden, 1973; FitzPatrick, 1987) and occurs on a wide range of lithologies. This phenomenon was studied in detail by Cresswell (1983), who documented overturning of weathered Devonian slate beneath head deposits at Constantine Bay in Cornwall. At this site, overturning involves slip along cleavage-parallel fractures, and Cresswell identified two distinct styles of shear strain. The first involves a progressive upwards increase in deformation and the second an upwards increase then decrease in shear strain so that at the top of the profile only slight deformation is evident. The former he ascribed to Pleistocene frost heaving and solifluction, with progressively greater rates of movement upwards through the profile. The latter he interpreted as indicating that the uppermost part of the weathered slate moved downslope as a rigid plug, with little or no internal deformation. In both cases the frictional strength of the fractured slate zone was probably reduced by high porewater pressures, and Cresswell believed that the former permafrost table lay close to the base of the weathered slate layer.

Coombe rock: head derived from periglacially weathered chalk

On the chalk of southern England, beyond the limits of maximum glaciation, periglaciation during successive glacial periods led to frost weathering of rock and the formation or modification of dry valleys (Chapter 8). Williams (1980) has demonstrated experimentally that chalk is highly susceptible to frost weathering, and his results suggest that the former active layer would have been reduced to fine calcareous mud in a few centuries. It seems likely that such material would have been rapidly eroded from slopes by solifluction and slopewash, causing significant modification of chalk landscapes during periglacial episodes. Several types of periglacial deposit occur in the English chalklands (Evans, 1968), the most important being *coombe rock*. This is genetically equivalent to head, and comprises a silty chalk mud containing chalk and flint clasts. An associated deposit, *brickearth*, commonly consists of a mixture of silt-sized quartz grains and numerous chalk pellets, and is widely interpreted as loessic material reworked by solifluction. Coombe rock is often thickest on the floors of dry valleys. The matrix consists predominantly of silt-sized chalk powder, and is highly frost-susceptible. Clasts tend to increase in size towards the base of the deposit, where they merge into the underlying weathered chalk bedrock (Shakesby, 1975), and often display a strong preferred downslope orientation.

The efficacy of Pleistocene frost weathering and gelifluction on the chalk is well illustrated by the considerable thicknesses of coombe rock in coastal exposures in Sussex (Hodgson, 1964; Williams, 1971; Jones, 1981). Between Chichester and Brighton the coastal plain is underlain by two interglacial raised beaches, the 30–35 m beach and the 10–15 m beach. The upper beach has been ascribed to the Hoxnian Interglacial, and the lower to the Ipswichian (Mottershead, 1977; Jones, 1981). Both are completely blanketed by coombe rock (Figure 7.16) and some brickearth (soliflucted aeolian silts), the coombe rock attaining a thickness of 20 m over the Ipswichian beach platform at Black Rock, Brighton (Williams, 1971). Farther to the west, the volume of coombe rock resting on raised beaches in the Portsmouth area suggests that at least 10.5 m of chalk was stripped from the southern side of Portsdown under periglacial

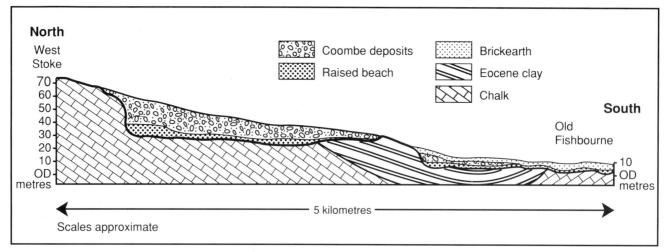

Figure 7.16 Coombe Rock mantling coastal raised beach platforms, Sussex. After Mottershead (1977).

Figure 7.17 Sarsen stones, Fyfield Down in Wiltshire. Photograph by Professor R.J. Small.

conditions (Edmunds, 1930). At Black Rock, the coombe rock contains the remains of mammoth (*Mammuthus primigenius*) and woolly rhinoceros (*Coelodonta antiquitatis*), both indicative of former cold-climate conditions (Gallois, 1965). The deposit here is well bedded, and becomes finer towards the top, where increasing quantities of loessic material are incorporated (Jones, 1981). Mottershead (1977) has pointed out that the 30–35 m Hoxnian beach platform in Sussex might be expected to be overlain by periglacial deposits of both pre-Devensian and Devensian age, but it has not yet proved possible to identify two distinct stratigraphic units and no intervening Ipswichian palaeosol has been found.

At Clatford Bottom in the Marlborough Downs, up to 3.5 m of solifluoted coombe rock mantles the lower parts of gentle south-facing slopes (Small, Clark & Lewin, 1970). The uppermost layer consists of about 0.5 m of flinty brown loam that is thought to represent decalcified coombe rock, and the presence of vertical loam-filled pipes that penetrate into the underlying unweathered coombe rock further attests to a fairly high rate of Holocene solutional weathering. The surface loam at this site is associated with concentrations of large boulders, known locally as 'sarsen stones', which form a stone stream along the valley bottom and can be traced up the valley sides to the clay-with-flint deposits of the adjacent plateau. Sarsens occur singly or in

Table 7.2: *Estimated maximum distance moved by block streams on the Wiltshire and Dorset chalk, and downslope from the granite outcrop of east Dartmoor.*

Bedrock	Locality	Distance moved by farthest-travelled blocks (m)	Average gradient (degrees)
Chalk	Clatford Bottom	4000	1.5
	Lockeridge Dene	3200	1.5
	Piggledene	3200	1.75
	Monkton Down	2300	2.75
	East Kennett	2000	2.25
	West Woods	1800	2.0
	Near Manton	1800	1.75
	Temple Bottom	1800	2.75
	Little Bredy	1800	2.0
	Portesham	1200	6.25
Granite	Leusdon Common	915	7.5
	Yarner Woods	760	11.5
	Bagtor Woods	700	6.0
	Buckland Beacon	520	11.5
	Poundsgate	520	8.5
	Ashburn Valley	400	11.0
	Lemon Valley	400	10.0
	Welstor	305	10.0
	Pinchaford	305	11.5
	Langworthy Brook	305	8.5

Source: Williams (1968).

sparse groups in the Chilterns and North and South Downs, but are more common in the chalk valleys of Wiltshire and Dorset (Figure 7.17). Frost heaving within the former active layer appears to have brought them to the surface, and on steeper valley sides they may have formed ploughing boulders that moved downslope more rapidly than the underlying soliflucting chalk regolith. The sarsens are thought to have originated as Tertiary silcretes (Summerfield & Goudie, 1980), and were interpreted by Small *et al.* (1970) as having been rafted to their present positions by solifluction. Williams (1968) used the distance moved by sarsen blocks in the Marlborough area as a guide to the amount of solifluction that has taken place. Table 7.2 summarises the distances involved in boulder movement from the interfluves, down the valley sides and along the valley floors. In general, these distances are about four times greater than those travelled by granite boulders that form similar stone streams on Dartmoor (Table 7.2). This difference led Williams to conclude that periglaciation has been responsible for much greater slope modification on the chalk of Wiltshire and Dorset than on the granite of Dartmoor.

Head deposits derived from till

To the north of the limits of maximum glaciation, drift deposits often show evidence of reworking by periglacial mass wasting, though often such reworking affects only a superficial layer no more than 1–2 m thick. Within this layer, elongate clasts show downslope orientation, and sometimes a crude stratification parallel to the ground surface is evident. Soliflucted till may form arcuate terraces with risers a metre or two high on lower

Figure 7.18 Soliflucted till forming coastal cliffs, Morfa-bychan, Dyfed, Wales.

Table 7.3: *Data from solifluction simulation experiment (Gallop, 1991).*

Bedrock	Mean grain size	Liquid limit	Plastic limit	Total volume displaced (cm³) in 13 cycles		Average volume displaced (cm³) per cycle	
	(phi units)	%	%	upper slope	lower slope	upper slope	lower slope
Granite	1.30	34		18.5	66.0	1.4	5.1
Limestone	0.62	29	20	12.0	18.5	2.1	4.5
Mudstone	1.90	28	20	106.0	135.5	8.2	10.4
Slate	3.90	34	31	93.0	175.7	7.2	13.5

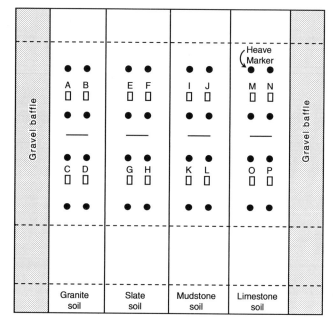

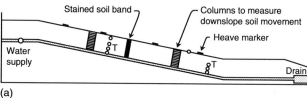

(a)

Figure 7.19 (a) Experimental design for simulation of periglacial mass movement. After Gallop (1991).

valley sides, and in valley bottoms reworked drift deposits may reach considerable thicknesses. In the valleys of the South Wales Coalfield, for example, up to 15 m of solifluctated till has been observed (Harris & Wright, 1980; Wright & Harris, 1980; Wright, 1983). Wright (1991) concluded that these sediments may be considered *paraglacial deposits* as they represent melt-out tills that were reworked by solifluction and mudflow immediately after deposition. A similar origin appears likely for some of the thick head deposits banked against the coastal bedrock hills of Cardigan Bay south of Aberystwyth (Figure 7.18). In this area, Watson & Watson (1967) observed a succession of 'Yellow Head', 'Blue Head' and 'Brown Head' units, all of which exhibit strong preferred downslope orientation of clasts. They concluded that all three units represent periglacial solifluction deposits. However the Blue Head (which constitutes the bulk of the exposed sequence) contains striated clasts and quartz grains with surface textures indicative of glacial crushing (Vincent, 1976). This unit may therefore result from slumping and flow of till down the sides of coastal hills immediately following withdrawal of the Late Devensian ice sheet. If so, it is better described as a paraglacial mass flow deposit rather than a periglacial solifluction deposit (Eyles, Eyles & McCabe, 1989). The Yellow and Brown Heads, however, are probably exclusively periglacial in origin.

The above examples suggest that paraglacial reworking of Devensian till was widespread in the latter part of the Dimlington Stadial. There is also evidence for a renewal of solifluction activity on low ground in Scotland and northern England during the Loch Lomond Stadial. This evidence takes the form of peat layers of Windermere Interstadial age overlain by solifluction deposits or solifluctated till. In the Glasgow area, excavations on the flanks of drumlins have revealed over 4 m of reworked till overlying interstadial peat deposits containing pollen indicative of a wet tundra environment (Dickson, Jardine & Price, 1976). At Springburn, interstadial peat buried beneath solifluctated till yielded a ¹⁴C age of 11 140 ± 110 yr BP, whilst a peat layer within the colluvial silt that overlies the displaced till gave an age of 5994 ± 50 yr BP, thereby demonstrating that resedimentation of the till occurred naturally, rather than as a result of anthropogenic disturbance. Similarly, peat buried by reworked till on the flanks of a drumlin at Robroyston yielded interstadial ages of between 11 650 ± 190 yr BP and 11 210 ± 150 yr BP. Dickson and his co-workers concluded that solifluction on the flanks of drumlins had resulted in the burial of wet tundra fenlands in the intervening

depressions. Solifluction during the Loch Lomond Stadial is also indicated by burial under solifluctated drift of Windermere Interstadial peats dated to 11 510 yr BP at Eston Beacon in the Tees Valley, 12 390 ± 100 yr BP at Brinziehill in Kincardineshire and 10 780 ± 50 yr BP at Woodhead in NE Scotland (Harkness & Wilson, 1979; Hall, 1984).

The relative importance of frost creep and gelifluction

It is generally assumed that granular head deposits moved downslope through a combination of frost creep and gelifluction under conditions of seasonal ground freezing and thawing (Harris, 1987a; Hutchinson, 1991). A novel simulation experiment has been carried out by Gallop (1991) to test the validity of this assumption. Regolith derived from granite, limestone, mudstone and slate was used to construct four adjacent slope units, each 0.35 m thick with a gradient of 12°, in a 5 metre square refrigerated box (Figure 7.19(a)). Water was supplied to the base of the soils via a layer of sand, and the slope was subjected to cyclic freezing and thawing from the surface downwards, each cycle lasting 20–25 days. Frost heave, thaw consolidation and resulting downslope displacement over 13 cycles were monitored for each lithology. Data relating to the experiment are summarised in Table 7.3, and the resulting vertical profiles of soil displacement (as indicated by columns of small tiles installed during slope construction) are illustrated in Figure 7.19(b). The maximum potential frost creep resulting from a single freezing cycle was calculated by assuming frost heave

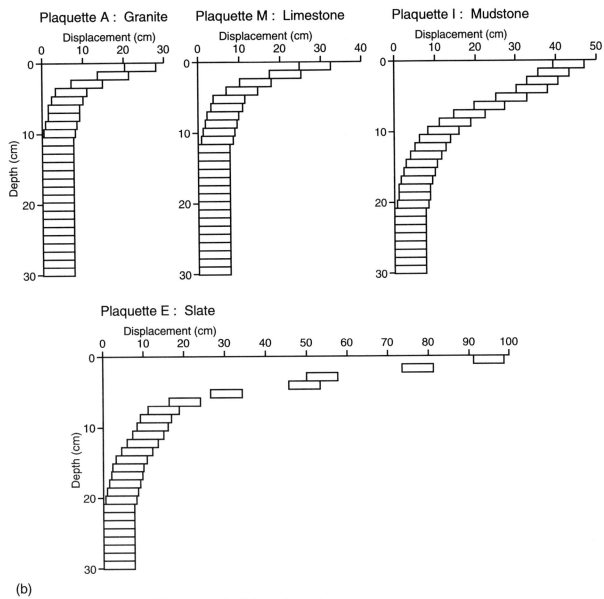

(b)

Figure 7.19 (b) Profiles of downslope soil displacement after 13 freeze–thaw cycles.

perpendicular to the slope and vertical resettlement. This calculation showed that the surface displacement of sandy granite and limestone regoliths can be largely accounted for by frost creep alone, but the greater movement rates observed in the silt-dominated mudstone and slate soils can be explained only by additional thaw-induced gelifluction.

This contrasting behaviour resulted mainly from the high permeability of the granite and limestone soils. Freezing of these soils was associated with the growth of near-surface needle ice, but little ice segregation at depth. Thawing of the needle ice led to significant downslope creep of the superficial layers, but pore-water pressures failed to develop as thawing progressed downwards through the granite and limestone soils since they contained little excess ice. In contrast, frost heaving of the mudstone and slate soils reflected extensive ice segregation at all depths. In these silt-rich soils, melting of excess ice was accompanied by the development of high porewater pressures that caused a loss

of frictional strength. Water contents immediately above the thaw plane generally exceeded the liquid limits of the soils, and in consequence viscous flow ensued. Gradual dissipation of pore pressures as soil drainage took place then led to stabilisation. Each freeze–thaw cycle was therefore accompanied by near-surface frost creep in the granite and limestone soils, and by both frost creep and gelifluction in the mudstone and slate soils. The remarkable similarity between the profile of granite soil displacement in Gallop's experiment (Figure 7.19(b)) and that of a weathered andesite vein within granitic head at Bellever, Dartmoor (Figure 7.20) suggests that the experiment provides a valid simulation of periglacial mass movement in such regolith. If so, it indicates that granular head with a dominantly silty matrix underwent flowage during seasonal ground thawing (gelifluction and, where the soil became very wet, possibly mudflow), whereas better-drained sandy head deposits moved downslope largely as a result of frost creep.

Figure 7.20 Deformation of an andesite vein by periglacial mass movement of sandy granitic head, Bellever Quarry, Dartmoor, Devon. Scale divisions 0.5 m.

this head are planar slip surfaces aligned roughly parallel to the ground surface and often extending continuously over considerable areas. Immediately beneath the head is periglacially-weathered clay, the top of which may also contain slip surfaces. This weathered clay is often severely brecciated and much softer than undisturbed clay at depth. On the surface, degraded lobate frontal banks marking the toes of landslides may sometimes be detected (Figure 7.21a), and in some cases it has proved possible to map slide tracks and headscarps (Hutchinson, 1991; Figure 7.21(a)).

The characteristics of slip surfaces within weathered clays and clay-rich head deposits have been studied by Spink (1991), who recognised six distinct types within weathered London Clay and Reading Beds clays (Figure 7.22). Three types of shear surfaces he attributed to active layer sliding: subhorizontal shears up to 2 m long at depths averaging 1.4 m below the top of the clay head (type 1); extensive shears, subparallel to the surface, that underlie the type 1 shears at the base of clay head (type 2); and small, closely-spaced, randomly-oriented shears often less than 100 mm long (type 3). Spink proposed that pri-

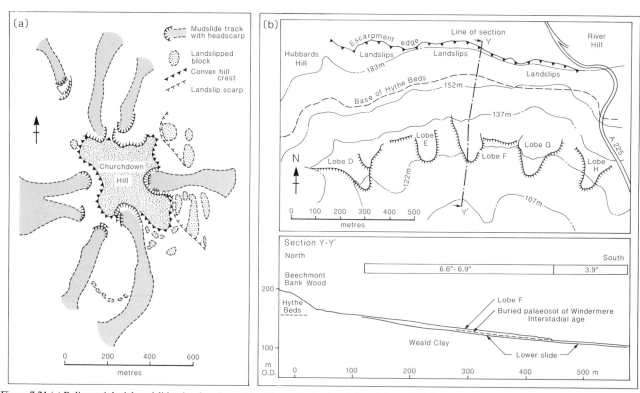

Figure 7.21 (a) Relict periglacial mudslides developed over Lias Clay on the flanks of Churchdown Hill, Gloucestershire. After Hutchinson (1991). (b) Relict periglacial mudslides developed over Lower Cretaceous Weald Clay near Sevenoaks in Kent. After Skempton & Weeks (1976).

Clay-rich head deposits

Hutchinson (1967) appears to have been the first to appreciate that many slopes developed in stiff clays owe their present form to periglacial slope processes and have undergone little modification under temperate conditions during the Holocene. Such slopes are often mantled by 2–3 m of head consisting of rock fragments in a clay or silty clay matrix. Underlying and within

mary shearing at the base of the former active layer is marked by type 2 shear surfaces while secondary shearing produced the type 1 discontinuities. The short type 3 shears he ascribed to internal deformation during movement. Deeper shear surfaces (types 4–6) were interpreted by Spink in terms of thaw subsidence during permafrost melting (Chapter 5), though one variety may reflect deep-seated landsliding during permafrost thaw. Shear structures, however, are not the only manifestation of

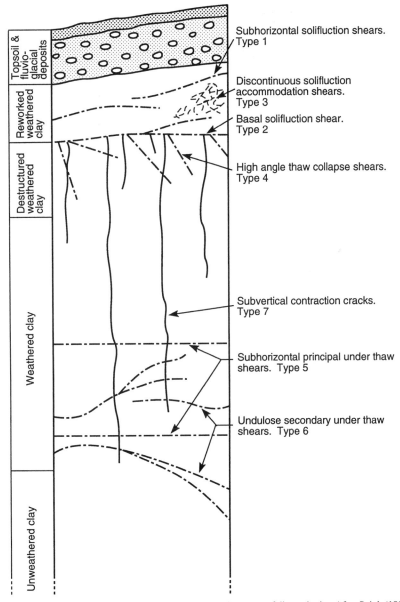

Figure 7.22 Schematic profile through periglacially-disturbed Eocene Clay, illustrating types of discontinuity. After Spink (1991).

freezing and thawing in clays. Modification of the upper 12 m of the Upper Lias Clay near Stamford in Lincolnshire has been described by Chandler (1972b), who observed small lumps of relatively unweathered clay (lithorelicts) surrounded by soft clay. In thin section the latter showed a 'turbulent' structure as if remoulded by rotation of the lithorelicts. In addition, frequent subhorizontal and occasional near-vertical fissures give the clay a blocky structure. This disturbed clay structure was interpreted by Chandler as resulting from the combined effects of ice segregation and subsequent thawing during permafrost aggradation and degradation respectively.

The practical significance of extensive shallow slip surfaces beneath clay head was highlighted in Kent when construction of the Sevenoaks by-pass road across the Hythe Beds escarpment caused unexpected landsliding (Weeks, 1969; Skempton & Weeks, 1976). Trial pits and boreholes revealed two 'solifluc-tion' sheets, the lower of which was around 3 m thick and underlain by Weald Clay. Its inclination decreased away from the escarpment from around 7° to little more than 3° (Figure 7.21(b)). The upper deposit was up to 4 m thick, consisted of similar clayey head and terminated around 400 m south of the escarpment in a lobate frontal bank. Separating the two deposits was an organic palaeosol that yielded a radiocarbon age of 12 200 ± 200 yr BP. The upper head must therefore have accumulated during the Loch Lomond Stadial, and the lower head presumably at some time earlier in the Devensian. Weeks also reported a much older head deposit of chert fragments in a sandy silty clay that mantles high ground in the area. A typical trial pit cut through the main Devensian head showed an upper 2 m of red-brown clay containing lumps of fissured grey-brown clay. Shear surfaces occur within and at the base of both units, the lowermost defining the top of the unweathered Weald Clay.

Table 7.4: *Data relating to shallow periglacial slides in Kent and Oxfordshire.*

Location	Slope (degrees)	Depth of shears (m)	Slope deposit	Parent material	c_r' (kPa)	$\varnothing_r'$ (degrees)
Sevenoaks, Kent	4	1.8–3.3	Chert and ragstone in a silty clay matrix.	Weald Clay	1.4	15.0
Tonbridge, Kent	4	0.9–3.0	Limestone clasts in a silty clay matrix.	Weald Clay	2.1	16.0
Tonbridge, Kent	7	1.05–4.5	Silty clay with siltstone fragments .	Wadhurst Clay	0 12.4	
Ditton, Kent	3	1.2–1.8	Silty clay with flints.	Gault Clay	0	12.4
Broughton, Kent	5	1.36	Clayey sand with flints. Shear in the upper few cm of London Clay	London Clay	0	14.0
Tetsworth, Oxfordshire	3.5	1.7–2.3	Silty clay with with flint.	Gault Clay	0	14.0

c_r' = effective residual cohesion; $\varnothing_r'$ = effective residual angle of friction. Source: Weeks, 1969.

It appears that the upper part of the main Devensian head moved farthest, since it contains material derived from the scarp, while the lower part consists only of displaced Weald Clay.

Similar shallow planar shear surfaces have been observed under or within clayey head at a number of other sites in Kent and Oxfordshire. Residual shear strength parameters relating to these shear surfaces are given in Table 7.4. Stability analyses for these sites predict that the critical slope for sliding when the ground is fully saturated is between 6° and 8°. As the surface slope in all instances but one is gentler than this, these analyses demonstrate that the slopes are stable under temperate conditions. As outlined earlier in this chapter, however, thaw of ice-rich soil may generate excess water contents and exceptionally high porewater pressures that effectively reduce soil strength and permit failure at much lower gradients. Weeks (1969) therefore concluded that these deposits accumulated as periglacial mudslides over a former permafrost table. Their modern equivalents might therefore be active layer detachment slides such as those illustrated in Figure 7.7. By utilising measured geotechnical parameters relating to clay head deposits in Kent, and by assuming a three-month period of summer thawing and a 2 m thick former active layer, Skempton & Weeks (1976) estimated that the thaw consolidation ratio (R) during periglacial slope failure was probably around 1.3. Allowing R to vary by ±25% to take account of the uncertainties in its estimation, they calculated the excess pore pressures generated during thaw using thaw-consolidation theory (see Harris, (1981b, pp. 96–104)), and showed that failure may have occurred on slopes of only 1.3–3.0°. They also showed that for failure on a slope of 2° the water content of the Weald Clay would have been approximately 55% and that of the matrix of the displaced material 50%. Since the liquid limit for Weald Clay ranges from 60% to 85% it is likely that failure by sliding during active layer thaw was initiated at water contents below the liquid limit, that is before the soil was prone to fluid flow.

The general pattern of shallow periglacial landsliding established in Kent probably applied, with local variations, over much of the outcrop of argillaceous rocks in southern Britain

(Hutchinson, 1991). Clayey head deposits underlain by planar shear surfaces are present, for example, on Lias Clay slopes in Northamptonshire and Lincolnshire (Chandler, 1970a; Penn, Royce & Evans, 1983; Figure 7.23). Larger-scale landsliding on Upper Lias Clay slopes in Northamptonshire has been investigated by Chandler (1976) in the Gwash Valley, and by Biczysko (1981) near Daventry. In both cases near-surface landslide deposits have been reworked by solifluction and shallow mudsliding. Organic clays buried by mudslide deposits in the Gwash Valley yielded ^{14}C ages of 10 860 ± 85 yr BP and 11 820 ± 85 yr BP, indicating that the final phase of instability at this site occurred under periglacial conditions during the Loch Lomond Stadial. Chandler concluded that the deep-seated landslides of the Gwash Valley developed when permafrost was absent, during a Middle Devensian interstadial, and that these were later modified by shallow periglacial mass wasting over permafrost during the Dimlington and Loch Lomond Stadials.

Also associated with Jurassic strata are the head-covered slopes of the Cotswold Hills in SW England. In the lower Swainswick Valley, near Bath, slopes of around 9° are mantled by 4–5 m of clay-rich head containing limestone fragments, and head also overlies the montmorillonitic Jurassic clay known as Fuller's Earth (Chandler *et al.*, 1976; Figure 7.24). Trial pits showed no basal shear surfaces in the head covering upper slopes, but revealed a continuous basal shear surface farther downslope where the head overlies a large, deep-seated landslide. In nearby Horsecombe Vale, a limestone caprock has foundered by sliding over the underlying Fuller's Earth. Farther downslope, up to 3 m of head consisting of limestone fragments in a clay matrix is overlain by clay-rich head derived from the Fuller's Earth. A palaeosol separating the two head deposits contains a Windermere Interstadial molluscan fauna, and a shear surface underlies the upper clay head. Chandler *et al.* (1976) interpreted the deep-seated landsliding at this site as a consequence of river valley incision and steepening of valley sides during the Devensian, and suggested that although some of the valley-side head may predate this landsliding phase, much is contemporaneous with it, while the Fuller's Earth head of Horsecombe Vale dates from the Loch Lomond Stadial.

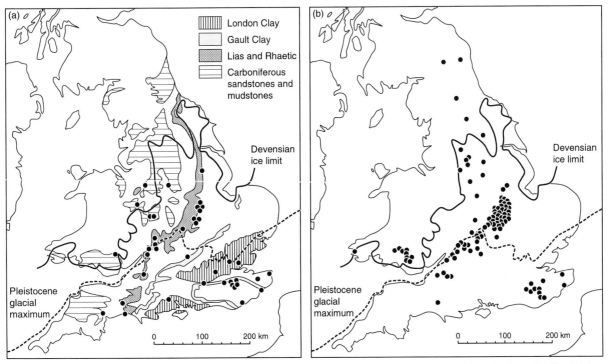

Figure 7.23 (a) Distribution of periglacial slope deposits derived from clay bedrock. (b) Location of main areas with cambering and valley bulging. After Hutchinson (1991).

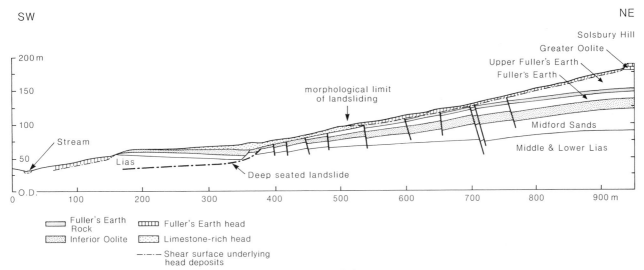

Figure 7.24 Section through Swainswick Valley near Bath. After Chandler *et al.* (1976).

Carboniferous mudstones in Staffordshire and Derbyshire are also mantled with Devensian clayey head deposits (Figure 7.23(a)). At Bury Hill in Staffordshire, planar polished slip surfaces occur at depths of up to 3.2 m within this head (Hutchinson, Somerville & Petley, 1973). In the broad valley of the Scow Brook, near Carsington in Derbyshire, head containing 62% clay is underlain by smooth, gently undulating shears at a depth of around 1.2 m (Skempton, Norbury & Petley, 1991). These shear surfaces are not continuous, but underlie approximately 40% of the slope. The presence of shallow shears beneath the lower slopes of the Scow Brook valley was implicated in the failure of

an earth dam during its construction in 1983 (Skempton, 1988). Gradients range from 4° to 6° and slopes are stable under present climatic conditions. Shear strength determinations on intact clay underlying the head indicated a peak effective cohesion value (c') of 10 kPa and a peak effective internal angle of friction (ϕ') of 20°. However, measurements across a shear surface indicated residual cohesion (c_r') of zero, and residual angle of friction (ϕ_r') of only 12°. This reduction of effective soil shear strength parameters to residual values by periglacial mudsliding reduced the factor of safety of the Carsington dam by about 22%, a fact that was to prove crucial to its subsequent failure.

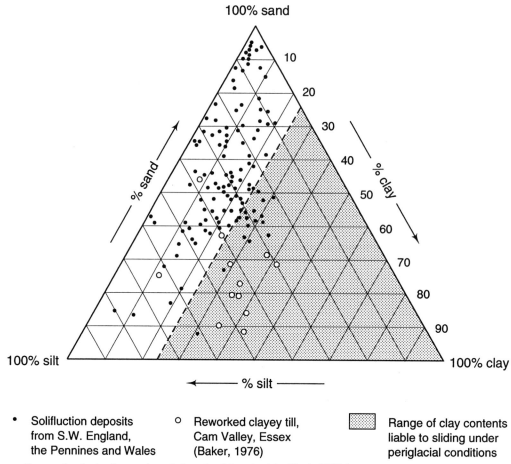

Figure 7.25 Ternary diagram of grain size for granular and clayey head deposits. After Harris (1987a).

Shallow landsliding under periglacial conditions is not confined to outcrops of argillaceous rocks, however, but also affects clay-rich tills. Those of East Anglia, for example, have experienced recurrent exposure to periglacial conditions since their deposition by the Anglian ice sheet. Baker (1976) has demonstrated that Devensian periglacial conditions promoted shallow landsliding in such tills, with little internal reorganisation of the displaced mass, though mudflow structures occur at sites where clay contents are below about 30% by weight. This example and those cited earlier therefore form part of a consistent pattern of shallow translational landslides in areas of the English lowlands where slopes of modest gradient are mantled by clay-rich soils. Stability analyses have shown that slope failure in these areas required the generation of porewater pressures higher than those likely to occur under a temperate climatic regime, and it is widely believed that such enhanced pore pressures were produced by thaw consolidation during episodes of Pleistocene periglaciation. In many respects the relict shallow clay slides of England resemble the active layer detachment slides of arctic environments (Figures 7.7 and 7.8). If this analogy is valid, it is likely that these landslides imply the former development of permafrost, as this would have impeded drainage and thus permitted high porewater pressures to be sustained in late summer when the former active layer reached its maximum depth.

Granulometry and geotechnical properties of head deposits

In general, the head deposits derived from periglacial weathering of non-argillaceous rocks in lowland Britain are matrix-supported sandy or silty diamictons containing only limited amounts of clay. Similarly, most soliflucted tills are silt-rich rather than clay-rich. Conversely, the argillaceous rocks of SE England and the Midlands have yielded clay soils, and, in the latter area, clay-rich tills (Figure 7.25). Silty and sandy soils derive much of their strength from intergranular friction and display little cohesion, while unremoulded clays have lower frictional strength, but are cohesive. Clay content is also the major factor influencing the Atterberg limits of a soil, such that increasing clay content is generally associated with higher liquid limits and plasticity indices (Figure 7.26). Geotechnical data from slopes subject to active gelifluction in present-day periglacial environments correspond closely with those obtained for poorly-sorted silty or sandy Pleistocene head deposits in lowland Britain, and simulation studies have confirmed that the latter are prone to thaw-induced downslope flow.

The low liquid limits and plasticity indices of silty and sandy head deposits make them susceptible to loss of strength and flow when water content is high. The release of meltwater from segregation ice, together with snowmelt inputs, must have generated high water contents in seasonally-thawing ground during

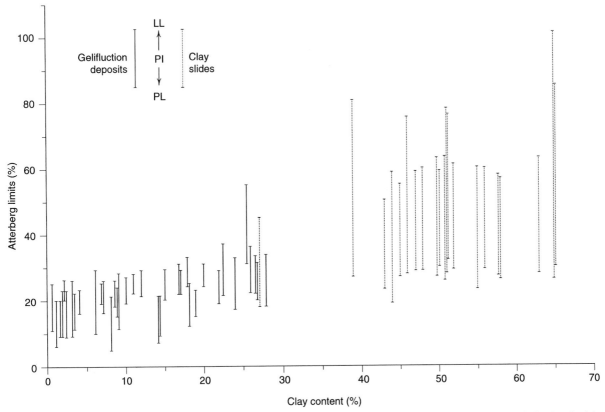

Figure 7.26 Relationship between Atterberg Limits and clay content in periglacial slope deposits. LL = liquid limit; PL = plastic limit; PI = plasticity index. After Harris (1987a).

periods of Pleistocene periglaciation. Where the soil matrix consisted of silt or very fine sand, permeability is unlikely to have been sufficient to allow immediate drainage of excess water, thereby favouring a build up of porewater pressures associated with water contents in excess of the liquid limit, and consequent downslope soil flow (gelifluction). Conversely, coarse sandy regoliths are not only less frost-susceptible, but are often also sufficiently free-draining to limit the build up of porewater pressures. In such soils, therefore, frost creep rather than gelifluction may have constituted the dominant process of periglacial mass wasting. In reality, however, some combination of both frost creep and gelifluction seems likely to have affected many of the poorly-sorted head deposits described above. In contrast, clay-rich soils with high liquid limits appear to have retained cohesion even when porewater pressures were sufficiently high to initiate instability. During the resulting slope failures, water contents appear generally to have remained below the liquid limit, so that the predominant mode of failure involved translational sliding over basal slip surfaces rather than gelifluction. In some present-day permafrost environments, high ice contents are common at the base of the active layer. During years of excessive thaw penetration, especially when rapid thawing occurs in late summer, high porewater pressures resulting from the melting of ice lenses at the base of the active layer are likely to trigger active layer detachment slides (Figure 7.7; Lewkowicz, 1990). The shallow planar slip surfaces underlying many clay-rich head deposits in lowland England almost certainly formed in this way.

Thaw-induced periglacial landslides in bedrock

Although shallow translational landsliding attributable to seasonal thaw of clay-rich soil is known to be widespread on the argillaceous rocks of lowland England, evidence for deep-seated landsliding associated with permafrost thawing is much less common. One possible example of slope failure below the depth of the former active layer occurs at Lyme Regis in Dorset, where Lias shales exposed in the side scarps of recent coastal landslides displays intense chevron folding, apparently associated with compression in the toe of a relict landslide (Hutchinson & Hight, 1987). The shales contain relatively brittle layers of fibrous calcite that display boudinage within the fold structures, and shale has entered the tension gaps between separate boudins, apparently by ductile flow. Hutchinson & Hight showed that a water content of 37–43% would have been necessary for ductile flow, which compares with an average present-day water content of 22.5% in disturbed Lias shales. They concluded that at the time of failure the undrained shear strength of the disturbed Lias was much lower than at present as a result of the thawing of ice-rich permafrost, and that this reduction in strength was responsible both for the landslide and for the ductile behaviour of the shale. Interestingly, the disturbed shales at Lyme Regis are overlain by clay-rich head around 2.3 m thick, derived from the shale and from nearby Cretaceous rocks. The head contains basal planar shears, and has moved over slopes as low as 4°. Hutchinson & Hight calculated that water contents of around 51% would have been necessary to mobilise the head on such gentle slopes, which implies that the base of the active layer probably contained more excess ice in its frozen state than the underlying permafrost. The average liquid limit for the head is 66%, which suggests that the active layer would have been prone to shallow planar sliding rather than gelifluction. Since the disturbed Lias shales are mantled with head, the phase of permafrost degradation and associated deeper landsliding must have been followed by renewed active layer mass movement, suggesting a return to periglacial conditions following the phase of permafrost degradation represented by the deep-seated landslide.

Relict ground-ice slumps

Since ground-ice slumps develop as a result of the melting of ice-rich permafrost, they generally form in areas prone to considerable thaw-subsidence and therefore have low preservation potential. However, in the Fenlands of eastern England a number of features interpreted as thermokarst depressions are bounded by arcuate microscarps, and both Burton (1987) and West (1991) have suggested that these miniature scarps may represent the remains of ground-ice slumps. Some of the best examples occur at Mepal Fen, 25 km north of Cambridge (Figure 5.23(c)). At this site a series of arcuate embayments surround a shallow thermokarst depression 2.2 km in diameter that is sunk below the level of the surrounding river terrace. The depression is floored by clay, thought by Burton (1987) to represent deposition of reworked Jurassic clay by 'thaw flowslides' or mudflows that emanated from ground-ice slumps around the periphery of the depression. Burton interpreted the arcuate microscarps on the margins of Mepal Fen as the relict headscarps of these ground-ice slumps, drawing an analogy with the 'thermocirques' described by Czudek & Demek (1970a). He envisaged that the initial thermokarst subsidence at this site was triggered by degradation of ice-rich permafrost underlying terrace gravels, and that the depression subsequently expanded to its present dimensions by headward retreat of the margins through retrogressive slumping. Stratigraphic evidence suggests that the thermokarst depression (and hence the ground-ice scarps at its margin) probably formed during the Dimlington Stadial (Chapter 5).

Slope evolution

As noted earlier in this chapter, the characteristic form of lowland slopes subject to repeated episodes of periglaciation is a smooth convexo-concave profile mantled by gelifluctate. Some slopes, however, also support erosional landforms attributable to differential weathering and mass wasting under periglacial conditions. These include upstanding rock residuals known as *tors*, and erosional benches termed *cryoplanation terraces*. As such forms are more frequently encountered in upland areas, however, consideration of their genesis and significance is postponed to later in this volume (Chapters 9 and 13 respectively). In the lowland zone, the considerable thickness of some valley-floor and coastal head deposits in areas that lie beyond the limits of Devensian glaciation contrasts markedly with an almost total absence of interglacial slope deposits, apart from localised superficial accumulations of Holocene colluvium. This contrast highlights the relict nature of many of the hillslopes in lowland

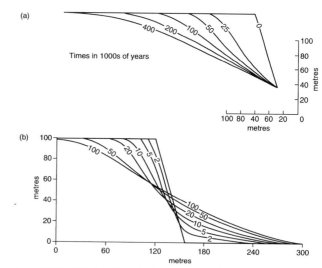

Figure 7.27 (a) Model of periglacial slope evolution with basal removal of debris. (b) Model of periglacial slope evolution with basal sediment accumulation. After Kirkby (1987).

Britain. Such considerations have led Scourse (1987) to support Te Punga's (1957) view that geomorphological activity in southern Britain has been negligible during interglacials except in the immediate vicinity of stream channels, and that the landscape owes its present form almost exclusively to cold climate processes.

The importance of periglacial slope processes in the development of the landscape in areas outside the limits of maximum glaciation was also emphasised by Kirkby (1984, 1987), who has attempted to construct a mathematical model of hillslope evolution for such areas. Kirkby estimated that over the past 500 000 years cold climates have prevailed for at least 80% of the time. It follows, he argued, that slope evolution in southern Britain should be simulated on the assumption of periglacial processes and transport rates, as slope processes during intervening interglacials have apparently had little effect on hillslope form. Kirkby also assumed that regolith undergoes landsliding at gradients above 26.5° and solifluction at lower angles, and he marshalled data from previous field measurements to justify an assumption that sediment transport by solifluction is at least an order of magnitude faster than transport by soil creep under humid temperate conditions. Figure 7.27(a) shows Kirkby's model simulation for a 100 m high slope with an initial 70° gradient and basal removal of debris. Under these assumptions the model predicts a progressive decline in gradient and the development of a marked convexity at the slope crest. Under conditions of basal sediment accumulation on a horizontal footslope (roughly equivalent to many coastal situations during glacial periods of low sea level), the model predicts the development of an extensive basal concavity as sediment (head) accumulates (Figure 7.27(b)).

The similarity in slope form between this second simulation and coastal areas where an apron of head has accumulated downslope from an abandoned Ipswichian cliffline (Figure 7.10) is striking, and provides strong support for Kirby's assumption that the present landscape of southern Britain was largely fashioned under periglacial conditions during successive Pleistocene glacial stages. In some coastal locations, how-

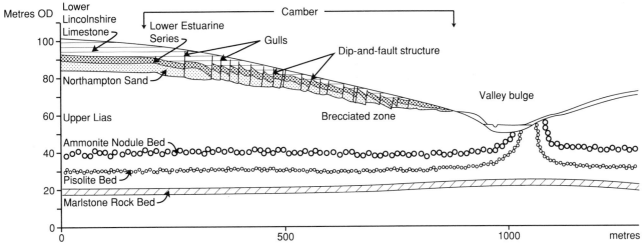

Figure 7.28 Cambered strata and valley bulging, Gwash Valley. After Horswill & Horton, 1976.

ever, Flandrian sea-level rise has resulted in removal of coastal head aprons and renewed erosion of the relict cliffline that formerly lay landward of such head deposits. The result of such renewed erosion is the formation of a *bevelled cliff*, consisting of a steep (30–35°) upper slope unit that is truncated at its downslope end by a near-vertical sea cliff. Such bevelled cliffs are common in both SW England and SW Wales (Steers, 1966). The upper 'bevelled' slope unit commonly displays downslope overturning of shattered strata, with bedrock merging upwards into a thin mantle of head.

In valley situations where erosion by rivers under periglacial conditions was responsible for removal of head deposits at the foot of slopes, Kirkby's first simulation (Figure 7.27(a)) provides a reasonable approximation of slope form. In many locations, however, head deposits accumulating on valley floors in southern Britain were not removed by fluvial erosion. Under such circumstances, Kirkby's model indicates the development of a pronounced footslope concavity associated with an aggrading valley floor, which again agrees well with observed slope forms in many parts of both lowland and upland Britain (Chapter 11). In locations where fluvial erosion under periglacial conditions has modified valley-side gradients and caused valley-floor widening, as is the case in many asymmetrical dry chalkland valleys (French, 1972, 1973) and in many of the larger river valleys of southern and central England (Castleden, 1977, 1980), the resulting gently inclined footslopes have been termed *cryopediments* or *periglacial pediments*. Although slope processes, particularly solifluction, are thought to have been instrumental in the development of these features, they are genetically dependent on fluvial processes, and are therefore considered in detail in the next chapter.

Cambering and valley bulging

The term *cambering* refers to the large-scale flexing and stretching of competent caprocks over the upper parts of valley-side slopes. Cambering normally occurs where valleys have been cut through a caprock into underlying softer strata, usually clays, and results from the deformation of the softer rocks under the weight of the overlying competent strata, causing the latter to extend down the valley sides as an attenuating 'drape'

(Figure 7.28). Bending and extension of the caprock is accompanied by the development of deep fractures termed *gulls* that run parallel to the valleyside contours and separate coherent blocks of rock (Figure 7.29). Such caprock blocks are often tilted towards the valley axis, so that they dip at an angle steeper than the general slope of the valley-side camber. Cambering is also often associated with *valley bulging*, which takes the form of broad anticlinal deformation of the argillaceous strata underlying the valley floor, with the fold axis running approximately parallel to the valley axis (Figure 7.28). It is generally considered that these large-scale structures developed under periglacial conditions, though at certain locations glacial erosion may also have been involved in the initial valley deepening. Cambering and valley bulging were first recognised in the Pennines (Miller, 1887; Watts, 1905; Lapworth, 1911), and similar features in Northamptonshire later formed the topic of a classic paper by Hollingworth, Taylor & Kellaway (1944). Both types of structure have subsequently been observed in association with almost every argillaceous formation of Carboniferous or later age (Arkell, 1947; Shotton & Wilcockson, 1950; Higginbottom & Fookes, 1971; Bristow & Bazley, 1972; Cooper, 1980). The Carboniferous rocks of the Pennines and south Wales, the Jurassic rocks of the Cotswolds, the east Midlands and Yorkshire and the Lower Cretaceous rocks of the Weald have all proved prone to the development of these structures (Figure 7.23(b)).

The characteristics of cambering and valley bulging have now been investigated at various sites in England. Much of this research has been carried out by engineering geologists responsible for assessing the suitability of such terrain for construction purposes. An excellent example is provided by site investigations for the Empingham Reservoir, in the Gwash Valley, Rutland (Horswill & Horton 1976; Figure 7.28). Here the geological succession consists of argillaceous Upper Lias mudstones overlain by competent Middle Jurassic ironstones and limestone. Beneath the cambered Middle Jurassic rocks, and within the valley bulge structure, the mudstone displays severe brecciation in the form of intact blocks embedded within a clay matrix. The depth of brecciation exceeds 25 m beneath the lower valley sides, but decreases upslope as the thickness of the overlying cambered ironstone and limestone increases. Also

Figure 7.29 Gull due to cambering of chalk, near Maidenhead, Surrey. Photograph by Dr D.M. Holyoak.

observed were extensive planar shear surfaces up to 50 m long within the valley bulge. The latter has been partly eroded by the river, but a tentative reconstruction suggests upward flexing of at least 25 m, and horizontal shortening of the mudstone strata of at least 60 m.

The Jurassic strata of the Cotswolds are also extensively cambered (Ackerman & Cave, 1967). In the Swainswick Valley near Bath, for example, Jurassic limestones are cambered over Lias clays (Chandler *et al.*, 1976; Figure 7.24). At this site the clays are brecciated to a total depth of 40 m, and gullies running along the valley sides mark the location of gulls separating blocks of limestone. At Radstock, also near Bath, excavation of foundation trenches on the upper slopes of the Avon Valley exposed a series of deep gulls in cambered Jurassic limestone (Hawkins & Privett, 1981; Figure 7.30). Most gulls at this site lay concealed under near-surface limestone strata in which extension during cambering had been accommodated through opening of closely-spaced joints, thus allowing the uppermost strata to bridge the open gulls below. Infilled gulls associated with cambering of Jurassic limestone above Lias clays and mudstones also occur to the west of Stow-on-the-Wold in Gloucestershire (Briggs & Courtney, 1972), and at Bredon Hill in Worcestershire (Whittaker, 1972). At these sites, troughs marking the locations of gulls run approximately parallel to the contours, and quarry sections reveal that the mouths of gulls are blocked with limestone rubble. Further examples of infilled

gulls occur on the sides of the Medway Valley in Kent, where the Lower Cretaceous Hythe Beds are cambered over the underlying Atherfield and Weald clays (Gallois, 1965). Here individual gulls extend laterally across the slope for as much as 150 m and are infilled with brickearth; the largest observed in section are 25 m deep and 16 m wide. Cambering and valley bulging have also been reported to occur in the Wealden and Purbeck rocks around Hastings and Lewes in Sussex (Lake *et al.*, 1987; Shephard-Thorn, 1987).

The origin of cambering and valley bulging has been discussed by several authors (e.g. Hollingworth *et al.*, 1944; Kellaway & Taylor, 1953; Hutchinson, 1991; Parks, 1991), and a variety of possible formative mechanisms have been proposed. It is generally agreed, however, that the main prerequisite is rapid incision of valleys through competent strata into underlying incompetent argillaceous beds (here for convenience referred to as clays). This releases large horizontal stresses in the latter, and under favourable conditions these may be sufficient to initiate lateral deformation of clays towards the valley axis (Vaughan, 1976). Such deformation may in turn produce some lowering of the valley sides and formation of a low-amplitude *proto valley bulge* on the valley floor (Parks, 1991). Deformation of clays under temperate conditions, however, appears unlikely to account for the dimensions of most valley bulges and associated valley-side cambering. The apparent inadequacy of the simple explanation outlined above

Figure 7.30 Gull in Jurassic Limestone, Radstock, near Bath. Photograph by Dr K.D. Privett.

Perhaps the most important mechanism of cambering and valley-bulging, however, becomes effective during climatic amelioration, when the ice-rich frozen clay undergoes thawing. Thaw consolidation of low permeability clay is liable to be accompanied by a marked reduction in shear strength, causing rapid deformation under the shear stresses set up by the weight of the overlying caprocks. Parks suggested that near the surface, the clay would have behaved like a viscous fluid, allowing cambered caprock blocks to settle and rotate downslope, thereby producing the characteristic steepening of dip commonly observed on individual blocks. Hutchinson (1991) has amplified Park's explanation, noting that permafrost thawing may have taken place from the bottom upwards as well as from the surface downwards. He suggested that basal thawing of ice-rich permafrost may have caused displacement of thaw-softened clay towards the valley floor, beneath the still-frozen ground above. Such lateral displacement may have been responsible for increasing the amplitude of valley bulging.

Most cambering and valley bulging in Britain apparently pre-

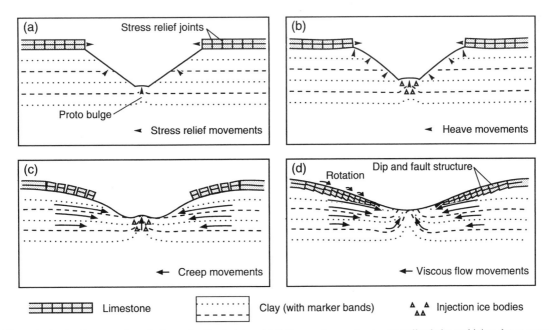

Figure 7.31 Stages in the development of cambering and valley bulging. (a) Valley erosion creates a proto valley bulge and joints due to stress release. (b) Frost heave increases the size of the bulge and the effects of stress relief in the caprocks. (c) Erosion of the valley and creep of ice-rich strata cause cambering and increase the size of the bulge. (d) Thaw of the strata creates high porewater pressure and consequent viscous flow, enhancing the effects of cambering and valley bulging. Rotation of caprock blocks gives rise to dip-and-fault structures. After Parks (1991).

prompted Parks (1991) to propose a general model of cambering and valley bulging under periglacial conditions (Figure 7.31). Parks argued that the downward penetration of permafrost into the clays was probably accompanied by ice segregation and consequent frost heave, which would have further stressed the overlying competent strata on valley sides. He suggested that additional valley incision and valley-side steepening by periglacial rivers may have initiated permafrost creep, causing additional lateral displacement of the clay and concomitant downslope extension of the overlying caprock (Figure 7.31).

dates the Devensian glacial stage. Chandler et al. (1976), for instance, showed that Devensian erosion and gravel terrace formation in the Cotswolds postdate cambering and valley bulging in this area. Similarly, Horswill & Horton (1976) concluded that the development of cambering and valley bulging in the east Midlands may even predate Anglian glaciation at some sites, while at others it postdates withdrawal of the Anglian ice sheet. It seems most likely that these large-scale phenomena actually developed over successive episodes of Pleistocene periglaciation.

Periglacial slope processes and landscape evolution: concluding comments

The last two decades have witnessed growing awareness that the present form of hillslopes in southern Britain predominantly reflects the operation of periglacial slope processes during successive glacial stages. This heightened appreciation of the importance of periglaciation in fashioning the landscape of this area reflects advances in several distinct areas of research. Three are particularly worthy of mention. First, field measurements in present-day periglacial environments have highlighted the relative rapidity of cold-climate mass wasting processes (Table 7.1), and indicate that slope evolution may be at least an order of magnitude faster under periglacial conditions than under temperate conditions. In Britain, this proposition is supported by the abundance and considerable thickness of periglacial slope deposits in areas not affected by Devensian glaciation, compared with the limited extent and thickness of Holocene sediments in such areas. A second major contribution to our appreciation of the importance of periglacial processes in dictating slope form in lowland Britain stems from the work of engineering geologists investigating the characteristics and evolution of low-gradient slopes on argillaceous rocks. It has now been demonstrated that the form of such slopes owes little to present-day processes, and essentially reflects failure of clay-rich sediments during thaw consolidation of a former active layer. Investigations by engineering geologists have also contributed considerably to our understanding of the influence of periglaciation in the development of large-scale deformation of bedrock in the form of cambering, valley bulging and deep-seated landsliding. Finally, our understanding of the importance of periglacial processes in the evolution of hillslopes in unglaciated parts of Britain has been furthered by research on Quaternary climatic fluctuations. The ocean core record in particular demonstrates that throughout much of the Quaternary the cumulative duration of cold stages greatly exceeded that of temperate interglacials and interstadials. Given the relative rapidity of periglacial mass wasting during cold stages, it is not surprising that the imprint of periglacial slope processes is so dominant in many parts of lowland Britain.

8

Lowland landscape modification by fluvial and aeolian processes

A consideration of the factors affecting sediment entrainment suggests that both running water and wind action are likely to be important in periglacial environments, particularly where incomplete vegetation cover exposes soil to slopewash and deflation. Appreciation of the geomorphic role of periglacial rivers, however, is a fairly recent development. Pioneering studies of this topic were carried out over two decades ago (Rudberg, 1963; St Onge, 1965; Cook, 1967; McCann, Howarth & Cogley, 1972), but even today fluvial processes tend to be under-represented in the literature on periglacial geomorphology (Clark, 1988b). In contrast, appreciation of the importance of wind action as a geomorphological agent under periglacial conditions is long established, originating with the recognition of extensive Quaternary *loess* (windblown silt) deposits in Europe, China and North America. It appears, however, that the cold dry zones outside the margins of the major Pleistocene ice sheets experienced much greater aeolian activity than the great majority of present-day periglacial environments, and that there are few (if any) appropriate modern analogues for Quaternary loess deposition (French, 1976). This chapter outlines and evaluates the consequences of Pleistocene periglacial fluvial and aeolian processes in lowland Britain; the effects of such processes in the upland zone are considered in Chapter 13.

Periglacial rivers and river valleys

Most of the large river valleys in southern England that have not been affected by glacial meltwater contain spreads of fluvioperiglacial sediments. In many cases Holocene fluvial activity has resulted merely in the trimming of these sediments and deposition of a superficial veneer of alluvium across them. Where glacial meltwaters entered the catchment, as was the case for instance in the Severn basin, fluvioglacial outwash gravels underlie Holocene alluvium at shallow depths (e.g. Shotton, 1977; Dawson & Bryant, 1987). Proglacial outwash systems will not be considered in detail here, since they represent part of the glacial rather than periglacial geomorphic system. However, there were times during Quaternary cold stages when the nival runoff regime of nonglacial tributaries in large basins such as that of the Thames was supplemented by glacial meltwater discharge. In many arctic environments there is a continuum between catchments that are dominated by glacial meltwater discharge, catchments fed both by glacial melt and snowmelt and catchments that contain no glacier ice and are therefore dominated by snowmelt (Marsh & Woo, 1981).

Rivers in periglacial environments differ significantly from those of the temperate zone in terms of their hydrological regimes. This difference largely reflects the importance of snow in their annual precipitation budgets. In North America, for instance, snow accounts for 30–35% of total annual precipitation near the southern limit of discontinuous permafrost, and over 50% in the high Arctic. Meltwater from the winter snowpack is mainly released during spring and early summer to produce a pronounced *nival flood* (McCann et al., 1972; Woo, 1983, 1986, 1990). The shallowness of the active layer in the early part of the melt season promotes surface runoff over slopes and this, combined with an incomplete vegetation cover, guarantees a flashy response to rainstorms (Cogley & McCann, 1976), and high runoff ratios. Church (1974) has identified four categories of arctic and subarctic river regime, as follows.

(1) *Subarctic nival.* This relates to rivers in the discontinuous permafrost zone in which the spring snowmelt flood dominates, with low summer discharges punctuated by occasional rainstorm-induced floods.

(2) *Arctic nival.* This regime is characteristic of the continuous permafrost zone, and is also dominated by the spring snowmelt flood, with low summer discharges interrupted by occasional rainstorm peaks. In smaller catchments flow may cease between summer rainstorm floods, but more gradual snowmelt and persistent snowbanks often sustain continuous summer discharge in larger systems (Marsh & Woo, 1981). Diurnal fluctuations in discharge are also common, reflecting diurnal cycles in the rate of snowmelt.

(3) *Proglacial.* Where such regimes pertain, rivers are fed by glacier melt, which maintains continuous summer flow, produces a longer snowmelt flood and generates short-term discharge fluctuations in response to variations in the rate of melting of snow and ice.

(4) *Muskeg.* The muskeg regime is characteristic of catchments in which extensive wetlands increase water storage, reduce flow rates and attenuate flood peaks.

Under Quaternary periglacial conditions in Britain, unglacierised catchments probably experienced runoff regimes broadly similar in character to the subarctic and arctic nival regimes described above. Examples of representative hydrographs for subarctic and arctic rivers are illustrated in Figure 8.1. The importance of the spring snowmelt flood tends to be greater in small systems where it may persist for only a few days. For instance, Umingmak Creek in Ellesmere Island, which has a catchment area of only 2.3 km², discharges around 95% of its water yield in the first two weeks of its annual summer flow period (Figure 8.1(c)). For the much larger Colville River in Alaska, which has a catchment area of c. 50 000 km², around half the annual discharge occurs in a one month period from late May to late June (Arnborg, Walker & Peippo, 1966). The geomorphological significance of such distinctive runoff regimes lies largely in the high stream competence and capacity that

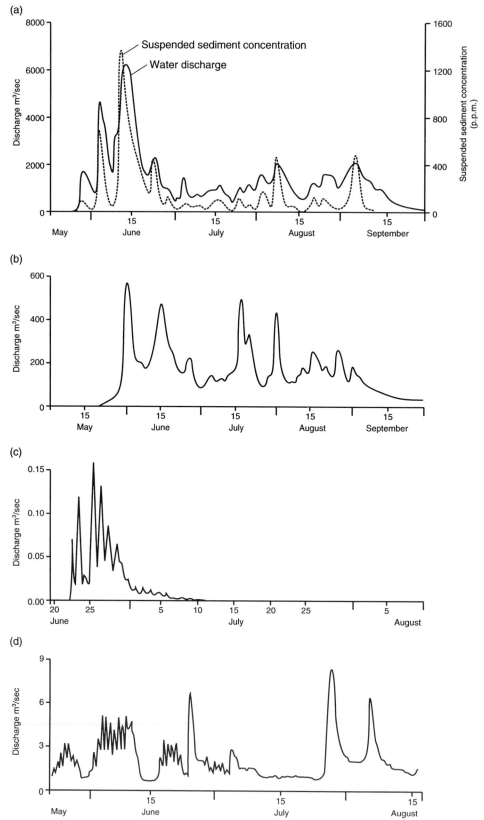

Figure 8.1 Examples of runoff hydrographs for arctic and subarctic rivers. (a) Colville River, Alaska, 1962. Arctic nival regime. Drainage area 50 000 km[2]. Note dominance of nival flood period in the plot of suspended sediment yield. After Arnborg *et al.* (1966). (b) Sagavanirktok River, Alaska, 1972. Arctic nival regime. Drainage area 5719 km[2]. After Scott (1978). (c) Umingmak Creek, Ellesmere Island, NWT, Canada, 1974. Arctic nival regime. Drainage area 2.3 km[2]. After Marsh & Woo (1981). (d) Dietrich River, Alaska, 1979. Subarctic nival regime. Drainage area 45 km[2]. After Onesti & Walti (1983).

Figure 8.2 Small river valley near Resolute, Cornwallis Island, NWT, Canada, showing frost-weathered bedrock surfaces and late-lying snow.

Figure 8.3 First order river, Ellesmere Island, NWT, Canada. The permafrost table lies at a depth of 0.4–0.6 m. The north-west-facing valley sides (in shadow) have developed steeper gradients than the south-east-facing valley sides.

prevail during the nival flood period. However, rivers in which autumn flow is shallow become frozen solid in winter, the ice adhering firmly to their beds. This *fast ice* provides protection from erosion during the early part of the succeeding spring flood (Scott, 1978; Woo, Marsh & Steer, 1983; Clark, Gurnell & Threlfall, 1988). Such protection is likely to be most significant in smaller river basins where the spring flood is of short duration, although in many small snow-fed catchments discharge ceases following early summer snowmelt, so that no water is available to form a protective ice-cover in the autumn. In larger rivers, fast ice is likely to clear before the period of peak discharge. Lateral erosion and channel instability are also more likely in larger rivers, since thermoerosion of frozen channel-side sediments can take place over a longer time period, becoming increasingly effective later in the nival flood when water temperatures begin to rise (Scott, 1978).

Data on sediment and solute transport by periglacial streams are limited, but those that are available suggest that suspended sediment concentrations rarely exceed 500 mg l^{-1}, while solute concentrations are generally between 25 and 100 mg l^{-1} for carbonate terrain and much lower on crystalline rocks. The importance of high discharge associated with the nival flood in transporting suspended sediment is clearly illustrated by the Colville River (Figure 8.1(a)), where daily suspended sediment yield exceeds 500 000 tonnes at the flood peak. In many periglacial river valleys the combined effect of frost weathering and rapid mass movement on valley sides leads to an abundant supply of coarse-grained sediment to river channels. The high competence and capacity of nival flood runoff causes rapid evacuation of this sediment from headstreams, so that periglacial rivers are often bedload-dominated. For rivers in Baffin Island, for example, Church (1972) estimated that bedload comprises more than 75% of sediment yield. Bedload transport is notoriously difficult to quantify, however, and data are therefore not widely available. McCann *et al.* (1972) noted that in the streams they studied in arctic Canada, the entire boulder-filled bed became mobile during the spring snowmelt flood, but it proved impossible to measure bedload transport for these sites. Insufficient data are currently available for generalisations to be made concerning total sediment yield from periglacial catchments, though Clark (1988b) has suggested that such yields are not markedly different from those of rivers of comparable size in humid temperate environments.

McCann *et al.* (1972) provided excellent illustrations of fluvial dissection on Devon and Cornwallis Islands, which has produced steep-sided V-shaped valleys up to 150 m deep. Here permafrost is continuous, the active layer rarely exceeds 0.5 m in depth, and valley sides are mantled with angular frost-weathered debris (Figure 8.2). Since snow tends to drift into the valleys, nivation processes probably enhance weathering and sediment delivery (Chapter 13). A much-discussed characteristic of periglacial river valleys is their tendency to develop *asymmetrical cross profiles* unrelated to underlying geological control. In his review of the topic, however, French (1976) noted wryly that *symmetrical* valleys are also common in present-day periglacial environments, but fail to attract the attention of geomorphologists. French identifies 'normal' asymmetry as occurring where north-facing slopes are steeper than south-facing (Figure 8.3). Such asymmetrical valley profiles have been described in Alaska, arctic Canada, Greenland and Siberia, and appear to result from unequal radiation intensities associated with slopes of opposing aspect. The greater intensity of insolation on south-facing slopes is thought to cause deeper active layer development and more effective gelifluction, which in turn is believed to cause more rapid slope decline than is the case for north-facing slopes. However, of possibly greater importance is the tendency of rivers to migrate away from south-facing slopes as a result of the greater sediment supply from that side. This leads to lateral erosion and steepening of north-facing valley-side slopes. However, other microclimatic factors may also lead to differences in thermal regime on slopes of different aspect. In north-west Banks Island, for instance, the prevailing winds cause preferential snow accumulation on north-east-facing slopes (French, 1971). This insulating snow cover raises winter ground temperatures and promotes prolonged active layer saturation in summer, and hence results in more rapid slope decline and greater sediment supply to the river. Lateral migration of the river towards the north-east steepens south-west-facing slopes, producing valley asymmetry opposite to the 'normal' pattern for the northern hemisphere. Clearly, therefore, a variety of local conditions may combine to promote contrasting types of periglacial valley development.

It was noted above that periglacial rivers often have gravel

Figure 8.4 Braided periglacial river channels: (a) Boulder Creek, Yukon Territory, Canada: a subarctic nival stream. (b) Black Top Creek, Ellesmere Island, NWT, Canada: an arctic nival stream.

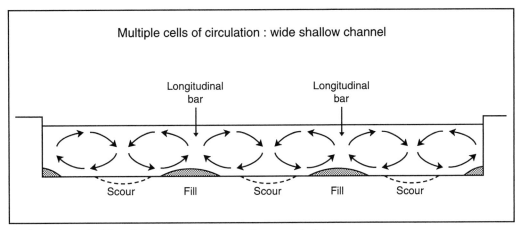

Figure 8.5 The role of secondary helical flows in forming braid bars in a shallow gravel-bed stream.

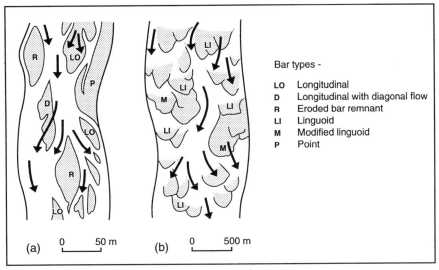

Figure 8.6 Classification of braid bars. (a) Gravel-bed stream. (b) Sand-bed stream. After Miall (1977).

beds, and as is common with bedload-dominated streams on wide valley floors, their channels are usually broad and shallow (Schumm, 1963, 1968). During the nival flood much of the river course is inundated, but as flood levels fall the channel becomes divided into a complex braided pattern with longitudinal bars separating individual channel segments (Miall, 1977; Bryant, 1982, 1983a). Figure 8.4 illustrates examples of braided channels in arctic and subarctic nival river systems. However, single-channel meandering streams also occur in arctic areas and are mainly associated with rivers that carry a relatively large component of their load in suspension. In areas of continuous permafrost a *talik* (zone of unfrozen ground) commonly underlies large rivers. Such taliks develop due to the release of latent heat during winter freezeback of the overlying water. Taliks are generally no more than a few metres thick beneath shallow braided water courses, but may be much thicker beneath deeper single-channelled rivers.

Miall (1977) has shown that the most important factors contributing to braiding are the presence of abundant coarse cohesionless bedload, a strongly fluctuating discharge regime, steep long profile gradients, and a lack of stabilising vegetation. All these factors frequently apply to rivers in periglacial environments. Braid bars in gravel-bedded rivers are initiated during periods of high discharge when both bed scour and deposition occur. Of crucial importance are secondary flows in which the water follows a helical spiralling path (Figure 8.5). Several secondary flows develop across the channel, producing zones of scour, where flow is directed towards the bed, and zones of deposition, where flow is directed towards the surface (Richards, 1982). In zones of surface flow convergence the velocity at the bed is raised, resulting in high bed shear and erosion, while in zones of surface flow divergence the opposite is the case, promoting deposition and the development of braid bars. These are classified according to their location, form and sedimentary characteristics (Figure 8.6). *Longitudinal bars* are diamond- or lozenge-shaped, and comprise mainly massive or crudely-bedded imbricate gravels. Imbrication is the tendency for flat pebbles to dip upstream and is characteristic of gravel deposited from traction. *Linguoid bars* occur mainly in sand-bedded rivers, and comprise cross-bedded sands, often with smaller bedforms superimposed on the bar platform during falling stage.

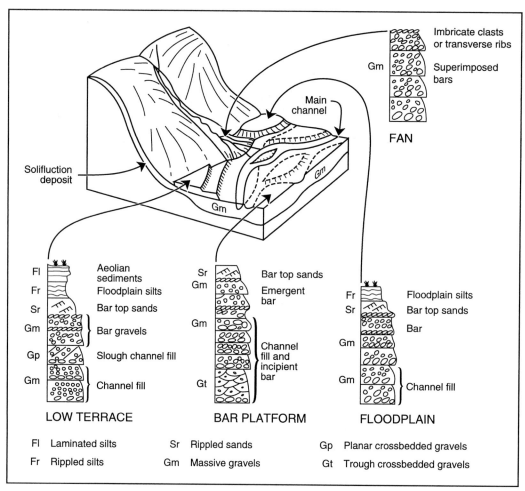

Figure 8.7 Model of periglacial alluviation. After Bryant (1982).

As we have seen, longitudinal bars are initiated during peak flow. As a flood wanes, progressively finer gravels are deposited on bar surfaces. When the bars break the water surface, secondary channels are cut across them, and these are in turn filled with cross-bedded sands. If flow velocities continue to fall, wedge-shaped sand units may accumulate in the lee of longitudinal bars in the main channels. Finally, as flows reach very low levels, areas of standing water are left as pools within which suspended sediments are deposited to form mud drapes. Although renewed erosion during the next major flood event is likely to remove much of the finer sediment deposited during waning discharge, some may be buried by the associated influx of gravels, and hence preserved. In aggrading gravel floodplains the resulting sedimentary sequence therefore consists of an irregular series of truncated fining-upward cycles. The bar gravels generally form planar sheets bounded by erosional disconformities, but larger lenses of imbricate coarse gravels and smaller lenses of cross-bedded sands mark the locations of primary and secondary channel fills respectively.

The characteristics of braided periglacial rivers in Spitsbergen and Banks Island have been studied by Bryant (1982, 1983a). The Spitsbergen catchments are partly glacierised, but Sachs River on Banks Island has an arctic nival regime. A depositional model developed by Bryant is illustrated in Figure 8.7. Two important elements are evident at these sites: first, the accumulation of aeolian deposits on inactive bar platforms and terraces away from the active channel (Good & Bryant, 1985), and second, the development of ice-wedge polygons on such inactive surfaces. On both Banks Island and Spitsbergen, deflation of fine alluvium occurs during summer when flow is low. This windblown sediment is then redeposited as laminated fine sands and silts, forming a fine-grained sediment cover over fluvial gravels (Figure 8.7). Examples of this interaction between wind and water have also been described by Fahnestock (1969), Pissart, Vincent & Edlund (1977) and Nickling (1978). Ice-wedge growth is associated with permafrost aggradation in areas of the floodplain abandoned by active channels, but subsequent channel migration may lead to truncation of the ice wedges and development of wedge casts as the permafrost thaws locally to form a new subchannel talik (Chapter 4). In this way ice-wedge growth followed by subsequent casting and burial may result in the formation of intraformational ice-wedge casts that mark migrations of active channels and their associated taliks rather than climatic fluctuations (Bryant, 1982, 1983a; Seddon & Holyoak, 1985). Such periglacial structures are often of critical importance in the identification of relict periglacial fluvial sediments deposited in a continuous permafrost environment.

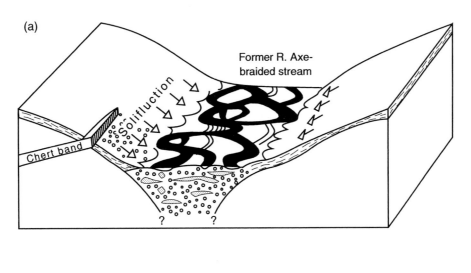

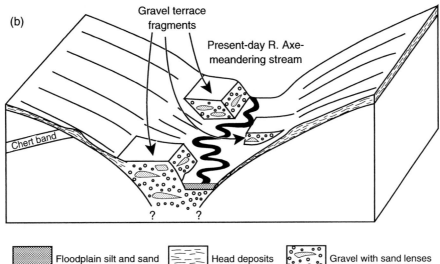

Figure 8.8 Two-stage model of the evolution of the Axe Valley, Devon: (a) valley floor alluviation under periglacial conditions; (b) present-day valley form. After Shakesby & Stephens (1984)

Pleistocene fluvial landforms and deposits

The significance of Pleistocene cold-climate fluvial processes in NW Europe has been emphasised by Gibbard (1988), who stressed the importance of accelerated rock weathering and associated niveofluvial dissection in providing large quantities of coarse sediment that were then transported and deposited by braided rivers. The resulting fluvial sands and gravels today make up the overwhelming bulk of alluvial deposits over much of NW Europe. Not all periglacial sands and gravels are confined to the floors of valleys, however. The Chelford Sands of eastern Cheshire, for example, were deposited in a broad fan complex emanating from the Pennines (Worsley, 1970). Fluvial deposition in this area occurred in shallow braided water courses, but wind action also added an aeolian component to the sedimentary sequence. The Chelford Sands Formation incorporates organic deposits of Early Devensian age, and numerous intraformational ice-wedge casts attest to the presence of permafrost during fan accumulation (Worsley, 1966b).

The role of periglaciation in river valley evolution

Beyond the maximum limits of the Quaternary ice sheets, valley sides are often mantled with periglacial solifluction deposits and show little evidence of significant modification during the present (Holocene) interglacial (Chapter 7). It appears, therefore, that valley development was significantly accelerated by periglaciation. Many smaller valley systems were apparently dominated by solifluction processes, but rivers in larger catchments were able to transport and rework solifluction debris, spreading gravel across aggrading valley floors and carrying away much of the finer-grained sediment in suspension. A good example of such a periglacially-modified river valley is the Axe Valley in Devon, which contains a considerable thickness of alluvial gravels (Shakesby & Stephens, 1984). These have an imbricate structure and comprise angular to rounded clasts, mainly of chert and greensand. The deposits are crudely bedded, with bedding emphasised in places by lenses of laminated sands. Similar laminated sands also occur within the gravels as small angular lumps that were apparently eroded, transported

Morphology   Process

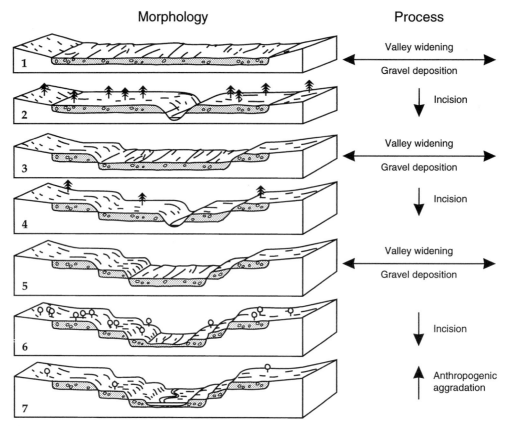

Figure 8.9 Model of the evolution of the Nene Valley during the Devensian and Flandrian. After Castleden (1980). 1. Early Devensian stadial. 2. Chelford Interstadial. 3. Middle Devensian stadial. 4. Upton Warren Interstadial. 5. Dimlington Stadial. 6. Early Flandrian. 7. Mid-Flandrian period of anthropogenic forest clearance.

and deposited as frozen 'clasts'. The gravels are extensively cryoturbated in their upper 2–3 m, and contain pre-Ipswichian Acheulian hand axes. At one exposure Shakesby & Stephens observed a 2 m thick layer of sand, silt and clay within the gravel sequence. This fine sediment lay at a depth of 10 m, and was inferred to represent overbank deposition by a meandering stream, in contrast to the braided network implied by the gravels. Pollen recovered from this finer bed indicated discontinuous boreal forest vegetation at the time of accumulation, characteristic of a late interglacial or interstadial period. A model of the evolution of the Axe Valley based on the observed sedimentary sequence is shown in Figure 8.8.

Many large river systems in southern England contain Pleistocene alluvial deposits similar to those in the Axe Valley. Those in the Nene Valley of Cambridgeshire and Northamptonshire consist of horizontally-bedded gravel and sand units, channel fills of very coarse rounded boulders, braid bar lenses of massive poorly-sorted gravels and subhorizontal surfaces delimited by organic material and fine-grained sediments (Castleden, 1977, 1980). These surfaces are associated with involutions and intraformational ice-wedge casts, and appear to represent temporarily-abandoned bar platforms beneath which permafrost became re-established. Radiocarbon dates from the intercalated organic beds indicate sediment aggradation during the Dimlington Stadial between about 30.0 ka BP and 18.0 ka BP. Similar gravels containing intraformational ice-wedge casts and involutions are exposed in gravel pits in the valley of

the River Bain in Lincolnshire (Bryant, 1983a), and probably underlie most of the larger valleys of the lowland zone. In the case of the Nene, and indeed many other rivers, the valley floor gravels underlying Holocene alluvium extend along the flanks of the valley as low terraces. Such terraces are often underlain by the remnants of a planed-off rock floor, referred to by Castleden as *gallery pediments*. The development of the Nene Valley through successive stadial and interstadial phases of the Devensian, as envisaged by Castleden, is illustrated in Figure 8.9. It is likely that similar cycles of periglacial valley widening (pedimentation) and gravel accumulation have affected many rivers, not only in southern England, but also beyond the limits of the last ice sheet in western and central Europe.

The largest of the lowland rivers affected by recurrent Pleistocene periglaciation is the River Thames, which lay largely beyond the limits of glaciation, though it received inputs of fluvioglacial sediment at least twice in the Early Pleistocene (Hey, 1965; Rose, 1989a,b). Fluviatile sand and gravel deposits are widespread in the Thames Valley system, occurring as complex terrace sequences that increase in height and age away from the present river channel. The terminology, correlation and chronology of the Thames terrace sequence is complex and will not be discussed in detail here. The reader is referred to reviews by Briggs & Gilbertson (1980), Green & McGregor (1980) and Gibbard (1985).

On the northern side of the lower Thames basin the higher terrace gravels may be traced beneath Anglian till in a broad

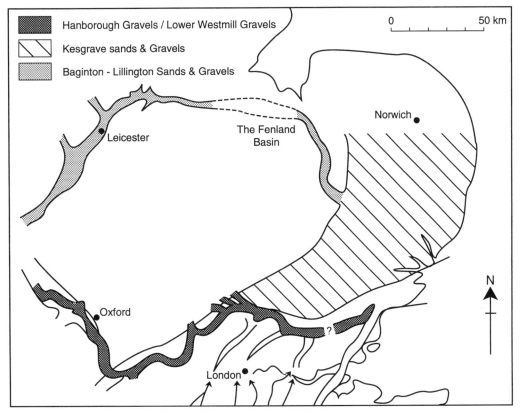

Figure 8.10 Fluvioperiglacial gravels of the pre-Anglian proto-Thames (after Rose, 1989b). Rose suggested that the course of the river migrated towards the south-east during the accumulation of these gravels in East Anglia.

belt extending through the Vale of St Albans and across East Anglia (Figure 8.10). This zone of sand and gravel marks the Early Pleistocene lower course of the proto-Thames. It appears that the advance of the Anglian ice sheet blocked the Vale of St Albans and forced the river to flow along a more southerly route (Rose, Allen & Hay, 1976; Gibbard, 1977, 1985; Baker & Jones, 1980; Green & McGregor, 1980; Green, McGregor & Evans, 1982; Bridgland, 1988). The sands and gravels deposited in East Anglia by the proto-Thames were termed the Kesgrave Formation by Rose et al. (1976). A temperate palaeosol ('The Valley Farm Soil') occurs in the upper parts of the Kesgrave Formation (Rose et al., 1985a,b; Whiteman & Kemp, 1990). This soil was subsequently disrupted by cryoturbation and by ice-wedge and sand-wedge development during the early Anglian to produce the arctic 'Barham Soil', a prominent palaeosol developed under particularly severe periglacial conditions (Chapter 6). The Valley Farm palaeosol therefore reflects pedogenesis during a temperate phase, while the superimposed Barham palaeosol marks a later period of periglaciation that affected the same surface before it was buried under Anglian till.

Quartz and quartzite pebbles were apparently supplied to the Thames catchment by rivers draining a pre-Anglian ice sheet in Wales and the Midlands. These pebbles may be traced eastwards along the discontinuous high-level Westland Green Gravel terrace of the middle Thames Valley to the Kesgrave Formation. There is also evidence that a major tributary of the proto-Thames drained the East Midlands at this time, depositing sands and

gravels contemporaneous with the Kesgrave Formation (Rose, 1989b; Figure 8.10). The Kesgrave Formation is up to 20 m thick and consists of sand and gravel beds with lenses of silt (Rose & Allen, 1977). The sands and gravels incorporate channel scour and fill structures, planar beds and cross-bedded units that are collectively indicative of sediment accretion within a large braided river of rapidly fluctuating discharge (Figure 8.11). The maximum water depth is estimated to have been at least 4.5 m, and flow was towards the north-east. The presence of intraformational ice-wedge casts indicates sedimentation under permafrost conditions. The Kesgrave Formation therefore provides evidence for prolonged periglaciation in East Anglia prior to the advance of the Anglian ice sheet.

The model of Quaternary river valley development shown in Figure 8.9 assumes valley widening under periglacial climates and incision during temperate episodes. The Thames Valley, which escaped glaciation, experienced a complex history of Quaternary climatic change. It is likely, therefore, that valley deepening, valley widening and sediment aggradation occurred repeatedly and under a variety of climatic regimes. Most present-day terrace surfaces are probably in part erosional, so that a single terrace fragment may incorporate sediments of different ages deposited under differing environmental conditions (Figure 8.12). Assuming interglacial hydrology and sediment supply to have been similar to that of the Flandrian, the Thames and its tributaries are likely to have been confined to single meandering channels during temperate interglacials (Green & McGregor, 1980; Gibbard, 1985), and indeed organic deposits

Figure 8.11 Kesgrave Formation fluvioperiglacial gravels, Coggeshall Quarry, Essex. Note planar beds of massive imbricate gravels (longitudinal bar deposits) cut by sand-filled secondary channels.

attempted by Bryant (1982, 1983a,b), who studied the Summertown–Radley Terrace gravels near Oxford. Here planar sheets of poorly-stratified gravel up to 1.2 m thick are cut by broad channels filled with cross-bedded gravel and sand. The deposits closely resemble those of the Nene Valley described earlier, and like the Nene gravels, contain horizontal erosion surfaces below which are intraformational ice-wedge casts and involution structures. Stratigraphic evidence and thermoluminescence dating suggest that the upper Summertown–Radley gravels are of Early Devensian age, though the Summertown–Radley Terrace also includes older fluvioperiglacial gravels at greater depths (Gibbard, 1985). Very similar terrace gravels occur in the valley of the Warwickshire Avon (Dawson, 1987), where they are mantled by lateral spreads of aeolian sands. Such evidence indicates that depositional processes closely analogous to those depicted in Figure 8.7 were common in the larger valleys of the British lowlands during periods of periglaciation.

In the valley of the lower Kennet, a south bank tributary of

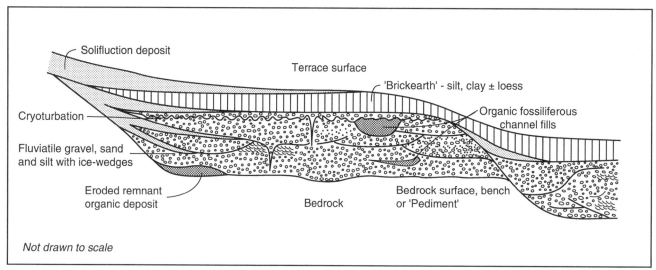

Figure 8.12 Typical sedimentary sequence associated with gravel terraces in the Thames Valley. After Gibbard (1985).

indicating temperate conditions generally occur within fine-grained channel fills. Again using the Holocene river as an analogue, it is probable that only limited valley widening or deepening occurred during most interglacials. Consistent with this inference is sedimentological evidence which indicates that sand and gravel aggradation took place under periglacial conditions within a braided river system (Gibbard, 1985). However, deepening of the Thames Valley also apparently took place in periglacial periods, either in response to changes in runoff regime and sediment supply, or as a consequence of eustatically-lowered sea levels. Hence valley widening, gravel deposition and valley incision may all have occurred during individual cold stages (Figure 8.13), followed by floodplain alluviation during ensuing interglacials. Given the complexity of climatic changes and associated sea-level fluctuations during the Quaternary, it is not surprising that the stratigraphy of the Thames terrace sequence is extremely complex and remains incompletely understood.

Detailed reconstruction of the depositional environment under which the Thames Valley gravels accumulated has been

the Thames, Cheetham (1975, 1976, 1980) identified three terrace levels: the Beenham Grange Terrace, 1–3 m above the alluvial floodplain; the Thatcham I Terrace, 7–9 m above the floodplain; and the Thatcham II Terrace, 10–12 m above the floodplain. Evidence provided by molluscan fauna and ^{14}C dating indicates that the Beenham Grange Terrace is of Late Devensian age. The relatively undisturbed nature of the terrace surface is shown by the preservation of braided channels now filled with fine-grained Holocene sediments, but still visible as crop marks (Figure 8.14). A radiocarbon date on organic material from within the Thatcham Terrace gravel indicates that sediment accumulation continued through the Middle Devensian into the Dimlington Stadial (Bryant, 1982; Bryant, et al., 1983). Cheetham employed an empirical equation relating channel gradient and bankfull discharge in a reconstruction of the palaeohydrology of the Kennet during deposition of the Thatcham and Beenham Grange terrace gravels. His calculations suggested that the bankfull discharges associated with terrace aggradation were at least twice those of the modern river. Cheetham also considered the modern drainage density of the

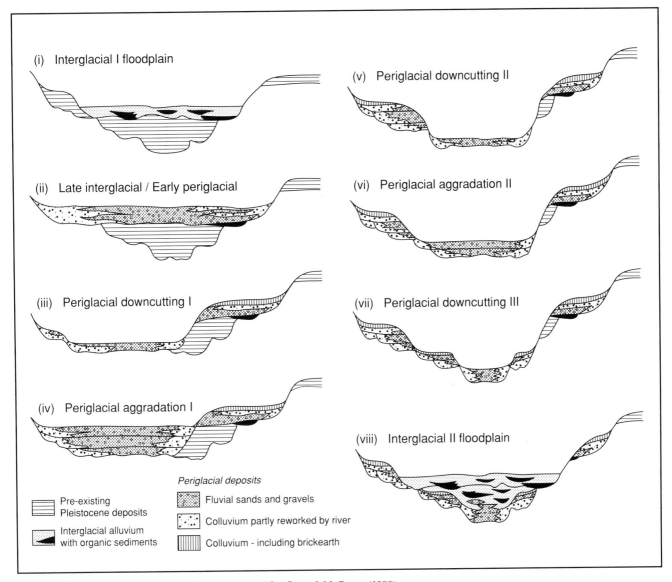

(i) Interglacial I floodplain

(ii) Late interglacial / Early periglacial

(iii) Periglacial downcutting I

(iv) Periglacial aggradation I

(v) Periglacial downcutting II

(vi) Periglacial aggradation II

(vii) Periglacial downcutting III

(viii) Interglacial II floodplain

Periglacial deposits

Pre-existing Pleistocene deposits

Interglacial alluvium with organic sediments

Fluvial sands and gravels

Colluvium partly reworked by river

Colluvium - including brickearth

Figure 8.13 Model of the evolution of the Thames terraces. After Green & McGregor (1980)

Kennet above Theale, and compared it with the drainage density obtained by plotting the valley axes of all dry valleys within the catchment. He showed that the present-day Kennet has a drainage density of 0.45 km km^{-2}, compared with a 'valley axis' drainage density of 2.34 km km^{-2}. As most dry valleys grade to the Beenham Grange Terrace surface rather than the modern river, it seems likely that they last carried significant surface drainage during the period of terrace accumulation. By assuming that discharge is proportional to the square of the drainage density (cf. Gregory & Walling, 1968), and an empirical relationship between mean annual flood discharge and drainage density (Carlston, 1963), Cheetham calculated that the discharge of the mean annual flood during deposition of the Beenham Grange Terrace probably lay between 153 m^3 s^{-1} and 418 m^3 s^{-1}, as compared with 40 m^3 s^{-1} for the modern Kennet. Such high magnitude palaeofloods were probably engendered by spring snowmelt, with meltwater running off the chalk along valleys that today no longer support surface drainage. In a simi-

lar analysis of the first and third terraces of the Blackwater Valley in Berkshire and Hampshire, Clark & Dixon (1981) estimated that peak palaeodischarges were between 19 and 37 times the maximum flow recorded for the River Blackwater today. Their conclusion therefore supports the notion that the gravel terrace deposits of many rivers in southern and central England accumulated under periglacial conditions when river regimes were dominated by high seasonal snowmelt-fed flood peaks.

The hydrological and geomorphological response of a lowland river system to the relatively rapid climatic fluctuations of the Late Devensian Lateglacial has been admirably demonstrated by Rose *et al.* (1980) in a study of the sands and gravels of the Gipping valley at Sproughton in Suffolk. Here the river sediments are underlain by chalk head and mudflow deposits in which a palaeosol is developed. The palaeosol is dated to shortly after the thermal maximum of the Windermere Interstadial. Lacustrine silts draped over the palaeosol are of

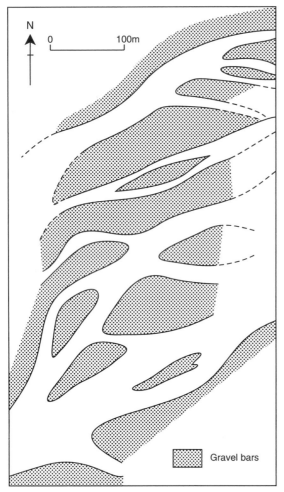

Figure 8.14 Braided palaeochannels in the Kennet Valley near Theale, Berkshire. After Cheetham (1980).

late Windermere Interstadial age. This suite of sediments was affected by river channel incision to a depth of at least 9 m (to -7 m OD), apparently as a result of a sudden increase in flood discharges. The incised channel subsequently filled with sand and gravel deposits incorporating a lower unit of large-scale cross-set beds representing point bar deposition by a meandering stream, and an upper unit of small-scale cross-sets formed as longitudinal bars in a braided system. Significantly, the bar deposits at this site contain a high proportion of unresistant chalk clasts, suggesting local valley erosion and only limited transport distances. Lenses of peat within the gravels contain coleoptera of arctic or alpine character, which indicate that the gravels accumulated mainly during the Loch Lomond Stadial, after which a meandering channel depositing fine-grained alluvium was re-established. Channel incision was considered by Rose *et al.* (1980) to have resulted from climatic deterioration and the establishment of very high nival flood discharges. Decreasing discharges, together with increasing sediment supply from valley-side solifluction, appear to account for subsequent sediment accumulation, initially in a large gravel-bed meandering river then later within a wide and relatively shallow braided system.

Although the most obvious effects of river action under periglacial conditions concern the erosion and deposition of sediment sequences, there is some evidence also for planation or incision of underlying bedrock. Castleden (1977, 1980), for example, reported periglacial gravels overlying broad flat bedrock valley floors in several river valleys in southern and central England. These planed-off bedrock surfaces are referred to as *periglacial pediments*. While the subgravel rockhead is certainly broadly planar in many valleys, in others periods of localised channel incision also apparently occurred, giving steep-sided scour hollows such as those underlying the lower Kennet Valley (Bryant, 1982, 1983a,b) and the lower Thames (Berry, 1979). In the dry valleys of the chalklands, valley asymmetry is associated with broad, gently-concave footslopes below south-west-facing valley-side slopes. These footslopes have gradients of between 5° and 9° and are up to 100 m wide. They were termed *cryopediments* by French (1973), and apparently reflect parallel retreat of adjacent valleyside slopes, as outlined in the following section.

Chalkland dry valley formation

The chalklands of southern England are characterised by steep scarp faces and gentle dip slopes, both of which are extensively dissected by dry valleys (Figure 8.15). Those cut in scarp faces exhibit a variety of forms, including steep-sided valleys such as those of the Chiltern escarpment near the Goring Gap (Jones, 1981), relatively short, steep-sided gashes such as the Devil's Dyke near Brighton and amphitheatre-shaped forms with steep headwalls and flat floors, which are common on both the North and South Downs (Bull, 1936, 1940). Dip-slope valleys are larger, often dendritic in form, and frequently display asymmetric cross-profiles (French, 1972, 1973). The term *coombe* was used by Bull (1936) specifically to denote scarp-slope dry valleys, but has since been used to describe dry valleys on both the scarp slopes and dip slopes of the Downs. Small semicircular hollows cut in Eocene sands and clays in the steep valley sides of the New Forest area were termed *dells* by Tuckfield (1986), and shown to be morphologically similar to many chalkland escarpment coombes as well as to the shallow dry valleys or *dellen* of central Europe (Czudek & Demek, 1971).

The formation of chalkland dry valleys has been attributed by some researchers to 'normal' fluvial erosion under temperate conditions, with only slight modification by periglacial processes. It has been argued that rivers were able to flow over the chalk at a time when water tables were much higher than today, and that falling water tables caused by beheading of dip-slope streams (Chandler, 1909), scarp recession (Fagg, 1923) or falling base-levels due to changing coastline position (Sparks, 1949) led to progressive drying up of the chalkland valleys. The role of spring-sapping in the development of escarpment dry valleys has been emphasised by Sparks & Lewis (1957) and Small (1961, 1964). For the Yorkshire Wolds, where glacial erosion also played a part in landscape evolution, detailed morphological analysis led Lewin (1969) to favour headward extension of valleys by surface streams as the primary cause of dry valley formation. This was thought to have caused lowering of the water table and consequent drying up of water courses. Lewin concluded that under periglacial conditions the valleys simply acted as 'gutters' that infilled with frost-weathered chalk as a result of valley-side mass movement.

Figure 8.15 Dry valleys: (a) scarp-slope valley, Barton in the Clay; (b) dip-slope valleys, South Downs. Cambridge University Collection of Aerial Photographs.

A second body of opinion has regarded dry valley formation as a consequence of ground freezing during Pleistocene cold stages. It is argued that at such times the normally permeable chalk would have been rendered impermeable, and that rivers fed by snowmelt and armed with the products of frost weathering would have been effective agents of erosion. Spurrell (1886) and Reid (1887) were the first to propose such an origin.

They also recognised that the chalky head or *coombe rock* that mantles the dry valley bottoms is the product of frost weathering and mass wasting, an interpretation that supports the notion of valley development under periglacial conditions. Bull (1936, 1940) stressed the role of snowmelt in generating high stream discharges even within relatively short escarpment valleys, and suggested that the arcuate headwalls of many escarpment coombes may have resulted primarily from nivation processes. The strongest evidence that periglacial processes played an important role in dry valley formation comes from the deposits they contain. Almost all are mantled with coombe rock, which tends to thicken downvalley and to be deepest on valley floors. Low gradient aprons or fans of chalky debris extend beyond the mouths of some escarpment dry valleys in both the South Downs (Bull, 1936) and the North Downs (Kerney, Brown & Chandler, 1964), though Williams (1980) showed that this is not always the case.

Kerney *et al.* (1964) described coombes in the North Downs escarpment near Brook in Kent. These have steep straight sides with gradients of 20–34° and gently sloping floors. The valley floors support thin deposits of chalky drift that thicken and coalesce downslope, forming broad fans which extend onto the clay vale below the scarp (Figure 8.16). These fans are generally over 2 m thick and contain up to 0.5 m of poorly-sorted chalky head overlain by fine-grained calcareous muds. The latter are thought to represent colluvial material deposited by surface wash. The muds are subdivided by a palaeosol dated on the basis of its molluscan fauna to the Windermere Interstadial. Kerney *et al.* attributed much of the coombe formation at Brook to rapid frost weathering, niveo-fluvial erosion and solifluction, though as Small (1964) pointed out, the large volume of fan debris reflects not only coombe formation, but also erosion of the escarpment itself. The last phase of active coombe erosion was considered by Kerney *et al.* to have been during the Loch Lomond Stadial.

Morphometric analysis of dry valley systems in the North Downs, South Downs and Berkshire Downs has shown them to be dominated by first order valleys. This may reflect rapid valley development through headward extension (Morgan, 1971), possibly associated with a history of episodic valley erosion under periglacial conditions with intervening temperate phases when geomorphological activity was largely limited to removal of chemically-weathered chalk in solution. Morgan stressed the consistently high drainage densities of these dry valley systems, citing a mean density of 2.26 km km^{-2} for the South Downs and Berkshire Downs. This greatly exceeds the present surface drainage density of 0.1 km km^{-2}. He considered that these high values reflect valley development when infiltration into the chalk was restricted by the presence of permafrost.

The conflicting theories of dry valley development discussed above were reconciled by Williams (1980), who emphasised the large amount of denudation of adjacent clay vales during the Quaternary. He concluded that erosion of the clay vales was accompanied by progressive lowering of water tables in the chalklands. According to Williams, during early Quaternary interglacials the chalkland valleys probably contained perennial streams fed by groundwater and capable of significant valley enlargement. He proposed that lowering of water tables resulted in loss of surface drainage under temperate conditions, so that periglacial valley enlargement (when surface water flowed over

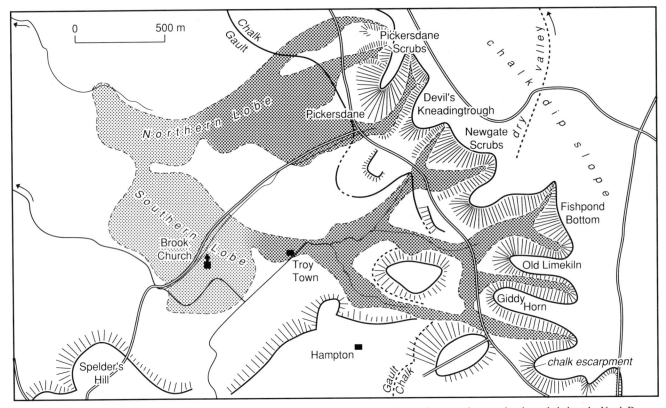

Figure 8.16 Escarpment coombes and their associated lobes of solifluction and colluvial sediments that extend across the clay vale below the North Downs near Brook in Kent. After Kerney *et al.* (1964).

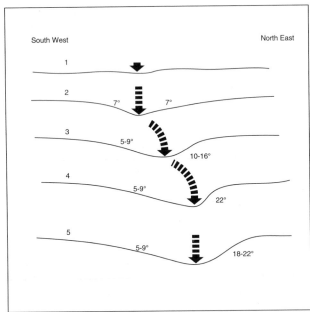

Figure 8.17 Model of the development of asymmetric valleys in chalklands under periglacial conditions. After French (1973).

impermeable frozen ground) became progressively more important through middle and late Quaternary times.

The development of valley asymmetry has also been attributed to periglacial processes (e.g. Ollier & Thomasson, 1957; French, 1972, 1973), though French (1976) conceded that such

processes may merely have modified pre-existing chalkland valleys. Moreover, local factors such as lithology, dip of bedding and lateral erosion by meandering streams may cause steepening of either valley side. It is nevertheless the case that in many downland valleys the south- and west-facing sides tend to be steeper (typically 16–22°) than the north- and east-facing valley sides (typically 5–9°). This pattern is the opposite of that widely reported from present-day periglacial environments in the northern hemisphere (French, 1976). In attempting to explain such reverse asymmetry, French (1973) argued that, during Quaternary cold stages, warmer south- and west-facing slopes experienced more rapid frost weathering due to more frequent and deeper periods of thawing. He suggested that this in turn favoured channel migration towards the more readily eroded south- and west-facing slopes (Figure 8.17). Fluviothermal erosion may also have encouraged such lateral migration (stages 3 and 4, Figure 8.17). French envisaged eventual cessation of lateral migration of streams as their capacity to transport debris was balanced by the rate of delivery of frost-weathered detritus to the slope foot. Thereafter, he suggested, south- and west-facing slopes would have evolved by parallel retreat, leaving a gently-concave basal cryopediment.

Williams (1980) attempted to calculate the volume of rock removed during the development of asymmetric valleys under periglacial conditions. For valleys in the Chilterns, Marlborough Downs and the North Dorset Downs he calculated that between 14% and 38% of valley volume may have been eroded asymmetrically under periglacial conditions, though there is clearly difficulty in deciding the form and dimensions of the valleys prior to

Figure 8.18 Deflation surface, Ellesmere Island, NWT, Canada.

periglacial modification. Williams nevertheless argued that his results indicate very significant chalkland valley modification under periglacial conditions, due largely to the development of an impermeable frozen substrate and the generation of high discharges during the annual snowmelt flood.

The evidence outlined so far in this chapter highlights the vital role that fluvioperiglacial processes played in the evolution of valleys in lowland Britain outside the limits of Devensian glaciation. Palaeohydrological reconstructions of flood discharges associated with rivers in this zone suggest that floods were much more spectacular under the cold climates of successive glacial stages than during intervening temperate interglacials, even though precipitation totals were probably lower during the former (Lockwood, 1979). This apparent anomaly reflects the rapid release of meltwater that must have accompanied melting of winter snow accumulations in spring. As in high arctic areas at present, rapid spring snowmelt during glacial stages must have generated powerful annual snowmelt floods of high capacity and competence. Coupled with the effects of frost weathering and periglacial mass movement in supplying sediment to river channels, such high magnitude snowmelt floods were responsible for valley widening, establishment of braided channel networks and the spreading of sediment across valley floors. Intimately associated with such redistribution of sediment across floodplains was the reworking of finer-grained material by wind, the geomorphological consequences of which we shall now examine.

Wind action

The presence of windblown silt or *loess* in the Quaternary drift cover of Britain was first recognised at a few localities in the early part of this century (e.g. Trechmann, 1920; Palmer & Cooke, 1923). Its much wider occurrence as a thin superficial layer, or as a component of other near-surface sediments such as head deposits, was not appreciated until much more recently. Since 1970 there has been extensive detailed mapping of loess in Britain by geologists and soil scientists, and as a result its known extent has increased considerably (Catt, 1977, 1985a,b). Of less widespread occurrence are periglacial windblown sand deposits known as *coversands*. These are often associated with fluvioperiglacial deposits.

Periglacial aeolian processes

The entrainment of fine particles by wind is only effective where a protective cover of vegetation is absent. Hence, dust storms and sand storms are common not only in many hot deserts but also in the sparsely-vegetated polar deserts of high latitudes. In such areas, winnowing of fine-grained material from exposed sediments often leaves a lag of pebbles covering the ground surface (Figure 8.18). Those pebbles that exhibit polishing and facetting by the abrasive effect of windblown sand are referred to as *ventifacts*. Periglacial wind action is, however, not restricted to the polar desert zone. Many periglacial rivers, whether niveo-fluvial or glaciofluvial in origin, deposit large quantities of silt and sand over extensive braided outwash plains, fans and valley floors. Since high discharge events are often restricted to the period of the nival flood in early summer, unvegetated braid bars and sandflats are exposed to deflation in late summer and winter, and in consequence extensive aeolian deposition may occur over adjacent land surfaces. In central Alaska, for instance, loess has been deposited along the margins of braided rivers (Péwé, 1968), and in south-western Greenland aeolian sand sheets border proglacial outwash plains while adjacent hills and plateaux are mantled by a thin veneer of loess (Hansen, 1970; Dijkmans, 1988; Koster, 1988).

Deflation during winter results in the transport of both sediment particles and snow, and these are often deposited in interdigitating layers to form *niveo-aeolian deposits* (Samuelsson, 1926; Cailleux, 1978; Koster, 1988). Melting of snow may destroy bedding in niveo-aeolian silts and sands (e.g. Pissart, Vincent & Edlund, 1977) or may produce complex deformation features, including folds and tension cracks, known as *denivation structures* (Cailleux, 1972, 1976; Koster & Dijkmans, 1988). Cold semiarid conditions rather than thick snowcover appear most favourable for niveo-aeolian transport and deposition (Koster, 1988). Wind action also occurs during the summer, however, especially in the vicinity of sparsely-vegetated braided river channels. On Banks Island, for instance, Pissart *et al.* (1977) and Good & Bryant (1985) observed extensive sand flats up to 5 km wide bordering the Thomsen and lower Sachs Rivers, both sites of intense summer aeolian activity. Good & Bryant noted that a network of shallow streams crosses the sandflats as the snow melts during the early part of the runoff season. At this time the ground is thawed to only a few centimetres depth, and the streams scour shallow channels in which rippled fluvial sand beds are deposited. Later in the summer, when the active layer is up to 0.5 m deep and snowmelt is complete, aeolian sand transport dominates, resulting in the deposition of finely-laminated beds of well-sorted sand (Figure 8.19). These are generally around 0.1 m thick, and bounded by extensive deflation surfaces on which lag deposits of ventifacts occur. The similarity of these fluvio-aeolian sediments to Quaternary coversands in Europe suggests that the processes observed adjacent to Sachs River may have been active over wide areas in the periglacial zone beyond the Pleistocene ice sheets. Deflation of sand is not restricted to alluvial deposits, however. Riezebos *et al.* (1986) have described rapid deflation of sand and silt from recently-deposited till adjacent to the glacier Holmströmbre in West Spitsbergen. Aeolian sediment transport at this site occurs both in summer, when strong katabatic winds blow from the

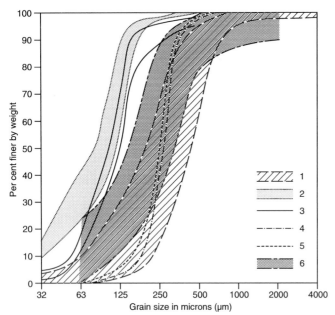

Figure 8.19 Granulometry of periglacial aeolian sands. 1. Thomsen River, Banks Island (Pissart *et al.*, 1977). 2. Holmströmbre, West Spitsbergen (Riezebos *et al.*, 1986). 3. Coversands, Vale of York (Matthews, 1970). 4. Coversands, Messingham, South Humberside (Buckland, 1982). 5. Shirdley Hill Sand, Lancashire (Wilson , Bateman & Catt, 1981). 6. Coversands from the Brecklands, East Anglia (Chorley *et al.*, 1966).

glacier, and in winter, when drifting of snow exposes bare ground to deflation. Transport of silt at Holmströmbre greatly exceeds that of sand in terms of both volume and distance, with sand being redeposited close to its source area. The grain-size characteristics of the sands at Holmströmbre (Figure 8.19) are remarkably similar to those of certain Quaternary coversand deposits in the Netherlands, suggesting that aeolian sediment transport processes at this site offer a valid analogue for those that operated across wide expanses of unvegetated till during the waning of successive Pleistocene ice sheets.

The various studies summarised above demonstrate the importance of wind erosion and deposition in present-day periglacial environments where vegetation cover is sparse. There is reason to believe, however, that periglacial wind action must have been much more widespread and effective in the past. During Quaternary glacial stages, strong winds generated by the development of steep pressure gradients around northern hemisphere ice sheets swept across the adjacent periglacial zone. Within this zone, vast areas of sparsely-vegetated till and outwash were exposed to wind erosion, particularly during ice-sheet retreat. Such periglacial wind action led to the entrainment and deposition of loess on a massive scale. In Europe, loess and aeolian sands mantle a broad area stretching from the Urals in the east to northern France in the west. Pye (1984) considered that deflation during Quaternary cold stages occurred mainly in late summer and autumn in a belt across Europe between the Scandinavian ice sheet to the north and the Alpine glaciers to the south. The evidence provided by contemporaneous pollen and faunal evidence indicates deposition of loess in a cold, fairly dry environment. Britain lies on the (relatively moist) western margin of this zone, and in consequence loess and coversand deposits are both thin and patchy in comparison with those of mainland

Europe. Indeed, as Williams (1975) pointed out, in all but the south-eastern corner of England, loess is more often a component of other periglacial deposits than a deposit in its own right.

Loess deposits in lowland Britain

Loess typically consists of a homogeneous, structureless, highly porous buff-coloured silt, with a grain-size distribution showing a pronounced mode in the range 20–40 μm (Pye, 1984). Mineralogically, loess deposits are usually dominated by quartz, but may also contain feldspar, mica, clay minerals, carbonate grains and various heavy minerals. Often present too are carbonate concretions and the shells of land snails (Russell, 1944), though in Britain postglacial pedogenesis has led to decalcification of loess to a depth of 0.7–0.9 m (Catt, 1985a,b). Since most loess deposits outside south-east England are less than 0.7 m thick, they are often completely decalcified but retain a grain-size distribution closely resembling that of calcareous loess (Figure 8.20). In many areas an aeolian silt component has been added to and mixed with other superficial deposits, such as the head deposits of south-west England (Mottershead, 1971) and west Wales (Watson & Watson, 1967). Catt (1985a,b) has also pointed out that many soils in south and east England, together with some in south Wales, the Midlands, the Pennines and north-west England, show a marked upward increase in silt content. This is particularly evident where the underlying bedrock has weathered to produce a parent regolith deficient in silt-sized particles. Catt reported that silt enrichment may be detected in soils from all parts of the landscape except floodplains that support a cover of Holocene alluvium. This suggests a pre-Holocene aeolian origin for the silt. Burrin (1981), however, has argued that even the silt in the floodplain deposits of the southern Weald consists largely of reworked loess that has been washed into valleys then redeposited by rivers.

The distribution of loess in England and Wales has been discussed by Catt (1977, 1985a,b). Areas with a loess cover greater than 1 m in thickness correspond with the Hamble, Hook and Park Gate soil series on maps produced by the Soil Survey of England and Wales (Figure 8.21), and a loessic origin has been inferred for many of the superficial deposits shown as 'brickearth', 'head brickearth', 'river brickearth' or 'loam' on British Geological Survey maps (Dines *et al.*, 1954). Not all brickearths are of loessic origin, however; some in the lower Thames Valley, for instance, were deposited as floodplain alluvium under interglacial conditions (Catt, 1979; Gibbard, Wintle & Catt, 1987).

Loess deposits thicker than 1 m and sometimes up to 4 m in depth occur throughout north Kent, on the coastal plain of west Sussex, and in parts of the lower Thames valley. One of the first sites to be described in detail is that at Pegwell Bay, south-west of Ramsgate in Kent (Pitcher, Shearman & Pugh, 1954). Here the loess is up to 4 m thick and overlies cryoturbated chalk and Tertiary rocks. In the weathered upper parts of the loess profile, less than 5% by weight is acid-soluble, but below a depth of about one metre the unweathered loess contains around 25% acid-soluble carbonate material. The loessic silt consists largely of quartz particles and the sediment has a strongly prismatic structure. The Pegwell Bay loess is typical of such deposits in south-east England and closely resembles the extensive loess found across the English Channel in continental Europe.

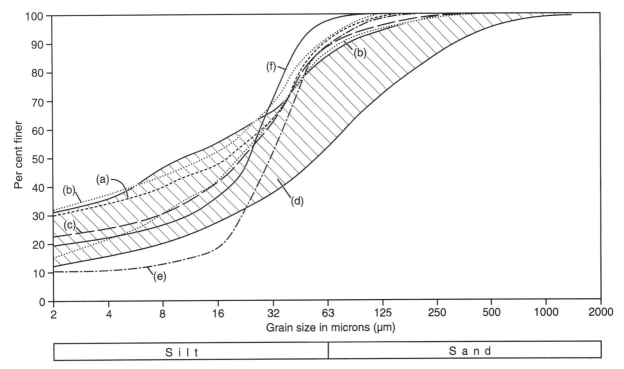

Figure 8.20 Granulometry of loess deposits.
(a) Binbrook, Lincolnshire (Catt, Weir and Madgett, 1974).
(b) Various sites in the Yorkshire Wolds (Catt *et al.*, 1974).
(c) Ford, Kent (Catt *et al.*, 1974).
(d) North Wales (Lee & Vincent, 1981).
(e) The Wirral (Lee, 1979).
(f) North-west France (Lautridou, Sommé & Jamagne, 1984).

The silty loessic components of soils from Yorkshire to Sussex and as far west as east Devon exhibit two regional trends: first, a progressive westward increase in the proportion of flaky heavy minerals such as chlorite and biotite, and second, a progressive westward decrease in modal particle size. Both changes may be due to progressive aeolian sorting by easterly winds transporting the loess from a source in the North Sea Basin (Catt 1978, 1985a,b). Lill & Smalley (1978) also argued that easterly winds driven by a large anticyclone centred over the Scandinavian ice sheet may have carried quartz silt into England. This argument may explain the relative abundance of loess in south-east England, which lies close to potential source areas in continental Europe. However, Eden (1980) demonstrated a north–south reduction in the grain size of certain heavy minerals in loess sampled along a transect from Norfolk through Essex to Kent. This pattern is consistent with silt transport from the North Sea Basin by north to north-easterly winds. The presence of sponge spicules in the loess of these areas also supports the contention that its source was the exposed bed of the North Sea.

The thin loessic silts of west Cornwall and the Isles of Scilly are somewhat coarser in texture and mineralogically distinct from loess deposits farther east, and were probably derived from glacigenic sediments in the southern Irish Sea Basin (Catt & Staines, 1982). Similarly, the silt-rich deposits that mantle the Carboniferous Limestone hills of north-west England and north Wales were probably derived from glacigenic sediments deposited in and around the Irish Sea Basin (Lee & Vincent, 1981; Vincent & Lee, 1981). These sediments are up to 1 m thick, unstratified, decalcified, light yellowish brown to brown in colour and generally stone-free. Silt grains have surface tex-

tures consistent with a loessic origin and some also show evidence of glacial abrasion. At Wylfa Head, on the north-west coast of Anglesey, similar silts overlie Late Devensian tills (Harris, 1991). Here the sediment is indurated, reveals a cryogenic microstructure and contains small-scale involutions. Smithson (1953) and Ball (1960) have shown that the silty loams overlying Carboniferous Limestone in north Wales contain suites of exotic heavy minerals that indicate derivation from glacigenic deposits, and Harris (1991) has demonstrated a mineralogical affinity between the silts at Wylfa Head and the Late Devensian tills at this site. These findings support the idea that such western loessic deposits reflect deflation of local glacigenic sediments.

The presence of loessic silt within Quaternary stratigraphic sequences has only rarely been reported for sites in the British lowlands. One example, at Warren House Gill in County Durham, was described by Trechmann (1920), though the section is no longer exposed. Trechmann observed a layer of silt sandwiched between two till units, and concluded that this represented an interglacial loess. Francis (1970), however, argued that this layer is a true periglacial loess of 'Wolstonian' age that immediately postdates the lower till. In contrast, Catt considered the Warren House Gill loess to be stratigraphically equivalent to the loess-enriched solifluction deposits that separate two Late Devensian till units (the Skipsea Till and the Withernsea Till) on the Yorkshire coast. Similar in stratigraphic context is a 2 m thick layer of aeolian silt that separates two tills on the Wirral Peninsula in north-west England (Lee, 1979). As in Yorkshire, both of the Wirral till units are considered to be of Devensian age. It appears that deposition of the lower till was followed by ice withdrawal and deflation of the exposed ground

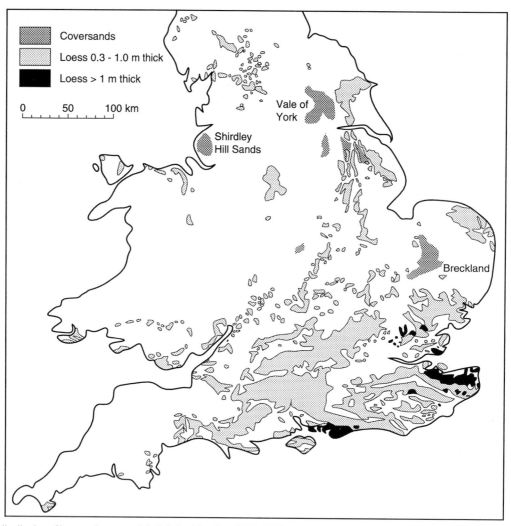

Figure 8.21 The distribution of loess and coversands in Britain. After Catt (1977, 1985a) with additions.

surface, prior to ice readvance and the emplacement of the upper till.

On the basis of stratigraphic, archaeological and mineralogical evidence, Catt (1977, 1978) considered the surficial loess deposits of the British lowlands to be of Late Devensian age. Thermoluminescence (TL) dating of samples of loess from various sites in southern and south-eastern England has generally supported this conclusion, but has also demonstrated the survival of isolated pockets of older loess at some sites (Wintle, 1981; Parks & Rendell, 1988, 1992). TL dating of a loess-enriched solifluction deposit underlying Late Devensian till at Eppleworth in east Yorkshire gave an age of between 17.5 ± 1.6 ka BP and 16.6 ± 1.7 ka BP (Wintle & Catt, 1985). This indicates loess deposition during the Dimlington Stadial, immediately prior to the advance of the last ice sheet across east Yorkshire. In Essex, the occurrence of a loess infill in an ice-wedge cast has also been interpreted as evidence for loess accumulation during the Dimlington Stadial (Eden, 1980). Parks & Rendell (1992) have suggested that earlier loess deposits in southern and south-eastern England represent two main phases of silt deposition, namely > 170 ka BP and 125–50 ka BP. The former age applies to a 'brickearth' horizon at a site on the

Sussex coastal plain at Boxgrove; two other sites, at Northfleet on the Thames Estuary and Holbury near Southampton, have yielded probable pre-Ipswichian TL ages; and a group of three widely-spaced sites gave Early Devensian TL ages within the range 62–50 ka BP. Such sites probably represent the remnants of formerly much more extensive loessic deposits. Even older aeolian silts occur in association with the extensive Early Anglian Barham palaeosol in East Anglia (Rose & Allen, 1977).

Coversands

Periglacial aeolian sand deposits are less widely distributed than loess in the British lowlands (Figure 8.21). The main areas with extensive coversands are the southern part of the Vale of York and the lower Trent Valley (Matthews, 1970; Gaunt, Jarvis & Matthews 1971; Buckland, 1982), Lincolnshire (Straw, 1963), SW Lancashire, (Jones, Tonks & Wright, 1938; Wilson *et al.*, 1981) and the Brecklands of East Anglia (Chorley *et al.*, 1966; Williams, 1975). In addition, Douglas (1982) has described well-sorted sands of possible aeolian origin that are incorporated into near-surface cryoturbation structures in the English Midlands. The coversands in the Vale of York and SW

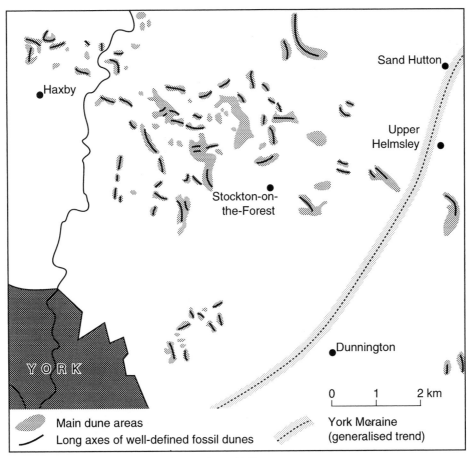

Figure 8.22 Late Devensian sand dunes in the Vale of York. After Matthews (1970).

Lancashire lie within the limit of Late Devensian glaciation, but those of the Brecklands and the Midlands lie to the south of the Devensian limit, and overlie pre-Devensian tills. Granulometrically, the coversands from these areas are remarkably similar (Figure 8.19), and resemble in this respect both aeolian sands in present-day periglacial environments and the Pleistocene coversands of continental Europe. In the Vale of York, low dune-like mounds of sand up to 6 m high are aligned approximately north-west to south-east and north to south (Matthews, 1970; Figure 8.22). Crescentic barchan-like dunes also occur in places. The mineralogy and variable degree of rounding of sand grains in these deposits indicate an essentially local derivation, from fluvioglacial and glaciolacustrine deposits in the Vale of York and lower Trent Valley. Deposition was by winds blowing from a direction slightly south of west. The presence of blown sand on the eastern side of the Trent Valley in Nottinghamshire (Edwards, 1967) also suggests deflation of sediments on the valley floor by west to south-west winds (Williams, 1975). Near Messingham and Flixborough in South Humberside, coversands show thin, irregular, sub-horizontal bedding and often pass laterally into shallow channel structures infilled with waterlain sand (Buckland, 1982). This suggests an environment of deposition similar to that found today adjacent to braided rivers in the Canadian Arctic (Good & Bryant, 1985). On the dip slope and across the lower parts of the escarpment of the Lincolnshire Jurassic limestone, coversands reach a thickness of up to 7 m.

The Shirdley Hill Sand in SW Lancashire covers an area of about 200 km². Near Southport, cross-bedded dune-like hillocks occur, but in many areas the sand forms a layer a few metres thick that lacks surface expression but exhibits horizontal stratification and cross-bedding where exposed in section (Wilson et al., 1981). Individual grains are generally rounded or subrounded, and scanning electron microscope analyses have revealed surface textures consistent with aeolian reworking of fluvioglacial sands. This interpretation is supported by the striking similarity between the mineralogy of the Shirdley Hill Sand and that of local Late Devensian fluvioglacial deposits. Since no directionally-consistent changes in sorting or grain size could be detected in the Shirdley Hill Sand, Wilson et al. (1981) concluded that they reflect local redistribution of fluvioglacial sand by winds blowing in a variety of directions, with no predominant direction of transport.

In East Anglia much of the chalk outcrop supports only a thin patchy cover of blown sand, but in the Breckland coversand is more or less continuous, reaching depths of up to 3 m (Williams, 1975). Such sand is often incorporated into large-scale polygons and stripes considered to represent nonsorted patterns developed over permafrost, probably under the severe climatic conditions of the Late Devensian glacial maximum (Watt et al., 1966). The coversand surface is hummocky in places, but Holocene wind action has destroyed any regular dune forms that may have been present at the time of accumulation (Williams, 1975). Surface grain size varies in a complex manner, though with a tendency to become finer towards the south-west. Regular spatial variations in grain-size with a periodicity of 125–1000 m have been explained by Chorley *et al.* (1966) as being possibly related to sand accumulation within dune systems, but additional smaller-scale variability in grain size probably relates to cryoturbation during formation of the large-scale patterned ground. Like those in the Vale of York and Lancashire, the Breckland coversands were probably derived locally, in this case from fluvial and fluvioglacial sediments, and possibly from decalcified till.

There is abundant evidence from radiocarbon dating of organic layers associated with the coversands of Yorkshire, Lincolnshire and Lancashire to indicate that the main period of sand accumulation was later than that of surficial loess accumulation. At Messingham in South Humberside, the coversand is underlain by an extensive peat horizon dated to 10 280 ± 120 yr BP, a date that implies coversand accumulation in the final stages of the Loch Lomond Stadial. Analysis of associated beetle faunas suggests that the peat accumulated in a marshy environment similar to the muskeg of the Canadian and Alaskan tundra (Buckland, 1982). Near Sutton-in-the-Forest in the Vale of York, a peat horizon 1.5–1.7 m below the coversand surface and 0.3 m above its base yielded a radiocarbon age of 10 700 ± 190 yr BP, which also indicates sand accumulation in the Loch Lomond Stadial. This site lies within the Late Devensian ice limits and the coversands fill wedge casts developed in the underlying till (Matthews, 1970). These casts are interpreted as thermal contraction cracks developed in permafrost after the retreat of the last ice sheet, and may imply that sand accumulation commenced in the wake of ice-sheet deglaciation. At an adjacent site in the Vale of York, a humose clay loam within coversand yielded a radiocarbon date of 9950 ± 180 yr BP (Gaunt *et al.*, 1971). Since much of the overlying near-surface sand apparently accumulated in recent times, this date may indicate cessation of coversand accumulation at the end of the Loch Lomond Stadial. Similarly, pollen studies by Godwin (1959) indicated that the Shirdley Hill Sand in Lancashire also dates largely to this period. Further pollen analyses by Kear (1968, 1977; Tooley & Kear, 1977), and a radiocarbon date of 10 455 ± 100 yr BP obtained on organic material from within the sand confirms this conclusion. The coversands of northern England therefore represent some of the youngest periglacial deposits present in the British lowlands.

In contrast, the coversands of the Breckland area in East Anglia are apparently somewhat older, since they are incorporated into the large-scale chalkland patterned ground structures assumed to have formed at the time of the Late Devensian glacial maximum of *c.* 18 ka BP (Perrin, Davies & Fysh, 1974). According to West *et al.* (1974), coversand accumulation in the Breckland occurred during the Early Devensian, prior to patterned ground formation. If this was so, the Breckland coversands are contemporaneous with the Older Coversands of the Netherlands (Catt, 1977).

The significance of periglacial fluvial and aeolian activity

Within the past two decades, it has become increasingly apparent that the morphology of many lowland floodplains is largely determined by coarse alluvial sediments deposited under periglacial conditions. River systems that drained the land beyond the Quaternary ice sheets, and indeed glacially-fed meltwater systems farther north, generated broad spreads of sand and gravel that infilled many valley floors, and Holocene alluvium often forms only a thin veneer covering these earlier deposits. In the chalklands, and to some extent in other calcareous areas, the dry valleys that form an important element of the landscape probably owe their formation in large part to fluvial processes operating under former periglacial conditions. Moreover, the rapidly-accumulating, poorly-vegetated alluvial deposits laid down under periglacial conditions across wide areas of southern Britain were susceptible to deflation, and provided the source areas for the coversands that mantle the surface in certain localities. The more widespread loess deposits of lowland Britain were apparently derived from deflation of seabed exposed by eustatic lowering of sea levels during glacial stages. Although not as thick as that in adjacent areas of northern Europe, the Late Devensian loess of England and Wales constitutes an important component of many soils, particularly those developed over calcareous bedrock. It has therefore had a marked influence on the natural vegetation of these areas, and indeed continues to influence land-use today. The presence of periglacial alluvial and windblown sediments also provides us with important evidence for the reconstruction of Quaternary palaeoenvironments, particularly in relation to the Devensian stage (Chapter 14). The episodic nature of aeolian deposition and the complexity of periglacial alluviation (as revealed for instance in the Thames Valley) not only serve to emphasise the complexity of Quaternary environmental changes, but also provide the means of reconstructing and understanding such changes. In this respect periglacial fluvial and aeolian sediments complement those periglacial features indicative of former cryogenic activity and the presence of permafrost (Chapters 4–7).

PART 3
The periglaciation of upland Britain

Upland Britain is largely underlain by resistant igneous and metamorphic rocks that have been subject to repeated episodes of glacial erosion. In consequence it is characterised by steep slopes and supports only a thin and discontinuous drift cover, though wide areas are covered by peat. Some periglacial processes still operate on the highest ground, but the most prominent periglacial features in upland Britain developed under much more severe conditions in the past, particularly during the Late Devensian (26–10 ka BP). The most widespread manifestation of former periglaciation is frost-weathered rock and the mantle of mountain-top detritus that cloaks upper slopes and plateaux. Chapter 9 considers the origin and age of this detritus and of the tors that protrude through it, as well as the significance of frost-weathered debris for delimiting the dimensions of former glaciers and for constraining the distribution of other periglacial features. These include both active and relict patterned ground features (Chapter 10) and a wide range of solifluction phenomena, some of which ceased to move some 10 000 years ago whilst others are active at present (Chapter 11). Chapter 12 is devoted to a consideration of the talus deposits that accumulated after deglaciation below steep rockwalls, and examines how these have been modified by avalanches, debris flow activity and the development of protalus ramparts and rock glaciers. The final chapter in this section considers the effects of various other geomorphic agents that have operated under periglacial conditions during the Late Devensian or under the milder conditions of the Flandrian. These include nivation processes, erosion and deposition by upland rivers, aeolian activity on high ground and the effects of periglacial processes in enhancing coastal erosion during the Late Devensian Lateglacial.

9
Frost weathering and mountain-top detritus

Of the great range of periglacial phenomena found on high ground in Britain, by far the most widespread is the frost-weathered *mountain-top detritus* that mantles many mountain slopes and plateaux. For although most mountains in Britain have been buried under one or more former ice sheets, glacial drift is rare on upper slopes, and the regolith mantle in such locations is often attributable to rock weathering under periglacial conditions. Such mountain-top detritus is, however, by no means uniform in its characteristics. On some mountains it takes the form of *blockfields* consisting of a complete cover of boulders with no visible fine soil and consequently little or no vegetation cover (Figure 9.1). More commonly, frost-riven boulders are embedded in a matrix of sand and silt. Vegetation cover on such terrain is often complete, even at altitudes exceeding 1000 m (Figure 9.2). Such contrasts usually reflect differences in the manner in which the underlying bedrock has responded to frost weathering. The characteristics and distribution of mountain-top detritus are important for two reasons. First, the composition of such detritus plays a fundamental role in governing the mode and rate of certain types of periglacial activity, and consequently exercises a powerful control on the distribution of many periglacial phenomena. Second, its distribution is often intimately related to the upper limit of glaciation during the Loch Lomond Stadial of *c.* 11–10 ka BP, and in some areas may even define the upper limits of the last ice sheet. The lower limits of *in situ* mountain-top detritus therefore offer valuable evidence for reconstructing the dimensions of former ice masses.

This chapter examines the nature and significance of frost weathering and mountain-top detritus on British mountains. The first part of the chapter deals with the effects of periglacial weathering, the second considers the characteristics of mountain-top detritus, and the third discusses the implications of rock residuals (tors) that stand proud of the weathered regolith mantle. The chapter concludes with an account of the significance of mountain-top detritus as an indicator of the dimensions of former glaciers and a summary of the ways in which its characteristics influence the distribution of periglacial landforms.

Periglacial weathering

Traditionally, the breakdown of rock in periglacial environments has been attributed to *frost weathering*, the mechanical disintegration of well-lithified rock as a result of repeated freezing and thawing. Many terms have been used to describe such rock breakdown, for example frost wedging, frost shattering or gelifraction. This multiplicity of terms, however, masks (or perhaps emphasises) a lack of understanding of the underlying processes, for although geomorphologists have long been aware of the dramatic effects of frost action on rocks, the mechanics of

frost weathering are still incompletely understood. Moreover, it is important to appreciate that rock breakdown by frost operates over a wide range of scales, producing granular as well as clast-sized detritus. Tricart (1956) subdivided frost weathering processes into 'microgélivation', which he considered to operate independently of rock texture, producing material of silt size or finer, and 'macrogélivation', which exploits structural and textural weaknesses in rocks to produce coarser debris. More recently, Lautridou (1988) has identified four types of frost weathering effects: frost scaling, due to formation of thin plates of internal ice parallel to the rock surface; frost splitting of intact rock; frost wedging, which exploits pre-existing cracks in rock; and granular disaggregation. In this account, the term *macrogelivation* is used to refer to the breakup of bedrock to produce clast-sized detritus. *Microgelivation* is considered to exploit textural weaknesses in intact rock to produce much finer debris by a combination of splitting, flaking, granular disintegration and comminution of grain aggregates or individual grains and crystals. The products of microgelivation include small rock flakes and coarse *gruss* (disaggregated rock) containing particles up to about 32 mm in size as well as sand and silt. The use of these two terms thus reflects mainly the size of the weathered products, and has no genetic implications.

Frost-weathering processes

Frost weathering has traditionally been attributed to stresses set up by expansion of water as it undergoes freezing in joints, cracks or bedding planes (resulting in macrogelivation) or within intergranular pores or microcracks (resulting in microgelivation). This simple explanation has been increasingly questioned, notably by White (1976a) and Walder & Hallet (1985, 1986). Alternative mechanisms of rock breakdown associated with freezing have been reviewed by McGreevy (1981), whose terminology is followed here.

The *volumetric expansion hypothesis* described above has its roots in the experimental work of Bridgeman (1912), who showed that freezing at 0 °C is accompanied by a 9% increase in volume, and that a maximum stress of 207 MPa could be developed at -22 °C. The maximum tensile strength of rock is much lower, around 25 MPa. However, Grawe (1936) pointed out that stresses set up by freezing of porewater are likely to be much less than the theoretical maximum, particularly because air trapped within a rock, being compressible, may greatly reduce effective freezing stress. McGreevy (1981) identified further constraints, such as the necessity of a high degree of rock saturation, rapid freezing and freezing to temperatures below -5 °C. He also amassed field data which suggest that many sites thought to favour frost weathering, such as snowbed

Figure 9.1 Blockfield developed on microgranite at 500 m on Sron an t-Saighdeir, Isle of Rhum, Scotland.

Figure 9.2 Vegetation-covered mantle of mountain-top detritus on the Fannich Mountains, northern Scotland.

margins, rockwalls and bergschrunds, often do not fulfil these requirements. However, even those who have questioned the effectiveness of frost weathering by volumetric expansion concede that under propitious circumstances the process may play some role in rock breakdown. What remains in doubt is whether the process is as widespread as previously believed and whether volumetric expansion of water on freezing is the only or even the dominant agent of microgelivation. Moreover, as White (1976a) observed, macrogelivation resulting from the expansion of water that freezes in cracks or joints (often termed 'frost wedging') has yet to be demonstrated. The efficacy of volumetric expansion as an agent of macrogelivation has nonetheless been assumed by many researchers, probably because it is difficult to envisage most other proposed 'frost weathering' mechanisms operating at anything but a microscale.

An alternative to the volumetric expansion hypothesis is the *capillary theory of frost damage*, which invokes rock breakdown through water migration to ice bodies within microcracks and other loci of weakness such as bedding planes. This model has been developed by Walder & Hallet (1985, 1986), who demonstrated that crack growth is likely to be most effective with slow cooling and within a temperature range of -4 °C to -15 °C. At higher temperatures, ice pressure is likely to be inadequate to promote crack growth, whilst at temperatures below -15 °C migration of unfrozen water within rock is inhibited. A possible ancillary effect is that movement of porewater through the rock may in itself engender internal stresses, and thus fracture or disaggregation (White, 1976a; Fukuda & Matsuoka, 1982). Experimental tests have recently confirmed the importance of segregation ice growth in microfractures as a mechanism of rock breakdown (Akagawa & Fukuda, 1991; Hallet, Walder & Stubbs, 1991). Although this model does not rule out the possibility of rock breakdown by other mechanisms, it appears to resolve many of the inconsistencies inherent in the volumetric expansion model, and offers a potentially very rewarding approach to the understanding of frost-weathering phenomena and their significance.

Finally, the *ordered water hypothesis* or *hydration hypothesis* (White, 1976a) has been widely accepted as an alternative to the traditional explanation of weathering by volume expansion of freezing water (e.g. Thorn, 1979b; Konishchev, 1982). An account of the mechanisms involved has been given by Fahey

(1983, pp.535–6). Interest in this hypothesis was kindled by the experimental research of Dunn & Hudec (1966), who found that carbonate rocks most susceptible to damage during freezing and thawing are those in which most of the water does *not* freeze, and concluded that this unfrozen water took the form of water molecules adsorbed onto clay surfaces. Such water is rigid but non-freezable and capable of exerting pressure against pore walls. Moreover, the presence of like (negative) charges at the 'free' ends of adsorbed layers across pore spaces probably enhances the disruptive effects of adsorption by creating repulsive forces, and additional stresses may be generated by the expansion of adsorbed water when the temperature falls below 4 °C. The degree of alignment or 'ordering' of water molecules increases as temperature is lowered, and increasing humidity results in more layers becoming ordered. As reduction in temperature results in an increase in relative humidity, both the degree of ordering and the thickness of adsorbed layers (and hence the stress generated by adsorption) should increase as temperature falls towards and below 0 °C. Thus failure results from 'terminal fatigue caused by expansion and contraction of the sorption sensitive rock by either wetting and drying or cycling through the freezing point' (Hudec, 1973, p.334). Adsorption and desorption may also generate certain important and possibly crucial ancillary effects. Adsorption reduces the surface tension of water films coating clay mineral surfaces, causing some relaxation of internal stress and consequent dilation, which in turn may enhance the polarity of the ordered water molecules (Dunn & Hudec, 1972). Moreover, sorption may result in osmotic movement of water through the rock, possibly generating pressures of up to 200 MPa (White, 1976a).

The efficacy of ordered water effects *vis à vis* volumetric expansion of freezing water is uncertain. Hudec (1973) maintained that the former is dominant in sorption-sensitive argillaceous rocks, and that the latter affects only coarser-grained rocks that are critically saturated (i.e. >91% of pores are water-filled), a proposition supported by Fahey & Dagesse (1984). Experiments carried out on schists by Fahey (1983) led to the tentative conclusion that volumetric expansion on freezing is 3–4 times more effective than hydration in yielding granular detritus, though his experimental methods have been strongly criticised (Hall, 1986). It is not clear whether adsorption–desorption is effective only for argillaceous rocks (Dunn & Hudec, 1972) or whether the process occurs more widely, as proposed

Figure 9.3 Effects of macrogelivation (frost wedging) in opening the joints between blocks of metamorphosed limestone at 1000 m altitude in the western Grampian Highlands, Scotland.

Figure 9.4 Macrogelivation and frost thrusting of boulders, Glyder Fawr, Snowdonia.

by White (1976a). Some authorities remain unconvinced of the importance of hydration as a frost-weathering process (Lautridou, 1988).

In his review of frost weathering processes, McGreevy (1981) concluded that it is unlikely that any single mechanism functions in isolation to cause rock breakdown, and several researchers have identified the concurrent operation of two or more modes of weathering (e.g. Keeble, 1971; Konishchev, 1982; Mugridge & Young, 1982; Fahey, 1983; Williams & Robinson, 1991). According to this view, the dominance of a particular process appears to be dependent on the complex interaction of three sets of variables (Hall, 1988; Matsuoka, 1991), namely thermal regime (amplitude of freezing cycles, rate of freezing), moisture conditions (degree of rock saturation, distribution of porewater, solute content and availability of unfrozen bulk water) and lithology (rock texture and chemistry). Walder & Hallet (1985, 1986), however, offer a fundamentally different perspective, claiming that the above 'fragmented view' of frost weathering can be reduced to a single model based on the thermodynamics of water in porous media and the mechanisms of crack growth. It is difficult to evaluate or reconcile these competing viewpoints. Moreover, the three hypotheses outlined above have different environmental implications. According to the volumetric expansion model, rock breakdown is favoured by wet conditions, rapid cooling and frequent freeze–thaw cycles (Matsuoka, 1990). The capillary damage theory predicts that breakdown will be optimal under wet conditions with slow cooling and sustained subzero temperatures in the range -4 °C to -15 °C. The ordered water hypothesis holds that breakdown will be favoured in argillaceous rocks (or chemically-weathered rocks) under conditions of fluctuating temperature and humidity.

A further problem is that both theoretical modelling and laboratory experimentation have focused on the processes of microgelivation acting on intact rock. Little attempt has been made to investigate the more dramatic phenomenon of macrogelivation, which involves not only the prising apart of joint-bound blocks (Figure 9.3), but also their subsequent chaotic realignment (Figure 9.4). Macrogelivation *may* simply be equivalent to frost wedging of rock (i.e. expansion of water that freezes in joints and cracks), and possibly the explosive release

of strain energy induced by rapid freezing (Michaud, Dionne & Dyke, 1989), but neither the process nor its environmental significance is adequately documented. The possibility that stress release may contribute to macrogelivation has been suggested by several authors, but though this seems a strong possibility it has not been rigorously investigated.

Chemical weathering in periglacial environments

Traditionally, cold environments have been viewed as being dominated by mechanical weathering processes, with minimal chemical activity (e.g. Peltier, 1950). As Caine (1974) has pointed out, there are good reasons for considering this to be the case, as low temperatures retard chemical reactions. Slow rates of chemical activity are likely to be accentuated by sparse vegetation cover, the brevity of the period over which nutrient extraction and plant metabolism occur and slow humic decay. However, Rapp's (1960a) finding that solutional loss is the principal agent of denudation in the mountains of northern Sweden suggests a very different picture, as does the occurrence in some periglacial areas of small-scale weathering features that are often associated with chemical activity, such as tafoni, oxide rinds and carbonate coatings (Cailleux & Calkin, 1963; Czeppe, 1964; Cailleux, 1968; Washburn, 1969; Woo & Marsh, 1977).

Several studies suggest that chemical weathering is indeed significant in some cold environments. The importance of carbonation and solution of limestones in high arctic areas is evident both in the levels of dissolved solids recorded in groundwater and streams (Cogley, 1972; Smith, 1972) and in the occurrence of groundwater precipitates (Tedrow & Krug, 1982), although the effectiveness of these processes appears to be extremely dependent on levels of biogenic carbon dioxide and thus on vegetation cover (Woo & Marsh, 1977). Reynolds & Johnson (1972) have stressed the effectiveness of chemical weathering in alpine environments, emphasising the importance of water and hydrogen ion supply rather than temperature as fundamental controls, and other studies appear to support their arguments. Dixon, Thorn & Darmody (1984), for example, established that gruss and soils on a nunatak in Alaska have experienced extensive chemical alteration through solution, loss of mobile constituents and formation of secondary clay

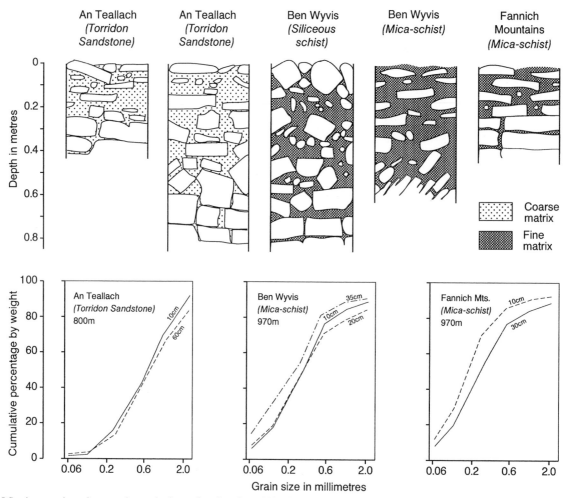

Figure 9.5 Sections cut through mountain-top detritus at five sites above 800 m in the northern Highlands of Scotland. The lower diagram depicts the granulometry of the matrix at three of these sites, showing the depths from which samples were obtained. Near-surface vertical sorting of clasts is evident in the first three sections. At the base of each section, *in situ* detritus merges into unweathered bedrock.

minerals. Although low temperatures probably ensure that such processes operate more slowly in periglacial areas than in other (humid) environments, it seems they may nonetheless play an important role in rock breakdown because of the tendency for chemically-altered rock to disintegrate more readily than sound rock when subjected to freezing and thawing (Lautridou & Ozouf, 1972; Lautridou, 1988).

Periglacial weathering in upland Britain

The importance of frost weathering on British mountains has long been recognised. Harker (1901) attributed the formation of scree on Skye to frost shattering, and the quartzite and conglomerate detritus that covers the hills of Caithness was considered by Crampton (1911) to result from 'the splitting action of frost'. A similar origin was envisaged by Marr (1916) for the blockfields and scree of Scafell Pikes in the English Lake District. Peach *et al.* (1912) described 'plateau frost debris' on mountains in northern Scotland, and noted that changes in the character of this detritus are intimately related to changes in underlying lithology. All of these authors attributed the production of frost-weathered debris to a period of periglacial conditions during downwastage of the last ice sheet.

Almost all later researchers have also adopted the view that the debris which mantles most British mountains is the product of frost weathering, though a few (e.g. Romans, Stevens & Robertson, 1966) have interpreted mountain-top detritus as frost-modified till. Whilst it is true that erratics are sometimes incorporated within mountain detritus, other considerations indicate that most debris covers result from *in situ* weathering. In particular, the composition of mountain-top detritus generally reflects that of the underlying bedrock, and tends to change abruptly at geological boundaries. Furthermore, sections cut through mountain-top detritus often exhibit an upwards transition from intact rock through loosened rock fragments to chaotic detritus (Figure 9.5), a transition irreconcilable with a glacial origin. Other authors have speculated that preglacial chemical weathering may locally have played some part in the formation of mountain-top detritus. Sugden (1971), for example, suggested that large boulders in the granite blockfields of the Cairngorms may have originated as corestones exhumed from an ancient saprolite cover. However, boulders embedded within mountain-top detritus on the Cairngorms (and elsewhere) are normally markedly angular, a feature indicative of mechanical rather than chemical weathering. The view adopted here is that mountain-top detritus on British mountains is pre-

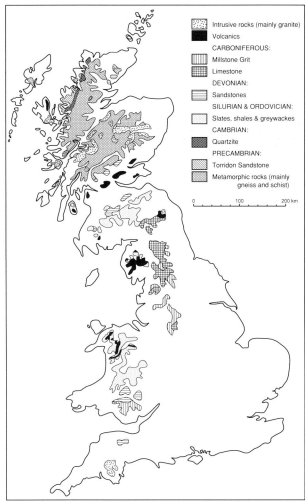

Figure 9.6 Generalised geological map, showing the range of lithologies that underlie ground above 400 m in Great Britain.

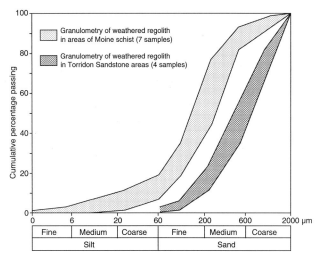

Figure 9.7 Grain-size envelopes for matrix material in mountain-top detritus developed on micaceous Moine schist (left) and Torridon Sandstone. The former is frost-susceptible, the latter is not.

dominantly the product of frost weathering, though the significance of chemical activity deserves further investigation and it is possible that some mountain-top regolith is of glacial, interglacial or preglacial rather than periglacial origin.

Response of different lithologies to frost weathering

The mountains of Britain are underlain by a great diversity of rock types. These range in age from the ancient basement rocks of Lewisian Gneiss in the Outer Hebrides and parts of NW Scotland to the Tertiary lavas and intrusive rocks on Skye, Rhum, Mull and Arran, and incorporate a rich variety of sedimentary, igneous and metamorphic types (Figure 9.6). The wide range of rock types that underlie high ground in Britain presents an opportunity for comparison of the effects of periglacial weathering on different lithologies, and for comparison of the results of laboratory experiments of rock breakdown with field evidence. Neither opportunity, unfortunately, has been fully exploited. Comparisons of the susceptibility of different lithologies to frost weathering by examination and analysis of weathered products have been limited in scope and descriptive in character, and the relationship between field and experimental data has been only tentatively explored. The following account summarises available information on the response to frost weathering of some of the principal upland lithologies, but is of necessity both qualitative and incomplete.

The Moinian and Dalradian metamorphic rocks that underlie most of the Scottish Highlands comprise mainly schists and gneisses that vary enormously in composition and habit, and consequently in response to periglacial weathering. In general, the susceptibility of such rocks to macrogelivation seems to be closely related to fissility. Slates, phyllites, mica-schists and pelitic schists often show signs of extensive fracture along cleavage and foliation planes, yielding platy clasts of variable size and often marked angularity, whereas more massive siliceous schists, gneisses and granulites tend to have fractured along joint sets to give more equidimensional clasts. Micaceous rocks also appear to have been generally more susceptible to microgelivation, as the clastic component of weathered mica-schists and related lithologies is often embedded in an abundant matrix, generally dominated by the fine sand fraction (60–200 μm) with an appreciable component of silt (Stevens & Wilson, 1970; Ballantyne, 1984; Figures 9.5 and 9.7). The more massive schists, gneisses and granulites appear to have been more resistant to microgelivation, and sometimes form blockfields. A good example of the contrast between the two types is seen on Ben Wyvis in northern Scotland. The main part of the Wyvis massif is underlain by mica-schists that form a smooth, vegetation-covered debris mantle, rich in fines. In contrast, the subsidiary summit of Carn Gorm is underlain by massive but well-jointed granulite that forms a blockfield from which fines are absent even at depths of 1.6 m or more. In ranking the susceptibility to frost action of various rocks in the Glencoe area of Scotland, Thorp (1981) rated Dalradian mica-schists as 'moderately resistant' to macrogelivation, but massive schists as 'highly resistant'.

Slates, shales and greywackes of Lower Palaeozoic age also underlie extensive upland tracts, including most of the Welsh uplands, the Lake District and the Southern Uplands of Scotland. Such rocks have proved to be amongst the most susceptible to frost weathering. Ball (1966) noted that the shales and slates of

Figure 9.8 Section cut through a Cambrian Quartzite blockfield at 960 m on An Teallach, northern Scotland.

Figure 9.9 Mountain-top detritus and summit tors on Bheinn Mheadhoin (1182 m) in the Cairngorm Mountains. Note the rounded nature of exposed boulder surfaces, a consequence of prolonged granular disintegration.

Wales possess strong cleavage that has facilitated disintegration to platy fragments and Caine (1963a) described the slates of Grassmoor in the Lake District as 'very susceptible to frost shattering'. Similar comments have been made about such rocks in Scotland and on the fringes of Dartmoor by Thorp (1981) and Waters (1964a) respectively. The composition of mountain-top detritus developed from weathering of such rocks in the Southern Uplands of Scotland has been described in detail by Tivy (1962) and Ragg & Bibby (1966). The characteristic regolith consists of small angular clasts, generally platy in form, embedded in an abundant matrix in which coarse silt and fine sand (30–250 μm) form the dominant fractions, although areas of more massive grit have yielded larger clasts and a coarser matrix. Most plateaux underlain by slates or shales support a complete vegetation cover, and bedrock outcrops are rare.

Quartzite and metamorphosed limestone are of much more limited distribution on British mountains, but have yielded some of the most spectacular frost-weathered regolith. The most conspicuous occurrences of such rocks are the Cambrian Quartzite strata that cap Torridon Sandstone mountains such as An Teallach and Liathach in NW Scotland and the similar but older Dalradian Quartzites that form the summits of some mountains in the Grampian Highlands, such as the Mamores and Schiehallion. Such fine-grained quartzites and metamorphosed limestones form extensive blockfields that completely bury the underlying bedrock. Although the well-jointed nature of such rocks has apparently made them exceptionally susceptible to macrogelivation, they have proved relatively resistant to microgelivation (Whyte, 1970). However, on some mountains blocks of shattered quartzite are embedded in a matrix of sand and silt (Galloway, 1961a), or excavation has revealed an infill of silt-rich fines some depth below the blockfield surface (Crampton & Carruthers, 1914). This suggests that the resistance of quartzite to granular disintegration has been variable.

Although exposed boulders in quartzite blockfields tend to be slightly rounded by weathering (Figure 9.8), fine-grained quartzite nevertheless appears to have proved in most cases exceptionally resistant to microgelivation, a feature attributed by Pye & Paine (1983) to its low porosity and the strength of the silica bonds in such rock. Coarser-grained quartzites, however, seem to have been more susceptible to disaggregation (Ballantyne, 1981) and it is these types that have often weathered to produce diamictons rather than blockfields.

Granite also exhibits a very varied response to weathering. Most of the granite mountains of Great Britain occur in the Scottish Highlands, where numerous large granitic batholiths are intruded into the Moinian and Dalradian metamorphic rocks. Not all of these granite areas form high ground, but those that do constitute some of the most impressive mountain scenery in Britain. They include the extensive Cairngorm and Lochnagar massifs, together with the mountains at the head of Loch Etive in the Western Grampians, the Red Hills on Skye and the craggy ridges of northern Arran. Other granite uplands occur in the Southern Uplands of Scotland (e.g. Cairnsmore of Fleet), northern England (Cheviot and the western fringes of the Lake District) and south-west England (Dartmoor and Bodmin Moor; Figure 9.6). Mountain-top detritus on granite terrain typically consists of boulders embedded in a coarse felspathic sandy matrix. Regolith of this type covers extensive tracts of the Cairngorms and Lochnagar (King, 1968; Shaw, 1977) and is found on other granite uplands as far removed as Ronas Hill on Shetland (Ball & Goodier, 1974) and Dartmoor (Waters, 1964a). The combination of abundant large clasts embedded in thick gruss suggests that the granites in such areas have proved susceptible to both macrogelivation acting along joints and to granular disintegration. The occurrence of high-level aeolian deposits derived from weathering of granites on the Cairngorms, Ronas Hill and mountains in the Western Grampians also suggests highly effective microgelivation (Ball & Goodier, 1974) as does the well-rounded nature of exposed boulder surfaces and rock outcrops (Figure 9.9).

Not all upland granites, however, exhibit identical responses to past frost weathering. This has been illustrated by Waters (1964a, 1965) with reference to Haytor Rocks on Dartmoor, where well-jointed granite has been completely disintegrated by frost weathering whilst more massive granite has remained

Table 9.1: *Comparative resistance of different lithologies to frost weathering.*

Author:	Waters, 1964a	Thorp, 1981	Ballantyne, 1981
Area	Dartmoor	Glencoe	Northern Highlands
Weathering type:	'Frost breaking'	'Frost-wedging'	'Microgelivation'
Criteria	Degree of disruption; abundance and size of weathered debris.	Size of clastic fragments.	Rounding of exposed surfaces; abundance of fines; laboratory experiments.
MOST RESISTANT ROCKS	Massive granite Quartz-schorl	Rhyolite, Andesite Massive granite Massive schists	Fine-grained Cambrian Quartzite Siliceous schists
	Medium-grained granite Quartz-porphyry Diabase	Limestones Schists Quartzites	Mica-schists
LEAST RESISTANT ROCKS	Fine-grained granite Metasediments	Porphyrites Fine-grained granite Slates	Weathered quartzite Weathered sandstone

largely intact. Similar contrasts were observed by Thorp (1981) in the Glencoe area of Scotland. On some granites, too, the effects of microgelivation are much less conspicuous. Blockfields in which fines are absent at the surface occur on some of the higher granite summits of the Cairngorms, such as Derry Cairngorm, and superb openwork blockfields have formed on the microgranite and granophyre of Sron an t-Saighdeir on the island of Rhum (Figure 9.1). Indeed, Sugden (1971) observed that on some granite blocks on the Cairngorms it is possible to recognise 'the original glacially-polished surface', which suggests that microgelivation may locally have had very little effect. Such conflicting evidence for highly effective and very limited microgelivation may be attributable, as Galloway (1961a, b) maintained, to differences in the response to frost weathering of rock that had experienced chemical alteration compared with sound rock. This interpretation certainly accords with the results of laboratory experiments (e.g. Lautridou & Ozouf, 1982). If this is the case, it implies that chemical alteration has played a more important role in the disintegration of upland granites (and possibly other rock types) than has hitherto been appreciated.

Rather similar to granite in their response to frost weathering are the upland sandstones. The most extensive of these are the Torridon Sandstones, which form a group of impressive peaks along the north-west seaboard of Scotland (Figure 9.6). The typical form of weathered detritus developed on Torridon Sandstone consists of boulders embedded in a matrix of coarse sand (Figures 9.5 and 9.7). The boulders are often 'slabby' in form, reflecting the prising apart of bedrock along near-surface bedding and pseudobedding planes (Godard, 1965). The matrix is the product of the detachment of grains and aggregates of grains from the parent rock. The susceptibility of Torridon Sandstone to granular disaggregation is evident also in the presence of extensive deposits of aeolian sand, derived originally by weathering, on most Torridon Sandstone mountains (Ballantyne & Whittington, 1987), and in the marked rounding

of exposed surfaces on clasts and rock outcrops. The characteristics of Torridon Sandstone regolith are replicated on sandstone uplands elsewhere in Britain, for example on the Devonian sandstone plateau of Ward Hill in Orkney (Goodier & Ball, 1975) and in Carboniferous sandstone areas of the Northumberland fells in northern England (Clark, 1971). On some mountains, however, sandstones and particularly conglomerates form blockfields, for example on Morven in Caithness (Crampton & Carruthers, 1914) and in mid-Wales (Potts, 1971).

Few researchers have attempted comparison of the susceptibility of different lithologies to frost weathering in the field. Moreover, studies to date have been entirely qualitative, and based on differing criteria (Table 9.1). Given the great geological diversity of British mountains, such field comparisons could be profitably extended, provided that replicable criteria can be devised to assess rock susceptibility to both macrogelivation (e.g. size and form of clastic fragments, depth of operation) and microgelivation (e.g. abundance of granular matrix, degree of rounding of exposed surfaces). Researchers working in upland Britain have tended to emphasise the importance of rock texture and structure as controls on the susceptibility of different rocks to weathering (Waters, 1964a; Godard, 1965; Ragg & Bibby, 1966; Thorp, 1981). Whilst these factors are certainly important in determining the degree and nature of rock breakdown and the abundance, size, shape and granulometry of weathered products, they have been emphasised at the expense of two other controls. The first of these is the age of weathered detritus, a topic considered in detail below; the second is the influence of chemical activity in rendering different lithologies more susceptible to microgelivation.

Past and present frost weathering on British mountains

Conflicting opinions have been expressed concerning both the age of frost-weathered debris on British mountains and the pre-

Joint depths in bedrock outside the limits of the Loch Lomond Readvance

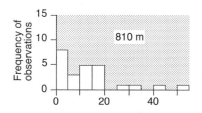
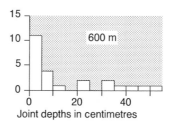
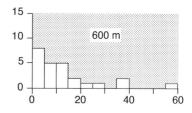

Joint depths in centimetres

Joint depths in bedrock within the limits of the Loch Lomond Readvance

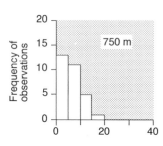
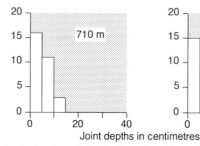
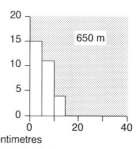
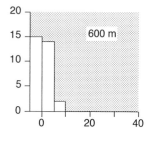

Joint depths in centimetres

Figure 9.10 Frequency histograms depicting the depths of open joints inside and outside the limits of the Loch Lomond Readvance on the Torridon Sandstone mountain of An Teallach in northern Scotland. The restricted depths of joints inside the area occupied by glacier ice during the Loch Lomond Stadial suggest limited operation of macrogelivation (frost–wedging) during the c. 10 000 years since the end of the stadial.

sent effectiveness of frost weathering. Several early commentators, such as Crampton & Carruthers (1914), interpreted mountain-top detritus as a relict deposit formed under severe periglacial conditions immediately after the last ice sheet downwasted below the level of mountain summits. Peach *et al.* (1912) noted that this interpretation explains the absence of *in situ* frost-weathered debris in high upland valleys that remained occupied by glacier ice when frost action was operating on exposed slopes and plateaux. Similarly, Galloway (1961b) believed that all coarse frost-weathered debris on Scottish mountains had formed during the Late Devensian Lateglacial, on the grounds that the present mountain climate is insufficiently severe to account for the 'disruption' he observed on high plateaux. Tivy (1962) added the telling comment that current frost weathering of shales and greywackes in the Southern Uplands of Scotland is unlikely to be effective, because both bedrock and overlying regolith are protected by an insulating cover of vegetation, peat and peaty humus, a situation common in many mountain areas. Conversely, however, some writers have described 'recent' fracturing, splitting or comminution of rocks on British mountains (e.g. FitzPatrick, 1958; Godard, 1959; Hills, 1969), which suggests that limited frost weathering may be active under present-day conditions.

The principal evidence concerning the age of frost-weathered debris on British mountains results from reconstructions of the limits of glaciers that formed in upland Britain during the Loch Lomond Stadial of *c.* 11–10 ka BP, the last period of intense cold to affect the British Isles. A marked contrast in the degree of frost weathering between adjacent areas inside and outside the limits of the Loch Lomond Readvance was first observed by Sissons (1967, pp. 223–4) on the Torridon

Sandstone rocks of the Applecross Peninsula and subsequently (Sissons, 1977a) on the metamorphic rocks of the Glen Moriston area. Sissons suggested that the absence of evidence for frost wedging of rock inside the limits of the readvance implies that macrogelivation has been insignificant since the end of the Loch Lomond Stadial. Later studies have supported this interpretation. On the island of Rhum, for example, *in situ* blockfields on Sron an t-Saighdeir (Figure 9.1) extend down to an altitude of 230 m, yet are absent from higher ground that was occupied by glaciers during the Loch Lomond Stadial (Ballantyne & Wain-Hobson, 1980). Similar relationships occur on Skye (Ballantyne, 1989a), in the Southern Uplands (Cornish, 1981) and in the Lake District (Sissons, 1980a). In the Western Grampians, Thorp (1981, 1986) identified a 'periglacial trimline' that corresponds to the upper limit of a major Loch Lomond Stadial icefield. On mountainsides above the level of this 'trimline', rocks commonly exhibit signs of marked disruption by frost weathering; below it, particularly on spurs, the rocks are ice-scoured and glacially polished. The preservation of features such as striae and friction cracks immediately downslope of the 'trimline' is a strong indication of the impotence of frost weathering (at least on some lithologies) since the disappearance of the last glaciers at around 10 ka BP.

Contrasts in the depths of open joints inside and outside the limits of Loch Lomond Stadial glaciers have also been interpreted in terms of limited operation of macrogelivation since the end of the stadial. In the Glencoe area, Thorp (1981) found that joint depths in schists do not exceed 0.15 m on rocks inside the glacier limits, yet reach 1.0 m immediately outside these limits. Similarly, at altitudes exceeding 600 m on Torridon Sandstone, open joints immediately outside the Loch Lomond

Figure 9.11 Contrasts in the roundness of exposed (right) and buried clasts from the Torridon Sandstone frost-weathered debris on An Teallach, NW Scotland.

Readvance limits commonly exceed 0.5 m in depth, but immediately within these limits achieve depths no greater than 0.17 m (Ballantyne, 1982b; Figure 9.10). Such evidence demonstrates that frost weathering was much more effective on British mountains before *c.* 10 ka BP than at any time during the Holocene. More direct evidence also indicates that macrogelivation is largely inoperative at present, except possibly as a cause of rockfall from cliffs. Rock exposed over the last century near the summits of mountains by the creation of paths displays no sign of joint enlargement or other forms of disruption, and areas of undisturbed frost-weathered detritus are often covered by moss, lichen, peat or a vegetation mat. Rounding of the exposed surfaces of bedrock outcrops and clasts indicates prolonged stability during which only superficial granular disintegration has been effective (Ballantyne, 1984). There is some evidence for recent splitting of intact rock on high ground in the form of clasts and outcrops exhibiting fresh, angular fractures (Godard, 1959; King, 1968; Hills, 1969). However, examples of recently-fractured clasts are rare (Kotarba, 1984), and hence do not contradict the general conclusion that macrogelivation has been of very limited importance since the end of the Loch Lomond Stadial.

Microgelivation also appears to have been much more effective under stadial conditions than under the milder climate of the Holocene. On many upland plateaux, particularly those underlain by fissile rocks such as mica-schists, slates and shales, the regolith mantle is protected from frost action by an insulating cover of vegetation, humus and in some areas peat. The frequent development of organic soils beneath the vegetation cover suggests that such regolith mantles are relict (Tivy, 1962; Romans *et al.*, 1966; Stevens & Wilson, 1970). The most likely explanation of the origin of the fine matrix within such vegetation-covered regolith is that it was produced by microgelivation under stadial conditions (Ragg & Bibby, 1966). That microgelivation was effective in upland areas during the Late Devensian Lateglacial has been demonstrated by Shaw (1977), who compared the weathering characteristics of granite boulders lying immediately inside the Loch Lomond Readvance limits on Lochnagar with those of boulders lying immediately outside these limits. Shaw found that, in terms of roundness,

resistance to rubbing and acoustic response, the boulders located outside the glacial limits were generally more weathered than those inside, which indicates that they had been affected by microgelivation (and other weathering processes) before the beginning of the Holocene.

Unlike macrogelivation, microgelivation has continued to operate during the Holocene. The effects of granular disintegration are apparent from the rounded appearance of exposed rock and clast surfaces, especially on granite and sandstone mountains (Figure 9.9). Such rounding contrasts strongly with the characteristic angularity of buried clasts and bedrock (Figure 9.11). Even individual boulders partly embedded in regolith are often rounded on their exposed surfaces yet sharply angular underneath. This contrast demonstrates that the granular disaggregation responsible for rounding exposed clast surfaces postdates the development of the regolith mantle itself, and therefore presumably occurred during the Holocene. It also indicates that the effects of Holocene microgelivation have been restricted to the surface, and that the regolith in which rounded clasts are embedded has been stable throughout the period of clast rounding. Not all researchers, though, have explained rounding of surface clasts and rock outcrops in terms of microgelivation. It has been proposed, for example, that the rounded granite tors and boulders of Dartmoor and the Cairngorms (Figure 9.9) are essentially rock residuals and corestones produced by preglacial or interglacial weathering under a deep saprolite cover (e.g. Linton, 1949, 1955; Sugden, 1968, 1971). Although the tors may indeed have evolved in this way (see below), analysis of the felspathic sand that surrounds such tors and supposed corestones indicates a dominantly mechanical origin, with limited chemical alteration (Eden & Green, 1971; Doornkamp, 1974; Paine, 1982). Moreover, the angular nature of most buried clasts comprising the 'clitter' (blockfields) of Dartmoor and the Cairngorms appears irreconcilable with the above hypothesis. There is, however, extensive evidence of kaolinisation of Dartmoor granite, and isotope studies by Sheppard (1977) suggest that this results from subaerial weathering rather than hydrothermal alteration.

The degree to which exposed boulders and rock outcrops on British mountains have been rounded by microgelivation is strongly conditioned by lithology, and has been shown by Ballantyne (1981) to be related to the rapidity of rock breakdown by granular disintegration in freeze–thaw experiments. In general, rounding is most pronounced on sandstones and granites, and much less so on rocks resistant to microgelivation such as fine-grained quartzites, siliceous schists and granophyre. Fissile rocks, such as phyllites, mica-schists, slates and shales, have often retained planar facets and sharp edges, as such rocks tend to break down by flaking rather than by granular disintegration.

Perhaps the most convincing evidence for the present-day operation of microgelivation on British mountains takes the form of thin accumulations of sand on the surfaces of ablating snowbanks in the lee of unvegetated plateaux. Such sand is originally derived by weathering of exposed rock surfaces upwind, and becomes incorporated in the snowpack during storms. On An Teallach, a sandstone mountain in northern Scotland, roughly six tonnes of windblown sand derived from the granular disintegration of boulder and bedrock surfaces upwind accumulated within the snowpack in a lee valley over

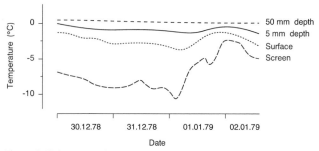

Figure 9.12 Screen, surface and subsurface temperatures at 665 m on An Teallach, northern Scotland, 30 December 1978 to 2 January 1979. The damped response of ground temperatures to a drop in air temperature to -10 °C reflects the insulating effect of a 5–10 cm thick cover of snow.

one winter (Ballantyne & Whittington, 1987). As the area of the plateau 'source' for the sand is *c.* 0.36 km², this implies a weathering yield of the order of 17 g m⁻² during that winter, roughly equivalent to an average ground surface lowering of 43 μm yr⁻¹ if it is assumed that summer losses are negligible. From analyses of the pollen stratigraphy of thick accumulations of such niveo-aeolian sands at the same site, Ballantyne & Whittington demonstrated that frost-weathered sand particles have been accumulating in this way throughout most of the Holocene. The weathering rates cited above probably represent maximal values for upland Britain, as conditions on An Teallach are particularly favourable for microgelivation. On most British mountains the current effects of this form of weathering are probably negligible.

In sum, the evidence outlined above indicates that frost weathering during the Holocene has had only a very limited effect in causing rock breakdown on British mountains. To some extent this is attributable to the protection afforded by an insulating cover of peat, humus or vegetation. This explanation, however, fails to account for the meagre evidence for Holocene frost weathering on mountains where rocks and clasts are exposed at the surface. A more general explanation for the cessation or near-cessation of disruptive frost weathering (particularly macrogelivation) at the beginning of the Holocene lies in the rapid and drastic warming that took place at that time. This proposition merits closer inspection in the light of the current controversy, outlined earlier, concerning the mechanisms of frost weathering. If intensity of frost weathering is considered to be primarily dependent on freeze–thaw cycling (the volumetric expansion hypothesis), there seems little reason to anticipate its near-cessation during the Holocene. For although under present conditions there may be only a small number of freeze–thaw cycles at the ground surface each year, moisture is certainly abundant and frost penetration (in soils) is known to reach at least 0.5 m at 900 m altitude. If frequency of freeze–thaw cycles represented the primary constraint affecting rate of rock breakdown, then similar conditions operating over the *c.* 10 000 years of the Holocene might be expected to have had much more effect on rocks exposed at high altitudes than is evident. Similarly, if ordered-water weathering (hydration) had been effective, the scale of Holocene rock breakdown again might be expected to be greater than it has been, as rock surfaces on British mountains experience repeated wetting by rain and meltwater and drying by wind in addition to freezing and thawing. If the views of Walder & Hallet (1985, 1986) are adopted, how-

ever, the meagre and superficial nature of Holocene frost weathering on British mountains becomes more readily comprehensible. Air temperatures lower than -10 °C are rare on British mountains at present, and there is no reason to suppose that markedly more severe conditions have pertained at any time during the Holocene. Since air temperatures of -10 °C or lower are most likely to occur when there is an insulating cover of snow on the upper slopes of British mountains, it seems probable that *ground surface* temperatures rarely drop more than a few degrees below freezing (Figure 9.12). Under such circumstances, rock temperatures are unlikely to enter the -4 °C to -15 °C range identified by Walder & Hallet as critical for crack growth in rocks, except possibly on cliff faces too steep to retain snowcover. The ineffectiveness of frost weathering on rock surfaces during the Holocene may thus reflect the infrequency with which ground temperatures have declined to levels at which such weathering becomes effective, regardless of the frequency of freeze–thaw cycles. Such an interpretation would appear to be supported by the restriction of rock rounding by Holocene microgelivation to the surfaces of exposed boulders and rock outcrops, which often protrude above a shallow cover of snow or ice on plateau surfaces, and are thus exposed to more extreme temperatures than the surrounding ground.

Chemical weathering in upland areas

The role of chemical weathering as an agent of rock breakdown on British mountains has received little attention. Kotarba (1984), however, noted pitting and grooving of rocks on the Western Hills of Rhum, from which he adduced that chemical weathering presently predominates over mechanical breakdown. Excavation of mountain-top detritus has sometimes revealed chemically-weathered clasts in near-surface horizons (e.g. Tivy, 1962), and King (1968) reported pockets of chemically-weathered granite high on the Cairngorms. Such observations are, however, difficult to evaluate in terms of the overall significance of chemical weathering on British uplands, and in this respect two other sources of data are of greater importance: studies of mineral weathering in mountain soils and measurements of the concentrations of dissolved material in streams draining upland catchments.

Studies of the effects of chemical weathering in upland regolith on the Southern Uplands of Scotland suggest that it has been of limited importance. Tivy (1962) found that despite the presence of weathered rock fragments in A₂ horizons on Lowther Hill, underlying C horizons display a lack of evidence for chemical alteration. Similarly, the fine fraction of soils above 600 m on Broad Law is characterised by low values of exchangeable base content, base saturation and pH in mineral layers. From this, Ragg & Bibby (1966) concluded that very little chemical weathering of minerals had taken place. Low values of exchangeable bases also proved characteristic of mountain soils from NE Scotland analysed by Romans *et al.* (1966), who interpreted an upslope convergence of surface and subsoil pH values in terms of decreasing rates of biological and chemical activity with increasing altitude. Perhaps the most telling evidence for limited chemical activity is that mountain soils in most glaciated upland areas contain abundant fresh biotite, amphibole and feldspar, minerals usually considered susceptible to chemical weathering (Wilson, 1985).

Figure 9.13 Cambrian Quartzite blockslope, Sàil Liath, An Teallach, NW Scotland. The more rounded boulders are erratics of Torridon Sandstone, believed to have been emplaced before formation of the quartzite block deposit.

A number of studies have focused on the clay mineral content of mountain soils. Pockets of mountain regolith have been found to contain appreciable quantities of gibbsite, kaolinite and halloysite, clay minerals traditionally associated with an advanced stage of weathering. The presence of such minerals, however, has been attributed to the survival of preglacial or interglacial pedorelicts either under a protective cover of cold-based glacier ice (Wilson & Bown, 1976; Hall & Mellor, 1988; Mellor & Wilson, 1989) or on nunataks that remained above the level of the last ice sheet (Stevens & Wilson, 1970; Reed, 1988). A contrary view has, however, been expressed by Hall (1983), who has interpreted the presence of gibbsite in granite regolith on the Cairngorms in terms of an early stage of chemical weathering at free-draining snowbed sites. Similarly, Green & Eden (1971) interpreted gibbsite in the granite gruss of Dartmoor as an initial product of the weathering process, rather than a relict of an earlier weathering phase. More typically, however, clay mineral development in mountain soils is restricted to vermiculitisation of micaceous minerals in A horizons under acid conditions (Wilson, 1985).

Use of concentrations of dissolved solids in streams draining upland basins as a measure of chemical weathering activity on mountains poses two problems. First, most measurements are made near the basin outlet, and are therefore unrepresentative of concentrations of dissolved solids on high ground. Secondly, studies in the Lake District (White, Starkey & Saunders, 1971) and NE Scotland (Reid, MacLeod & Cresser, 1981) have shown that a large component of the chemical composition of streams draining upland areas reflects atmospheric inputs rather than chemical weathering. For Ardessie Burn, which drains An Teallach in NW Scotland, Ballantyne (1981) calculated that the mean concentration of dissolved solids is c. 20 mg l^{-1}, equivalent to a yield of c. 50 t km^{-1} yr^{-1}. However, this figure includes both atmospheric inputs as well as dissolved salts resulting from chemical weathering, and as most of the basin lies below 600 m, reveals nothing of the yield from high-level regolith. Measurements of the conductivity of springwaters high on the same mountain indicated average concentrations of dissolved solids of less than c. 12 mg l^{-1} (Ballantyne, 1985), and, as such values include atmospheric inputs, suggest that chemical

weathering is of very restricted significance. A more detailed study of chemical weathering rates within an upland catchment in NE Scotland (Reid et al., 1981) revealed large outputs of both silicon (38.5 kg ha^{-1} yr^{-1}) and bicarbonate (48 kg ha^{-1} yr^{-1}) that do not reflect atmospheric inputs and hence indicate considerable chemical weathering. The same study showed that breakdown of plagioclase feldspar accounts for about 75% of the output due to weathering. However, as this research was conducted in a basin mantled by glacial drift and with a restricted altitudinal range (236–778 m), the implications of these results for chemical weathering under present periglacial conditions on high ground are uncertain.

In sum, assessment of the past and present role of chemical weathering on British mountains is hampered by a paucity of relevant data, and such evidence as is available is somewhat conflicting. It seems reasonable to conclude that chemical weathering has been of secondary importance in the *direct* breakdown of rocks, but of greater importance in upland pedogenesis. Its role in promoting more effective microgelivation (*cf.* Lautridou & Ozouf, 1982) is unknown. Such conclusions must remain provisional, however, pending research on chemical alteration of rock surfaces in mountain areas.

Mountain-top detritus

Mountain-top detritus is subdivisible into two main categories. The first is that of block deposits, which are characterised by a surface cover of boulders with little or no interstitial fine material. Such deposits are commonly referred to as *blockfields* (or *felsenmeer*) where they occupy broad expanses of level or gently-sloping terrain (Figure 9.1), *blockslopes* where they veneer moderate or steep slopes (Figure 9.13), and *blockstreams* where they occupy broad valleys or gullies (Caine, 1968; White, 1976b). The second category consists of diamictons in which clasts are embedded in a matrix of fines. Although such diamictons are far more common on British mountains than true block deposits, there is no generally-accepted term to describe them. In this account *debris-mantled surface* and *debris-mantled slope* are employed to refer respectively to plateaux and slopes veneered by regolith that comprises clasts embedded in a matrix of fines.

Blockfields and blockslopes

Early hypotheses of blockfield formation invoked various non-periglacial origins (Caine, 1968; Strömqvist, 1973), but it is now widely accepted that most mountain-top blockfields in areas that have experienced periglacial conditions are primarily the result of frost action. Two categories of blockfield have been recognised: *autochthonous blockfields*, which consist largely or entirely of the products of *in situ* weathering of bedrock, and *allochthonous blockfields*, in which the boulders have some other origin, such as glacial deposits or corestones eroded from saprolite covers. Widely-reported characteristics of block deposits include the presence of fines below the surface and a tendency for coarse debris to decrease in size with depth (Klatka, 1961, 1962; Caine, 1968; Potter & Moss, 1968; Clapperton, 1975; Iwata, 1983). The absence of surface fines may simply reflect eluviation (Svensson, 1967), but a downwards-fining sequence appears to imply vertical frost sorting

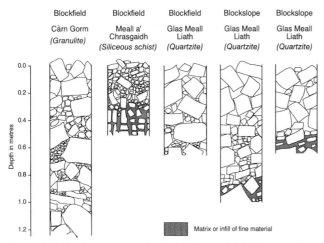

Figure 9.14 Sections cut through blockfields and blockslopes on three mountains in the northern Highlands of Scotland.

Figure 9.15 Section cut through a Cambrian Quartzite blockslope in northern Scotland. Note the slight edge-rounding of exposed clasts, the downward-fining within the deposit and the infill of silty fines at the base.

(Dahl, 1966; Strömqvist, 1973). The size and shape of the largest boulders in autochthonous blockfields usually reflect the joint density of the underlying bedrock (Söderman, 1980). The thickness of such block deposits is variable. On Scandinavian mountains, Strömqvist (1973) found a fairly uniform depth of 0.7 m, and Dahl (1966) reported an average depth of 1.0 m (maximum 1.7 m). Clapperton (1975), however, described block deposits up to 3 m deep in the Falkland Islands.

Andersson (1906) ascribed the formation of the block deposits of the Falkland Islands to macrogelivation of the underlying quartzite, solifluction of the weathered mass and immobilisation as a result of washing out of fines. This three-fold sequence has been repeatedly invoked to account for blockfield formation, and a fourth process, vertical frost sorting, has been added to account for downward-fining of clasts (e.g. Strömqvist, 1973). Variations on this model have concerned mainly the origin of bouldery debris and the nature of downslope movement. Caine (1968), for example, maintained that the boulders in Tasmanian blockfields may have been derived from saprolite, and others have described blockfields formed from frost-sorted till (Richmond, 1962; Dahl, 1966; Gangloff, 1983). Microgelivation is generally considered to have produced the fines within blockfields, though Klatka (1961, 1962) considered that Carpathian block deposits also contain fines derived from till and loess. Gelifluction within such fines is widely held responsible for *en masse* movement of block deposits, though frost creep may also have caused downslope displacement (Richmond, 1962, Rudberg, 1964; Dahl, 1966). There is general agreement that wash, possibly aided by piping (Smith, 1968) and 'subsurface mudflows' (Strömqvist, 1973) has been effective in removing fine material from block deposits, and thus contributing to their ultimate immobilization (Klatka, 1961; Dahl, 1966; White, 1976b). Strömqvist (1973) considered that the blockfields in Scandinavia were formed under permafrost conditions, and that blockfield depth on level ground reflects that of the former active layer. This attractive idea is supported by the presence of relict ice-wedge polygons (Svensson, 1967) and large-scale sorted patterned ground (Dahl, 1966) on blockfield surfaces, but its validity remains untested.

Although mountain-top detritus is widespread on the higher parts of British mountains, true blockfields and blockslopes are relatively rare, being confined to rocks that have proved more susceptible to macrogelivation than microgelivation. In central Wales, for example, Potts (1971) found that the distribution of block deposits is limited to areas underlain by grits, conglomerates and igneous rocks. In Scotland, blockfields and blockslopes are best developed on quartzite, microgranite, felsite, porphyrite, siliceous schists, metamorphosed limestones, granulite and (in some areas) granite (Kelletat, 1970a,b; Whyte, 1970; Ryder & McCann, 1971). Block deposits on British mountains have been widely interpreted as autochthonous products of periglacial weathering. Tufnell (1969) and Ball & Goodier (1970) proposed that some blockfields in northern England and Snowdonia respectively may be derived from till, but supplied no evidence in favour of this interpretation.

The structural characteristics of blockfields have been investigated by Ballantyne (1981), who excavated block deposits on three mountains in the Northern Highlands of Scotland: Càrn

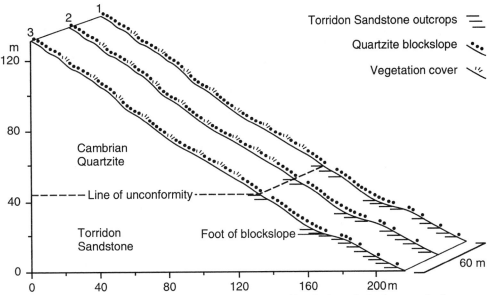

Figure 9.16 Downslope extension of Cambrian Quartzite blockslope deposits over Torridon Sandstone bedrock as a result of mass movement, Glas Mheall Liath, An Teallach, NW Scotland.

Gorm (Ben Wyvis), Glas Meall Liath (An Teallach) and Mheall a'Chrasgaidh (Fannich Mountains). He also excavated sections in a steep (31°) quartzite blockslope on the flanks of Glas Mheall Liath. The depth of the excavated block deposits proved very variable. Bedrock outcrops form 'islands' amid a sea of boulders on Càrn Gorm, yet a short distance from such outcrops Ballantyne excavated 1.6 m of boulders without reaching either bedrock or an infill of fine material. In contrast, the blockfield on Mheall a'Chrasgaidh barely reaches 0.4 m in depth, and has an infill of silt and sand in its lower parts. Both blockfields, however, show some downwards fining near the top of excavated sections (Figure 9.14), and a similar trend was evident in most pits excavated on Glas Mheall Liath (Figures 9.8 and 9.15). On Càrn Gorm, mean clast diameter declines from 198 mm at 0.1 m depth to 82 mm at 0.4 m depth and 71 mm at 0.8 m depth, but as bedrock is approached, large boulders (including some more or less *in situ* above intact bedrock) tend again to be dominant. The quartzite blockslopes of Glas Mheall Liath exhibit less marked vertical sorting than the blockfields, and all excavated sites contained an infill or matrix of fines at variable depth (Figures 9.14 and 9.15). These fines contain 13–17% silt (2–60 µm), and are therefore probably frost-susceptible. Another contrast between the blockslopes and the blockfields is that whereas clasts at the surface of the latter display no preferred orientation, a strong preferred downslope orientation of clasts is evident on most parts of the blockslope, indicative of former mass movement (cf. Caine, 1968; Paine, 1982). Downslope movement of this blockslope is also evident from the presence of boulder lobes on its surface. Even more impressive evidence for movement is found at the foot of the blockslope, which extends downslope for a distance of 40 m across Torridon Sandstone strata, burying the sandstone completely under a cover of quartzite boulders 0.4–0.5 m deep (Figure 9.16).

The structure of the excavated block deposits suggests that they are the product of macrogelivation of sound but well-jointed bedrock. In particular, the tetragonal regularity of many boulders and the presence of *in situ* joint-bound blocks at the base of some sections indicates the operation of macrogelivation along joint sets. Ballantyne interpreted the downwards variations in clast size in terms of repeated heaving and resettling of the regolith as a result of cyclic freezing and thawing. He argued that during such vertical movement the framework of larger boulders would have acted as a sieve, so that the smaller stones occupying the voids between boulders diminish in size downwards. Movement of blockslope deposits may have resulted from gelifluction within frost-susceptible fines at the base of such deposits, but as such fines fill the voids between tightly-wedged boulders this seems unlikely. Ballantyne favoured frost creep as the mechanism of blockslope movement, envisaging repeated heaving and downslope resettling of the regolith due to deep annual freezing and thawing of frost-susceptible basal fines in the manner suggested by Rudberg (1964). Such repeated heaving and resettling is consistent not only with downwards fining of clasts within block deposits, but also with the development of large-scale relict sorted patterned ground on block deposits where interstitial fines occur at shallow depth (Chapter 10). If valid, this interpretation suggests that blockslope movement may have occurred within the active layer above a former permafrost table. However, this does not necessarily imply that blockfield or blockslope depth is equivalent to the former depth of the active layer, as suggested by Strömqvist (1973), because creep of debris is likely to have resulted in the infill of depressions, giving rise to the observed marked variation in blockfield and blockslope depths. Only where the pre-existing surface was perfectly regular is Strömqvist's hypothesis likely to hold true.

There is no evidence to suggest that blockfields are currently forming on British mountains. Exposed boulders are edge-rounded as a result of granular disaggregation, a characteristic of prolonged stability, and many block deposits are covered by moss, lichen, thin soils or peat. Moreover, *in situ* periglacial

Figure 9.17 Section through vertically-sorted frost-weathered regolith developed on massive schists, Ben Wyvis, northern Scotland.

block deposits occur only outside the limits of the glaciers that formed in upland Britain during the Loch Lomond Stadial of *c.* 11–10 ka BP, which implies that neither blockfields nor blockslopes have developed during the Holocene.

Debris-mantled slopes and plateaux

Periglacial weathering mantles comprising clasts embedded in a matrix of fine material have excited much less interest than blockfields and blockslopes. Such neglect is unfortunate, because in many present or former periglacial environments the former are much more widespread than the latter. The importance of frost-weathered diamictons was, however, recognised by Lundqvist (1962, p. 64) who coined the term 'nonsorted field' to describe 'those widespread areas where...frost action asserts itself merely in an incomplete concentration of stones towards the surface'. Where the concentration of clasts at the surface is more complete, stone pavements (or *dallages*) may occur (Bout & Godard, 1973). These differ from blockfields in that clasts at the surface are firmly embedded in underlying fines. Similar debris mantles occur on slopes (e.g. Jahn, 1975; McArthur, 1975; Sukhodrovskiy, 1975; Church, Stock & Ryder, 1979), and have been described by French (1976, p. 152) as 'characterised by a relatively smooth profile with no abrupt breaks of slope, and a continuous or near-continuous veneer of frost-shattered or solifluction debris'.

Throughout upland Britain, debris mantles comprising matrix-supported clasts are widespread on high ground, particularly on mountains underlain by lithologies that are susceptible to granular disaggregation, such as shales, greywackes, mica-schists, sandstones and most granitic rocks. The distribution of debris-mantled slopes and plateaux is thus complementary to that of blockslopes and blockfields in terms of lithology. Though some debris mantles on high ground are undoubtedly of glacigenic origin, the structure of such regolith generally indicates formation as a result of weathering, sorting and mass movement under periglacial conditions. A thorough account of the characteristics of such debris was provided by Ragg & Bibby (1966), who excavated five pits on the higher parts of Broad Law (840 m) in the Southern Uplands of Scotland. All profiles above 600 m revealed a rubble layer of small angular

fragments of greywacke overlying a fine sand layer in which angular stones increased in size and frequency with depth. Ragg & Bibby interpreted this deposit as the product of frost weathering and vertical frost sorting associated with annual freezing and thawing of the active layer above permafrost. They attributed formation of the 'rubble layer' to eluviation of fines, an interpretation suggested by a coating of sand on the upper surfaces of stones, and interpreted a preferred downslope orientation of clasts in the upper parts of such deposits in terms of solifluction. Fairly similar structural characteristics were observed by Ballantyne (1981) in pits excavated on plateaux in the northern Highlands of Scotland (Figure 9.5). On An Teallach a surface concentration of edge-rounded sandstone slabs (which locally form splendid stone pavements) immediately overlies a sand-rich zone containing fewer and smaller stones. This in turn grades downwards into a zone dominated by angular boulders that increase in size with depth, and in places occur more or less *in situ* above unweathered bedrock. Pits excavated in regolith overlying fairly massive schists on nearby mountains also revealed a surface concentration of edge-rounded boulders overlying a matrix-dominant zone (Figure 9.17), but such vertical sorting was not evident in pits excavated in regolith derived from mica-schists. A more fundamental difference between the regolith derived from sandstone and that overlying schist plateaux lies in the granulometry of the matrix, which is strongly influenced by parent lithology. The matrix of regolith derived from sandstone is dominated by the medium and coarse sand fractions (0.2–2.0 mm), and contains negligible silt, but the matrix of regolith developed on schists comprises mainly fine and medium sand (0.06–0.6 mm), with an appreciable quantity (4–15%) of grains finer than 0.06 mm (Figure 9.7). As will be demonstrated at the end of this chapter, such differences in granulometry play a vital role in constraining the periglacial processes that have operated on these two contrasting types of regolith, and hence on the range of periglacial landforms that they support.

In all the excavations depicted in Figure 9.5, weathered regolith grades downwards into intact bedrock, and hence is clearly *in situ*. Like Ragg & Bibby (1966), Ballantyne interpreted this regolith as autochthonous frost-weathered debris, attributing the coarse fraction to macrogelivation, the fines to contemporaneous microgelivation, and the concentration of boulders at the surface to vertical frost sorting. Such plateau-surface debris mantles are clearly relict features relating to a past episode or episodes of much more extreme climate. As with block deposits, exposed clast surfaces are invariably edge-rounded, a sure indicator of prolonged stability. Most debris-mantled surfaces, apart from some developed on coarse-grained rocks such as granite or sandstone, tend to support a complete vegetation cover. Moreover, mature soils (alpine podzols and rankers) have developed within mica-schist debris mantles; such soils indicate regolith stability over several centuries or millennia. Like blockfields, debris-mantled surfaces in the northern Highlands are absent from plateaux that were actively glaciated during the Loch Lomond Stadial, which indicates that their formation antedates the Holocene.

Excavations in debris-mantled slopes have revealed structures similar to those of plateau regolith, namely a surface concentration of coarse debris, an underlying matrix-rich layer and a basal layer of angular clasts, some of which are *in situ* above

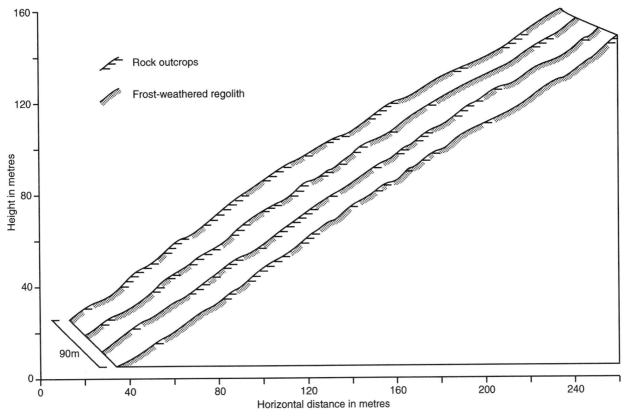

Figure 9.18 Profiles surveyed down a debris-mantled slope on resistant Torridon Sandstone. The sandstone beds have no apparent dip relative to the slope. An Teallach, northern Scotland.

intact bedrock (Tivy, 1957; Ragg & Bibby, 1966). The regolith texture of such slopes in Scotland has been analysed by Innes (1986), who found that most debris mantles are very poorly sorted and strongly fine-skewed, and that though particle size distributions are strongly influenced by parent lithology, there is also considerable variation between samples collected at different points across a single slope. The form of steep debris-mantled slopes also varies. On some mountains the debris mantle completely blankets the slope, and bedrock is absent at the surface; on others the debris cover is interrupted by step-like rock outcrops, with debris infilling the intervening treads. Such differences reflect the depth of regolith cover and thus the resistance of the underlying rock. On the schists of Ben Wyvis, for example, frost-weathered regolith forms a smooth, uninterrupted cover on slopes of up to 36°. In contrast, debris-mantled slopes of similar gradient on Torridon Sandstone mountains are interrupted by frequent outcrops of intact rock, and hence have a much less regular profile (Figure 9.18).

Unlike talus slopes (Chapter 12), debris-mantled slopes display no consistent downslope trend in terms of the size of surface clasts. Surface and near-surface clasts, however, generally exhibit a strong preferred downslope orientation indicative of mass movement (Ragg & Bibby, 1966; Ballantyne, 1981; Paine, 1982). Though soil development and the rounding of exposed clasts indicate that hillslope debris covers, like those on plateaux, are relict features formed under much more severe conditions in the past, the upper layers of such regolith are subject to downslope movement under present conditions. The higher parts of vegetated debris-mantled slopes often support

active solifluction landforms, though solifluction now affects only the uppermost few decimetres of regolith (Chapter 11). Unvegetated debris-mantled slopes frequently show signs of disruption by frost heave, and some support active sorted stripes (Chapter 10). Rates of surface clast movement on such striped surfaces are remarkable, and apparently reflect rapid downslope displacement of surface stones by the growth and collapse of needle ice. Caine (1963a) reported median rates of 70–254 mm over one winter on slopes of 9–19° in the Lake District, and marker clasts monitored by Ballantyne (1987a) over three years moved over 1 m downslope on a 23° slope on Tinto Hill in southern Scotland. On steep (>30°) unvegetated debris slopes on An Teallach, however, the same author recorded only modest rates of surface stone creep, with a median value of 11 mm yr[-1], and noted that measurable displacement (>1 mm yr[-1]) could not be detected at depths greater than about 10 cm (Ballantyne, 1981). Reported depths of diamictic frost-weathered debris mantles are variable. Romans et al. (1966) described maximum depths of 0.5 m on plateaux in NE Scotland, and Ballantyne (1981) found no evidence for plateau debris deeper than 0.8 m on mountains in the northern Highlands. Depths of 0.6 m (Caine, 1963a), 0.6–1.0 m (Tivy, 1962) and 1.0 m (Galloway, 1958) have been reported on slopes of up to 20° in the Lake District, Lowther Hills and Southern Uplands respectively. The sections examined by Ragg & Bibby (1966) on the slopes of Broad Law, however, ranged from 1.5 m to 3.5 m in depth, and they and others (Tivy, 1962; Watson, 1969a) have observed that debris mantles tend to thicken downslope, presumably as a result of mass movement.

Table 9.2: *Locations and lithologies of British tors.*

Area	Lithology	Sources
1. England		
Isles of Scilly	Granite	Scourse (1987)
Dartmoor (SW England)	Granite	Linton (1955)
		Palmer & Neilson (1962)
		Gerrard (1974, 1978, 1988a)
Exmoor (SW England)	Sandstone	Mottershead (1967)
Weald (SE England)	Sandstone	Robinson & Williams (1976)
Charnwood Forest (Midlands)	Granite, microdiorite and hornstone	Ford (1967)
Tabular Hills (Yorkshire)	Silicified grits	Palmer (1956)
Derbyshire	Dolomite	Ford (1969)
Pennines	Gritstone	Palmer & Radley (1961)
		Linton (1964)
		Cunningham (1964, 1965)
Stiperstones (Shropshire)	Quartzite	Goudie & Piggott (1981)
Cheviot Hills	Granite	Douglas & Harrison (1985)
2. Wales		
Central Wales	Igneous rocks, grits	Potts (1971)
Pembrokeshire	Flinty thyolite	Linton (1955)
Prescelly Hills	Dolerite	Linton (1955)
3. Scotland		
Cairngorm Mountains	Granite	Linton (1949, 1955)
		King (1968)
NE Scotland	Granite	Linton (1955)
Ochil Hills	Andesite	Linton (1955)
Ben Loyal (Sutherland)	Syenite	Linton (1955)
Caithness	Sandstones and grits	Linton (1955)
Trotternish, Skye	Basalt	Ballantyne (1990, 1991a)

Nevertheless, the restricted depths of *plateau* debris mantles make it tempting to relate the thickness of these to former active layer depths. If this interpretation is valid, it implies former active layer depths not exceeding 0.8 m on plateaus at 800–1000 m in northern Scotland, and of 1.0 m or less at altitudes of 700–800 m in the Southern Uplands.

Upland tors

A *tor* may be defined as a residual mass of bare bedrock that rises conspicuously above its surroundings, is isolated by free-faces on all sides, and owes its formation to differential weathering and mass wasting (cf. Pullan, 1959; Caine, 1967; Selby, 1972). Tors occur on high ground in many parts of Britain, particularly on massive, resistant rocks. Many of the most spectacular examples occur on granite massifs, particularly those of Dartmoor in SW England and the Cairngorm Mountains in NE Scotland, but the range of lithologies that support tors is considerable (Table 9.2). Reviewing the distribution of tors in Britain, Linton (1955) noted that they tend to occur on massive siliceous rocks, but not on argillaceous rocks. In fact tor-like rock residuals occur in mica-schist on many Scottish mountains, and indeed schistose tors have been widely reported outside Britain (e.g. Dahl, 1966; Martini, 1969; Wood, 1969). Tors are, however, absent on weak sedimentary rocks such as shales (Potts, 1971). According to Gerrard (1988a), tors range in height from less than 3 m to over

50 m. Those in upland Britain, however, occupy only the lower part of this range. The largest of the Cairngorms tors, the Great Barn of Bynack, measures 51 m by 48 m around the base and is approximately 15 m high (King, 1968; Figure 9.19). A similar height is reached by the tallest Dartmoor tors, Vixen Tor (16.5 m) and Hay Tor (15.2 m). The highest tor of the Stiperstones in Shropshire, the Devil's Chair, rises 12–20 m above the surrounding surface (Goudie & Piggott, 1981; Figure 9.20).

Tors occupy a number of topographic positions. Often the most prominent are summit tors, which crown the highest ground. Outstanding examples are Hay Tor, Hound Tor and Great Staple Tor on Dartmoor, the magnificent tors of Ben Avon and Beinn Mheadhoin in the Cairngorms (Figure 9.9), the Stiperstones in Shropshire (Figure 9.20) and the Eagle Stone in the southern Pennines. On granite domes, further impressive tors rise above spur ends and the break of slope between plateau surfaces and valley sides. Amongst the best examples are Vixen Tor above the River Walkham and Black Tors in the Meldon Valley, both on Dartmoor, and the Great Barn of Bynack (Figure 9.19) in the Cairngorms. Rather similar in location are scarp-edge tors, such as Brimham Rocks in the northern Pennines and the Bridestones in NE Yorkshire, which are detached from the margins of resistant caprocks that fringe upland plateaus (Palmer, 1956; Palmer & Radley, 1961). Tors also stand out from some valley sides, often marking the outcrop of more resistant rock, but these are generally less impres-

Figure 9.19 The Great Barn of Bynack, a 15 m high granite tor located at an altitude of 1070 m on the edge of a summit plateau in the Cairngorm Mountains.

Figure 9.20 Quartzite tors at the Stiperstones, Shropshire.

sive. Valley-side tors on Dartmoor, for example, do not exceed 7 m in height (Gerrard, 1974). Valley-floor and lowland tors are uncommon in Great Britain, though Palmer & Radley (1961) cite two examples: Plumpton Rocks, south-east of Harrogate, which result from dissection of a small cliff, and the Hemlock Stone of Triassic Sandstone west of Nottingham, standing isolated above the River Trent. Further examples occur on Exmoor and on the fringes of Dartmoor, and on granite outcrops in the Isles of Scilly (Scourse, 1987).

Tors are certainly not exclusive to periglacial environments, and several theories have been proposed to account for their formation. L.C. King (1948, 1958), for example, envisaged summit tors ('skyline tors') as resistant residuals produced by scarp retreat, and scarp-edge tors ('subskyline tors') as products of differential scarp retreat above a developing pediplain. Others, such as Czudek (1964), Demek (1968, 1969), and Martini (1969) have argued that some tors represent rock residuals produced by scarp retreat at the heads of cryoplanation terraces. Neither theory has found favour as an explanation for British tors. Pediments are absent around most tors, and though cryoplanation benches have been identified on Dartmoor (Te Punga, 1956; Waters, 1962) these are rarely associated with tors. Instead, explanation of tor development on British uplands became polarised around two rival hypotheses, the proponents of which periodically engaged in a singularly dogmatic and occasionally acrimonious exchange of viewpoints.

The first hypothesis, the two-stage hypothesis, was developed at length in a classic paper by Linton (1955). Writing mainly of the Dartmoor tors, Linton argued that these are 'the result of a two-stage process, the earlier stage being a period of extensive sub-surface rock rotting whose pattern is controlled by structural considerations, and the later being a period of exhumation by removal of fine-grained products of rock decay' (Linton, 1955, p. 472; Figure 9.21). Linton interpreted the rounded 'woolsack' form of many Dartmoor tors as evidence of their origins as corestones within a deep saprolite mantle, remnants of which he believed to form the widespread sandy gruss or 'growan' deposits of Dartmoor. He inferred that the initial prolonged period of deep weathering occurred under a warm, possibly subtropical climate, and that stripping of the saprolite was accomplished by solifluction and meltwater during the 'last glacial episode'. This interpretation was accepted by several authors as an explanation for the Dartmoor tors (Orme, 1964; Waters, 1964a, 1965, 1971) and those of the Cairngorms (King, 1968). It was also warmly embraced by geomorphologists studying tors in Central Europe (Jahn, 1962, 1974; Demek, 1964a,b), North America (Cunningham, 1969) and Australia (Caine, 1967). Within Britain, however, the two-stage hypothesis met fierce opposition. In a series of articles, Palmer (1967; Palmer & Radley, 1961; Palmer & Neilson, 1962) argued that the tors of upland Britain are the products of a single 'cycle' of periglacial denudation involving the action of macrogelivation on rocks of variable resistance and the removal of weathered products by solifluction. Significantly, Palmer's ideas were stimulated by his research in the Pennines, where the majority of tors are scarp-edge features associated with caprocks of resistant millstone grit. At such sites, tors are associated with landslides, blockfields and solifluction debris; evidence for deep-weathered rock is usually absent, and a 'periglacial' origin of the sort Palmer envisaged seems entirely feasible. Summit tors such as Eagle Rock pose a greater problem; these he considered, somewhat vaguely, to represent 'erosional relics' of minor gritstone beds.

Both sides then carried the debate into the enemy camp: Palmer & Neilson (1962) offered a radically different interpretation of the Dartmoor tors to that proposed by Linton, and the latter responded by offering 'an essay in analysis' on the origin of the Pennine tors (Linton, 1964). Palmer & Neilson (1962, p.336) claimed that Dartmoor 'preserves the finest set of periglacially-formed tors in Britain', arguing that the growan deposits, being deficient in clay, represent mechanically rather than chemically weathered bedrock, that tors do not rise out of a saprolite cover, and that the spreads of angular boulders or 'clitter' surrounding many Dartmoor tors are the product of frost weathering. They proposed an evolutionary model that invoked removal of regolith from upper slopes by solifluction and tor formation through differential frost weathering of the rock thereby exposed at the surface (Figure 9.21). Linton's response was to attempt to demonstrate the equivocal nature of the evidence favouring an exclusively periglacial origin for the gritstone tors in the Pennines. He also furnished examples of weathered bedrock in the Pennine area, evidence, he claimed, of a formerly more extensive saprolite cover. Linton concluded that the scarp-edge tors of the Pennines may be adequately explained in terms of deep weathering, subsequent exhumation and possibly valley-side cambering. Although he acknowl-

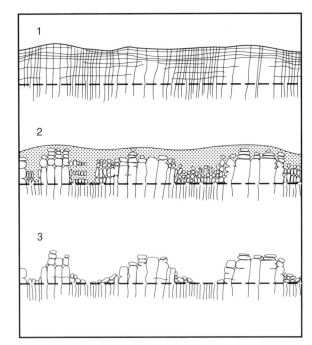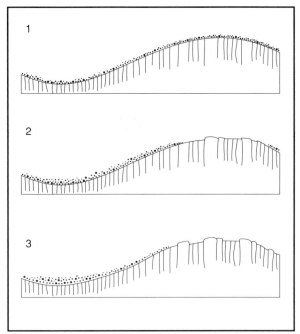

Figure 9.21 Two models of tor genesis on British uplands. Left: the two-stage model proposed by Linton (1955), which invokes preferential deep weathering of densely-jointed bedrock and subsequent stripping of the resultant saprolite cover under periglacial conditions. Right: the single-stage model of Palmer & Neilson (1962), which involves removal of regolith from summits by solifluction and differential frost weathering of bedrock thus exposed.

edged the association of the Pennine tors with blockfields, land-slips and solifluction deposits, he argued (somewhat perversely) that 'there is no reason...to suppose any connection between these processes and tor formation' (Linton, 1964, p.12).

In one respect proponents of both theories were in accord, namely the importance of joint spacing in determining the survival of rock residuals. Linton (1955) believed that corestones survived to form tors only in zones of low joint density that consequently remained immune to subsurface chemical weathering; Palmer & Neilson (1962) believed that zones of low joint density favoured tor formation because of their resistance to frost wedging (Figure 9.21). Low joint density has also been widely invoked as the fundamental reason for tor survival in many different environments, irrespective of lithology and proposed mode of tor evolution (Jahn, 1962, 1974; Demek, 1964a; Caine, 1967; Cunningham, 1969; Martini, 1969; Wood, 1969; Selby, 1972). This aspect of tor development has been investigated by Gerrard (1974, 1978) and Ehlen (1992) with reference to the Dartmoor tors. Gerrard found that summit tors occur only where relative relief is great, and postulated that incision by rejuvenated streams had resulted in release of compressive (lateral) stresses, thus permitting the opening of vertical joint sets and the penetration of weathering agencies. Release of compressive stresses may also explain the apparently anomalously *high* density of joint spacing in some Dartmoor tors, and may account for the fact that the location of tors on the Cairngorms shows little relationship to joint density (King, 1968).

In part, the debate concerning the origin of tors in upland Britain reflects the nature of tors themselves: being residual landforms, they may represent the end products of several possible evolutionary pathways (*cf.* Gerrard, 1984), and the origins of tors in a particular area must be inferred from their morphology, their distribution and the characteristics of associated landforms and deposits. Given that the Pennine tors tend to differ in

all of the aforementioned respects from those of Dartmoor, it may appear that the search for a single unifying model of tor evolution is fruitless, and that it is more realistic to evaluate the evidence for each area independently. For example, in their study of the quartzite Stiperstones in Shropshire, Goudie & Piggott (1981) pointed to the absence of 'corestones' or other evidence for deep weathering in this area, and to the angular nature of clasts in the sorted stripes that surround these tors (Figure 9.20). From this evidence they inferred that the Stiperstones tors were produced by frost weathering and mass movement under periglacial conditions, probably during the Devensian. Similarly, the association of tors with dated solifluction deposits in the Scilly Isles led Scourse (1987) to conclude that these tors were formed under periglacial conditions during the Middle and Late Devensian.

Conversely, however, the notion of a 'two-cycle' origin has continued to prove attractive as an explanation of the granite tors of Dartmoor and the Cairngorms (King, 1968; Waters, 1971; Gerrard, 1974, 1978), both sites where the appearance and location of summit and spur tors are consistent with a 'corestone' origin. Such tors are often massive and monolithic, and unlike frost-shattered bedrock outcrops on other lithologies (such as quartzite or schist) are not skirted with miniature taluses of derived frost-weathered debris. Often they display not the slightest evidence for reduction by frost wedging (Figures 9.9 and 9.19), although microgelivation has probably contributed to rounding of exposed surfaces. It now seems unlikely, though, that the initial phase of deep weathering of granite took place under a humid subtropical climate. The present rather thin covering of growan or sandy gruss on Dartmoor bears little chemical or textural resemblance to deeply-weathered granites in subtropical areas, as it exhibits a lack of feldspar alteration and a low clay content (Eden & Green, 1971; Doornkamp, 1974). Although gibbsite is present, this is considered by Green & Eden (1971) to be an initial product of the weathering process and not indicative of weathering under humid tropical conditions. Further pertinent evidence concerning the age of granite saprolites and the conditions under which they developed comes from the research of Hall (1985, 1986a, b) on the deeply-weathered crystalline rocks of NE Scotland. Hall demonstrated that the saprolite in this area consists of two types. The first is a widespread sandy gruss, which, like the Dartmoor growan, displays limited mineral alteration and reflects prolonged weathering under humid *temperate* conditions in the Pliocene and early Pleistocene. The second type is a much less widespread clay-rich gruss with advanced mineral alteration indicative of prolonged weathering under a pre-Pliocene subtropical climate. Hall's findings suggest that the growan of Dartmoor may also represent prolonged weathering under temperate conditions in the Pliocene and Pleistocene. Indeed, it raises the possibility that growan formation was resumed during successive interglacials, with progressive emergence of the tors as a result of saprolite stripping during intervening periods of periglacial climate throughout the later Pleistocene.

In sum, the available evidence appears to point to a polygenetic origin for tors in upland Britain. Some, such as the Stiperstones and the scarp-edge gritstone tors of the Pennines, may have evolved exclusively as a result of differential frost weathering and periglacial mass movement. Others, such as the summit and spur tors of Dartmoor, may indeed represent resistant residuals that were merely exhumed from a pre-existing sandy gruss under periglacial conditions. If this interpretation is correct, then the factors responsible for different modes of tor evolution in different areas require exploration.

One possible controlling factor is tor location. In the Bohemian Highlands and Tasmania, summit tors have been interpreted as 'two-stage' forms whilst adjacent scarp-edge tors have been attributed to scarp retreat under periglacial conditions (Demek, 1964a; Caine, 1967). Although this possibility cannot be excluded as an explanation for the Pennine tors, it also cannot be of general application: the Stiperstones, for example, are summit tors with no associated evidence for initial formation under a saprolite cover. Similarly, those granite tors located below the plateau-edge break of slope on Dartmoor and in the Cairngorms are morphologically similar to summit tors and are usually unrelated to a retreating scarp face. An alternative interpretation is that lithology has acted as the principal control on the nature of tor development. Clearly, 'two-stage' tors can only have developed on lithologies that experienced the widespread development of a thick saprolite cover during long periods of tectonic stability. The widespread development of 'corestone'-type summit and spur tors on granitic rocks, not only in the Cairngorms and Dartmoor but also on the Cheviot Hills and outlying hills in NE Scotland such as Bennachie and Ben Rinnes, suggests that granitic rocks have proven particularly favourable to the development of such tors. The abundance of sandy gruss deposits in the two main areas of granite tor development (Dartmoor and NE Scotland) lends support to this interpretation.

A final factor that may have played some role in the nature of tor evolution is glaciation. Linton (1949, 1955) initially argued that tors cannot have survived the passage of active glacier ice, hence are diagnostic of areas that remained ice-free throughout the Pleistocene. He supported this argument with the pertinent

observation that tors tend to be absent in uplands (including granite uplands) that have experienced intense glacial modification, such as those of Snowdonia, the Lake District and the western Highlands of Scotland. Later, however, he modified this view, suggesting that on the tor-bearing uplands of eastern Scotland 'the ice lay thinly, accumulated slowly and moved but little' thereby permitting the survival of tors 'genetically comparable' with those of unglaciated Dartmoor (Linton, 1959, p. 43). This later view was supported by Sugden, who found in the Cairngorms 'limited but positive evidence that parts of the preglacial surface actually bearing tors have been over-ridden by ice' (Sugden, 1968, p.85). Amongst the evidence cited by Sugden is the presence of schist erratics around the Argyll Stone and adjacent tors at 850 m altitude on the Feshie Hills west of Gleann Einich. He also noted that on Beinn Mheadhoin a tor with a roche moutonée form rises above a surface of ice-smoothed bedrock slabs. Such evidence suggests that the Cairngorm tors may indeed have survived glaciation, and indeed raises the possibility that the stripping of a saprolite cover from around the tors of NE Scotland may have been accomplished mainly by glacier ice rather than by periglacial mass movement (cf. Caine, 1967). On the other hand, the absence of tors from granite mountains in the Western Highlands (such as those around Loch Etive) suggests that the tors on the Cairngorms and other mountains in NE Scotland survived because the erosive potential of successive ice sheets in this area was rather limited. Support for this interpretation is provided by the preservation of preglacial weathering covers in NE Scotland, which suggests that the ice sheets that crossed this area were cold-based, frozen to the underlying substrate and consequently incapable of substantial erosion (Hall & Sugden, 1987; Hall & Mellor, 1988).

Elsewhere in upland Britain evidence for the survival of tors under glacier ice appears equivocal. Clapperton (1970) considered that the Cheviot tors had survived glaciation, but Douglas & Harrison (1985) opposed this idea in view of the great volume of solifluction deposits overlying till in this area. Similarly, Cunningham (1965) claimed that tors had survived glaciation on the Derbyshire Pennines, but Palmer & Radley (1961) believed that the tors in formerly-glaciated parts of the Pennines formed under severe periglacial conditions during and after ice-sheet retreat. The Stiperstones of Shropshire lay immediately outside the maximum reach of the Late Devensian ice sheet, and were therefore interpreted by Goudie & Piggott (1981) as having formed as a result of periglacial activity during the Devensian. On the Isles of Scilly, Scourse (1987) drew an interesting distinction between prominent tor forms outside the limits of maximum ice-sheet glaciation and rounded, eroded tors, some with a roche moutonée form, that occur just within this glacial limit. On balance, the available evidence suggests that tors occupying formerly glacierised parts of upland Britain were capable of surviving the passage of glacier ice, but only where such ice had limited erosional potential. This conclusion is broadly in accord with that reached by researchers working on 'glaciated' tors in other parts of the world (e.g. Caine, 1967; Sugden & Watts, 1977; Gangloff, 1983; Watts, 1983). It is likely, however, that *frost-shattered* tors above areas of *intense* glacial erosion may indeed be diagnostic of former nunataks that rose above at least the last ice sheet. One possible example of this occurs on Ben Loyal in the far north of Scotland, where shattered syenite tors on high ground contrast markedly with ice-abraded roches moutonées on the lower slopes of the moun-

Figure 9.22 Periglacial trimline in the Western Red Hills, Isle of Skye. The former upslope limit of the Loch Lomond Readvance is defined by the upper limit of glacial drift (which coincides approximately with the upper limit of vegetation cover) and the lower limit of scree and frost-weathered bedrock.

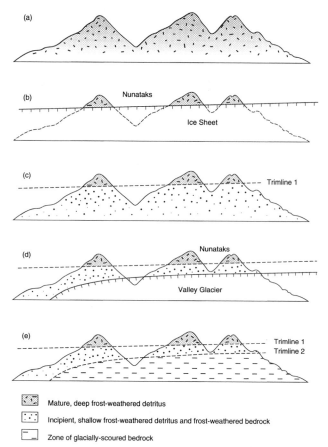

Figure 9.23 The formation of two periglacial trimlines separating three weathering zones. (a) Prolonged frost weathering results in the formation of a mature cover of deep mountain-top detritus. (b) The lower limit of this detritus is trimmed by an advancing and thickening ice sheet. (c) During and after ice-sheet downwastage, frost again attacks the bedrock below the trimline, modifying glacially-moulded surfaces and possibly producing a thin cover of frost-weathered regolith. (d) A later ice advance or readvance trims the lower frost-modified zone. (e) Retreat of the last glacier leaves a zone of unmodified glacially-scoured bedrock and glacial drift below the lower trimline.

tain (Linton, 1955). Similarly, the basalt plateau of the Trotternish Escarpment in northern Skye supports frost-shattered tors on higher ground, whilst intervening cols contain roches moutonées and display no evidence for macrogelivation, a contrast interpreted by Ballantyne (1990) as indicative of the maximum altitude achieved by the Late Devensian ice sheet in this area. Frost-shattered tors, however, form only part of the periglacial evidence for former nunataks in upland Britain, a topic examined in more detail below.

Periglacial trimlines

The term *periglacial trimline* describes the maximum level to which glacier ice has eroded or 'trimmed' a pre-existing zone of frost-weathered rock or debris on a hillslope. In practice, the trimline cut by the most recent glacier advance in any area is marked by the boundary between a downslope zone of glacial drift and glacially-abraded bedrock, and an upslope zone of frost-weathered bedrock or *in situ* frost-weathered detritus (Figure 9.22). This boundary may be fairly sharp, occurring within an altitudinal range of few tens of metres, or rather diffuse. The clarity of the transition depends, amongst other things, on the effectiveness of glacial abrasion and the extent to which frost-weathered debris moved downslope following glacier thinning. Where successive glacial advances of progressively diminishing extent have occurred, it may be possible to identify more than one trimline. In such cases the upper (and hence older) trimline(s) are often more difficult to distinguish, as they are represented only by contrasts in the degree to which the terrain has been modified by frost weathering (Figure 9.23). An important criterion for identifying such older trimlines is the thickness or maturity of *in situ* frost-weathered detritus on flat or gently-sloping ground, particularly on plateaux and cols.

Periglacial trimlines are associated with two further concepts, both of which have excited controversy. One is the *nunatak hypothesis*, first enunciated by the botanists Blytt and Sernander over a century ago to explain the distribution of certain plant species in Scandinavia. As originally conceived, the nunatak hypothesis is the proposition that certain mountain summits remained unglaciated, at least during the last glacial, thereby providing floral refugia. Periglacial trimlines may delimit the maximum altitude reached by former ice sheets against the flanks of such nunataks. A related concept is that of *weathering zones*: as successive trimlines delimit altitudinal zones that have been exposed to weathering processes for different lengths of time, then the degree of rock weathering and soil development should be more advanced in the zone above any trimline than in the zone below. The term *weathering limit* has been employed to describe the boundaries between such zones. In particular, autochthonous blockfields and other types of thick *in situ* mountain-top detritus have been widely accepted as diagnostic of summits and plateaux that have experienced prolonged periglacial weathering, uninterrupted by glacial erosion (e.g. Nesje, 1989).

Trimlines and weathering zones have been employed in reconstructions of the dimensions of former ice sheets both in Norway (Sollid & Sørbel, 1979; Sollid & Reite, 1983; Nesje *et al.*, 1987, 1988; Nesje & Sejrup, 1988) and in eastern Canada (Ives, 1957, 1958, 1975, 1978; Løken, 1962; Boyer & Pheasant,

1974; Grant, 1977). The association of blockfields with former nunataks in such reconstructions has attracted criticism, however. Dahl (1966) maintained that blockfields in the Narvik area of northern Norway have developed since the downwastage of the last ice sheet, and hence are not diagnostic of former nunataks, though his reasoning was strongly challenged by Ives (1966). The blockfields of the Torngat Mountains in Labrador-Ungava, previously interpreted by Ives (1958) as representing unglaciated enclaves, have been reinterpreted by Gangloff (1983) as glacial till reworked by periglacial processes, mainly on the grounds that blockfield matrix material has sedimentological characteristics similar to those of nearby moraines. A more serious problem in the use of periglacial trimlines to delimit former glaciers is posed by the possibility that *in situ* mountain-top detritus on high ground may have been preserved under shallow, cold-based glacier ice whilst lower slopes were scoured by active warm-based ice (Sugden, 1968, 1977b; Holdsworth & Bull, 1970; Boulton, 1979; Watts, 1983; Rasmussen, 1984). This possibility is often difficult to refute. Mature blockfields have been exposed by the shrinkage of thin, cold-based ice caps in northern Norway (Whalley, Gordon & Thompson, 1981), and Sugden & Watts (1977) have shown that tors and blockfields on Baffin Island survived the passage of the last ice sheet. The main argument against this possibility is the often abrupt nature of the boundary between glacially-abraded and frost-weathered terrain, as the transition from erosive warm-based ice to passive cold-based ice within an ice sheet is likely to have migrated during ice-sheet build up and decay. A consistent downglacier descent in trimline altitude is also indicative of a 'true' periglacial trimline that marks the altitude of a former glacier surface, as is the downvalley continuation of a weathering limit by lateral moraines (Grant, 1977; Nesje *et al.*, 1987, 1988).

Recognition of periglacial trimlines on British mountains has been somewhat belated. Inevitably, much of the work on this topic has concerned trimlines cut by the glaciers of the Loch Lomond Readvance, the final episode of glaciation in upland Britain. Marked contrasts between areas of glacially-abraded bedrock inside the limits of this readvance and frost-wedged bedrock, mountain-top detritus and other relict periglacial forms outside these limits were noted by Sissons (1967, pp. 223–4), and were subsequently employed by various researchers to delimit the higher reaches of Loch Lomond Readvance glaciers in the Scottish Highlands, the Hebrides, the Southern Uplands, the Lake District and Snowdonia (e.g. Sissons, 1972, 1974b, 1977b, 1980a; Ballantyne & Wain-Hobson, 1980; Cornish, 1981; Gray, 1982). Explicit development of the concept of Loch Lomond Readvance trimlines, however, began with the work of Thorp (1981), who established a methodology for the systematic mapping of relevant evidence and used this technique to reconstruct the former maximum dimensions of Loch Lomond Readvance glaciers in parts of the Western Grampian Highlands. Thorp (1981, p. 49) defined the trimline cut by these glaciers as 'a narrow zone (10–30 m) separating strongly ice-moulded from frost-affected bedrock', and by tracing this zone downvalley mapped former glacier limits over an altitudinal range of 300–800 m. He found that trimlines are best developed on steep truncated spurs, where glacial abrasion has often been most effective, and employed a wide range of periglacial evidence (including relict

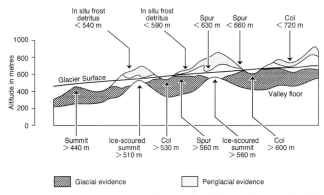

Figure 9.24 Criteria for trimline identification. The maximum and minimum altitudes constrain the position of the former glacier surface.

solifluction landforms, summit blockfields, debris-mantled slopes, relict talus, jagged frost-shattered rock and tor-like summits) to identify nunataks that had escaped glaciation during the Loch Lomond Stadial. He also noted, as did Ballantyne (1982b), marked contrasts in the depths of open joints inside and outside areas that were glaciated during the stadial (Figure 9.10). Finally, Thorp pointed out some of the difficulties posed by trimline identification, noting in particular that frost-resistant lithologies such as lavas, coarse-grained granites and massive schists tend to exhibit only limited evidence of Lateglacial frost action, and indeed preserve erosional forms (such as roches moutonées and even striae and chattermarks) produced by the passage of the last ice sheet over high ground, well above the readvance trimline.

Thorp (1986) subsequently employed similar techniques in a remarkable reconstruction of the vast (2000 km²) icefield that occupied the Western Grampians during the Loch Lomond Stadial. Besides using more 'conventional' geomorphological evidence to delimit the lateral margins of this icefield, he obtained its vertical dimensions by ascending and mapping trimline evidence on 198 mountain spurs, of which 171 yielded trimline altitudes within a vertical range of 60 m. The altitudes of high-level cols supporting frost-weathered bedrock and debris provided additional evidence limiting the maximum altitude of the former icefield. Conversely, ice-scoured cols and the glacially-abraded summits of lower hills imposed minimum altitudes on the ice surface (Figure 9.24). Trimline mapping using Thorp's approach has subsequently been employed in a reconstruction of the dimensions of a smaller (155 km²) Loch Lomond Stadial icefield on the Isle of Skye (Ballantyne, 1989a; Figure 9.25). The use of trimlines in this way has great potential for the accurate three-dimensional reconstruction of the vast icefield that lay athwart much of the Scottish Highlands during the Loch Lomond Stadial, and hence for the reconstruction of stadial palaeoclimate and glacier dynamics. The main obstacle to the completion of this task is the enormous demands that trimline mapping on high ground makes on the stamina of the researcher.

Attempts to identify trimlines older than those of the Loch Lomond Readvance are still in their infancy. For over a century, there was general consensus that the last ice sheet covered almost all of the high mountain summits of Great Britain. Geikie (1894, p.82), for example, believed that only the mountains of Harris in the Outer Hebrides protruded through a great

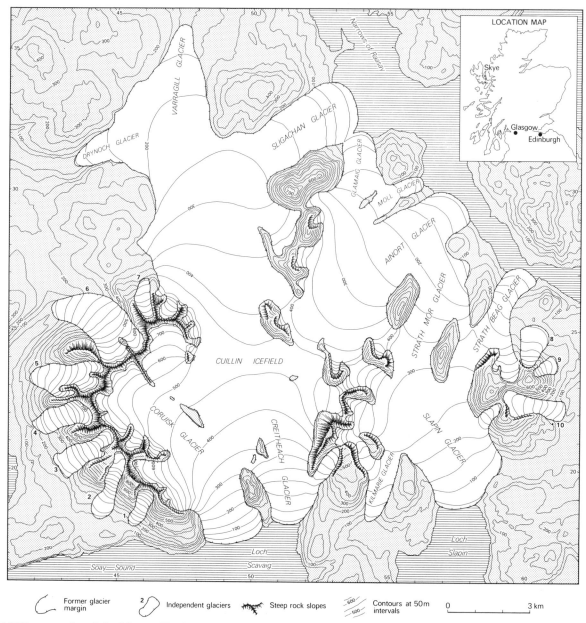

Figure 9.25 Reconstruction of a Loch Lomond Readvance icefield in south-central Skye. The upper limits of the icefield were delimited through identification of periglacial trimlines such as that shown in Figure 9.22 (Ballantyne, 1989a).

'mer de glace' that covered the entire Scottish mainland like an icy shield, a view accepted by Sissons (1967) over 70 years later. Occasionally, however, this dogma has been challenged. As noted previously, Linton (1949, 1951, 1955) argued that tor-bearing summits remained above the level of the ice as nunataks, but he subsequently moderated this view, conceding that the survival of tors may merely identify areas of limited glacial modification. The debate concerning the survival of nunataks above the level of the last Scottish ice sheet nevertheless continued to smoulder, fuelled by a succession of admittedly rather equivocal arguments. Ragg & Bibby (1966) maintained that the great volume of mountain-top detritus on Broad Law implied that the higher parts of the Southern Uplands probably remained above the level of the last ice sheet; Romans *et*

al. (1966) argued that cryogenic redistribution of the fine fraction within alpine soils on mountains in NE Scotland had occurred under periglacial conditions between 28 ka BP and 12 ka BP, which implies that such summits escaped ice-sheet erosion; and Stevens & Wilson (1970) suggested that the presence of kaolinite and halloysite in alpine soils on Ben Lawers indicates that the higher parts of this mountain remained above the Late Devensian ice sheet. The existence of nunataks above the Late Devensian ice sheet was also argued on geomorphological grounds. On the Trotternish Escarpment of northern Skye, Anderson & Dunham (1966) observed no evidence of glacial erosion above *c.* 500 m, and suggested that the higher parts of the escarpment remained above the ice-sheet surface. Credit for the first attempt to map the form of the ice-sheet surface belongs to Godard (1965), who based his tentative reconstruction of the dimensions of 'maximum glaciation' on the altitudes of block-fields in northern Scotland, though subsequent research showed his mapping to be erroneous (Ballantyne, Sutherland & Reed, 1987; Reed, 1988).

Despite the above arguments, the prevailing belief in complete burial of the Scottish mountains by the Late Devensian ice sheet remained unshaken. This view was reinforced by theoretical models of the form of the ice sheet, which predicted that even the highest mountains were buried under several hundred metres of glacier ice (Boulton et al., 1977; Gordon, 1979). Such models, however, were based on the assumption that the last ice sheet had covered the entire Scottish mainland, extending westwards to the edge of the continental shelf and eastwards into the central and northern North Sea area. The validity of this assumption was first questioned in a trenchant article by Sissons (1981b). Although there is strong evidence to suggest that some earlier ice sheet extended to these limits, stratigraphic research has now confirmed that the last Scottish ice sheet was laterally much less extensive than was traditionally believed (Sutherland, 1984; Bowen et al., 1986; Figure 2.1). Recognition that the dimensions of the last ice sheet had been exaggerated encouraged a renewed search on the Scottish mainland for evidence of former ice-sheet nunataks. Ballantyne (1984) observed that on mountains in NW Scotland outside the limit of the Loch Lomond Readvance, the transition with increasing altitude from ice-scoured bedrock to slopes mantled by thick in situ frost detritus is often not gradual, but occurs over a few tens of metres. This suggests that the boundary between the two is a trimline representing the upper limit of a former ice-sheet. Moreover, in Wester Ross the lower limit of in situ mountain-top detritus descends north-westwards, and is thus consistent with the former direction of regional ice-sheet movement as indicated by striae and erratics. Ballantyne (1984) suggested that this upper trimline may have been cut by the Wester Ross Readvance, but later mapping showed that it lies high above the level of the Wester Ross moraines, and therefore predates this readvance (Ballantyne et al., 1987).

Subsequent research by Reed (1988) confirmed that a high-level weathering limit occurs on mountains throughout Wester Ross, at altitudes well above those reached by glaciers during the Loch Lomond Readvance. This weathering limit tends to be best developed on Torridon Sandstone, a lithology little affected by macrogelivation since ice-sheet downwastage. Reed found that soils above the weathering limit contain clay mineral assemblages different from those represented in soils below this limit. Gibbsite was found only above the weathering limit, as was a significantly greater representation of kaolinite and halloysite. Reed interpreted these contrasts as indicative of the survival of pre-Late Devensian pedorelicts above the weathering limit. If correct, this implies that in situ mountain-top detritus above the weathering limit predates the last ice sheet maximum, and escaped ice-sheet erosion. As the westwards and eastwards descent of the weathering limit closely follows the directions of former ice-sheet movement as indicated by striae, ice-moulded bedrock and erratics, Reed concluded that it represents a trimline marking the maximum vertical extent of the last ice sheet in this area.

Support for this interpretation may be found on the Trotternish Escarpment of northern Skye. The higher parts of this basalt escarpment support a thick cover of in situ mountain-top detritus with blockfields, relict bouldery solifluction lobes and small frost-shattered tors. In contrast, the intervening cols provide unequivocal evidence for the passage of glacier ice in the form of ice-moulded bedrock and roches moutonées. Once again, the upslope transition from ice-scoured terrain to slopes covered by frost-weathered debris is not gradual, but occurs at most over about 30 m, and in places is marked by an abrupt lower limit to soliflucted debris. This frequently coincides with a springline, where water draining through the permeable frost-weathered deposits emerges at the surface, and consequently with the upper limit of drift gullies and the lower limit of xerophytic vegetation. As the Loch Lomond Readvance in this area is represented only by two cirque glaciers, Ballantyne (1990) proposed that the abrupt contrast between the ice-scoured and frost-weathered zones represents an ice-sheet trimline equivalent to that in Wester Ross. He tested this proposition by plotting the altitude of the weathering limit on a south–north projection plane. The result revealed a steady decline in the altitude of the weathering limit from 580–610 m at the southern end of the escarpment to 440–470 m at its northern end (Figure 9.26). As independent geomorphological evidence indicates that the regional movement of the last ice sheet across Trotternish was northerly, this trend is consistent with interpretation of the weathering limit as a trimline representing the maximum altitude achieved by the Late Devensian ice sheet.

The research outlined above implies that two periglacial trimlines and hence three weathering zones may be identified on some mountains in northern Scotland. It is widely agreed that the lower trimline was cut by the glaciers of the Loch Lomond Readvance whilst the upper, more contentiously, has been interpreted as delineating the surface of the last ice sheet at its maximum thickness. In a few locations, both trimlines may be clearly seen, for example at the western end of Loch Glascarnoch, on the road from Inverness to Ullapool. From here the view to the north-west reveals a thick cover of frost-weathered boulders on the high ground of Am Faochagach (954 m). This debris mantle has been interpreted by Reed (1988) as forming above the level of the last ice sheet. Immediately downslope, bedrock surfaces are frost-weathered but covered by only a thin discontinuous cover of debris; this zone, according to Reed, experienced periglacial weathering only after ice-sheet downwastage. The lowermost zone is one of ice-scoured bedrock and glacial drift, and lies within the area occupied by glacier ice during the Loch Lomond Readvance, here delimited by a massive end moraine.

The status of the high-level weathering limits of Wester Ross and Trotternish remains debatable, however. There are two points of contention. First, it is arguable that the mountains were completely overrun by the last ice sheet, and that these limits actually represent the former boundary between erosive, warm-based ice on low ground, and passive, cold-based ice on mountain summits. The main arguments countering this suggestion are first, the consistency of trimline gradients with former directions of ice movement and, second, the abruptness of the boundary in certain locations, even though elsewhere the clarity of the trimline has been obscured by mass movement. It is also feasible that independent cold-based ice caps covered and protected some debris-mantled plateaus whilst the surrounding valleys were occupied by erosive ice streams. It has

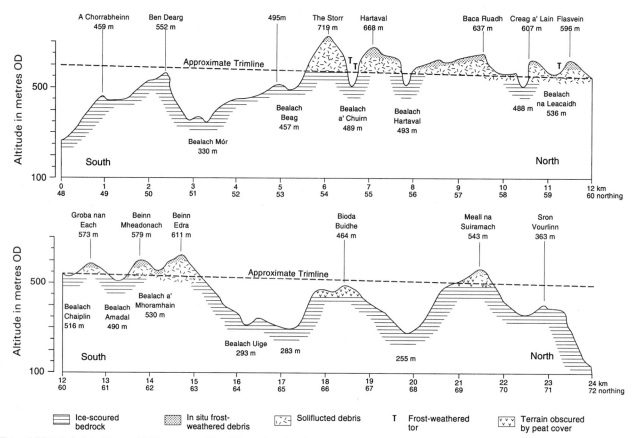

Figure 9.26 Relief of the Trotternish Escarpment, Isle of Skye, plotted against a south–north projection plane and showing the lower limits of *in situ* frost-weathered detritus and soliflucted debris, and the upper limit of glacially-scoured bedrock. The northwards descent of this weathering limit suggests that it represents a trimline cut by the last ice sheet at its maximum thickness. From Ballantyne (1990).

been shown that even deeply-weathered gruss on Scottish mountains apparently survived successive glaciations under the protective cover of plateau ice domes (Hall & Mellor, 1988). This scenario, however, seems less plausible where the debris cover occupies the crests of steep arêtes.

The second point of contention concerns the age of mountain-top detritus above the high-level weathering limit. On Torridon Sandstone and the basalts of Trotternish, ice-moulded bedrock downslope has experienced negligible macrogelivation, even though such rock surfaces were exposed to weathering under a severe cold climate during ice-sheet downwastage and again during the Loch Lomond Stadial. This indicates that the development of a thick regolith cover above the weathering limit required very prolonged exposure to severe periglacial conditions, which in turn favours the notion that the initial development of such detritus predates the Late Devensian glacial maximum. However, the presence of erratics within the high-level mountain-top detritus of Wester Ross implies that summits above the weathering limit were glaciated at some time in the past, though such erratic carry may, of course, be pre-Devensian (Ballantyne *et al.*, 1987). The clay mineral evidence presented by Reed is more difficult to assess. Although some authors have interpreted the presence of gibbsite, halloysite and kaolinite in mountain soils as due to the survival of interglacial (or even preglacial) pedorelicts (e.g. Stevens & Wilson, 1970; Wilson & Bown, 1976; Mellor & Wilson, 1979),

the weathering status of such clay minerals is uncertain; gibbsite in particular may be an early product of the weathering process (Green & Eden, 1971), especially at free-draining snowbed sites (A.M. Hall, personal communication).

Such considerations suggest that the evidence for ice-sheet nunataks in northern Scotland must be treated with caution. If this interpretation of the evidence is valid, however, it has exciting implications. First, mapping of the high-level trimline will permit a much more accurate reconstruction of the dimensions of the last ice sheet than has hitherto been possible, even though summits near to the main centres of glacier accumulation seem likely to have been completely buried under ice at the time of the ice-sheet maximum. Reed's (1988) research suggests that the former ice-shed in Wester Ross increased in altitude southwards, and the absence of thick *in situ* periglacial detritus on most mountains in southern Ross (where distinctly ice-moulded bedrock crowns some summits) suggests that the level of the former ice sheet rose above the highest peaks in this area (Gordon, 1979, 1981). This is consistent with total ice cover over the Western Grampians, still farther to the south (Thorp, 1987). The existence of ice-sheet nunataks in NW Scotland, however, suggests that others may have existed in peripheral uplands elsewhere in Britain, such as Snowdonia, mid-Wales, Sutherland and Caithness, the Outer Hebrides and the Eastern Grampians.

A second implication of the existence of ice-sheet nunataks

in NW Scotland is that it throws light on the age of blockfields and other thick covers of mountain top detritus. On lithologies resistant to macrogelivation, such as some lavas, Torridon Sandstone, coarse-grained granites and massive schists, glacially-abraded bedrock at high altitudes has often been only superficially modified by macrogelivation subsequent to ice-sheet downwastage (Sugden, 1968; Thorp, 1981; Reed, 1988; Ballantyne, 1990). Thick covers of *in situ* frost-weathered regolith on such rocks appear to imply a much longer history of development, dating back to before the Late Devensian ice-sheet maximum, and possibly as far back as the Ipswichian Interglacial or earlier. Conversely, well-jointed or fissile lithologies susceptible to frost wedging have in some areas yielded a thick cover of mountain-top detritus since the down-wastage of the last ice sheet. The Cambrian Quartzite block-fields and blockslopes of An Teallach provide an excellent example of such Lateglacial mountain-top detritus, as these occur well below the high-level weathering limit and have extended downslope over ice-scoured Torridon Sandstone out-crops (Figure 9.16). Such evidence indicates that blockfields and other periglacial regolith covers are not everywhere diag-nostic evidence for ice-sheet nunataks. Because of the variable susceptibility of different rock types to macrogelivation, moun-tain-top detritus on British mountains is not all of the same age; the regolith on rocks susceptible to frost-wedging, such as well-jointed quartzite, fissile schists and shales, may be of much more recent origin than that on rocks relatively immune to macrogelivation.

Mountain-top detritus and the distribution of periglacial landforms

Many small-scale periglacial landforms, such as patterned ground and solifluction features, are developed on the frost-weathered detritus that mantles the higher parts of British mountains. It follows from what is written above that the distri-bution of such landforms is strongly influenced by the extent of former glaciers. Not only are relict Lateglacial features absent from terrain occupied by glaciers during the Loch Lomond Readvance, but also periglacial landforms tend to be lacking on mountains that experienced erosion by the last ice sheet and did not subsequently develop an appreciable regolith cover. This is the case, for example, in the vast zone of 'areal scouring' identi-fied by Gordon (1979, 1981) in the southern part of the NW Highlands, and is exemplified by the rocky summits of Knoydart and the Kyleakin Hills of eastern Skye. On such mountains periglacial frost-action landforms are rare, simply because the terrain consists mainly of ice-moulded bedrock with at most a thin, disco tinuous cover of frost debris.

The distribution of periglacial phenomena on British moun-tains is also strongly influenced by the nature of the underlying regolith. As noted earlier in this chapter, the variable response of different lithologies to periglacial weathering has produced regolith covers with strongly differing structural and textural characteristics, ranging from bouldery blockfields to regoliths composed of small clasts embedded in a silt-rich matrix. Three broad categories of mountain-top detritus may be recognised:

(1) openwork block deposits developed on rocks such as quartzite, microgranite, granulite and siliceous schist;

(2) diamictons consisting of clasts embedded in a coarse cohesionless sandy matrix and typical of most sandstone and granite areas; and

(3) diamictons relatively rich in silt and fine sand, formed by the weathering of such rocks as mica-schist, slate and shale.

These categories are referred to below as types (1)–(3). In practice, however, the categories overlap, particularly as verti-cal sorting of clasts in types (2) and (3) may result in the forma-tion of a superficial block deposit (Figures 9.5 and 9.17). The critical difference between type 2 regolith and type 3 regolith is that the latter is frost-susceptible, but the former is not (Figure 9.7). Thus periglacial landforms whose development is depen-dent on ice segregation are usually absent on type (2) regolith.

In general, the assemblage of periglacial phenomena present on any mountain is strongly controlled by the nature of the regolith mantle, and thus, indirectly, by the nature of the under-lying bedrock (Ballantyne, 1984, 1987a). This strong litholog-ical control may be illustrated with reference to three contrasting mountain areas, each of which supports a different regolith type: Sron an t-Saighdeir on the Isle of Rhum, where the under-lying microgranite forms superb type (1) blockfields and block-slopes; An Teallach in Wester Ross, a Torridon Sandstone mountain mantled by a cover of type (2) sand-rich regolith; and Ben Wyvis in Easter Ross, a mountain composed largely of mica-schists that have weathered to produce a cover of silt-rich type (3) regolith (Figure 9.5). Most of the periglacial landforms developed on the type (1) block deposits of Sron an t-Saighdeir are relict Late Devensian features, such as boulder sheets and lobes, together with large-scale sorted circles and stripes (Figure 9.27). Apart from slope-foot forms, and occasional sorted circles where fines reach the surface, type (1) regolith rarely supports active periglacial features. The sand-rich type (2) regolith of An Teallach also supports boulder sheets and lobes, but here the most widespread forms are wind-related Holocene features that reflect the cohesionless nature of such regolith. Characteristic landforms include deflation surfaces, wind-patterned ground, small turf-banked terraces, niveo-aeolian sand deposits, sand hummocks and small active niva-tion hollows excavated within the sand deposits. Recent hills-lope debris flows are also particularly common on such terrain (Figure 9.28). In contrast, the frost-susceptible type (3) regolith of Ben Wyvis supports a rich range of sorted and nonsorted solifluction landforms, together with ploughing boulders, earth hummocks and nonsorted hummock stripes (Figure 9.29). Active and relict sorted patterned ground also occurs on such regolith. Though such features are not evident on Ben Wyvis, where the vegetation cover is unbroken, they occur on bare ground on type (3) regolith developed on the lavas of Orval (571 m) in western Rhum (Figure 9.27). Wind-patterned ground and small turf-banked terraces also occur locally on type (3) regolith, but are much less widespread than on cohesionless, sandy debris mantles.

The relationships between type of mountain-top detritus and the range of periglacial landforms present on any mountain are summarised in Figures 9.30 and 9.31. The consistency of these relationships suggests that lithology (or, more specifically, the response of different rock types to frost weathering) is the dom-inant influence on the *range* of periglacial phenomena present on any mountain. Thus nearby mountains of similar altitude and

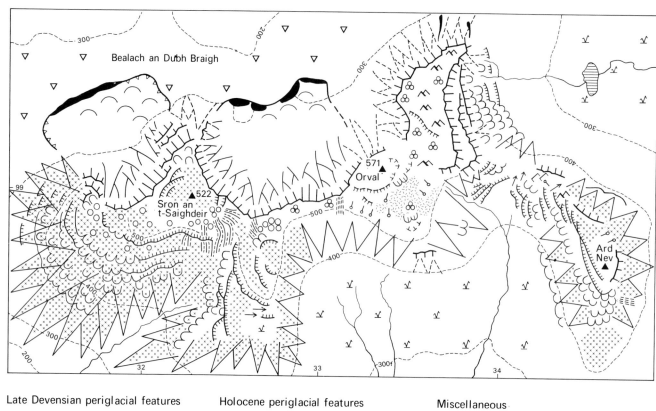

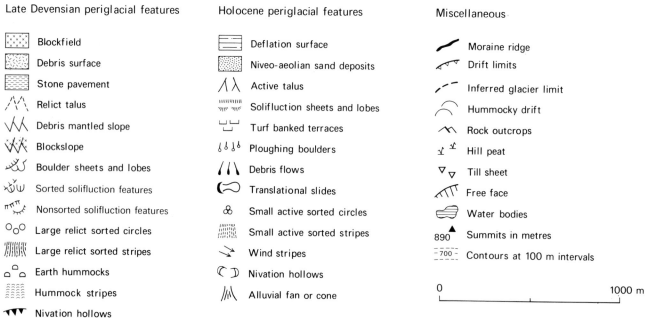

Figure 9.27 Periglacial features on the Western Hills, Isle of Rhum. The area around Sron an t-Saighdeir is underlain by microgranite that has yielded type (1) regolith (blockfields and blockslopes) on weathering. The plateau around Orval is underlain by Hawaiite lavas that have weathered to produce a shallow mantle of frost-susceptible type (3) regolith.

relief but different lithology, such as An Teallach and Ben Wyvis, may support utterly different assemblages of periglacial features. It follows that if the weathering response of the rock underlying a mountain is known, then the range of periglacial phenomena present may be predicted with considerable accuracy. The main difficulty in making such predictions arises from

the variable weathering response of some lithologies. Granite, for example, in places underlies blockfields, but more often forms a diamicton with a coarse sandy matrix. Locally, however, granitic matrix material contains up to about 40% silt and is thus frost-susceptible (Figure 9.32). For this reason granite massifs such as the Cairngorms often support a remarkably var-

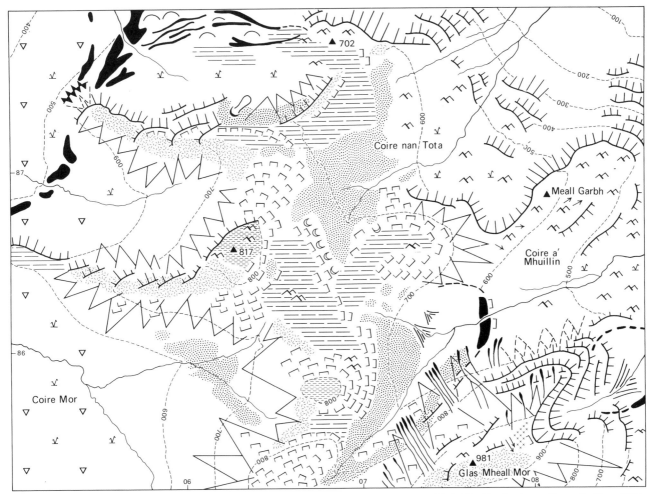

Figure 9.28 Periglacial features on the northern plateau of An Teallach, an area of type (2) regolith underlain by Torridon Sandstone. Key as for Figure 9.27.

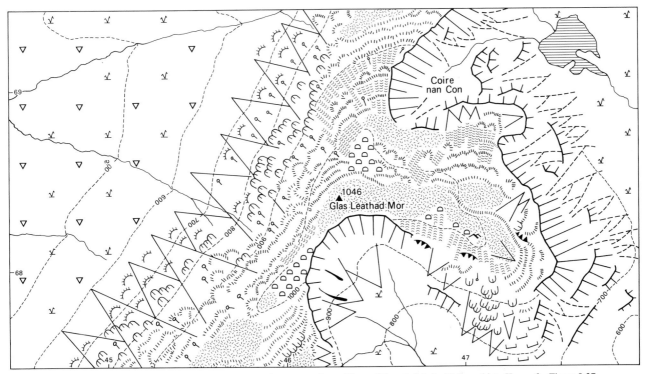

Figure 9.29 Periglacial features on the summit ridge of Ben Wyvis, an area of type (3) regolith underlain by Moine schists. Key as for Figure 9.27.

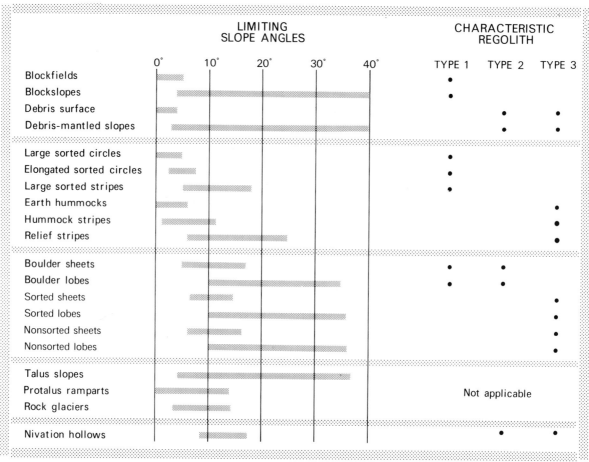

Figure 9.30 Local controls on the distribution of relict (Late Devensian) periglacial features on British mountains: limiting slope angles and characteristic regolith type.

ied assemblage of periglacial landforms. Mountain-top detritus on sedimentary rocks tends to be much more predictable in terms of the range of periglacial phenomena it supports.

The significance of mountain-top detritus on British mountains

In general, frost-weathered regolith has been somewhat neglected as a topic for study. Only blockfields have received much attention, probably because these are striking in appearance and dramatically different from other Quaternary deposits. In this chapter, an attempt has been made to demonstrate that, far from lacking interest, all types of mountain-top detritus are of major importance for our understanding of past and present periglacial processes, for explaining the distribution of upland landforms, and even for reconstructing the dimensions of former glaciers. Many uncertainties remain, however, and it is appropriate to conclude by identifying areas for further research as well as some of the main points that have been established. Perhaps the most interesting lesson to be learned from a detailed study of mountain-top detritus in upland Britain concerns the diversity of weathered products. This reflects the wide range of lithologies that underlie high ground, and the manner in which different rocks have responded to weathering under stadial conditions. The characteristics of quartzite blockfields in

northern Scotland, for example, have little in common with the matrix-rich regolith that has developed on shales in the Welsh uplands, yet both reflect the response of bedrock to a similar history of frost weathering. Because of their magnificent lithological diversity, the mountains of Great Britain present a unique natural laboratory for studying the response of rock to prolonged periglaciation. This laboratory, alas, has attracted few scientists.

A second point that emerges from this chapter is the feebleness of frost weathering on British mountains during the Holocene, despite the frequency of winter freezing at high altitudes under present-day conditions. This suggests that freezing to temperatures well below 0 °C is a prerequisite for both microgelivation and macrogelivation, and thus provides circumstantial support for the notion that capillary frost damage (ice segregation) may be the most important mode of rock breakdown (Walder & Hallet, 1985, 1986). Widespread belief in the ubiquity of effective frost weathering during the Loch Lomond Stadial of c. 11–10 ka BP also needs to be qualified. Although some fissile or well-jointed rocks yielded a cover of mountain-top detritus at this time, other lithologies proved so resistant to Lateglacial frost weathering that they have retained the morphology and microforms imposed on them by the passage of the last ice sheet. This is strikingly illustrated on the mountains of NW Scotland, where thick blockfields of shattered quartzite bor-

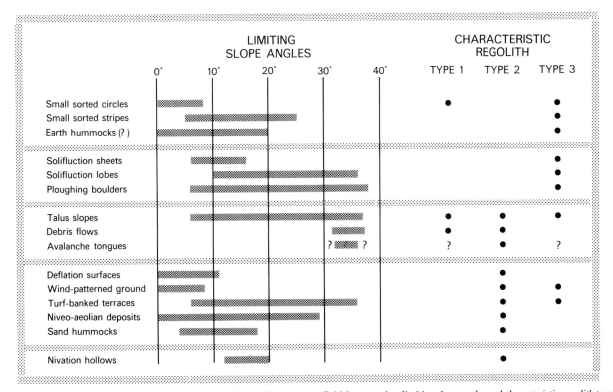

Figure 9.31 Local controls on the distribution of active periglacial phenomena on British mountains: limiting slope angles and characteristic regolith types.

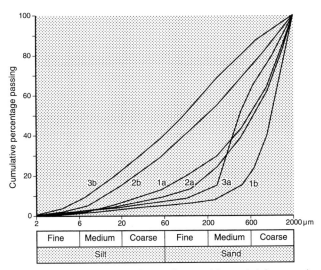

Figure 9.32 Grain-size curves for matrix material sampled from granite regolith at three sites on the Cairngorm plateau, Scotland. The 'a' samples were extracted from a depth of 10 cm, the 'b' samples from a depth of 30 cm.

der glacially-abraded slabs of Torridon Sandstone. A corollary of the resistance of some rocks to weathering during the Loch Lomond Stadial is that where these same rocks *do* support a thick cover of mountain-top detritus, a very prolonged period of periglacial weathering seems to be implied.

Despite the limited effect of Loch Lomond Stadial frost weathering on some lithologies, the contrasts between glacially-scoured and frost-weathered bedrock in upland areas have been fruitfully employed in the identification of periglacial

trimlines that mark the upper limits of Loch Lomond Readvance glaciers (Thorp, 1981, 1986). The significance of higher and thus older weathering limits, defined by the downslope limits of thick *in situ* mountain-top detritus on resistant rocks, is more controversial. Such high-level weathering limits have so far been identified only in northern Scotland, where they have been interpreted as trimlines marking the former altitude of the Late Devensian ice sheet. If this interpretation is valid, it implies that former nunataks may be identifiable on other mountain areas peripheral to the main centres of ice-sheet accumulation. Identification of such nunataks and associated trimlines would permit a much more accurate reconstruction of the dimensions of the last ice sheet than has hitherto proved possible. However, it is difficult to refute an alternative explanation of high-level weathering limits as representing the former boundary between passive cold-based ice on high ground and active erosive ice streams that occupied neighbouring valleys. A related controversy surrounds the interpretation of certain clay minerals (gibbsite, kaolinite and halloysite) in mountain soils. Although some authorities have interpreted these as pedorelics indicative of summits that escaped ice-sheet erosion, others have argued that such clay minerals may represent an early stage of the weathering process, hence are not diagnostic of pre-Devensian or preglacial weathering.

The interpretation of upland tors as exhumed pre-Devensian or even pre-Quaternary rock residuals also remains contentious. In part, this controversy seems to arise from the broad use of the term 'tor' to describe upstanding bedrock residuals irrespective of origin and history of development, and research in diverse areas is needed to establish if criteria can be devised to differentiate exclusively 'periglacial' tors that developed as a consequence of frost weathering from those that have a longer and more complex history of differential weathering and exhumation. Only when such criteria are established can the distribution of 'frost-weathered' tors be used with confidence as evidence for determining the lateral and vertical extent of the last and possibly earlier ice sheets.

Finally, it has been shown that the characteristics of mountain-top detritus control the distribution of many upland periglacial phenomena. Although the nature of this control has been established by relating different types of regolith to assemblages of periglacial landforms, the links between regolith type, periglacial processes and resulting landforms have not been rigorously established. One possible approach to this problem is through the use of laboratory simulations similar to those employed by Gallop (1991) to investigate the links between regolith type and solifluction processes (Chapter 7). The relationship between regolith type and periglacial processes is considered further in the following four chapters, which concern the nature and significance of individual periglacial phenomena in upland areas.

10
Patterned ground on British mountains

As outlined in Chapter 6, the term *patterned ground* refers to terrain that exhibits regular or irregular surface patterns defined by microrelief, vegetation cover, or alternation of fine and coarse debris. In plan, such patterns commonly take the form of circles, polygons or irregular networks on flat or gently-sloping terrain, and grade progressively into stripes as gradient increases. Although the origins of such features are disputed, it is generally agreed that they develop through mass displacement of soil as a result of seasonal freezing and thawing of the ground, both in the active layer above permafrost and in areas where permafrost is absent. Large-scale relict examples of both sorted and nonsorted patterned ground occur south of the limit reached by the last ice sheet, and have been widely interpreted as indicative of former permafrost conditions (Chapter 6). It seems likely that such patterns were once widespread beyond the margins of Late Devensian glaciation, but have been largely obliterated by cultivation. The higher parts of British mountains, however, have been spared the plough, and here both relict and active patterned ground features occur in abundance. This chapter considers the characteristics and significance of these landforms.

Sorted patterned ground on British uplands

It has long been recognised that British mountains support sorted patterns of recent origin as well as relict forms that developed under the severe periglacial climate of the Late Devensian. Active forms tend to be less than 1 m in pattern width, to be unvegetated and to show signs of recent superficial cryoturbation. Reported examples of relict sorted patterns, however, generally exceed 1 m in width and are distinguished by concentrations of clasts alternating with completely vegetated cells or stripes of fine debris (Figure 10.1). Prolonged inactivity of such large-scale sorted patterns is indicated by the development of mature soils within the vegetated cells of large-scale sorted circles and by the rounding of exposed clasts by microgelivation.

Relict sorted patterned ground

Although rarely described, large relict sorted circles, polygons and nets are common on mountain plateaux mantled by bouldery regolith, except where there is a cover of vegetation or peat. They tend to be best developed on rocks such as quartzite and microgranite that have weathered to produce abundant boulders embedded in a matrix of silt-rich frost-susceptible fines. According to Tufnell (1969), large sorted polygons with vegetated cells are widespread on the higher parts of the northern Pennines, and achieve a diameter of 15 m on Knock Fell. The great size of the features described by Tufnell, however,

suggests that these may represent relict thermal contraction polygons into which clasts have been concentrated by lateral sorting (cf. Holmes & Colton, 1960). Elsewhere, relict sorted patterns on mountain plateaux tend to be much smaller and to take the form of circles or irregular nets rather than polygons. Those on Beinn an Fhurain in Assynt, for example, range in diameter from 1.8 m to 4.0 m; those on Clisham in the Outer Hebrides are 1.0–4.0 m in width, and most of those on Sron an t-Saighdeir on the island of Rhum are 2.0–2.5 m wide (Godard, 1965; Kelletat, 1970b; Ryder & McCann, 1971; Figure 10.1). Relict sorted patterns in blockfield areas occasionally take the form of *debris islands* (cf. Washburn, 1979 p. 129), which are isolated cells of vegetated fines, usually 1–3 m in diameter, that sporadically interrupt the cover of boulders. Good examples occur on the Western Hills of Rhum and on the ridge north of Y Garn in Snowdonia. In places, relict circles or nets grade downslope via transitional elongate forms into stripes of similar width. On Beinn an Fhurain, for example, approximately equidimensional circles occur on gradients lower than 2.5°, elongate circles on slopes of 2.5–7.5°, and stripes on steeper slopes, whilst in the Cairngorms King (1971b) found that 'polygons' are limited to slopes of less than 5° and stripes to slopes of less than 18°. A similar slope-related transition from sorted polygons to sorted stripes occurs on the Stiperstones Ridge in Shropshire (Goudie & Piggott, 1981; Figure 6.18). Relict sorted stripes with pattern widths exceeding 2 m also occur on Great Dun Fell and Knock Fell in the northern Pennines, on Greatrigg Man and Nethermost Pike in the Lake District, and on the Western Hills of Rhum. The most impressive relict sorted stripes hitherto described are located on slopes of 12–18° at 425–490 m altitude on Rhinog Fawr in north Wales. According to Ball & Goodier (1968), these have a pattern width of 5–8 m, and the coarse stripes contain boulders of grit 0.6–1.5 m in length.

Although most reported examples of relict sorted patterns on British mountains exceed 2 m in width, this may reflect selective preservation of larger examples. The formation of large-scale sorted patterns is often accompanied by the development of smaller secondary sorted forms within the cells of primary circles, polygons or nets, and it seems likely that these have escaped notice simply because they are obscured by vegetation. A few authors, however, have described small-scale relict patterns. Godard (1965) reported a second 'generation' of sorted circles 0.4–1.0 m in diameter on mountains in NW Scotland. Although these are barely larger than active forms in the same area, their central cells are completely vegetated. Outstanding examples of small-scale relict sorted stripes occur in north Wales. In the Rhinog Mountains, vegetated sorted stripes with pattern widths of 0.7–1.0 m were described by Ball & Goodier

Figure 10.1 Relict sorted circles 2 m in diameter on a granulite blockfield, Sron an t-Saighdeir, Isle of Rhum.

Figure 10.2 Small relict stone stripes on Glyder Fach, Snowdonia.

(1969), and superb examples of relict stripes of similar dimensions occur at 910 m west-south-west of the summit of Glyder Fach in Snowdonia (Figure 10.2). Even smaller partly-vegetated stripes, 0.45–0.75 m in width, occur nearby at 760 m on the south-east shoulder of Y Garn. Ball & Goodier (1970) considered these active on the grounds that the fine stripes undergo heave in winter, but it is difficult to envisage how such vegetated stripes may experience lateral sorting.

The altitude of reported examples of relict sorted patterns rises irregularly eastwards across the Highlands and Islands of Scotland, from 500 m on Rhum and the Outer Hebrides, to 600 m on Quinag on the NW mainland, to over 1000 m in the Cairngorms and SE Grampians (Kelletat, 1970b). It seems likely that this pattern reflects that of blockfields with incomplete vegetation cover, and hence is largely a response to present exposure rather than past climate. If this interpretation is correct, it suggests that such features may have developed much more widely than their reported distribution would suggest, but are now obscured under the peat and vegetation cover that mantles many lower slopes and plateaux. Studies of the structure and sediments of large-scale relict sorted circles (Ballantyne, 1984; Figure 10.3) have revealed characteristics typical of frost-sorted patterned ground elsewhere, including (1) a tendency for clasts to occupy troughs surrounding cells of predominantly fine material, (2) coarsening-upwards of such clasts, (3) elevation of fine centres above the margins and (4) a frost-susceptible matrix in the cells (cf. Goldthwait, 1976). King (1971b) proposed that large sorted 'polygons' in the Cairngorms formed through selective *in situ* weathering of porphyritic granite, but this explanation appears irreconcilable with the structural characteristics noted above and cannot apply to other lithologies.

As noted earlier, the presence of undisturbed soil horizons under vegetation cover in the centres of large relict sorted circles and the rounding of exposed clasts by microgelivation suggests prolonged inactivity. These considerations, combined with the considerable depth of sorting (up to 0.6 m on Beinn an Fhurain) have led most researchers to conclude that large-scale sorted patterns on British mountains were produced under severe periglacial conditions in Late Devensian times (Godard, 1965; Tufnell, 1969; Kelletat, 1970b; Ryder & McCann, 1971; Ballantyne & Wain-Hobson, 1980). The absence of large-scale sorted patterns within the limits of the Loch Lomond

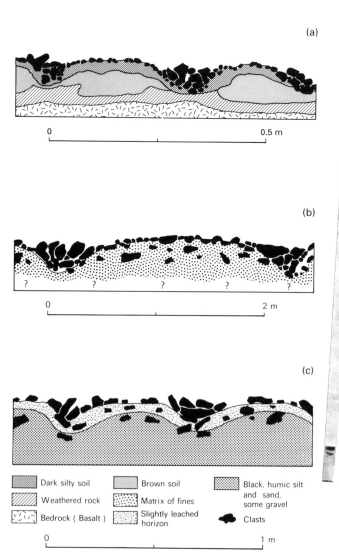

Figure 10.3 Sections excavated through sorted patterned ground on British mountains. (a) Active sorted circles on Orval, Rhum. (b) Relict sorted circles, Sron an t-Saighdeir, Rhum. (c) Active sorted stripes, Tinto Hill, southern Scotland.

Readvance is consistent with this view (Ballantyne, 1984). Such patterns are known to occur on plateaux in the Grampian Highlands that were covered by the Late Devensian ice sheet (cf. Thorp, 1987), and in these areas seem likely to have developed during the Lateglacial after the downwastage of glacier ice from high ground. It is conceivable, however, that some large-scale patterns may have been of earlier origin and were preserved throughout the Dimlington Stadial under cold-based glacier ice (cf. Whalley et al., 1981). Ball & Goodier (1968) considered that the low altitude and exceptional size of the very large sorted stripes on the Rhinog Mountains were indicative of formation under stadial conditions prior to the Lateglacial, but this suggestion is refuted by Foster (1970) on the grounds that the Rhinogs were completely covered by an ice cap during the Dimlington Stadial, so that the sorted stripes presumably developed subsequent to its disappearance. The age of smaller relict sorted forms is also contentious. Goodier & Ball (1969) suggested that the small vegetated stripes on the Rhinogs may have formed during the Little Ice Age. A similar age was advocated for relict stripes and 'polygons' on the Cairngorms by King (1971b), but as his supporting evidence consisted of lichenometric measurements unrelated to a suitable dated control curve this conclusion may be considered unfounded. Although it is possible that small-scale relict patterns such as those illustrated in Figure 10.2 developed during the Holocene, there is no conclusive evidence for this.

Regional occurrences of large-scale relict sorted patterns such as occur on British mountains are widely believed to be indicative of former permafrost (cf. Williams, 1975; Goldthwait, 1976; Washburn, 1979, 1980, 1985). If it is accepted that patterning results from density inversion in the former active layer with the resultant establishment of free convection cells (see Chapter 6), then the reported dimensions of most relict sorted circles (2.0–2.5 m) imply former active layer depths of 0.5–0.7 m on plateau surfaces. This is consistent with the maximum depth of sorting (0.6 m) recorded during excavation of large sorted circles on Beinn an Fhurain. However, on present evidence it is impossible to state when such conditions occurred. Most large relict sorted patterns may have developed either during the downwastage of the last ice sheet, or during the subsequent Loch Lomond Stadial; some, as argued above, may have been preserved from earlier periods of periglaciation. Moreover, the implied active layer depths may not relate to full-stadial conditions, but may reflect patterned ground formation within a rather deeper active layer during the development or degradation of the underlying permafrost.

Active sorted patterned ground

Of all the periglacial phenomena on British mountains, none has captured the attention of geomorphologists more than active sorted patterned ground. Small sorted stripes were first identified in the Lake District nearly a century ago (Ward, 1896), and by the 1930s researchers were enthusiastically describing active sorted patterning as confirmation that the higher parts of British mountains experience processes normally associated with arctic and alpine environments (Gregory, 1930; Simpson, 1932; Hollingworth, 1934; Hay, 1936). Active sorted patterned ground has subsequently been reported to occur in virtually all parts of upland Britain, including Snowdonia (Tallis &

Figure 10.4 Small active sorted circles, Beinn an Fhurain, NW Scotland.

Kershaw, 1959; Ball & Goodier, 1970), the Lake District (Hay, 1943; Caine, 1963a, b, 1972; Warburton, 1985, 1987), the Pennines (Tufnell, 1969), southern Scotland (Miller et al., 1954), the Hebrides (Godard, 1958; Ryder & McCann, 1971; Birks, 1973) and various parts of the Scottish Highlands (Godard, 1965; Kelletat, 1970a). From these accounts a number of common characteristics emerge. First, although many of the above authors have described active 'sorted polygons' on British mountains, all the supposed 'polygons' known to the authors or illustrated in the above accounts are circular, oval or irregular in plan and lack straight edges, hence are more accurately described as sorted circles or nets (Figure 10.4). Secondly, although reported dimensions vary considerably (Tables 10.1 and 10.2), active sorted circles rarely exceed 0.6 m and stripes 0.7 m in pattern width (repeat distance), and sorting is confined to the uppermost 0.2 m of soil or less. Most remarkable of all, however, is the consistency that emerges in all accounts regarding the maximum size of sorted clasts, which is invariably reported as 15–16 cm or less. Collectively, these data suggest that shallow ground freezing limits the maximum depth of sorting, which in turn dictates the dimensions of the resultant patterns and the maximum size of sorted clasts. If so, it follows that clast size alone imposes severe constraints on the occurrence of active sorted patterns in upland Britain; only regolith containing clasts that are predominantly smaller than c. 15 cm would appear to experience sorting into patterns under present conditions.

Structurally, active sorted patterns resemble miniature versions of the relict circles and stripes described in the previous section. Clasts tend to be concentrated in shallow troughs, 5–15 cm deep, that border fine cells or separate fine stripes (Figures 10.3–10.5). On striped slopes the depth of troughs tends to increase downslope, possibly as a result of erosion by surface wash that is concentrated in clast-filled furrows (Warburton, 1987). Some clasts are flat-lying, others are on edge, and on slopes many are aligned downslope. The transition from circles through elongate circles to stripes appears to occur at about 4–6°, but may be observed at only a few sites, such as the plateau between Foel Gach and Carnedd Llewelyn in Snowdonia (Tallis & Kershaw, 1959; Ball & Goodier, 1970). The maximum gradient on which stripes are reported to occur is 23°, on the southern flank of Tinto Hill in southern Scotland

Table 10.1: *Characteristics of some active sorted circles and polygons in upland Britain.*

Site	Lithology	Maximum sorting depth (cm)	Maximum cell width (cm)	Maximum pattern width (cm)	Maximum clast length[a] (cm)
The Storr Isle of Skye	Basalt		25	40	15
Orval Isle of Rhum	Basalt	7	25	30	12
Scùrr nan Gillean Isle of Rhum	Felsite	9	20	30	15
Tinto Hill Southern Scotland	Felsite	12	30	40	16
A'Mharconaich Drumochter Hills	Gneiss		30	40	15
Glen Lyon Southern Grampians	Schist		20	30	15
Lake District Northern England	Slate			50	
Carneddau Mountains North Wales	Slate	15	50	60	10

[a] Larger clasts are not incorporated in the network of the sorted pattern.
Sources: Godard, 1958; Tallis & Kershaw, 1959; Warburton, 1985; unpublished observations.

Table 10.2: *Characteristics of some active sorted stripes in upland Britain.*

Site	Lithology	Maximum sorting depth (cm)	Maximum cell width (cm)	Maximum pattern width (cm)	Maximum clast length[a] (cm)
Scùrr nan Gillean Isle of Rhum	Felsite		15	20	15
Ard Nev Isle of Rhum	Granophyre		12	25	7
Scaraben Caithness	Quartzite		20	40	
Cairngorm Mountains Scotland	Granite		10	20	10
Beinn a'Chuallaich Southern Grampians	Schist		25	40	15
Tinto Hill Southern Scotland	Felsite	20	45	70	16
Ben Cleuch Ochil Hills	Lavas		20	40	
High Pike Lake District	Slate		28	51	15
Helvellyn Lake District	Volcanics	13		52	13
Skiddaw Lake District	Slate	13		34	10
Grasmoor Lake District	Slate	6		27	15

[a] Larger clasts are not incorporated in the sorted pattern
Sources: Hollingworth, 1934; Miller *et al.*, 1954; Galloway, 1958; Caine, 1963a; Warburton, 1985; unpublished observations.

Figure 10.5 Active sorted stripes, Tinto Hill, southern Scotland.

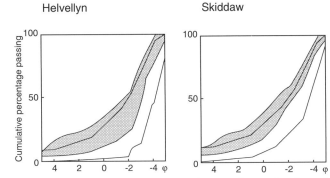

Figure 10.6 Envelope curves summarising the textural properties of sorted stripes and underlying parent materials for sites on Helvellyn (Borrowdale Volcanics) and Skiddaw (slate) in the Lake District. Each envelope is based on particle-size analysis of 7–8 samples. The stippled envelopes represent samples taken from fine stripes. From Warburton (1987).

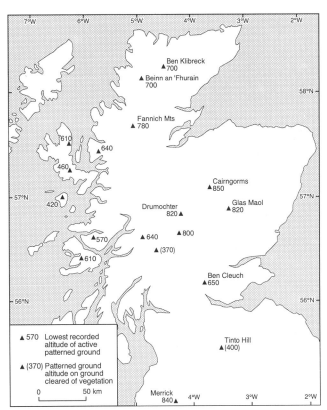

Figure 10.7 The lower altitudinal limits in metres of active sorted patterned ground at various sites in Scotland.

(Figure 10.5). In some areas, the lower limit of striped ground is marked by the lobate front of an active solifluction sheet (Warburton, 1985). During winter, fine stripes and the cells of sorted circles tend to experience preferential frost heave, and rise several centimetres above the level of the adjacent coarse debris. Excavation of frozen fine stripes in the Lake District and on Tinto Hill has revealed segregated ice lenses (Caine, 1963b; Ballantyne, 1981), and granulometric analyses of stripes in the Lake District suggests that these are confined to frost-susceptible soils containing appreciable quantities of silt (Warburton, 1985, 1987; Figure 10.6).

Two factors dictate the regional distribution of active sorted patterned ground in upland Britain: lithology and vegetation cover. We have seen above that patterning is apparently restricted to regolith in which clasts less than c. 15 cm long are embedded in a matrix of frost-susceptible fines. In consequence, patterns occur only on a limited range of rock types, particularly lavas, schists, shales and felsite (Tables 10.1 and 10.2), which tend to have yielded an abundance of fine soil and small platy clasts in response to periglacial weathering (Chapter 9). Equally notable is the poverty of active patterned ground on certain other lithologies, such as coarse-grained granite and Torridon Sandstone. Active sorted patterns are rare on the granite plateau of the Cairngorms, for example, and in the authors' experience

are absent on the Torridon Sandstone mountains of NW Scotland, even though these areas support wide expanses of unvegetated regolith. Lithology may also impose local constraints on pattern distribution: in the Lake District, occurrences of active sorted patterns are denser and more tightly clustered on Skiddaw slates than on the Borrowdale Volcanics (Caine, 1972).

Active sorted patterns occur only on sparsely vegetated terrain, hence are absent from mountains on which vegetation cover is complete. The shales that underlie many Welsh hills and those of southern Scotland, for example, support regolith ideal for pattern development, but are often completely covered by peat and vegetation; the same is true of many mountains underlain by mica-schist in the Scottish Highlands. Intriguingly, the development of sorted circles and stripes in many areas appears to have occurred following the relatively recent removal of an overlying cover of vegetation and peaty soil, and bare patterned areas are often bordered by eroding soil scarps (Hollingworth, 1934; Hay, 1936; Warburton, 1985). On Tinto Hill in southern Scotland, for example, superb sorted stripes (Figure 10.5) occupy regolith denuded of overlying peat and vegetation by burning and overgrazing (Miller et al., 1954). The influence of vegetation cover in preventing pattern development is evident in a marked westward decline in the lower limit of active patterning across Scotland, from 850 m in the Cairngorms to 420 m on Rhum (Figure 10.7), a trend that has been interpreted in terms of increase in exposure westwards and a corresponding decline in the altitude at which unvegetated ground occurs (Galloway, 1958; Kelletat, 1970a). Significantly, the 'anomalous' (bracketed) values of 400 m and 370 m in

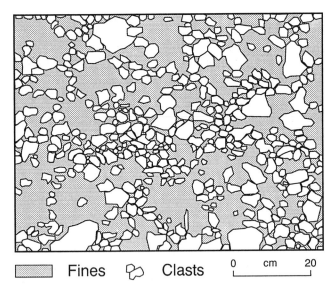

Fines   Clasts   0   cm   20

Figure 10.8 Sorted net developed on frost-susceptible felsite regolith in an experimental plot 30 m above sea-level in St Andrews, in eastern Scotland. The net formed over a few weeks during the severe winter of 1981–2, when ground freezing reached depths of up to 15 cm.

Figure 10.7 represent sites where vegetation has been artificially removed. This suggests that the dominant *regional* altitudinal control on active patterning is not temperature or precipitation, but simply the availability of vegetation-free sites on suitable regolith. This conclusion is supported by Spence's (1957) discovery of active frost-sorted patterned ground near sea level in Shetland, on serpentine soils inimical to vegetation colonisation. The occurrence of active patterns at such low levels suggests that frost sorting may, under favourable circumstances, operate almost anywhere in Great Britain. This proposition was investigated by one of the authors, who succeeded in 'growing' small sorted nets in an experimental plot of frost-susceptible felsite regolith in his garden at St Andrews, in eastern Scotland, during the winter of 1981–2 (Figure 10.8).

The recency of some pattern development is attested not only by the development of circles, nets and stripes on denuded regolith, but also by the appearance of sorted patterns on spoil heaps in the Lake District (Hollingworth, 1934) and in the southern Grampians (Ballantyne, 1981). There is evidence that small patterns may form or re-form rapidly under present winter conditions. Miller *et al.* (1954) reported complete re-formation of sorted stripes over disturbed ground on Tinto Hill over two winters, and further observations at the same site have revealed that both circles and stripes can re-form over disturbed plots within a few winter weeks. Tallis & Kershaw (1959) have observed that more fragile sorted patterns on the Carneddau in Snowdonia may be obliterated during the summer months by strong winds and driving rain, only to re-form during the succeeding winter. Reported rates of downslope clast movement on striped ground are rapid. On Grasmoor in the Lake District, Caine (1963a) recorded median rates of 70–254 mm over a single winter on gradients of 9–19°, and on a 23° slope on Tinto Hill in southern Scotland, marked clasts travelled over 1 m downslope between November 1977 and May 1980 (Ballantyne, 1987a). Such rates are comparable to the most rapid recorded (cf. Washburn, 1979, p.154). At both sites stone movement was restricted to periods of freeze–thaw activity, and

negligible displacement was recorded during the summer months when temperatures remained above freezing. Subsurface soil movement, however, tends to be much slower, and declines approximately exponentially with depth from a few centimetres per year near the surface to zero at *c.* 10 cm depth (Grasmoor) or *c.* 14 cm depth (Tinto Hill). Caine concluded that frost creep and sliding of surface clasts during thaw are the main agents of movement, but on Tinto Hill creep due to needle ice growth and collapse is apparently responsible for the accelerated movement of surface debris. Surface wash may also contribute to stone movement.

The processes underlying the development of small sorted patterns on British mountains have given rise to much speculation but little hard evidence. Following the lead given by Simpson (1932), almost all commentators have agreed that recent patterned ground develops as a result of 'freeze–thaw of water-soaked soil', but most accounts exhibit a calculated vagueness as how regular patterning is actually initiated. Hay (1936) noted that small sorted stripes in the Lake District merge upslope into a subsurface system of parallel rills, and postulated that such rills are of fundamental importance in generating the troughs in which clasts accumulate. He envisaged lateral sorting into such troughs as a result of needle ice growth and collapse. This explanation, however, cannot be extended to the formation of nets and circles on level ground. Caine (1963b) also ascribed the initiation of small stripes in the Lake District to differential frost heave, but attributed this to small-scale variations in the frost susceptibility of the underlying soil. Some support for this mechanism is provided by the frequent occurrence of adjacent sorted circles of widely-different cell diameters, which Ballantyne (1987a) interpreted as favouring the notion of pattern formation through upfreezing and outfreezing of clasts from randomly-spaced concentrations of frost-susceptible fines. It is difficult, however, to envisage how this process may ultimately produce regularly-spaced stripe patterns on sloping regolith.

A completely different approach to the explanation of pattern initiation has been adopted by Warburton (1987), who investigated the possibility that the formation of sorted stripes in the Lake District may reflect density inversion and consequent porewater convection during thaw of frozen ground, as proposed by Ray *et al.* (1983; see Chapter 6). This model predicts width to depth-of-sorting ratios of *c.* 2.7–3.8 for sorted stripes. Although Warburton found that the mean width to depth ratios of samples of stripes on the Borrowdale Volcanics and Skiddaw Slates fall within this range (*c.* 3.5 in both cases), many stripes proved to have width to depth ratios outside the range predicted by the convection model. Warburton suggested that such 'anomalies' may reflect the effects of modification of the initial sorted framework by mass movement and rillwash. His results, therefore, neither refute nor confirm convection as a possible mechanism for stripe initiation. Moreover, as argued in Chapter 6, the porewater convection model is of dubious validity for soils of limited hydraulic transmissibility, and subsequent research has provided evidence favouring pattern initiation through convection of liquefied soil rather than porewater (Hallet & Prestrud, 1986; Harris & Cook, 1988; Harris, 1990). The two models appear indistinguishable in terms of predicted width to depth ratios. Interestingly, Hollingworth (1934) long ago rejected convection as the mechanism for pattern initiation

Figure 10.9 Earth hummocks near the summit of Glas Maol (1068 m) in the SE Grampian Highlands of Scotland.

which take the form of domes of predominantly fine soil separated by a network of depressions (Figure 10.9). On slightly steeper slopes, earth hummocks sometimes become aligned downslope in parallel bands to form *hummock stripes* (Figure 10.10). With further increases in gradient, individual hummocks may merge to form nonsorted *relief stripes* consisting of alternating ridges and furrows aligned downslope. All three varieties of nonsorted patterned ground are restricted to ground that supports a complete vegetation mat. It must be emphasised, however, that not all earth hummocks grade into stripe forms with increasing gradient; hummocky microrelief has been reported to occur on gradients as steep as 14–17° in the northern Pennines (Tufnell, 1975; Pemberton, 1980). By contrast, on Ben Wyvis in northern Scotland, hummocks are replaced by hummock stripes on slopes of 1–6°, and these in turn grade into relief stripes on slopes of 6–11° (Ballantyne, 1986c). Hummocky microrelief is widespread in the British uplands, particularly on lithologies that have weathered to produce a cover of silt-rich frost-susceptible regolith. Hummocks are abundant on the cryoplanation terraces of Cox Tor in SW

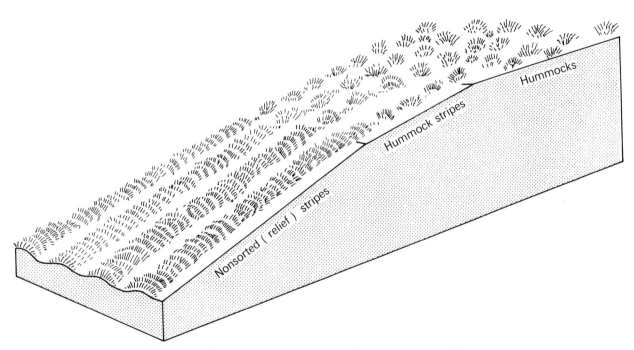

Figure 10.10 The transition from earth hummocks via hummock stripes to nonsorted relief stripes. Based on field sketches.

in the Lake District, on the ground that surface concentrations of clasts do not penetrate the undulating surface of the underlying clay loam, and hence provide no evidence for circulatory soil displacement (Figure 10.3(a)). It is arguable, though, that the structures excavated by Hollingworth represent stable equilibrium forms, in which redistribution of clasts by convection has effectively halted further lateral sorting.

Nonsorted patterned ground in upland Britain

Three types of nonsorted patterned ground occur on the mountains of Great Britain. By far the most common are *earth hummocks* (thufur), which generally occupy low-gradient sites and

England (Te Punga, 1956; Waters, 1962) yet appear to be absent from the coarser soils of the adjacent Dartmoor uplands (Gerrard, 1988b). In the Scottish Highlands, they are widespread on frost-susceptible montane soils derived from schists, but rare or absent on sandstone or granite regolith (Ballantyne, 1986c). Hummocks have also been described in the northern Pennines (Tufnell, 1975, 1985) and in Snowdonia (Ball & Goodier, 1970). Hummocky microrelief is not, however, confined to high ground, and has been mapped by Pemberton (1980) down to 150 m in northern England.

Measured dimensions of earth hummocks in Britain vary widely, but few exceed 0.3 m in height or 1.0 m in diameter (Table 10.3); they are therefore generally smaller than hum-

Table 10.3 :*Dimensions of earth hummocks on British mountains.*

Site	Height (cm)			Diameter (cm)		
	min	mean	max	min	mean	max
Great Dun Fell, Pennines (Tufnell, 1975)		20			35	
Knock Ore Gill, Pennines (Tufnell, 1975)		15			37	
Underbarrow and Scout Scars, N England (Tufnell, 1975)		28			71	
An Cabar, Ben Wyvis, Scotland (Ballantyne, 1981)	6	16	32	32	63	127
Glas Leathad, Ben Wyvis, Scotland (Ballantyne, 1981)	9	18	44	28	65	118
Drumochter Hills, Scotland (Chattopadhyay, 1982)	18	27	50	40	78	150

mocks in arctic and subarctic environments (Schunke & Zoltai, 1988; Chapter 6). They also differ in form. Earth hummocks in arctic lowlands and the thufur of Iceland are often knob-like in shape and closely spaced, with narrow interhummock depressions (cf. Tarnocai & Zoltai, 1978, figure 2; Washburn, 1979, figure 5.29; Schunke & Zoltai, 1988; Figure 10.1). In such areas, hummocks generally have width:height ratios of the order of 2:1. In contrast, the hummocks described on British mountains are characteristically dome-like and separated by rather broad depressions (Figure 10.9) with width to height ratios of 3:1 or 4:1. The low-level 'hummocks' described by Pemberton in northern England are even more subdued, taking the form of gentle undulations with mean width:height ratios of about 10:1.

Nonsorted hummock stripes and relief stripes appear to be rarer than hummocky microrelief. Small-scale relief stripes are well developed on steep (15–30°) slopes above 670 m on Y Lethr, in the Rhinog Mountains of north Wales. Here they take the form of vegetated ridges and troughs with a height difference of 0.1–0.2 m and a pattern width (from one ridge crest to the next) of 0.7–1.0 m (Goodier & Ball, 1969). Rather broader stripes occur on some Scottish mountains, for example on Glas Maol in the SE Grampians and on Merrick in the Southern Uplands. The most outstanding examples yet documented are those on Ben Wyvis in northern Scotland. Here hummock stripes with ridges typically 1.0 m in width are separated by furrows 0.4–0.7 m across, and grade downslope into relief stripes with a pattern width of around 2 m and a height difference of 0.1–0.3 m between adjacent ridges and troughs. The nonsorted stripes on Ben Wyvis tend to follow the direction of maximum slope, though they often meander slightly on gentle slopes and divide where the slope is convex (outward) in plan or join where the slope is concave in plan, thus maintaining a roughly equal spacing. Individual stripes on Ben Wyvis are traceable for up to 110 m downslope (Ballantyne, 1986c).

The literature on nonsorted patterned ground in upland Britain reveals little consensus regarding the age, genesis or significance of these landforms. Indeed, it seems likely that no single model is applicable in all cases, and some of the hummock forms that have been described may even be of nonperiglacial origin. Broadly, three sets of hypotheses have been proposed to account for hummocks and related nonsorted stripes on British mountains. First, it has been suggested that these features represent nonsorted patterns that initially developed under permafrost conditions during the Late Devensian Lateglacial. Second, some authors have advanced the view that supposed 'nonsorted' hummocks and stripes represent modification during the Holocene of initially sorted patterns. Finally, researchers working in the north of England have proposed that some hummocks at least may be of recent origin and are capable of developing on disturbed ground even under present conditions. These three views are examined below.

Investigation of the hummock and stripe patterns on Ben Wyvis led Ballantyne (1986c) to suggest that these are inherited from patterns initially developed during the Lateglacial. On this mountain there is a clear transition from hummocks to hummock stripes to relief stripes with increasing gradient, hence any explanation of hummock development must also account for the formation of nonsorted stripes on adjacent slopes. Trenches dug across both hummocks and stripes revealed well-developed podzols with a solum depth of up to 0.6 m. Significantly, clearly-defined soil horizons follow the undulations of the ground surface (Figure 10.11). This indicates that cryoturbation and solifluction are presently ineffective, as podzolic soil horizons are unlikely to survive internal mass displacement of sediment during thaw consolidation. Moreover, trenching showed that the sizes and concentrations of clasts underlying ridges and hummocks are identical to those underlying adjacent depressions, so inheritance from a pre-existing sorted pattern can be ruled out. Particle-size analyses suggest that both stripes and hummocks are restricted to particularly frost-susceptible regolith (>20% silt and clay); adjacent hummock-free regolith proved relatively deficient in silt. In view of the transitional nature of hummocks, hummock stripes and relief stripes on Ben Wyvis, it is reasonable to assume that all three are the products of the same underlying processes. The fact that stripes occupy slopes whilst hummocks are restricted to level or nearly level ground indicates either that stripes represent hummocks subsequently modified by mass movement (solifluction), or that the formative processes affected slopes and level areas in different ways. The former possibility seems unlikely. Although it is possible to envisage downslope 'stretching' of individual hummocks by solifluction, the development of regularly-spaced relief stripes demands that irregularly-spaced hummocks somehow migrate laterally across the slope and line up to form the ridges of the stripe pattern, a

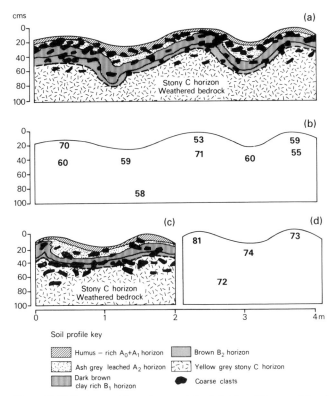

Figure 10.11 Sections cut through nonsorted relief stripes (diagrams (a) and (b)) and earth hummocks (diagrams (c) and (d)) on Ben Wyvis in northern Scotland. The figures on diagrams (b) and (d) refer to the mean intermediate axis lengths, in millimetres, of clasts sampled at the points shown.

(a) Lateglacial

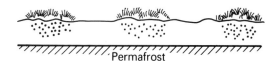

(b) End of Lateglacial/Early Flandrian

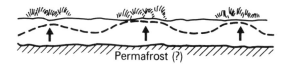

(c) Early Flandrian

(d) Flandrian

Figure 10.12 Possible evolution of earth hummocks, hummock stripes and nonsorted relief stripes as envisaged by Ballantyne (1986c). (a) Under stadial conditions, irregularly-spaced vegetated circles and regularly-spaced vegetation-defined stripes develop above the active layer. (b) The insulating effect of vegetation cover results in more rapid frost penetration under bare ground. The resulting undulating freezing plane induces mass displacement. (c) Hummocks and relief stripes formed by subsurface injection of sediment may have continued to develop as surface microrelief perpetuated irregular frost penetration. (d) Podzolisation results in depletion of clay and silt from the uppermost part of the soil, rendering it less frost-susceptible. Cryoturbation is consequently diminished.

colourful but somewhat implausible notion. Furthermore, clast fabrics in the stripe patterns lack the strong downslope preferred orientation characteristic of soliflucted debris. It therefore appears that the stripes developed *ab initio*, and do not represent hummocks modified by slope processes.

As outlined in Chapter 6, hummock formation is believed to result from differential frost heaving (Van Vliet-Lanoë, 1988a,b), cryostatic pressures set up during ground freezing (Nicholson, 1976) or, more likely, from mass displacement of liquefied soil during thaw (Mackay, 1979b, 1980). All three models invoke uneven ground freezing and thawing. To explain the transition between hummocks and stripes, Ballantyne (1986c) suggested that uneven freezing during the Lateglacial was produced by the development of vegetation-defined nonsorted patterns, with irregularly-spaced clumps of vegetation on level ground but regularly-spaced vegetated stripes following percolines (zones of concentrated subsurface water movement) on slopes. Such vegetation-defined patterns are common, for example, on the Canadian arctic archipelago (e.g. Washburn, 1947). Once established, the insulating effect of such vegetation cover would cause more rapid frost penetration under adjacent bare ground, which would in turn induce mass displacement and updoming of vegetated areas during freezing or thawing, in the manner envisaged by Nicholson (1976) and Mackay (1979b, 1980) respectively. In this way both hummocks and relief stripes would develop simultaneously, grading into each other (as hummock stripes) on gentle slopes. A similar evolution was envisaged by Washburn (1979, p. 153) for the Breckland stripes of eastern England. Ballantyne speculated

that the hummocks and stripes may have ceased to be active as a result of progressive eluviation of clay and silt from upper soil horizons during the Flandrian (Figure 10.12). Three further points of evidence favour a Lateglacial origin for nonsorted hummock-stripe assemblages in upland Britain. The first is the absence of such assemblages on ground occupied by glacier ice during the Loch Lomond Stadial. The second is the low dome-shaped form of the hummocks, which suggests degradation of originally higher and more sharply defined forms. The last is the maturity of the podzols developed within such features, which indicates prolonged stability for a period of at least several centuries and probably millennia.

In contrast to the model outlined above, some researchers have viewed hummocks and relief stripes as relict sorted patterns that have been modified by mass displacement and overgrown by vegetation during the Flandrian. Ironically, this hypothesis also had its origins on Ben Wyvis. Galloway (1961a) noted the similarity in planform between the 'ring' (hummock) and stripe patterns on this mountain and sorted patterns he had

Figure 10.13 Sand hummocks on An Teallach, NW Scotland.

observed on Greenland, and on this basis inferred that the former had evolved from sorted nets and stripes that themselves developed during a period of 'much harsher climate'. In support of this conjecture, Galloway claimed that larger stones tend to be concentrated under the depressions that separate hummocks and stripes, but later excavations on the same mountain (Figure 10.11) revealed no evidence for such lateral sorting. Galloway's hypothesis was revived by Chattopadhyay (1982) on the basis of excavation of three hummocks on the Drumochter Hills in the SE Grampians. Chattopadhyay suggested that these hummocks evolved from sorted circles that had become covered by a vegetation mat and experienced updoming because of more rapid frost penetration under the stony borders. The sections he depicted, however, show that the stony layers that underlie the depressions extend under the hummocks themselves, and are unlike those typical of sorted circles (e.g. Figure 10.3). It is notable, though, that the small relief stripes investigated on the Rhinog Mountains in Wales by Goodier & Ball (1969) also display slight lateral sorting. Given the similarity in dimensions between these features and partly-vegetated sorted stripes elsewhere (Figure 10.2), it is not unlikely that such 'nonsorted' patterns simply represent vegetation-covered sorted stripes. Goodier & Ball suggested that these stripes were formed during the Little Ice Age, between the sixteenth and eighteenth centuries AD, but their evidence for this is circumstantial.

Finally, Tufnell (1975) and Pemberton (1980) have advanced the view that some hummocks in northern England are of recent origin. It is worth noting, however, that those investigated by Tufnell in the northern Pennines tend to be smaller than the features represented on Scottish mountains (Table 10.3), whilst the features described as 'hummocks' by Pemberton are very subdued undulations of the ground surface. Neither type grades downslope into nonsorted stripes. Tufnell considered that hummocks may form under present conditions on the grounds that 'small hummocks occur on ground that must have been disturbed when the radio station and masts [on Great Dun Fell] were built' (1975, p. 367), but he offered little explanation of hummock formation besides concluding that those in the northern Pennines 'are similar to the Icelandic thufur and ... consequently of periglacial origin' (p. 365). As noted earlier, however, the similarity between earth hummocks in Britain and mature Icelandic thufur is questionable. Some support is offered for Tufnell's views by reports of hummocks of apparently recent origin on relatively low ground in the central

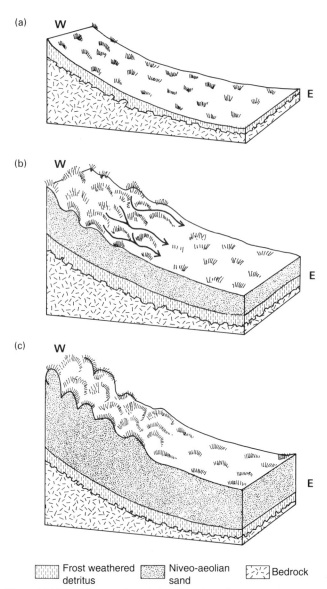

| | Frost weathered detritus | | Niveo-aeolian sand | | Bedrock |

Figure 10.14 Sand hummock evolution through preferential accumulation of niveo-aeolian sand on tussocks and flushing of sand from between tussocks by nival meltwater (schematic).

and eastern Pennines (Cotton, 1968) and the Vale of Eden in northern England (Pemberton, 1980). The latter author documented instances where subdued hummocks occur on formerly cultivated ground, and concluded that all of the hummocky microrelief in this area had developed within historical times, during intervals of climatic severity. Again, however, this research throws little light on hummock formation. Pemberton rejected the possibility that the hummocks he examined may simply be degraded molehills and though he concluded that they are 'clearly the products of frost penetration' (1980, p. 500), he provided no direct evidence for this assertion.

In sum, little can be said for certain concerning the age and origin of many nonsorted patterned ground features in upland Britain. The development of mature soils in some clearly indicates prolonged inactivity, but whereas it is attractive to ascribe a Lateglacial age for such features this cannot be verified on present evidence. Similarly, although there are clearly strong genetic affinities between some hilltop hummock fields and nonsorted stripes on adjacent slopes, insufficient research has been carried out to distinguish between the rival claims of Ballantyne (1986c), who interpreted such features as derived from vegetation-defined nonsorted patterns, and those authors who prefer to interpret such forms as having evolved from sorted patterns. The case for recent development of hummocks by cryoturbation also remains inconclusive. Whilst the work of Tufnell (1975) and Pemberton (1980) shows that some hummocky microrelief may be of recent origin, it fails to demonstrate that these forms result from freezing and thawing of the ground. Closer examination of the internal structures of such recent hummocks may prove rewarding in determining whether these reflect cryogenic mass displacement of soil or have some completely different origin.

Of interest in this context are spectacular hummocks that occupy recent niveo-aeolian sand deposits on An Teallach and other Torridon Sandstone mountains in NW Scotland (see Chapter 13). These are rounded rather than dome-shaped and much more closely spaced than, for example, the relict hummocks on Ben Wyvis (Figure 10.13). They are also much larger, averaging 0.5 m in height and 0.7 m in diameter, and occur on slopes of 4–18° rather than level ground. They are not associated with stripe development. Such hummocks, although superficially similar to earth hummocks, do not appear to be frost-action phenomena, and were interpreted by Ballantyne (1986c) as having developed over the last century as the result of trapping of niveo-aeolian sand by vegetation tussocks on slopes that are sufficiently steep to permit eluviation of sand from between the tussocks by nival meltwater (Figure 10.14). Ballantyne suggested that such features be termed *sand hummocks* to distinguish them from 'true' earth hummocks formed by cryoturbation.

Summary

The mountains of Great Britain support a wide variety of sorted and nonsorted patterned ground phenomena. Large-scale sorted circles and stripes are relics of stadial and probably permafrost conditions. Most small-scale patterned ground, in contrast, is actively developing even under the much milder maritime periglacial conditions of the present. The age and status of nonsorted hummocks and relief stripes, however, remain debatable. All of these forms invite further research. The age and palaeoenvironmental significance of relict patterned ground forms are still uncertain, and although active forms offer a superb and readily accessible 'laboratory' for the study of ground patterning processes, only a few researchers (e.g. Caine, 1963b; Warburton, 1987) have made fruitful use of this facility. It is widely agreed, however, that the development of patterned ground represents a particular response to freezing and thawing of the soil. Another outcome of this process has been the slow downslope movement of soil on mountain slopes. The significance of this process, and the landforms it has produced, are examined in the following chapter.

11
Solifluction landforms in upland Britain

Both upland and lowland landscapes in Britain have been substantially modified by *solifluction*, the slow downslope movement of soil under conditions of seasonal freezing and thawing. In soils with little cohesion, the principal mechanisms of solifluction are *frost creep* resulting from repeated heaving and resettling of the soil, and *gelifluction*, which involves slow saturated flow during thaw consolidation of ice-rich frozen soil. In clay-rich cohesive soils, however, thaw consolidation and particularly the melt of ice lenses near the base of the active layer may result in localised translational sliding of the soil mass (Harris, 1987b). In lowland Britain, the effects of solifluction include the lowering of valley-side gradients, the reworking of soil or drift and the accumulation of head deposits on valley floors and shoreline platforms (Chapter 7). In southern England in particular such effects have been renewed during successive glacial stages, so that solifluction may be considered the principal agent of hillslope evolution. In contrast, the solifluction phenomena of most upland areas have developed only since exposure of the land surface by downwastage of the last ice sheet. Upland Britain nonetheless supports a fascinating range of solifluction landforms and deposits. Some of these are relict features that relate to conditions much colder than those of the present, whilst others, on the higher parts of mountains, are currently active.

In many upland areas it is possible to identify a catena of solifluction phenomena (Figure 11.1). Gentle upper slopes are occupied by smooth *solifluction sheets* composed of frost-weathered regolith. These often form a continuous debris mantle that completely buries the underlying bedrock. As gradient increases beyond about 5°, such solifluction sheets often terminate downslope in regular steps or *risers*, which run continuously along the slope for tens or hundreds of metres. Successive sheets descending downslope in this manner give a 'staircase' appearance to the upper parts of some mountains. With further increases in gradient, the planform of such risers becomes increasingly crenulate or lobate in response to local variations in the rate of movement of debris, forming *lobate solifluction sheets*. Many of the features referred to as *solifluction lobes* represent the lobate extensions of such sheets. Farther downslope still, lobate sheets tend to override one another, isolating individual lobes from their neighbours. In addition, upper slopes currently subject to solifluction invariably support *ploughing boulders*, which move downslope at a rate exceeding that of the surrounding soil, leaving depressions or furrows in their wakes.

On the steep central portions of mountain slopes, solifluction lobes tend to die out downslope and to be replaced by a smooth, unbroken sheet of solifluction debris. On upper slopes this may consist of solifluction frost-weathered detritus of the sort described in Chapter 9 (cf. Tivy, 1962; Ragg & Bibby, 1966; Watson, 1969a; Innes, 1986). Farther downslope, however, it more often comprises glacial drift that has been smoothed and displaced by solifluction to form a *solifluction till sheet*. In areas that were reoccupied by glacier ice during the Loch Lomond Readvance, such smooth solifluction sheets or soliflucted till sheets frequently terminate abruptly downslope against thicker, gullied drift deposited during this final phase of glaciation. Conversely, in some upland areas that escaped glaciation at this time, soliflucted drift acccumulated extensively on valley floors (Figure 11.1). Such valley floor accumulations now take the form of drift terraces as a result of subsequent river incision, and hence are usually referred to as *solifluction terraces*. The same term has been applied to several different phenomena, but is retained here as the process of 'terracing' through river incision is analogous to that responsible for the formation of true alluvial terraces. Some valley-floor solifluction terraces support a secondary microrelief of risers and lobes, particularly near the break of slope between the terrace tread and the slope behind.

Superimposed on this relatively simple picture are a number of complexities. Although it is widely agreed that valley-floor solifluction terraces are relict phenomena of Late Devensian age, both relict and active solifluction sheets and lobes occur on the upper parts of mountains, sometimes in juxtaposition. Further, even amongst the relict solifluction phenomena on upper slopes there is considerable structural diversity: at least three varieties of solifluction lobe may be distinguished. Finally, some upper slope solifluction sheets grade into stepped *turf-banked terraces* that reflect retardation of creeping debris by bands of vegetation. Such terraces have developed in response to the combined action of mass-movement and the effects of wind on vegetation cover, and are considered more fully in Chapter 13. Here, attention is focused first on the characteristics of valley-floor solifluction deposits, then on relict periglacial phenomena on the upper parts of mountains, and finally on the nature and effects of active solifluction at high altitudes.

Valley-floor solifluction terraces

Valley-floor solifluction terraces are characterised by two components: a steep frontal scarp or bluff, 3–20 m high, and a gently-sloping smooth tread that merges upslope with soliflucted slope deposits (Figure 11.2). Such treads commonly slope towards the valley axis at gradients of less than 15°, though valley-floor accumulations with gradients of up to 30° have been identified in the Cheviot Hills (Douglas & Harrison, 1987) and shallow depressions occur on some terraces in south Wales (Crampton & Taylor, 1967). Solifluction terraces may occupy one or both sides of valley floors, are commonly 20–300 m wide, and may extend continuously upvalley for hundreds of

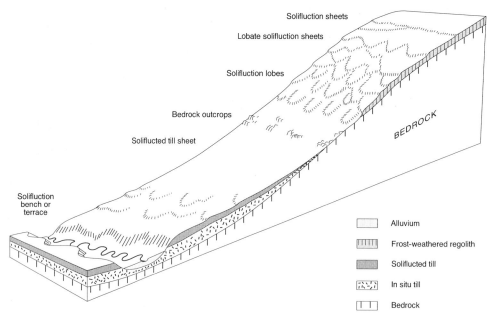

Figure 11.1 A catena of solifluction phenomena typical of those found on the slopes of British mountains (schematic).

Figure 11.2 Valley-floor solifluction terraces near Sourhope in the Cheviot Hills, northern England. Photograph by Dr T.D. Douglas.

metres (Figure 11.3). They occur in south and central Wales (Crampton & Taylor, 1967; Watson, 1969b, 1970; Potts, 1971), Snowdonia (Ball & Goodier, 1970), the Howgill Fells (Harvey, Alexander & James, 1984), Derbyshire (Wilson, 1981), the Isle of Man (Thomas, 1976; Dackombe & Thomas, 1985), the Cheviot Hills (Douglas & Harrison, 1984, 1985, 1987) and parts of the Southern Uplands of Scotland (Galloway, 1961a). The 'rubble-drift' terraces that occupy upland valleys on Dartmoor, though smaller, are probably of similar origin (Waters, 1965). Certain aspects of terrace distribution are interesting. First, although most of these areas are within the limits of the last ice sheet, all lie outside those of the subsequent Loch Lomond Readvance, which suggests that terrace development ceased with the onset of warmer conditions during the Flandrian. Secondly, terraces are most commonly developed on lithologies that have weathered to produce drift or regolith containing an abundance of silt, particularly mudstones, slates and shales in Wales, Derbyshire, the Isle of Man and the Southern

Uplands, and andesites in the lower parts of the Cheviots. Thirdly, terraces are well developed only on broad valley floors: they tend to be absent or poorly represented in steep, incised valleys, and for this reason rarely occur above 300–500 m. Finally, various researchers have claimed to identify systematic asymmetry in terrace development. Watson (1969b), for instance, found that terraces in the Aberystwyth area are preferentially developed below slopes with northerly aspects, and attributed this to the effects of more pronounced frost weathering on such slopes. Conversely, Crampton & Taylor (1967) envisaged a more complex cyclical sequence of events, with an initial preferential development of terraces at the foot of slopes with southerly or westerly aspects, arguing that such aspects favour the development of a deeper active layer and thus more effective frost creep and gelifluction. Potts (1971) and Douglas & Harrison (1987), however, detected no systematic asymmetry in terrace development, and concluded that terrace distribution primarily reflects the availability of drift or regolith upslope.

In general, terrace sediments consist of massive, poorly-sorted diamictons that display only crude stratification, though embedded clasts exhibit the strong preferred downslope orientation typical of solifluction deposits. Diamicton units are sometimes interbedded with thinner units of sorted gravel indicative of reworking by slopewash (Dackombe & Thomas, 1985). In terraces near Aberystwyth, Watson (1969b) identified a recurrent threefold sequence comprising washed gravels sandwiched between units of massive stony clay. The platy texture of terrace sediments in south Wales, together with the presence of indurated horizons (fragipans), led Crampton & Taylor (1967) to conclude that terrace sediments had accumulated as a result of solifluction over a former permafrost table. The sedimentological properties of valley-floor solifluction deposits in the Vale of Edale (Derbyshire) have been investigated in some detail by Wilson (1981). This site lies just outside the limits of the last ice sheet, and the surrounding slopes and plateaux support a range of relict periglacial phenomena, including tors and

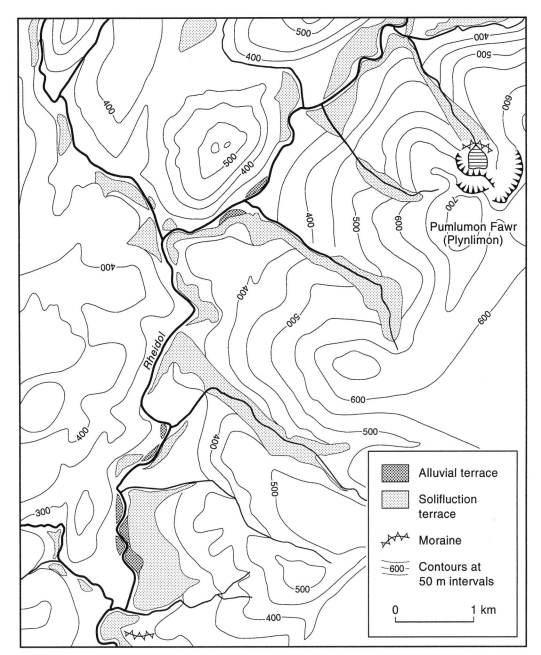

Figure 11.3 Solifluction terraces of the upper Rheidol basin, near Plynlimon, mid-Wales. Based on a map by Watson (1969b).

cryoplanation terraces. The soliflucted valley fill is derived from Namurian shales and mantles the valley bottom to a depth of 10 m or more. Although the valley fill has no distinctive surface morphology, variations in sediment thickness on the south side of the valley suggest deposition in the form of a number of contiguous fan-like spreads. As in similar deposits elsewhere, clasts exhibit a strong preferred downslope orientation. The deposits are poorly sorted, with the clay content for bulk samples ranging from 4.6% to 17.6%, and gravel content ranging from 30% to 72%. The matrix, though predominantly sandy, contains 11–27% clay. Significantly, Wilson demonstrated that the valley-floor solifluctate at Edale is frost-susceptible and has low plastic and liquid limits.

The source of soliflucted sediment has attracted some controversy. Watson (1969b) maintained that solifluction terraces near Aberystwyth are largely composed of sediments derived from frost-weathered regolith, but Potts (1971) demonstrated convincingly that terraces elsewhere in mid Wales are composed of soliflucted till. Detailed work by Douglas & Harrison (1984, 1985, 1987) in the Cheviots showed that terrace sediments may have had more than one source. In granite areas, exposures in solifluction terraces exhibit a tripartite sequence of deposits overlying *in situ* till. The lowermost unit comprises soliflucted till with a high proportion of fines. This is covered by up to 5 m of soliflucted growan (granite gruss), which in turn is overlain by up to 1 m of soliflucted gelifractate derived from frost weather-

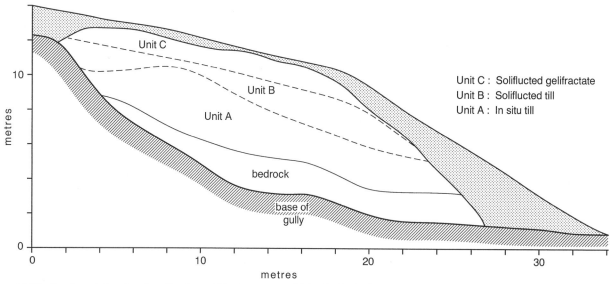

Figure 11.4 Section through the edge of a valley-floor solifluction terrace at Makendon in the Cheviot Hills. After Douglas & Harrison (1987).

ing of bedrock outcrops upslope. In andesite areas, soliflucted gelifractate directly overlies soliflucted till, with no intervening growan unit. At Makendon, for example, a section cut through a terrace revealed *in situ* till up to 5 m thick overlain by soliflucted till up to 2.5 m thick, which in turn is overlain by a third unit comprising up to 2 m of soliflucted andesitic gelifractate (Figure 11.4). Progressive stripping of slope deposits from the surface downwards may be indicated by these sequences. In granite areas, first till, then growan, then frost-weathered regolith appear to have been removed in succession, and redeposited in an inverted stratigraphic sequence on lower slopes and valley floors. This inversion of the original sedimentary sequence is similar to that proposed for head deposits on granite terrain in SW England (Waters, 1964a; Scourse, 1987; Chapter 7).

There is general consensus that valley-floor solifluction terraces in upland Britain accumulated as a result of solifluction during the Late Devensian. In areas occupied by the last ice sheet, the build up of solifluction deposits on valley floors presumably occurred during the Loch Lomond Stadial of *c.* 11–10 ka BP, though accumulation may have commenced following ice-sheet retreat in the waning stages of the Dimlington Stadial prior to *c.* 13 ka BP (Douglas & Harrison, 1984, 1987). The terraced form of these deposits apparently results from the formation of steep frontal bluffs by subsequent river incision and lateral erosion. The thickness of deposit implied by bluffs up to 20 m high, however, may give a misleading impression of the efficacy of Lateglacial solifluction; as the Cheviot sections show, the bulk of some terraces may comprise *in situ* till veneered by only a few metres of solifluccted debris. Moreover, the predominance of solifluccted till within many such veneers suggests that much of the solifluccted debris may be of fairly local provenance.

Relict solifluction phenomena on the upper slopes of British mountains

A remarkable range of solifluction landforms cover the convex upper slopes of British mountains, and such diversity has engendered much terminological confusion. In this account, the

Figure 11.5 Lobate boulder sheets in the Cairngorm Mountains, Scotland.

term 'sheet' refers to a detritus mantle that is terminated downslope by a long, low step or riser. This term is preferred to 'terrace', as the latter has connotations of horizontality. As noted earlier, on low-gradient slopes sheets normally possess fairly straight risers, but with increasing gradient the planform of the risers becomes increasingly crenulate and irregular (Figure 11.1). Most features referred to as 'lobes' actually represent the U-shaped frontal parts of lobate detritus sheets (King, 1972), though isolated lobes may occur as a result of the overriding of one sheet by another.

Types and characteristics of relict solifluction landforms

Sheets and lobes have frequently been classified as 'stone-banked', 'turf-banked' and 'vegetation-covered' (e.g. Galloway, 1961a; Mottershead & White, 1969; Ball & Goodier, 1970) but such terms vary in usage and do not necessarily distinguish landforms of different structure, significance or age. More recent research has employed internal structure as a basis for distinguishing different classes of solifluction landform. Three categories have been differentiated on British mountains:

boulder sheets and lobes, sorted solifluction sheets and lobes, and nonsorted sheets and lobes (Ballantyne, 1984). All three types occur on slopes of between 6° and 36°.

Boulder sheets and lobes (Figures 11.5 and 11.6) are com-

Figure 11.6 Boulder lobe, Cairngorm Mountains, Scotland.

posed of very coarse detritus, and are amongst the most impressive of all upland periglacial landforms. They occur on block-slopes and debris-mantled slopes where the regolith contains a high concentration of large boulders. Superb examples festoon the slopes of granite massifs such as the Cairngorms, Lochnagar and Dartmoor, whilst rather smaller examples occur on quartzite, gneiss and Torridon Sandstone. In the Lake District they cover the long west-facing slope of Scafell, and are widespread on Helvellyn. In Snowdonia, good examples occur on Y Garn and Carnedd Llywelyn above 835 m. Boulder lobes often extend far down the flanks of Scottish mountains. In the Cairngorms they descend to 540 m, whilst in the Hebrides they occur at 350 m or lower (Figure 11.7). Although in some areas the distribution of these features is truncated downslope by the upper limits of the Loch Lomond Readvance, elsewhere their lower limits frequently coincide with a break of slope or the upper limit of thick peat cover. Boulder sheets and lobes are generally more massive than the other types identified above, with risers up to 6 m high on granite mountains in Scotland but rarely exceeding 3 m on other lithologies, a contrast that appears to be related to the size of component boulders (Shaw,

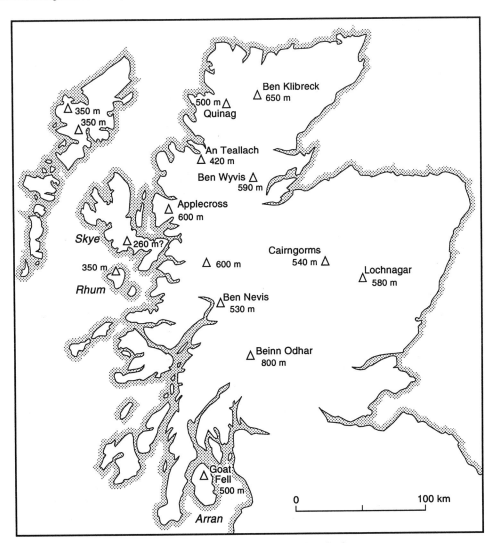

Figure 11.7 Lower limits of boulder sheets and boulder lobes in the Highlands and Islands of Scotland.

Table 11.1: *Characteristics of solifluction lobes on some Scottish mountains.*

Location	Lithology	Altitude range (m)	Gradient range (degrees)	Riser height (m)			Lobe width (m)		
				min	mean	max	min	mean	max
Boulder lobes									
Lochnagar area	Granite	580–1110	10–34	0.3	1.5	5.9	3.9	14.8	33.3
Cairngorms	Granite	540–1210	7–35	0.4	1.2	3.5	3.5	8.2	16.5
An Teallach	Sandstone	420–890	10–31	0.5	2.2	3.9	4.9	9.5	18.7
Sorted solifluction lobes									
Glas Maol	Schist	–	18–33	0.2	0.6	1.3	3.9	6.3	10.4
Drumochter Hills	Schist	640–900	8–35	0.3	0.7	1.3	1.7	5.3	14.0
Creag Meagaidh	Schist	650–1100	8–35	0.3	0.8	1.3	1.8	6.8	16.5
Relict nonsorted solifluction lobes									
Ben Wyvis	Schist	650–950	10–36	0.4	0.8	1.9	4.5	7.6	14.5
Active nonsorted solifluction lobes									
Drumochter Hills	Schist	750–950	6–27	0.3	0.4	0.7	1.9	3.4	6.7
Ben Wyvis	Schist	800–1000	10–36	0.2	0.5	0.9	2.0	5.7	10.3
Fannich Mountains	Schist	800–990	10–34	0.3	0.5	1.0	1.9	6.1	10.2

Sources: Shaw, 1977; Ballantyne, 1981 and unpublished data; Chattopadhyay, 1982.

1977; Table 11.1). Some are entirely vegetated, but more often vegetation cover is restricted to the treads (thus 'stone-banked lobes'). Other widely-reported characteristics of boulder sheets and lobes include gentle riser angles (typically 20–30° except on steep slopes) and a weak preferred downslope orientation of boulders. Individual lobes vary greatly in width, from as little as 4 m to over 20 m (Table 11.1). The most remarkable attribute of these landforms is the coarseness of constituent debris. According to Shaw (1977), the average diameter of granite boulders in some lobes on Lochnagar exceeds 1.0 m. More representative, perhaps, are lobes in the Cairngorms sampled by Chattopadhyay (1982), who measured mean boulder diameters of 0.2–0.5 m (Figure 11.6). Material of this calibre represents a formidable deterrent to excavation, but Shaw nonetheless succeeded in trenching five boulder lobes. In each of these he found a layer of peat overlying '...a mass of boulders without interstitial fines' (1977, p. 111) and no evidence of size sorting. Other researchers, however, have found an infill of sand or fine gravel between the boulders and some evidence of vertical and lateral sorting (Metcalfe, 1950; King, 1972; Ballantyne, 1981; Figure 11.8).

Sorted solifluction sheets and lobes are those in which a surface concentration of boulders is underlain at shallow depth by fine soil with a variable stone content. These tend to occur on lithologies that have weathered to produce a mixture of boulders and fine-grained soil. Good examples are developed on Moine and Dalradian schists in the Scottish Highlands, for example on Ben Wyvis, the Fannich Mountains, the Drumochter Hills and Glas Maol in the SE Grampians. None have been reported below 650 m. On average, sorted solifluction lobes are about half as thick as boulder lobes. Risers are steep (characteristically 30–50°), rarely exceed 1.3 m in height and are sometimes vegetated, though treads are often vegetation-free (thus 'turf-banked lobes'). Excavations by Ballantyne (1981) and Chattopadhyay (1982) have shown that sorted lobes

comprise an openwork concentration of boulders, 0.2–0.4 m thick, that merges downwards into a stony diamicton of similar or greater depth. This in turn frequently overlies an organic-rich buried soil horizon (Figure 11.8). Grain-size analyses of the fine matrix within four such lobes have demonstrated that it contains 14–26% silt and is consequently frost-susceptible.

In contrast, *nonsorted sheets and lobes* are those composed of a fairly homogeneous mass of soil and stones, with little or no evidence for lateral or vertical sorting. Such features usually support a complete vegetation cover (thus 'vegetation-covered lobes'). Some nonsorted sheets and lobes on the higher mountains of north Wales, the Lake District and Scotland are active under present conditions; others, particularly at lower alitudes, are relict. Active forms tend to be smaller, with riser heights not exceeding 1.0 m, have steep risers, and usually overlie buried organic material of Flandrian age. Conversely, relict nonsorted sheets and lobes are up to 2.0 m thick, are fronted by degraded risers, and do not overlie organic deposits (Table 11.1; Figure 11.8). Both commonly occur on lithologies, such as mica-schist, slate and shale, that have weathered to produce abundant small clasts embedded in a fine-grained, frost-susceptible matrix. Large relict nonsorted lobes occur, for example, immediately outside the former margins of a Lateglacial ice cap on the schists of the Gaick plateau in the central Grampians (Sissons, 1974b), and downslope of active lobes on the flanks of Ben Wyvis (Ballantyne, 1984). Somewhat degraded nonsorted lobes of inferred Late Devensian age are also found on mountains underlain by shales in the Southern Uplands (Tivy, 1962; Ragg & Bibby, 1966) and on the Carneddau, in Snowdonia, above 760 m (Ball & Goodier, 1970). Relict nonsorted sheets and lobes, however, are not restricted to high-level sites. Features of this type descend to only 170 m on the remote island of St Kilda, in areas that escaped glaciation during the Late Devensian (Sutherland et al., 1984). Even more remarkable are assemblages of relict nonsorted sheets and lobes described by

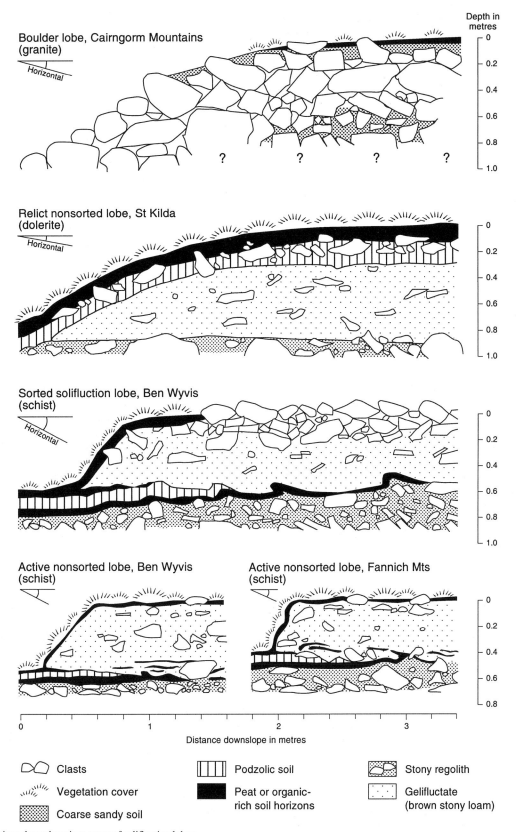

Figure 11.8 Sections through various types of solifluction lobes.

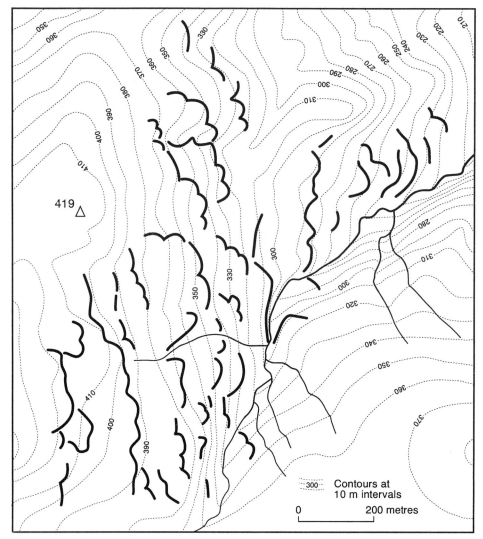

Figure 11.9 Large relict solifluction sheets and lobes in the valley of the Afon Cerniog, central Wales. After Pissart (1963a).

Pissart (1963a, b) at altitudes around 300 m in the Cerniog, Cledan, Colwyn and Nodwydd valleys in central Wales. Here, vast solifluction sheets 50–500 m wide form broad flights across gentle slopes with easterly and north-easterly aspects (Figure 11.9). Tread gradients range from only 1.5° to 7°, riser angles from 12° to 27°, and riser heights from 2 m to no less than 15 m. Pissart believed that under stadial conditions the drift of snow onto lee slopes resulted in the formation of transverse snowbeds, and that the huge solifluction landforms in these valleys reflect slow downslope movement of drift saturated by nival meltwater.

Age and formation of relict solifluction landforms

Although it is widely accepted that boulder sheets and lobes are relict features relating to a more severe climate than that of the present, the age of these landforms has been subject to debate. Sugden (1971) and King (1972) proposed that the boulder lobes of the Cairngorms may have formed during the 'Little Ice Age' of the sixteenth to eighteenth centuries AD. This view, however, is supported only by lichenometric measurements of dubi-

ous validity (King, 1972), and appears to have been conditioned by Sugden's (1970) belief that much of the Cairngorm plateau was occupied by glacier ice during the Loch Lomond Stadial, a view that implies that all periglacial features on the plateau are of later (i.e. Flandrian) age. Sissons (1979a), however, has argued convincingly that only corrie and valley glaciers of restricted extent developed in the Cairngorms during the Loch Lomond Stadial, so that sites presently occupied by boulder sheets and lobes were exposed to the severe periglacial conditions of the time. Strong support for a Lateglacial age for these features is provided by numerous observations that boulder lobes extend right down to the upper limits of Loch Lomond Readvance glaciers, but never occur within these limits (e.g. Sissons, 1972; Sissons & Grant, 1972; Shaw, 1977; Chattopadhyay, 1982). Such observations imply that features of this type ceased to develop under the relatively mild conditions of the Flandrian. So consistent is the mutual exclusivity of boulder lobes and areas occupied by the Loch Lomond Readvance that presence of the former is often employed as a criterion for delimiting the vertical dimensions of the latter (e.g. Thorp, 1981, 1986).

The same argument is valid for large relict nonsorted sheets and lobes. As noted above, superb examples occur immediately outside the limits of an ice cap that developed on the Gaick plateau during the Loch Lomond Stadial. On ground formerly occupied by this ice cap, however, relict solifluction landforms are completely absent. This dramatic contrast strongly suggests that these features developed before the end of the stadial (Sissons, 1974b). Moreover, numerous miniature meltwater channels that cross the area formerly occupied by the Gaick ice cap are fresh and unaffected by solifluction. From this evidence, Sissons argued that solifluction has been of negligible importance on the plateau since the ice cap disappeared, which reinforces his conclusion that the relict solifluction features here must have developed before the end of the Loch Lomond Stadial. In mountain areas that were occupied by active, warm-based (i.e. erosive) glaciers during the last ice-sheet maximum, the above observations imply that both boulder sheets and lobes and large relict nonsorted solifluction sheets and lobes developed in the interval between ice-sheet downwastage and the end of the Loch Lomond Stadial, and it seems likely that the forms now visible were last mobile during the latter period. Similar landforms in areas that remained unglaciated during the Devensian (such as the boulder lobes of Dartmoor and the relict solifluction lobes of St Kilda) may, however, have developed over much longer timescales. Sugden (1971) has mooted the interesting possibility that some of the larger periglacial features (such as boulder sheets and lobes) on the Cairngorms may have survived the Late Devensian under the protective cover of a thin, cold-based ice cap, but the validity of this suggestion has not been established.

Although the close relationship between boulder lobes and relict nonsorted lobes and the limits of Loch Lomond Stadial glaciers indicates that such lobes experienced negligible downslope movement after the end of the stadial, this does not appear to have been the case with sorted solifluction landforms. These also occur only outside the limits of the Loch Lomond Readvance, and display a depth of sorting (up to about 0.8 m) compatible with development under stadial conditions. Both of these characteristics indicate a Lateglacial origin. However, as noted earlier, excavations have shown that some sorted lobes overlie buried organic soils (Figure 11.8). A soil horizon buried under a sorted lobe on Ben Wyvis was found to contain pollen assemblages indicative of a mid or late Flandrian age, which indicates that lobe movement continued, at least intermittently, long after the end of the Loch Lomond Stadial (Ballantyne, 1984). The steep risers of many sorted lobes also suggest recent activity. Chattopadhyay (1982) reported that markers installed on the surfaces of sorted lobes exhibited downslope shift averaging 10–16 mm yr^{-1}, but it is unclear whether this reflects *en masse* movement of these landforms or merely the effects of superficial frost creep.

The sorted and nonsorted solifluction lobes and sheets of upland Britain are morphologically, sedimentologically and structurally similar to lobes and 'terraces' studied in other periglacial environments and attributed to movement dominated by frost creep and gelifluction respectively (e.g. Benedict, 1970, 1976). The former movement of massive boulder sheets and lobes is more difficult to explain. It has been variously attributed to 'flow of debris during snowmelt over frozen substrate' (Watt & Jones, 1948), 'gelifluction flow in a non-uniform sheet' (Ball & Goodier, 1970) and 'viscous flowage'

(King, 1972). However, these interpretations encounter a powerful objection: as pointed out by Waters (1964a) with reference to lobes on Dartmoor, gelifluction *sensu stricto* is likely to have been impossible because of the lack of fines amid the boulder mass. Galloway (1958, 1961a, b) and Kelletat (1970b) attempted to overcome this difficulty by suggesting that boulder lobes formerly contained a matrix of fine soil that has subsequently been washed out, and indeed that lobe immobilisation resulted from eluviation of fines. Although plausible, this idea also encounters difficulties. First, boulder lobes occur on some lithologies (such as quartzite) that are resistant to the production of fines by microgelivation. Second, the interstitial fines that occur within some sandstone and granite boulder lobes appear to be too coarse to be frost-susceptible, in which case gelifluction cannot have occurred. Finally, this explanation suggests that accumulations of washed fines should occur downslope of boulder sheets and lobes, but these are not evident.

Recognising these difficulties, Shaw (1977) proposed that boulder lobes moved downslope through deformation of internal ice ('rock glacier creep'), an explanation also favoured by Chattopadhyay (1982). This interpretation, however, raises insuperable difficulties. To permit deformation, ice could not simply fill interstices (as suggested by Shaw), as contacts between boulders would be maintained and the lobes would thus remain rigid and immobile. Even if it is assumed that the boulders were embedded within a *matrix* of deforming ice, lobe movement would still appear to be mechanically impossible. To permit movement, the yield strength of polycrystalline ice must be overcome. Research on glaciers suggests that this is of the order of 100 kPa (Paterson, 1981). The maximum shear stress generated within an ice-filled lobe is approximated by

$$\tau = \rho \, g \, h \sin \alpha \qquad (10.1)$$

where τ is shear stress at the base of the lobe, ρ is the density of the lobe, g is gravitational acceleration, h is the thickness of the lobe and α is gradient. If we assume an 'average' granite boulder lobe 1.5 m thick on a gradient of 20° (Table 11.1), a void ratio of 25%, and densities of 2700 kg m^{-3} for boulders and 900 kg m^{-3} for ice, then basal shear stress (τ) works out at *c.* 11 kPa, barely one-tenth of the value required to induce deformation of internal ice.

An alternative solution to the problem of boulder lobe movement is that this was accomplished by frost creep and possibly gelifluction operating in fines at the base of the openwork boulders, so that the latter were simply transported passively downslope (cf. Rudberg 1962, 1964). In this context it is worth recalling that blockfields and blockslopes are often underlain by a zone containing an infill of fine soil that would enable such movement to occur, provided that this zone experienced annual freezing and thawing (Chapter 9). It is pertinent to note that Shaw's (1977) excavations did not reach the base of the boulder lobes he investigated, and hence do not rule out the possibility of frost creep and gelifluction operating in subsurface sediments. An attractive feature of this explanation is that it implies that boulder sheets and lobes may be genetically indistinguishable from sorted solifluction landforms, except that they contain coarser debris and that the openwork layer is thicker. A wider implication is that all types of solifluction sheets and lobes on British mountains (and presumably elsewhere) may form part of a continuum of forms, in which differences in structure and

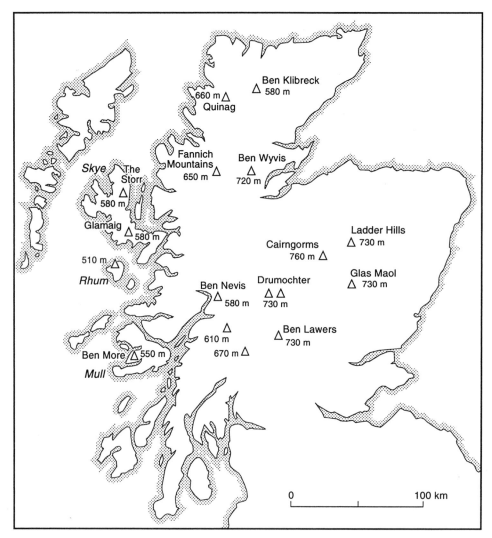

Figure 11.10 Lower limits of active solifluction in the Highlands and Islands of Scotland. Partly based on data in Kelletat (1970a).

dominant process are determined simply by the properties of the regolith on which they have developed, and hence ultimately on the response of the underlying bedrock to frost weathering. In terms of this model, nonsorted sheets and lobes may be seen as characteristic of regoliths comprising small clasts embedded in an abundant, frost-susceptible fine matrix, and dominated by gelifluction rather than frost creep, so that evidence for sorting is slight. Sorted sheets and lobes may be considered intermediate forms developed in regolith containing abundant boulders embedded in fine soil. Preservation of vertical sorting within such features suggests that both frost creep and gelifluction have been effective (cf. Benedict, 1970). Finally, boulder sheets and lobes may represent large-scale sorted solifluction landforms that have developed on lithologies which have weathered to produce an abundance of very large boulders and a limited amount of fine soil. This model is supported by the close association of different types of solifluction features with particular lithologies, and by the difficulty of drawing clear distinctions between types in the field, even after excavation. Its validity, however, requires further testing, particularly through further analysis of the struc-

ture and sedimentological characteristics of representative landforms.

Active solifluction phenomena

Evidence for active solifluction is widespread in upland Britain, but generally limited to the higher parts of mountain slopes. According to Kelletat (1970a), the lower limit of active solifluction tends to decline northwards and westwards across the Scottish Highlands, possibly in response to increasing wetness. For example, on Mull, Skye and Ben Nevis, Kelletat detected active solifluction phenomena down to 550–580 m; farther south and east, on the Drumochter Hills, Ben Lawers, the Cairngorms and the SE Grampians, he found that similar evidence extends no lower than 730–760 m (Figure 11.10). Similarly, landforms indicative of active solifluction are rare below 600–700 m in northern England and north Wales. Current solifluction activity is manifest in two closely-associated groups of landforms, namely active nonsorted solifluction sheets and lobes, and ploughing boulders. As discussed above, however, some sorted solifluction lobes of Lateglacial origin

Figure 11.11 Active solifluction sheets and lobes at 900 m near the summit of Scúrr Breac in the Fannich Mountains, northern Scotland.

Figure 11.12 Trench cut through an active solifluction lobe in the Fannich Mountains, northern Scotland, showing a soil horizon buried by downslope movement of the lobe.

continued to move downslope during the Flandrian and may still experience intermittent activity under present conditions.

Active solifluction sheets and lobes

The presence of active solifluction phenomena in Great Britain was first noted by officers of the Geological Survey working on the mountains of northern Scotland, and perceptively attributed by them to 'soil creep aided by the movement of snow' and 'freezing and thawing of water in the lower layers' (Peach *et al.*, 1912, 1913; Crampton & Carruthers, 1914). Subsequent research has shown that active nonsorted solifluction sheets and lobes differ from their relict counterparts in four ways. First, they are generally smaller, with risers 0.2–1.0 m high (Table 11.1; Figure 11.11). Second, active lobes sometimes occur on slopes that were occupied by glacier ice during the Loch Lomond Stadial, for example in Strath Dionard in northern Scotland and in the Drumochter Hills (Sissons, 1977b; Chattopadhyay, 1982). Third, whereas the risers of relict lobes are usually degraded, those of active lobes are characteristically steep (often around 45° or more) and sometimes bulge downslope (Figure 11.11). Finally, excavation of apparently active lobes has usually revealed a rather structureless brown soil consisting of angular stones embedded in a dominantly sandy but frost-susceptible matrix and overlying organic soil horizons or peat fragments containing Flandrian pollen grains (White & Mottershead, 1972; Mottershead, 1978; Ballantyne, 1981; Chattopadhyay, 1982; Figures 11.8 and 11.12). Active sheets and lobes are most widespread on argillaceous rocks, such as schists, slates and shales, that have weathered to yield abundant frost-susceptible fines; they tend to be rare on lithologies, such as granite and sandstone, that are characteristically overlain by coarse sandy (and hence nonfrost-susceptible) regolith.

Radiocarbon dating of organic matter buried by the advance of small solifluction lobes has demonstrated that such features have been moving intermittently downslope throughout the late Flandrian. Vegetation layers from depths of 0.4–0.7 m beneath small lobes in the Cairngorms yielded ^{14}C ages of 4480 ± 135 yr BP and 2680 ± 120 yr BP (Sugden, 1971), and four radiocarbon determinations obtained on peat buried by a solifluction lobe on Ben Arkle in northern Scotland gave ^{14}C ages ranging from

5541 ± 55 yr BP to 3984 ± 50 yr BP (Mottershead, 1978). Sugden suggested that the Cairngorm dates may be explained by lobe movement during periods of climatic deterioration around 2500 yr BP and during the Little Ice Age, but gave no reason for rejecting the simpler hypothesis that the ages obtained on the vegetation fragments actually reflect the age of burial. Mottershead believed that the Arkle dates indicate progressive burial of peat by solifluction after *c.* 4000 yr BP in response to increased slope erosion under conditions of deteriorating climate, and suggested that recent lobe movement has been limited.

An entirely different picture is suggested by five radiocarbon dates obtained from a continuous *in situ* soil horizon buried by the downslope advance of a solifluction lobe in the Fannich Mountains, also in northern Scotland (Figure 11.12). These yielded ^{14}C ages that range from 890 ± 120 to 530 ± 90 yr BP, all of which are statistically indistinguishable from a date of 660 ± 70 yr BP obtained from the same soil horizon 0.6 m downslope from the lobe front (Figure 11.13). This implies that the ^{14}C ages obtained from soil buried under the lobe may simply approximate the age of organic material in the soil *before* burial, rather than the timing of burial itself. Ballantyne (1986d) interpreted this evidence as indicating very rapid, very recent advance of this lobe over a distance of at least 3 m at a *minimum* rate of 22 mm yr^{-1}. He demonstrated that the rapidity of lobe movement implied by this sequence of dates cannot have been sustained throughout the Flandrian, but appears to reflect greatly accelerated recent activity. This, he suggested, may have been triggered either by climatic deterioration during the Little Ice Age of the sixteenth to nineteenth centuries AD, or by local vegetation degradation caused by overgrazing.

In other periglacial environments, radiocarbon dating of organic matter buried by the advance of solifluction lobes has provided interesting information on the timing and rate of solifluction movement. In particular, downslope sequences of dates obtained from intact buried soils or peat layers have permitted identification of phases of accelerated movement that apparently coincide with episodes of Holocene climatic deterioration (e.g. Benedict, 1970; Alexander & Price, 1980; Gamper & Suter, 1982; Gamper, 1983; Reanier & Ugolini, 1983; Matthews, Harris & Ballantyne, 1986). Radiocarbon dates obtained from organic matter buried by solifluction on British

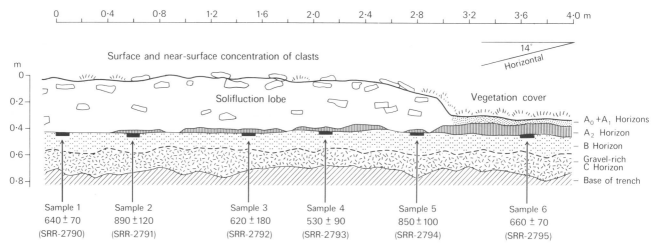

Figure 11.13 Stratigraphy of the solifluction lobe illustrated in Figure 11.12, showing the location of ^{14}C-dated samples and the radiocarbon ages obtained for each. From Ballantyne (1986d).

mountains have hitherto failed to yield a complete picture of such events, though they suggest that accelerated movement may have occurred around 4000 yr BP (Mottershead, 1978), around 2500 yr BP (Sugden, 1971), and within the past few centuries (Ballantyne, 1986d). Assessment of the validity of these conclusions, however, requires a much larger data base. The three studies described above nevertheless demonstrate the potential for dating solifluction activity in upland Britain, and thus increasing our understanding of Holocene climatic fluctuations in this area. Interestingly, accelerated solifluction during the Little Ice Age has been inferred from some intriguing historical evidence. On the upper slopes of Y Lethr, in the Rhinog Mountains of north Wales, a linked series of boulder-fronted lobes that run obliquely across a 15–20° slope was interpreted by Goodier & Ball (1969) as having developed through the collapse and downslope displacement of an ancient drystone wall. The lobes now appear stable. This fascinating evidence, however, is undermined by doubts as to whether the proposed wall ever existed. The lobes closely resemble sorted solifluction lobes of Lateglacial origin, and it is not clear why medieval Welshmen should construct a wall that starts and finishes on open slopes and runs obliquely uphill. It is also notable that walls constructed on Y Lethr in the early nineteenth century are completely intact, as are numerous drystone boundary walls at higher altitudes in the Scottish Highlands. The proposition is tantalising, but on present evidence unconvincing.

Despite the impression of rapid recent lobe movement suggested by bulging risers and some radiocarbon dates, measured rates of current solifluction activity are low. Downslope movement of markers installed on the surfaces of two steep lobes on Ben Wyvis over a single year (1976–7) did not exceed 18 mm and 11 mm (Ballantyne, 1981). Such rates could reflect the operation of near-surface frost creep alone. On a 16° slope in the Drumochter Hills, Chattopadhyay (1982) recorded 50 mm of surface displacement on a nonsorted lobe over three winters, declining to zero at 20 cm depth, and again explicable entirely in terms of superficial frost creep. Such rates are low in comparison with those measured on lobes in other periglacial environments (cf. Benedict, 1976), and indeed when compared with the displacement of surface clasts on unvegetated slopes in the Lake District and on Tinto Hill (Caine, 1963a; chapter 9).

However, it is equally possible that these rates are unrepresentative. Solifluction is an intermittent process, favoured particularly by slow, deep freezing of saturated soil (Harris, 1973), and therefore likely to be much more effective in some years than others, especially in areas of marginal activity.

Ploughing boulders

Intimately associated with active solifluction sheets on British mountains are *ploughing boulders*, otherwise known as ploughing blocks or gliding blocks. Ploughing boulders are boulders located at the downslope end of vegetation-covered furrows (Figure 11.14). Such furrows, together with the characteristic 'bow-waves' of turf pushed up immediately downslope of boulders, indicate downslope boulder movement at a rate exceeding that of the surrounding soil. Ploughing boulders are amongst the most common manifestations of recent periglacial activity in Britain. First noted in the Lake District by Hollingworth (1934) and Hay (1937), they have subsequently been observed in many parts of the Scottish Highlands (Galloway, 1958; King, 1968; Kelletat, 1970a; Shaw, 1977; Ballantyne, 1981; Chattopadhyay, 1982, 1983, 1986), the Southern Uplands (Galloway, 1961a; Tivy, 1962; Ragg & Bibby, 1966), northern England (Johnson & Dunham, 1963; Tufnell, 1969, 1972, 1976) and north Wales (Goodier & Ball, 1969; Ball & Goodier, 1970). Isolated examples have been observed at altitudes as low as 450–500 m in the northern Pennines, on Lochnagar, and on Ben More on the island of Mull, but the lower limit of ploughing boulders on most mountains is much higher. In north Wales, for instance, they extend down to 610 m on the Moelwyn Mountains, 700 m on the Rhinogs, 760 m on the Glyders but only 885 m on the Carneddau (Goodier & Ball, 1969). It is pertinent to note that ploughing boulders have been interpreted elsewhere as representing the lowest limit of current periglacial slope activity (e.g. Furrer, 1965; Höllermann, 1964, 1967).

Two aspects of the distribution of ploughing boulders are striking. First, they appear to be confined to areas of vegetated, frost-susceptible regolith, and hence are rare on lithologies that have weathered to yield blockslopes or debris with a matrix of coarse sandy fines. In the northern Highlands, for example, they

Table 11.2: *Characteristics of ploughing boulders on some British mountains.*

Area	Boulder Width (cm)			'Bow-wave' height (cm)			Furrow length (cm)		
	min	mean	max	min	mean	max	min	mean	max
Northern Pennines (Tufnell, 1972)	<30	50	150						800
Lochnagar: Granite (Shaw, 1977)	23	70	200	2	17	83	15	72	317
Lochnagar: Schists (Shaw, 1977)	20	52	150	2	15	83	13	128	888
Northern Highlands (Ballantyne, 1981)	23	70	185				24	187	1890
Grampian Highlands (Chattopadhyay, 1983)	20	77	190	9	23	55	50	238	560

Figure 11.14 Ploughing boulders on the Fannich Mountains, northern Scotland.

are abundant on mountains underlain by schists, such as the Fannichs and Ben Wyvis, but absent from the nearby sandstone massif of An Teallach. Secondly, ploughing boulders show a remarkably close association with active solifluction sheets and lobes. On the Fannichs and Ben Wyvis they extend downslope only as far as the lower limit of active solifluction, and Kelletat's (1970a) diagrams show that the distributions of active lobes ('Rasenloben') and ploughing boulders ('Wanderblöcke') are virtually identical on most of the mountains he visited in the Scottish Highlands. Similarly, Tufnell (1976) mentioned a close association in the northern Pennines between ploughing boulders and 'gelifluction terraces', and Chattopadhyay (1983) noted that in the Drumochter Hills they occur on vegetated solifluction sheets. The range of gradients on which ploughing boulders occur is also very similar to that (6–36°) reported for active solifluction on British mountains. In the northern Highlands a range of 10–34° was recorded by Ballantyne (1981), similar to that of 9–38° measured by Shaw (1977) on Lochnagar and consistent with Tufnell's (1972) observation that few in the northern Pennines occur on slopes of less than 10° or greater than 30°. Galloway (1958), however, observed ploughing boulders on slopes as low as 6° in the

Southern Uplands, and Chattopadhyay (1982) recorded a gradient range of 4–37° in the Grampians.

Various authors have measured the morphological characteristics of ploughing boulders on British mountains (Table 11.2), and have attempted to relate these to controlling variables. The results have been somewhat conflicting, but three significant features have emerged. The first is that only boulders larger than about 20–25 cm in diameter show evidence of downslope movement faster than that of the surrounding soil (Table 11.2). This suggests that a certain critical mass is required for the initiation of 'ploughing'. Secondly, measurements of boulder orientation have shown that the great majority are aligned directly downslope, or nearly so. Finally, both Ballantyne (1981) and Chattopadhyay (1983) found that furrow length is positively related to boulder size. Given that furrow length reflects both the rate of boulder movement and the rate at which furrows are infilled by solifluction, this relationship is interpretable in two ways. The simplest explanation is that large boulders leave large furrows that are more likely to survive encroachment and hence tend to be preserved for a greater distance upslope. Alternatively, large boulders may move faster relative to the surrounding regolith than small boulders. Chattopadhyay also found that furrow length is positively related to altitude on the Drumochter Hills, and adduced from this that increasing climatic severity favours more rapid boulder movement.

The majority of ploughing boulders on British mountains appear to experience at least intermittent movement under present conditions, though those that are now deeply embedded in the soil often show little sign of recent activity. Immediately upslope of most boulders, however, there is a deep niche a few centimetres wide that often extends to the base of the boulder. Such niches are probably diagnostic of recent activity. Recorded rates of boulder movement vary widely. On the northern Pennines, Tufnell (1976) measured average rates of 0.4–64.0 mm yr⁻¹ on five boulders over a ten-year period and showed that boulder movement during the summer months is negligible. Rates recorded over shorter periods in the Scottish Highlands are rather lower. On Lochnagar, Shaw (1977) measured average rates of 0.3–8.7 mm yr⁻¹; on the Fannich Mountains, Ballantyne (1981) recorded a range of 6.0–34.5 mm yr⁻¹; and on the

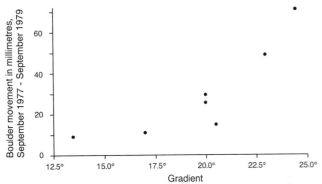

Figure 11.15 The relationship between downslope movement of seven ploughing boulders and gradient, Fannich Mountains, northern Scotland.

Drumochter Hills, Chattopadhyay (1986) noted displacements averaging 1.5–7.0 mm yr^{-1}, and found that the timing of movement apparently coincides with the spring thaw. The latter two researchers also found that rates of movement are strongly related to gradient (Figure 11.15). Although the displacement of most boulders seems to take the form of sporadic and possibly seasonal slow creep, it appears that rapid sliding may very occasionally occur: Johnson & Dunham (1963) recorded a downslope shift of one boulder of about 1.5 m within a single year.

Although abundant data have been collected on the characteristics of ploughing boulders in Britain (Tufnell, 1972; Shaw 1977; Chattopadhyay, 1982, 1983), attempts to explain the cause of boulder 'ploughing' have been largely speculative and divorced from the evidence available. Tufnell (1972, p. 259–260) suggested four mechanisms of boulder movement, namely heating and cooling (insolation creep *sensu* Statham (1977)), frost creep, freezing and expansion of water trapped upslope of a boulder and 'sliding of a block on the upper surfaces of frozen ground which lies directly beneath a rather wet and fluid layer '. He implied that all four mechanisms contribute to boulder displacement. Ballantyne (1981), however, challenged the adequacy of these mechanisms and proposed that boulder movement results primarily from gelifluction. He suggested that ice lenses develop underneath boulders during winter freezing, and that melt of such lenses in spring is accompanied by a critical rise in porewater pressures and consequent reduction in shearing resistance. As a result, boulders may be capable of sliding through the surrounding soil, but in doing so allow dissipation of excess porewater so that movement is short-lived. Movement may be facilitated by the water content of surrounding soil during thaw exceeding its liquid limit.

Several pieces of evidence favour gelifluction as the primary mechanism responsible for accelerated boulder movement. This explanation accounts for the close association of ploughing boulders with active solifluction sheets and lobes, and their absence on soils that are too coarse to permit the formation of segregation ice. It also explains the absence of ploughing boulders below a critical size or mass, as small boulders presumably exert insufficient stress on soil downslope to allow 'ploughing' to occur. The explanation proposed above may also account for the observed relationships between boulder size and furrow

length, as boulders with large mass may be expected to exert greater stress against soil downslope and hence move faster relative to the surrounding regolith. Similarly, the relationship between boulder displacement and gradient (Figure 11.15) may be accounted for in terms of increasing shear stesses as the slope steepens. The preferred downslope orientation of most ploughing boulders, like that of clasts within soliflucted soils, is interpretable in terms of boulders rotating to adopt an alignment of least resistance. The 'bow-waves' of soil downslope and at the margins of many ploughing boulders may, in terms of the above hypothesis, represent pushing aside of soil liquefied during thaw consolidation of the surrounding ground. Finally, the above model is consistent with Chattopadhyay's (1983) observation that movement takes place during thaw.

Although the mechanism proposed above awaits more direct testing, it is supported by work on the movement of a giant ploughing boulder in the Finse area of south-central Norway. Reid & Nesje (1988) have shown that the thermal diffusivity of this boulder is an order of magnitude higher than that of surrounding moist soil during freezing. They proposed that this allows preferential freezing at the base of the boulder during winter, with associated formation of ice-rich sediment. They also suggested that thaw of such sediment in spring causes extrusion of liquefied mud and boulder movement under the influence of gravity. Although this explanation was developed with regard to an exceptionally large boulder, it offers a general model for accelerated boulder movement, and appears to be consistent with the abundant data that have been collected concerning the characteristics of ploughing boulders on British mountains.

Solifluction landforms in upland Britain: problems and potential

The picture of past and present solifluction activity sketched out above is necessarily a provisional one. There remain many unknowns. In particular, the mechanism and palaeoenvironmental significance of boulder lobe development and movement are still uncertain, and must remain so until accurate information becomes available on the subsurface structure of these intriguingly intractable landforms. Representative rates of present-day solifluction await measurement, and the potential of radiocarbon dating of buried organic matter to identify episodes of enhanced solifluction activity (and thereby, indirectly, Holocene climatic deterioration) has barely been tapped. Despite the relative accessibility of active solifluction lobes and ploughing boulders in some parts of the British uplands, little use has been made of British sites in investigation of solifluction processes. In short, large parts of this research field remain untilled.

The last three chapters of this book have concentrated mainly on the effects of periglaciation, and particularly frost action, on the upper parts of British mountains. On valley floors below steep rockwalls, however, periglacial processes have resulted in the development of an equally fascinating suite of landforms that reflect the accumulation and modification of debris derived from the cliffs above. It is to consideration of these that we now turn.

12
Talus slopes and related landforms

Talus slopes are steep valley-side slopes formed by the accumulation of debris at the foot of rockwalls. The term *talus* is used to describe both the slope form and its constituent material. *Scree* has often been used as a synonym, but in common usage generally refers to any slope cover of predominantly coarse debris, irrespective of location, and this distinction is maintained here. Talus slopes are not restricted to present or former periglacial environments, but occur in all areas where the products of rock weathering have accumulated at the foot of a cliff. Taluses are nevertheless particularly prominent features of mountain areas that experience or formerly experienced a periglacial climate. One reason for this is that such areas have, with few exceptions, previously experienced glacial erosion, which resulted in the formation of steep rockwalls on the flanks of glacial troughs and at the heads of corries. Another is the manifest effectiveness of macrogelivation (Chapter 9) in dislodging blocks of rock that fall to the valley floor and accumulate as talus. Rockfall, whether in the form of discrete falls of individual clasts or more catastrophic events involving failure of large parts of a cliff, is normally the primary process responsible for talus accumulation. Much of the interest of periglacial talus slopes, however, arises from modification of 'primary' rockfall talus by a series of 'secondary' processes, some of which occur only in cold climates. The main processes of modification are fourfold: (1) snow and slush avalanche activity; (2) debris flow activity; (3) protalus rampart formation; and (4) protalus rock glacier formation. All four may involve major modifications in the form and sometimes the sedimentological characteristics of rockfall talus slopes (Figure 12.1). This chapter considers first the characteristics, age and significance of rockfall talus in Britain, then examines the nature and effects of talus modification under periglacial conditions.

Rockfall talus slopes

In planform, rockfall talus slopes are of three types: talus sheets, where debris delivery has been fairly uniform across the slope; talus cones, where rockfall has been concentrated or funnelled down a major rock gully; and coalescing talus cones, formed where talus cones intersect laterally (Figures 12.1–12.3). The three types are transitional, and the dominance of one or other is simply a reflection of the irregularity of the source rockwall. In long profile, talus slopes produced by rockfall and unmodified by other processes have been shown to comprise two units: an upper straight slope and a basal concavity (e.g. Howarth & Bones, 1972; Statham, 1973, 1976a; Kotarba, 1976), though a small convexity has also been observed at the crest of some taluses (Church *et al.*, 1979), and the basal concavity may be absent where there is basal removal

of debris (Andrews, 1961; Howarth & Bones, 1972). Where basal erosion is ineffective, the upper straight slope often stands at a overall gradient of *c.* 35–36° (Chandler, 1973).

Two models of rockfall talus accumulation have been proposed to account for this consistent gradient. The traditional view is that this angle represents the *angle of repose* of coarse talus material, in other words the gradient at which accumulating talus debris comes to rest after dry avalanching. This interpretation implies that debris is periodically redistributed over the surface of a talus by shallow debris slides. However, Statham (1973, 1976a) maintained that the *c.* 35–36° gradient characteristic of the straight slope is actually a few degrees less than the angle of repose as defined above. He also pointed out that talus redistribution by dry avalanching (the 'angle of repose' model) cannot account for the basal concavity at the foot of most rockfall talus slopes, nor for the widely-reported phenomenon of *fall sorting*, represented by a general downslope increase in the size of surface clasts. As an alternative, Statham proposed that the form of rockfall taluses represents a response to rockfall travel distances and the movement of surface clasts triggered by rockfall impact. He pointed out that clasts falling onto a talus slope possess a certain amount of impact energy and hence are capable of movement down slopes with gradients lower than the repose angle. In terms of this model, the gradient of the talus steepens slightly through time as rockwall height (and therefore impact energy) diminishes; only when the talus has covered most of the contributing rockwall does its gradient approach that of the 'angle of repose', so that dry avalanching may occur only in the late stages of talus development. According to this model, therefore, the gradient of rockfall talus is related to the input energy of falling clasts, rather than the frictional strength of the accumulating debris. The basal concavity was considered by Kirkby & Statham (1975) to form early in the buildup of talus, when the impact energy of accumulating particles is generally high. They demonstrated experimentally that as the upper straight segment of a talus increases in length, fewer falling clasts reach the base of the slope, so that the straight slope grows at the expense of the basal concavity.

Carson (1977), however, argued that the upper parts of talus slopes stand at or close to the angle of repose for talus material, and supported his argument with observations of dry avalanching on accumulating gravel stockpiles. Dry avalanching on natural talus was also inferred by Church *et al.* (1979) from their work on Baffin Island taluses. The importance of dry avalanching on rockfall talus has been particularly emphasised by Whitehouse & McSaveney (1983), whose studies of the age of different parts of talus slopes indicated a progressive buildup of material near the talus crest rather than even distribution of

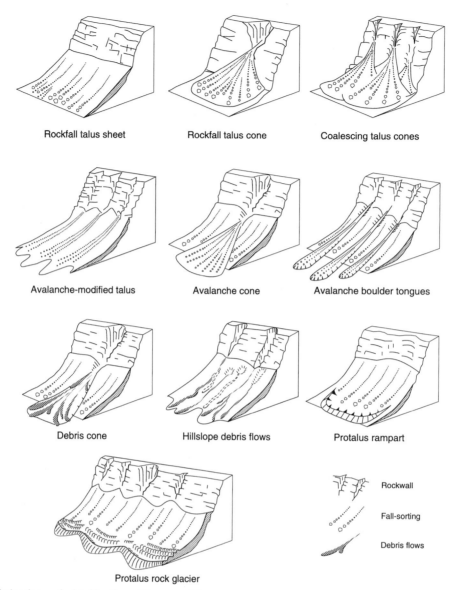

Figure 12.1 Types of talus slope and related landforms (diagrammatic).

clasts across the talus surface. The relative validity of the 'angle of repose' and 'discrete rockfall' (Statham) models thus remains in dispute. Although the latter is attractive in that it provides an explanation for fall sorting and the presence of a basal concavity, it appears to underrate the importance of dry avalanching as a redistributive process (Church *et al.*, 1979).

Talus in periglacial environments gives the impression of great thickness, but this is rarely the case. Rapp (1960a) summarised data showing the range of thickness to be 1–35 m, but both Young (1972) and French (1976) considered talus to be generally less than 5 m thick. In arctic areas, talus may be underlain by permafrost at shallow depth. Structurally, taluses consist of a shallow layer of coarse openwork debris that overlies a deeper layer in which fines are abundant, either as an interstitial infill or as a true matrix (Church *et al.*, 1979). Such fines may be exposed at the surface at the sites of debris slides or gullying. Sorting of surface clasts at any point on a talus is usually poor, but many authors have recorded a general increase

in clast size with distance downslope from the source rockwall. Such fall sorting has been explained in terms of the generally greater kinetic energy of larger clasts (Rapp, 1960a; Bones, 1973), though Statham (1973, 1976a) intepreted this phenomenon in terms of higher frictional losses for small particles. Once fall sorting has been established, however, it tends to be self-perpetuating, as the talus acts as a kind of sieve with individual clasts tending to settle amongst others of similar calibre (Statham, 1973; Church *et al.*, 1979). Fall sorting is particularly diagnostic of unmodified rockfall talus, and is liable to be absent on taluses reworked by extensive avalanche activity or debris flows. On such unmodified talus, systematic across-slope variations in clast size may reflect variable joint density in the source rockwall (Wilson, 1990).

In Great Britain, talus slopes occupy three types of location. First, substantial taluses have accumulated below some sea cliffs, particularly in areas where glacio-isostatic recovery has resulted in retreat of the sea from the cliff foot. Excellent exam-

Figure 12.2 Lateglacial talus sheets in the Lairig Ghru, a steep-sided glacial breach in the Cairngorm Mountains, Scotland. The light-toned lines down the talus represent the tracks of recent debris flows.

Figure 12.3 Holocene talus cone, Coire Lagan, Isle of Skye. The lobate structures on the talus surface may represent dry avalanching or debris flow activity.

Table 12.1: *Minimum and maximum gradients recorded on the upper part of talus slopes.*

Site	Lithology	Gradient (degrees)		Source
		minimum	maximum	
Wasdale, Lake District	Andesite	30	36	Andrews, 1961
Eglwyseg Escarpment, Wales	Limestone	32	38	Tinkler, 1966
Cader Idris, Wales	Granophyre	-	38–39	Statham, 1973
Cader Idris, Wales	Mudstone	-	35–36	Statham, 1973
Cader Idris, Wales	Volcanics	-	39–40	Statham, 1973
Isle of Skye, Scotland	Gabbro, Granite	32.5	38	Statham, 1976a
An Teallach, Scotland	Sandstone	35.1	36.5	Ballantyne & Eckford, 1984
Lomond Hills, Scotland	Quartz dolerite	35.2	36.3	Ballantyne & Eckford, 1984
Lochnagar, Scotland	Granite	31.5	39.2	Ward, 1985

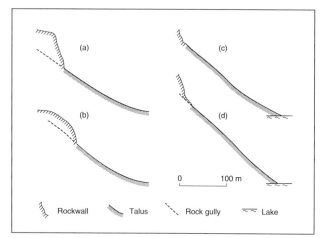

Figure 12.4 Rockfall talus profiles, Cader Idris, Wales. The upper slope gradients of profiles (a) and (b) have probably been lowered by debris flow activity. Note absence of basal concavity in the profiles of slopes terminating in the lake. From Statham (1973).

ples fringe the coasts of the islands of the Inner Hebrides, such as Mull, Rhum and Skye, as well as the cliffed coasts of eastern Scotland, Northumberland and parts of Wales (Savigear, 1952). Second, talus slopes occur below structural escarpments in lowland areas, for example at the foot of limestone scarps in Yorkshire and Wales (Tinkler, 1966) and below major sills in northern England and central Scotland (Ballantyne & Eckford, 1984). Finally, talus mantles the lower slopes of many glacial troughs and corries in upland areas; it is with this last group that this chapter is primarily concerned.

One point that emerges strongly from examination of the literature on talus slopes in upland Britain is that virtually all of those investigated bear the hallmarks of unmodified rockfall talus: a steep, straight upper slope, a basal concavity, and marked fall sorting. To some extent, however, this picture may be misleading, as researchers investigating talus slopes are liable to avoid areas that have experienced obvious modification. Even so, the consistency of form reported in the literature is notable. A steep upper slope appears to be absent only from talus slopes that have been reworked by avalanching snow (Ballantyne, 1989b), severely eroded by debris flows, or obliterated by major landslides (Ballantyne & Eckford, 1984). A basal concavity is similarly a ubiquitous characteristic, except where talus terminates downslope in a lake. This is the case, for

example, with the great coalescing talus cones that flank Wastwater in the Lake District (Andrews, 1961) and those of Cader Idris in Wales that end in Llyn Cau (Statham, 1973); in such cases only a straight slope may be evident above the water level (Figure 12.4). The gradients of the upper portions of talus slopes in upland Britain also reveal a measure of consistency. Almost a century ago, Lord Avebury (1902) observed of mountain rockwalls in England that 'frost detaches fragments, which eventually slide down and form a scree or talus at an angle [of] about 35°'. Whilst the overall (toe to crest) mean gradient of talus may be markedly less, this figure appears to be about average for the gradient of the straight upper slopes of taluses in upland Britain, though the range of reported values is wide, from 30° to 40° (Table 12.1). Again, however, this wide range may be deceptive, as much depends on the length of the slope facet measured. For example, the narrow range (35.1-36.5°) reported by Ballantyne & Eckford (1984) reflects the *average* values for the entire straight slope section, whereas the much broader range (31.5-39.2°) measured by Ward (1985) includes individual slope facets within the upper straight zone. Such differences highlight the fact that although the average gradients of the upper straight slopes may fall within a fairly circumscribed range, steeper and gentler facets occur within this zone. The causes of such irregularities include debris piling up at low gradients behind exceptionally large boulders and, where the talus is shallow, steps in the underlying bedrock.

Irrespective of the effects of surface irregularities, there appear to be two systematic controls on the overall gradient of the upper straight slope. One is lithology: some rock types, such as slate or mica-schist, tend to yield platy or slabby debris, whilst others, such as quartzite, form 'blocky' or approximately equidimensional boulders. Although such boulders may roll downslope more readily than platey debris, (Statham, 1973, 1976a), they also tend to form a more stable slope through interlocking of clasts, and thus to be more resistant to impact-induced movement. Although no studies have been carried out on the influence of particle form on the gradient of the upper parts of talus, the differences between the maximum gradients on taluses composed of different rock types on Cader Idris (Table 12.1) suggests a measure of lithological control. The other (and potentially more significant) control on gradient is talus maturity. This may be expressed as the ratio H_o/H_i, where H_o is the vertical height of the talus and H_i is the height of the entire slope (talus plus rockwall). According to Statham's model of talus accumulation by discrete rockfall events, imma-

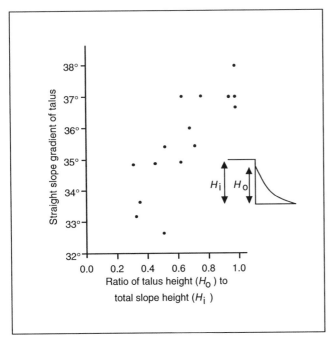

Figure 12.5 Increase in straight slope gradient with increase in talus height (Ho) relative to that of the entire slope (Hi). From Statham (1976a).

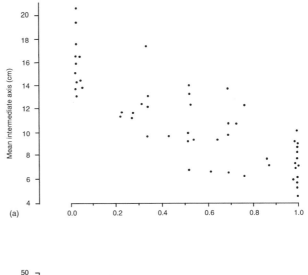

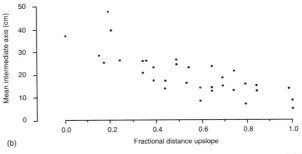

Figure 12.6 Variation in mean clast size with distance upslope on rockfall talus: (a) unvegetated talus slopes in the Cuillin Hills, Skye (from Statham (1976a); (b) vegetation-covered talus on An Teallach, NW Highlands.

ture talus (low H_o/H_i) should have a relatively low straight slope gradient, as the energy of impacting clasts will be high. Progressive burial of the rockwall, however, will reduce the impact energy of falling particles and thus permit the development of a steeper straight slope. Statham's (1976a) observations on the Isle of Skye taluses indicate that this is the case (Figure 12.5), though the relationship is at best a very general one. Ballantyne & Eckford (1984) reported H_o/H_i ratios of 0.28 and 0.67 for an active talus cone and a relict talus sheet in the same area, yet average straight slope gradients for both are c. 35°.

Fall sorting, another attribute of talus that appears to favour accumulation through discrete rockfall rather than through periodic redistribution of material by dry avalanching, has been observed on talus slopes in the Lake District (Hay, 1937; Andrews, 1961), Wales (Tinkler, 1966; Statham, 1973) and Scotland (Shaw, 1977). In an exhaustive study of this phenomenon, Statham (1976a) investigated variations in mean clast size down 21 talus profiles on the Isle of Skye (Figure 12.6). He found that 14 displayed a progressive increase in clast size, 4 an overall increase, and 3 no trend, the last being attributed by him to 'processes other than rockfall'. Fall sorting he attributed primarily to the 'sieving' effect of increasing clast size downslope. However, Ballantyne (1981) found equally strong evidence of fall sorting amongst samples of clasts resting on a vegetated talus in northern Scotland (Figure 12.6). This appears to indicate that fall sorting is dependent not only on particles being trapped amongst interstices, but reflects also the greater success of larger particles in overcoming frictional losses as they travel downslope.

In contrast to taluses in high alpine environments, the great majority of those abutting rockwalls in upland Britain are largely or entirely vegetation-covered. At lower elevations in particular a grassy vegetation mat often completely obscures the underlying debris, as is the case with the talus that sweeps down from Salisbury Crags in Edinburgh (Figure 12.7). Even at

Figure 12.7 Vegetation-covered Lateglacial talus slopes, Salisbury Crags, Edinburgh.

600 m altitude on An Teallach in NW Scotland, large talus slopes support a complete vegetation cover, and the magnificent screes of Wasdale, perhaps the finest in England, are partly vegetated with grass and shrubs. Only in high corries in the remote fastnesses of the Scottish Highlands are unvegetated talus slopes to be found, and even on these, surface boulders frequently support an irregular cover of moss and lichen. The persistence of vegetation cover on most British taluses suggests that these are predominantly relict slopes that experience very limited rockfall accumulation under present conditions. This impression is reinforcd by observations of widespread erosion

on many taluses, particularly by debris flow activity, but also by stream gullying and shallow translational landslides. Some talus slopes have also been disrupted or buried as a result of large-scale rock slope failures (Ballantyne & Eckford, 1984).

Such observations have prompted commentators to infer that talus slopes in upland Britain were formed under a much more severe climate than that of the present. This was the view of Harker (1901), who noted that current rockfall from the cliffs of the Cuillin Hills in Skye is infrequent, and who ascribed the great taluses of that island to a period of sustained periglacial conditions following the wastage of the last glaciers. Later researchers have followed Harker's lead, attributing relict taluses in Wales, northern England and Scotland to formation during the Late Devensian Lateglacial (Ball, 1966; Tufnell, 1969; Ball & Goodier, 1970; Ryder & McCann, 1971; Kotarba, 1984; Ballantyne & Eckford, 1984; Ballantyne & Kirkbride, 1987). Andrews (1961) suggested that the Wasdale screes in the Lake District last received a significant input of debris during the Little Ice Age of the seventeenth and eighteenth centuries AD, but provided no evidence in support of this suggestion.

Pronounced rockfall activity during the Lateglacial is to be anticipated on *a priori* grounds. Retreat of the Late Devensian ice sheet exposed in many upland areas a landscape of glacially-steepened and thus potentially unstable rockwalls, and the severe periglacial climate that persisted during the early stages of ice-sheet withdrawal and during the Loch Lomond Stadial of *c.* 11–10 ka BP resulted in widespread shattering of exposed rock surfaces on high ground (Chapter 9). Indirect evidence for pronounced rockfall activity during and before the Loch Lomond Stadial is provided by the huge quantities of debris, apparently of rockfall origin, that were reworked and redeposited by Loch Lomond Readvance corrie glaciers. Thus bouldery moraines in corries on the Island of Rhum were described by Ryder & McCann (1971, p. 299) as 'ablation moraine deposits consisting largely of scree debris deposited on the surfaces of melting corrie glaciers'. More recently, Benn (1989) has found that lateral moraines deposited by Loch Lomond Readvance corrie glaciers in NW Scotland are significantly larger on the eastern and northern margins of the former glaciers than on the western and southern margins. Studies of the aggregate shape characteristics of the debris within these moraines allowed him to demonstrate that this asymmetry reflects glacial entrainment of pre-stadial talus from below east- and north-facing rockwalls, coupled with the direct supply of abundant rockfall debris onto the former glacier surfaces.

More direct evidence favouring a Lateglacial age for most upland taluses comes from comparison of the maturity of taluses inside the limits of the Loch Lomond Readvance glaciers with that of taluses in areas which escaped glaciation during the Loch Lomond Stadial. The former have accumulated only under the relatively mild conditions of the Holocene; the latter have accumulated since ice-sheet deglaciation, during both the Lateglacial and the Holocene. On the Western Hills of Rhum, Kotarba (1984) observed contrasts between the well-developed, vegetation-covered talus sheets that occur outside the limit of the Loch Lomond Readvance, and the more active, less vegetated talus cones that occur inside these limits. He concluded that the main period of slope development terminated at the end of the Loch Lomond Stadial, and that Holocene activity has been largely restricted to modification of relict slopes by debris slides and

Figure 12.8 Immature Holocene talus cones in Glas Tholl, a corrie on An Teallach in the NW Highlands of Scotland. Modification of some cones by debris flows is evident from the deposition of parallel levées aligned downslope.

debris flows. Similarly, on An Teallach in northern Scotland, Ballantyne & Eckford (1984) located mature talus sheets only outside the maximum extent of the Loch Lomond Readvance glaciers. They found that whereas relict talus outside the glacial limit had buried most of the rockwall upslope ($H_o/H_i = 0.67$), an active talus cone below a much higher and steeper rockwall inside the glacial limit was at a much less advanced stage of development ($H_o/H_i = 0.28$; Figure 12.8). Some caution is necessary, however, in attributing maturity of talus development to age alone, as topographic and lithological factors may play an equally important role in determining rate of talus accumulation (cf. Olyphant, 1983). For example, Kotarba (1987) has presented data which indicate that taluses in western Rhum have a H_o/H_i index of 0.57–0.61, whilst those in the northern Cairngorms are much less mature ($H_o/H_i = 0.21$–0.32). Both sites lie inside the limits of the Loch Lomond Readvance. Kotarba has attributed this contrast to differences in climate, but it seems more likely that it reflects the much higher joint density of the Rhum granophyre compared with that of the Cairngorm granite.

The notion that talus slopes outside the limits of Loch Lomond Stadial glaciation are essentially Lateglacial relicts receives strong support from measurements of present rockfall

Table 12.2: *Present-day and historical rockwall retreat rates in Great Britain and elsewhere.*

Location	Lithology	Rockwall retreat rate (mm yr^{-1})			Source
		min.	mean	max.	
Great Britain, present–day					
An Teallach	Sandstone	0.013	0.015	0.016	Ballantyne & Eckford, 1984
Lomond Hills	Quartz-dolerite	0.009	0.015	0.063	Ballantyne & Eckford, 1984
Snowdonia	Volcanics	0.010	0.015	0.021	Stuart, 1984
Great Britain, Loch Lomond Stadial					
See Table 8.3	Various	1.64		3.29	Ballantyne & Kirkbride, 1987
Arctic environments, present day					
Lappland	Schist	0.04		0.15	Rapp, 1960a
Spitsbergen	Limestone	0.05		0.50	Rapp, 1960b
Yukon	Various	0.007		0.17	Gray, 1972
Arctic environments, Holocene					
West Greenland	Volcanics	0.5		2.4	Frich & Brandt, 1985
NW Spitsbergen	Quartzite	0.10		0.72	André, 1985
Ellesmere Island	Limestone	0.30		1.30	Souchez, in French, 1976
Alpine environments, present day					
Tatra Mountains	Dolomite	0.1		3.0	Kotarba, 1972
Front Range	Various		0.76		Caine, 1974
French Alps	Various	0.05		3.0	Francou, 1988
Alpine environments, Holocene					
Swiss Alps	Various	1.0		2.5	Barsch, 1977
Austrian Alps	Gneiss, schist	0.7		1.0	Poser, 1954

activity. Ballantyne & Eckford (1984) calculated a mean rock-wall retreat rate of 0.015 mm yr^{-1} for the cliffs overlooking two vegetated talus slopes in Scotland, and an identical figure was calculated by Stuart (1984) for sites in Snowdonia. Although these average values conceal considerable variations from one slope transect to another and do not take into account infrequent large-magnitude rockfalls, they indicate negligible talus accumulation at present: recalculated as talus accumulation rates, the figure for An Teallach (Table 12.2) implies addition of new debris at an average rate of only 0.004 mm yr^{-1} across the talus surface, whilst that for the Lomond Hills in Fife implies an averaged accumulation of only 0.001 mm yr^{-1}. If such rates are representative of accumulation throughout the 10 000 years of the Holocene, they imply average talus accumulation of only 40 mm and 10 mm respectively. Whilst these figures represent only rough approximations, they certainly indicate that much more frequent rockfall must have occurred in the past to produce extensive taluses at these sites. Comparison of measured rates of current rockwall retreat above relict talus slopes reveals that these are generally lower than those recorded in arctic environments, and very much lower than those characteristic of alpine areas (Table 12.2). An even more instructive comparison, however, is with rockwall retreat rates for the Loch Lomond Stadial in upland Britain. These have been calculated by Ballantyne & Kirkbride (1987) on the basis of the volumes of protalus ramparts that accumulated at the foot of talus slopes during the stadial, and average 1.64–3.29 mm yr^{-1}, two orders of

magnitude greater than the estimates of current retreat rates (Table 12.2). These figures are similar in magnitude to estimates for high-alpine environments at present, and lend credence to the view that relict taluses in Britain accumulated very rapidly during the Lateglacial and have since experienced only a limited addition of debris.

The similarity of the reconstructed stadial rockfall rates with those measured in alpine environments prompted Ballantyne & Kirkbride (1987) to suggest that during the Loch Lomond Stadial the climate was particularly conducive to release of rock by frost wedging. They argued that conditions for freeze–thaw activity were probably favoured by a combination of strong insolation during the spring, summer and autumn months with much cooler air temperatures than at present, and noted that mid-latitude alpine mountains offer the closest modern analogue for this situation. They also pointed out, however, that the instability of oversteepened rockwalls following ice-sheet deglaciation probably also enhanced Lateglacial rockfall activity. The importance of rockwall instability in accelerating rockwall collapse during and immediately after deglaciation has been demonstrated by André (1985, 1986), who found that the most rapid rates of rockwall retreat that she measured on cliffs in Spitsbergen (up to 0.7 mm yr^{-1}) resulted from pressure release following glacier surges. This raises the possibility that the large relict Lateglacial taluses of upland Britain may be interpreted as essentially *paraglacial* landforms (resulting from rapid disintegration of oversteepened rockwalls immediately after deglaciation) rather than *periglacial* features produced by enhanced frost action operating on cliffs during the Lateglacial. However, the relative immaturity of most Holocene taluses below steep rockwalls that were occupied by glacier ice during the Loch Lomond Stadial suggests that paraglacial effects have generally been of secondary importance, and that the primary impetus for the development of mature talus slopes during the Lateglacial was climatic.

No information is available on current rates of rockwall recession or debris accumulation associated with Holocene talus that has built up within the limits of the Loch Lomond Readvance. However, observations on the frequency of rock-falls in Glas Tholl, a spectacular sandstone corrie on An Teallach (Figure 12.8), suggest that steep, high rockwalls over-looking immature but active talus cones experience a much greater incidence of rockfalls than those above relict Lateglacial talus slopes. At this site Ballantyne & Eckford (1984) reported an average frequency of 0.67 falls per hour, which contrasts with an average of 0.12 rockfalls per hour for the quartz-dolerite scarp above the Lateglacial talus that skirts the Lomond Hills in Fife. By comparison, Luckman (1976) recorded fre-quencies of 0.20–0.94 rockfalls per hour in the Canadian Rocky Mountains; Gardner (1983) calculated an average of 0.49 falls per hour for the Highwood Pass areas in the same range; and Åkerman (1984) noted frequencies averaging 1.1–1.5 falls per hour from steep cliffs in Spitsbergen. However, the observa-tions on An Teallach and the Lomond Scarp were of brief dura-tion, confined to the summer months, and yield no information concerning the magnitude of individual falls.

In sum, present information on the evolution of talus slopes in upland Britain suggests that a broad distinction can be made between those that began to accumulate immediately after ice-sheet deglaciation and those that date back only to the end of the Loch Lomond Stadial and are thus essentially of Holocene age. Both groups are largely the products of rockfall accumulation, being characterised by an upper straight slope at a gradient of around 35°, a basal concavity and some degree of fall sorting. Taluses that accumulated throughout the Lateglacial tend to have reached a more mature stage of development than those of Holocene age. However, whereas the latter may still be actively accumulating under present conditions, the former are essen-tially relict periglacial (and to a lesser extent paraglacial) land-forms, products of enhanced rockfall during the Lateglacial, on which current accumulation is negligible. Some taluses, how-ever, have been markedly modified by processes other than rockfall. The remainder of this chapter examines the nature of these processes, their significance and their effects.

Avalanche-modified talus

In many alpine environments, snow and slush avalanches have played a major role in modifying both the form and sedimento-logical characteristics of talus slopes. Avalanches can generate enormous impact stresses (of the order of 0.1–1.4 MPa accord-ing to Mellor (1978)), but their effectiveness as agents of ero-sion and debris transport is not a simple function of size and velocity. For effective avalanche transport there must be an abundance of loose, entrainable debris that is unprotected by overlying snowcover or vegetation (Gardner, 1970; Ackroyd, 1986). The most effective avalanches are therefore either full-

depth avalanches that extend to the underlying ground surface or spring avalanches that travel from rockwalls upslope over snow-free talus (Luckman, 1977). Individual avalanches over talus slopes have been observed to contain up to 2000 m³ of debris (Ackroyd, 1986). Although there is some evidence that avalanches may directly erode rockwalls, producing funnel-shaped avalanche chutes (Luckman, 1977), their most impor-tant geomorphic role is in eroding debris from the upper parts of talus slopes and redistributing it farther downslope. The actual process of avalanche erosion is poorly understood, though Gardner (1970) observed entrainment of debris through the 'balling up' of advancing snow containing boulders. The effect of such redistribution of debris is twofold: an overall reduction in talus gradient (particularly in the upper part of the slope, which is often reduced well below the 35–36° characteristic of unmodified rockfall talus) and the formation of a long, sweep-ing basal concavity that may even extend a short distance uphill against the opposite slope (Rapp, 1959; Luckman, 1971, 1977, 1978; Huber, 1982). Avalanches also tend to destroy fall sorting by depositing poorly-sorted debris on the lower parts of the slope. Avalanche-transported clasts are often sharply angular (as a result of collisions during transport) and many are perched in unstable positions on top of other boulders. Such *perched boulders*, together with 'drapes' of fine material across boulder surfaces, reflect deposition from an ablating snowcover, and thus contrast with rockfall boulders, which tend to be wedged in stable positions amid adjacent clasts. The microrelief of avalanche-modified talus is often characterised by *debris tails*, which are ridges of debris that extend downslope in the lee of large boulders and reflect erosion of the surrounding unpro-tected talus surface (Gardner, 1970; Luckman, 1971, 1977, 1978; Gray 1973).

Snow avalanches also produce a number of distinctive ero-sional and depositional landforms. *Avalanche impact pits* are widespread in avalanche-prone areas: they are depressions of variable size, often water-filled, excavated in unconsolidated debris by the impact of repeated powerful avalanches where the ground levels off at the foot of an avalanche track (Davis, 1962; Peev, 1966; Corner, 1980; Fitzharris & Owens, 1984). Equally widespread in alpine environments are *avalanche boulder tongues*, which comprise avalanche deposits that extend beyond the toe of the talus. At their most distinctive they take the form of what Rapp (1959) called 'roadbank tongues', which as his term suggests are tongue-shaped embankments of avalanche debris that may rise several metres above the level of the valley floor, are markedly concave in long section but tend to be flat-topped (and sometimes asymmetric) in cross section (Luckman, 1977, 1978; Huber, 1982). Less well documented are what Rapp termed 'fan-type avalanche boulder tongues' or *avalanche cones*. These have a fan-shaped planform and closely resemble debris cones that accumulate as a result of repeated debris flow activity. Rapp (1959) suggested that they reflect more powerful avalanche activity than roadbank tongues, but they may simply indicate a less 'channelled' mode of avalanche deposition, or a progressive buildup of debris; as Rapp stressed, the two types are part of a developmental contin-uum of avalanche landforms.

In Great Britain, most avalanches occur unseen and unheard in remote mountain areas, and structures and communications are rarely affected. As a result, data on the nature of avalanches are scant for most upland areas. A survey of avalanche activity in Scotland by Ward (1980, 1984), however, provides a reasonable picture of the nature and frequency of avalanches in the Grampians and Cairngorms. Ward has shown that nearly all mountain avalanches in these areas are cornice falls, sluffs of 'flowing' snow or shallow slab avalanches, with fracture depths of 0.5–1.0 m; both full-depth avalanches and airborne avalanches are rare. The great majority of recorded avalanches have runout lengths of less than 250 m. Much larger avalanches are, however, not unknown; Ward (1984) observed one in the Cairngorms that had travelled 2 km and 'part of the way up Beinn Bhrotain', and he noted that around 1959 a large avalanche from Coire na Ciste in the northern Cairngorms travelled at least 500 m beyond the position of the car park and skiing facilities that have subsequently been built at this site. In part, the rarity of large avalanches reflects a lack of steep slopes of alpine dimensions, and the relative mildness of present conditions; incursions of warm air may lead to thaw at any time during the winter and thus prevent the accumulation of a thick snowpack. Ward, Langmuir & Beattie (1985) have also shown that the structure of the Scottish mountain snowpack differs from that in many polar and alpine environments where winter temperatures are lower. Typical profiles show masses of wind-slab above an equigranular layer of old snow, with thin ice layers at all depths. Because of the rarity of prolonged cold periods with temperatures below -10 °C, depth hoar (which forms a potential failure plane in alpine snowpacks) is virtually unknown. Slab avalanches tend to occur in the surface layers when temperature and snow density are low (less than -4 °C and 200–250 kg m^{-3} respectively). Thaw periods, however, may result in the development of slush zones above ice layers, giving rise to wet snow avalanches.

Irrespective of magnitude, few of the avalanches observed by Ward contained significant amounts of debris, which suggests that despite their frequency (many Cairngorm sites avalanche annually), avalanches are currently of limited importance as agents of talus modification in upland Britain. Even the full-depth avalanches observed by Davison & Davison (1987) on Meall Uaine in the SE Grampians resulted only in the movement of a few boulders a short distance downslope and localised uprooting of turf, though admittedly the avalanche track at this site supports a protective vegetation cover. Very limited talus slope modification is also suggested by Ward's (1985) study of avalanche activity in a corrie on Lochnagar (1155 m). The taluses surveyed here by Ward displayed only slight concavity, increasing from 29° at the base (where the talus enters a small loch) to almost 35° at the crest, with individual facet angles exceeding 38°, and are effectively indistinguishable from those of unmodified rockfall talus. Persistence of fall sorting on the Lochnagar taluses also indicates minimal avalanche modification. Indeed only small-scale avalanche effects were evident: a small number of perched boulders; bare patches on lichen-covered surfaces; small pits in vegetated talus, where boulders have been removed by avalanches; and scratch marks attributable to the passage of avalanche-entrained boulders. The geomorphic impotence of avalanches at this site was attributed by Ward to their small volume, the massiveness of the granite boulders in the talus and the presence of a protective cover of old snow.

The rarity of recent avalanche landforms on British mountains suggests that Ward's findings are representative for most areas. At a small number of particularly favourable high-level sites, however, the geomorphic role of recent avalanches is much more prominent. One such site is the Lairig Ghru, a high-level glacial breach that cuts through the western Cairngorms. Here a number of small roadbank-type avalanche tongues sweep down from markedly concave talus slopes and extend across the valley floor, and in a few cases angular avalanche debris from the east side of the valley has been transported some distance up the opposite slope (Figure 12.9). Another outstanding site is Coire na Ciste, a magnificent high-level (900–1300 m) corrie scalloped out of the north face of Ben Nevis, Britain's highest mountain. Here Ballantyne (1989b) identified classic examples of active avalanche talus, with average gradients of less that 30°, markedly concave profiles and a complete absence of fall sorting (Figure 12.10). At the foot of these taluses lies an apron of sharply-angular avalanche debris, some of it lichen-free and clearly recent, that is spread for a distance of up to 50 m across the floor of the corrie (Figure 12.11). A small tarn dammed by a 3 m high rampart of avalanche debris occurs at the break of slope at the foot of one avalanche chute, and appears to be an avalanche impact pit (Figure 12.12), though not necessarily of recent origin. The Pools of Dee, near the highest part of the Lairig Ghru, may have a similar origin.

Though evidence for such pronounced recent avalanche modification of talus is limited to a small number of favourable sites, it might be expected that avalanches played a more prominent role in talus slope modification in the past, particularly during the Loch Lomond Stadial and possibly even during the Little Ice Age of the sixteenth–nineteenth centuries AD, when snowcover on some mountains was perennial (Sugden, 1971) and when avalanche activity was enhanced in southern Norway (Grove, 1972). In Scotland, relict, overgrown avalanche boulder tongues occur at the feet of vegetated talus slopes in Glen Feshie and Glen Taitneach as well as in the Lairig Ghru and on Skye (Ballantyne, 1991a), but the age of these features is unknown. Possible examples of relict avalanche cones also occupy the lower parts of some taluses, for example in the Bealach Dubh, a glacial breach that separates Aonach Beag from Ben Alder in the central Grampian Highlands, but none have been investigated in detail. As noted earlier, most relict talus slopes surveyed in Britain have profiles typical of unmodified rockfall talus. It seems unlikely, though, that avalanches were not of much greater geomorphic importance during colder and snowier episodes in the past, and it is probable that only lack of research has inhibited identification of their effects. One promising approach might be to examine exposures in the lower parts of the debris cones that flank many upland valleys, to establish whether the sharply angular, unsorted debris characteristic of avalanche deposits is represented stratigraphically beneath or intercalated with debris flow deposits.

Debris flow activity

Of all the processes that have modified the form of talus slopes in Great Britain, by far the most widespread has been debris flow activity. The term *debris flow* is primarily employed here to refer to the rapid downslope flow of poorly-sorted debris mixed with water, but it is also used to refer to the landforms produced by individual flows. Brunsden (1979) classified debris

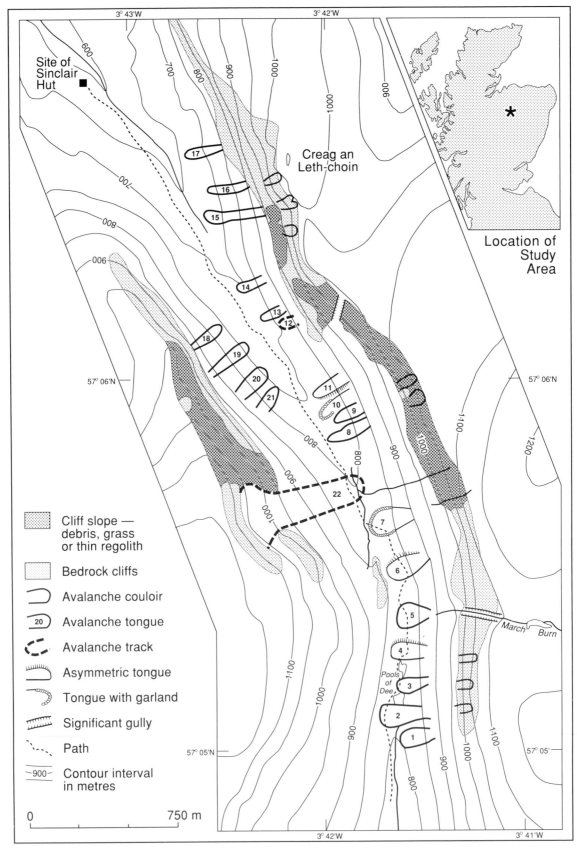

Figure 12.9 Avalanche landforms in the Lairig Ghru, Cairngorm Mountains, mapped by B.H. Luckman.

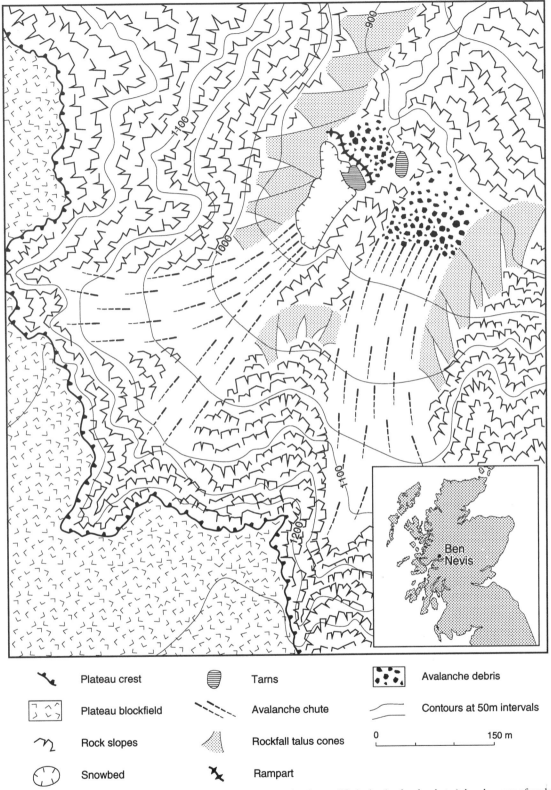

Figure 12.10 Geomorphological map of Coire na Ciste, Ben Nevis, showing avalanche-modified talus (avalanche chutes), basal aprons of avalanche debris and an avalanche rampart that has been deposited at the downslope end of a water-filled avalanche impact pit.

Figure 12.11 Poorly-sorted angular avalanche debris on the floor of Coire na Ciste, Ben Nevis. Note the sharply-angular perched boulders in the foreground.

Figure 12.12 Water-filled avalanche impact pit on the floor of Coire na Ciste, Ben Nevis. The tarn is dammed by an avalanche rampart on the far side of the impact pit.

Figure 12.13 Hillslope debris flow on a Lateglacial talus slope in the Lairig Ghru, Cairngorm Mountains. The source area of the flow is a gully incised in the talus by earlier debris flow erosion, and the flow terminates downslope on an incipient debris cone. The levées of older, partly-vegetated debris flows can be seen to the right of the recent flow track.

flows by scale and the nature of source area into three types, namely catastrophic flows, hillslope flows and valley-confined flows. The first are not represented in upland Britain. *Hillslope flows* are those that flow down an open hillslope, and are not topographically constrained; *valley-confined flows* are those that are confined for much of their length to a pre-existing valley or gully. The two categories, however, are transitional. Hillslope flows often follow shallow gullies cut in drift, talus or regolith by previous flow events (Figure 12.13).

Debris flows are restricted neither to talus slopes nor to periglacial environments. As agents of mass transport they may have most significance in hot semiarid environments (Johnson & Rodine, 1984) rather than in periglacial areas. The process is nonetheless of widespread importance in modifying slopes in cold mountainous environments, including both high-latitude mountains such as those of Spitsbergen and northern Scandinavia (e.g. Jahn, 1976; Rapp & Nyberg, 1981; Larsson, 1982) and mid-latitude mountains such as the Alps and Tatras (e.g. Kotarba & Strömquist, 1984; Van Steijn, de Ruig & Hoozemans, 1988). In such areas the prime effect of debris flow activity is the reworking of talus, though flows may also occur on steep slopes mantled by frost-weathered regolith or glacial

drift. The overall effect of debris flows operating on talus is to erode material from the upper part of the slope and deposit it near or beyond the talus foot, thus lowering the overall gradient of the slope and creating a long, sweeping basal concavity, similar in many ways to that produced by avalanche modification (Figure 12.2). Repeated valley-confined debris flows ultimately result in the accumulation of a relatively low-gradient (12–25°) *debris cone* near the talus foot (Figure 12.1). Another point of similarity is that debris flows disrupt the fall-sorting characteristic of rockfall talus. Slopes modified by debris flows, however, support a very different microrelief from that of avalanche-modified talus. Individual flows on the surface of talus are characterised upslope by erosion of a gully, which is continued downslope by two parallel debris levées that mark the path of the flow and which terminate downslope in one or more lobes of bouldery debris (Figure 12.13). Debris flow sediments also differ greatly from those deposited by snow avalanches, for the most part consisting of a diamicton in which clasts of variable size are embedded in a sand-rich matrix.

The great majority of debris flows are initiated by slope failure in the form of shallow landslides (Innes, 1983b; Johnson &

Rodine, 1984). Such slides are usually triggered by intense rainstorms that cause a rise in porewater pressures and consequent reduction in shearing resistance. Caine (1980) attempted to define boundary conditions of rainfall intensity and duration that control the onset of such sliding, but these offer only rough guidelines, as propensity for failure is influenced by other factors such as antecedent moisture conditions, gradient, and the depth and mechanical properties of the talus or regolith. The nature of the transition from sliding to flow is incompletely understood, but the essential mechanism appears to be one of progressive liquefaction and remoulding of the sliding mass, with the addition of further water increasing intergranular distances and thus decreasing its effective strength (Campbell, 1974; Rodine & Johnson, 1976; Johnson & Rodine, 1984). The nature of the flow itself is also contentious. Johnson & Rodine (1984) have advocated a form of Bingham flow, in which a raft or 'plug' of debris is carried downslope by laminar flow acting along the sides and base of a channel. Innes (1983b), however, has suggested that cohesionless grainflow operates where the clay content of the debris is low, and considered that this mode of flow is likely in the context of hillslope debris flows in Scotland. The transport of large boulders within debris flows, sometimes over gradients as low as 5°, has been variously attributed to buoyancy, the viscosity of the fluid phase, excess porewater pressures and dispersive pressures within the flow (e.g. Hampton, 1975, 1979; Rodine & Johnson, 1976; Pierson, 1981; Johnson & Rodine, 1984). Coarse debris tends to move upwards within a mobile flow, then is carried to the front by greater surface velocities (Takahashi, 1981). Concentrations of boulders may temporarily dam the flow, so that debris flows tend to move in a series of surges (Campbell, 1974; Okuda et al., 1980), with velocities of up to 16 m s^{-1} (Curry, 1966). The deposition of levées at the margins of flow tracks probably reflects low velocities at the edge of flows, though it has also been attributed to bulldozing by the advancing flow (Sharp, 1942) or dispersive sorting (Pierson, 1980).

Both hillslope and valley-confined debris flows occur throughout upland Britain, though their frequency is much greater in some areas than in others. Both types are endemic on talus, but hillslope flows are common also on steep slopes mantled by frost-weathered detritus or glacial drift. Individual flows tend to have a concave long profile, but closely follow local changes in the direction of maximum surface slope, curving around the margins of talus cones, debris cones and even moraines (Statham, 1976b; Addison, 1987). Both Ballantyne (1981) and Innes (1983b) have identified a lower threshold gradient of 30° for flow initiation in the Scottish Highlands, and the latter has found that the great majority originate on slopes of 32–42°. As elsewhere, the steep upper parts of debris flows in upland Britain comprise zones of net erosion, whilst net deposition (in the form of marginal levées and terminal lobes) is characteristic of the lower zone. The gradient at which deposition succeeds erosion is variable. On the Black Mountain in Carmarthen (Wales), Statham (1976b) found that this transition occurred at c. 16° for all of the recent flow tracks he surveyed, and he interpreted this as the gradient of the transport slope on which flows neither erode nor deposit debris. Ballantyne (1981), however, found that the transition occurred on much steeper (20–28°) slopes on An Teallach in northern Scotland. Similarly, whereas the Black Mountain flows came to rest on a slope of c. 8°, those surveyed on An Teallach were found to

have stopped on gradients of 11–23°. These differences may reflect greater flow viscosity at the latter site. One intriguing feature observed in both areas is that in the depositional zone upslope of the terminal lobe, vegetation between the levées has often been totally undisturbed by the passage of the flow, an indication of very low basal shear stresses. The dimensions of debris flows vary widely. On the Black Mountain, the tracks are typically 1.0–1.5 m wide and levées are 0.3–0.4 m high; on An Teallach, most tracks are 1–3 m wide, with levées 0.3–0.5 m high. Miniature flows also occur, as do much larger specimens. A giant rockslide-generated flow in Knoydart, for example, is over 400 m long and 50 m wide, with levées up to 20 m high (Ballantyne, 1992). In an extensive survey of hillslope debris flows in the Scottish Highlands, Innes (1985) found that the volume of transported material ranged up to c. 230 m^3, but that only a small proportion of the flows surveyed involved movement of more than 30 m^3 of debris. Ballantyne (1981) observed the presence of boulders more than 1 m long in the levées of large hillslope flows on An Teallach. The largest measured c. 1.8 m^3, equivalent to a mass of approximately 4.5 tonnes.

Debris cones that have accumulated through repeated debris flow deposition at the feet of talus or other slopes are also widespread throughout upland Britain, though much less common than individual flows. Magnificent cones flank many glacial troughs in the Scottish Highlands, such as Glencoe, Glen Etive and Glen Feshie (Innes, 1983a; Brazier, 1987; Brazier, Whittington & Ballantyne, 1988; Brazier & Ballantyne, 1989) and other fine examples have been described by Harvey et al. (1981) in the Howgill Fells and Statham (1976b) on the Black Mountain. However, by no means all valley-confined debris flows terminate on debris cones. Conversely, small debris cones sometimes occur at the foot of hillslope flows. Good examples of such cones occur near the highest point of Drumochter Pass, in the eastern Grampians. Here, gullies cut in thick drift on a featureless hillslope by past debris flows have become favoured sites for later flows, with a resultant accumulation of cones at the slope foot. It appears that, given a sufficient thickness of talus or drift, hillslope flows may gradually evolve into valley-confined flows.

Debris flows and cones are not restricted to the more mountainous parts of Great Britain. On the scarp slope of the Lomond Hills in Fife, for example, flows occurred in 1928 and 1985, and the vegetated scars and deposits of earlier flow events occur across the entire talus; others occurred nearby on the lower slopes of the Ochil Hills in November 1984 (Jenkins et al., 1988); a small flow took place in 1977 on Arthur's Seat (251 m) in Edinburgh; and Beven, Lawson & McDonald (1978) have documented a translational slide and debris flow that occurred on the North York Moors during an exceptional rainstorm in 1976. Even in mountainous areas of high relief, however, debris flow activity is much more marked in some locations than others. A map by Innes (1983a) based on an airphoto search for the presence of debris flows within 100 km^2 grid squares provides a rough indication of the main areas of debris flow activity in Scotland (Figure 12.14). Innes considered that his map represents only the minimal distribution of debris flows, but though imprecise it highlights areas of widespread debris flow activity, such as the NW Highlands, Skye, Rhum, Rannoch–Glencoe–Lochaber and the Cairngorms. Innes did not record any debris flows in certain areas of steep relief, such as the

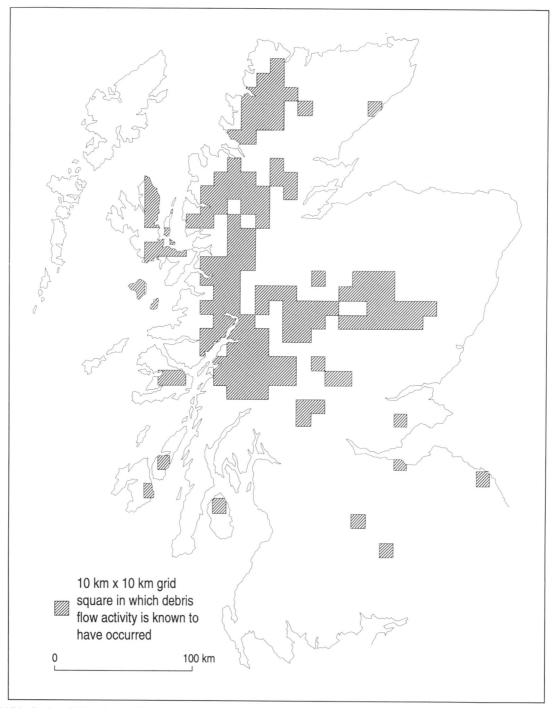

Figure 12.14 Distribution of debris flow activity in Scotland based on presence or absence of debris flows within 10 km x 10 km grid squares. Adapted from Innes (1983a).

mountains of the Outer Hebrides, Morvern, Knoydart, the SE Grampians and the Southern Uplands.

Such contrasts appear to reflect differences in lithology. Hillslope flows in particular are markedly more abundant on slopes mantled by talus, regolith or drift with a relatively coarse matrix. Innes (1983b) has suggested that flow-susceptible slopes in Scotland are mantled by debris with a clay content no higher (and often much less) than 3%, and that such debris tends to be both coarser and less well sorted than that on slopes

not prone to debris flows. Debris flow activity therefore tends to be most prominent on rocks that yield sand-rich regolith on weathering, such as the Torridon Sandstone of the NW Highlands and the granite of the Cairngorms. In such areas, hill-slope debris flows may cover wide areas: Strachan (1976), for example, mapped 58 individual flows across a 4 km long stretch of talus on Baosbheinn, a sandstone mountain in the NW Highlands, and Ballantyne (1981) reported 45 across 1 km of slope on An Teallach in the same area. Similar frequencies

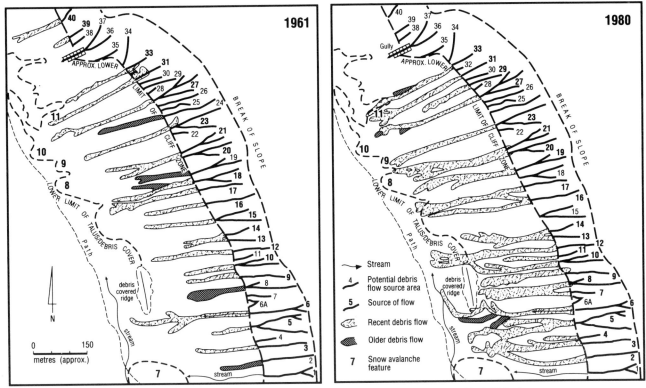

Figure 12.15 Increase in recent debris flows, 1961–80, in part of the Lairig Ghru, Cairngorm Mountains. Mapped by B.H. Luckman.

occur on the talus of the Lairig Ghru in the Cairngorms (Figure 12.15). In contrast, areas underlain by schists, which usually yield a much finer, silt-rich regolith on weathering, tend to support far fewer flows. The reason for the susceptibility of sand-rich debris to debris flow activity has not been explored in detail. Ballantyne (1986a) suggested that it may reflect the high infiltration rates associated with such debris, as these permit a rapid rise in the water table during periods of intense precipitation, leading to increased porewater pressures, reduction in effective normal stresses and consequent failure and flow. The susceptibility of sand-rich debris cover to sliding and flow is probably also enhanced by the cohesionless nature of such debris and by the absence, particularly on high ground, of a complete vegetation cover. A further important control on the distribution of debris flow activity is sediment availability. Some mountain areas are zones of 'areal scouring' (Gordon, 1981), in which glacially-polished bedrock crops out widely and there is little drift or regolith on slopes and very limited talus development. The mountains of Morvern, Morar and Knoydart in western Scotland are typical of this type of terrain. In these areas, debris flow activity is limited simply by the lack of sediment on steep slopes.

Debris flows occur more frequently in Great Britain than is commonly appreciated. Recent flow activity has been documented throughout the British uplands, for example in south Wales (Statham, 1976b), Snowdonia (Addison, 1987), the Howgill Fells (Wells & Harvey, 1987), the northern Pennines (Carling, 1987), the NW Highlands (Ballantyne, 1986a) and the Cairngorms (Baird & Lewis, 1957). The frequency of flow events varies greatly from area to area, but under the most propitious circumstances may be remarkably high. Innes (1983a),

for example, found that 71 individual flows were mobilised in the Lairig Ghru (Cairngorms) between 1970 and 1980. One reason why the high frequency of debris flow events in upland Britain is rarely appreciated is that most occur in remote areas, and cause little damage to structures or communications, but there are exceptions. Following a thunderstorm in 1953, debris flows were mobilised over a wide area of the Western Highlands of Scotland, and caused extensive damage to fencing, ditches, forestry and roads (Common, 1954). A similar event in 1968 closed the Achnasheen–Kinlochewe road in NW Scotland (Strachan, 1976). More recently, a flow of c. 350 m³ of debris engulfed a house at the foot of the Ochil scarp in central Scotland (Jenkins et al., 1988). Another reason why the frequency of debris flows in Britain is often underestimated is that they occur during torrential rainstorms, and hence are rarely witnessed. Intense rainstorms, usually convectional or occlusional in nature, appear to have been responsible for initiating all documented recent flows in upland areas. Rainfall intensities responsible for triggering recent flows include that of c. 40 mm in one hour for a flow event in Snowdonia (Addison, 1987), 55–80 mm in 2.5 hours for another in the Howgills (Wells & Harvey, 1987) and c. 105 mm in two hours for a third in the northern Pennines (Carling, 1987). However, not all exceptional rainstorms promote debris flow activity, even on flow-susceptible slopes. Much seems to depend on antecedent moisture conditions. In September 1981, for example, over 140 mm of rain fell in 24 hours over the catchment of the Ardessie Burn, which drains the western slopes of An Teallach. The storm generated specific river discharges of 4.1–4.8 m³ s⁻¹ km⁻² (Acreman, 1983), but on the mountain only two small debris flows occurred. This lack of debris flow activity appears to reflect very dry conditions during the preceding month, so that despite the intense rainfall most slopes remained stable.

Little research has been carried out on debris flow processes in Britain. On the Black Mountain, Statham (1976b) found that the volume of debris that accumulated in one year on the floor of a flow gully (8.4 m³) was approximately balanced by that removed by debris flows from that and adjacent gullies in the same year (8.3–11.5 m³). This suggests that in this area debris flows act as a cathartic, regularly removing the sediments that accumulate on the floors of gullies incised into the upper parts of talus. His findings also highlight the importance of sediment availability as opposed to rainstorm intensity as a major control on debris flow activity. Rather more attention, however, has been focused on the nature of debris flow sediments and their interpretation. Wells & Harvey (1987) identified three distinct facies associated with recent (valley-confined) debris flow deposits in the Howgill Fells, and interpreted these as indicative of viscous debris flows, dilute debris flows with a slightly higher water content and flows transitional to streamflows. The grain-size composition of the matrix in the deposits they examined proved to be relatively clay-rich (11–15%), from which they inferred that movement had taken the form of laminar, non-Newtonian slurry flow. This contrasts (but does not necessarily conflict) with Innes' (1983b) conclusion that most hillslope flows in Scotland, which are deficient in clay (<3%), have moved by cohesionless grainflow. Innes' views are supported by work by Carling (1987), who concluded that structures in bouldery lobes deposited within a steep upland valley in the northern Pennines indicate movement by low-viscosity grainflow with frequent particle collisions. Evidence for low viscosity flows was also outlined by Brazier & Ballantyne (1989), who found that the form of flow deposits revealed in section in Glen Feshie indicate that, on leaving the source gully, flows had spread widely over convex cone surfaces rather than following a single linear track to the slope foot. They suggested that such flows were less viscous (i.e. had a greater water:sediment ratio) than is the case with most hillslope flows. In sum, the available evidence suggests that although the majority of flows in British uplands, particularly hillslope flows, travel by a process of cohesionless grainflow, others with a higher clay content may involve laminar non-Newtonian flow of the type proposed by Johnson & Rodine (1984). Viscosity, and therefore the form of the flow deposit, would appear to be at least partly dependent on water content.

Most debris cones in upland Britain are completely vegetated, and show little sign of recent activity except for fluvial modification. Attempts to establish the age and significance of these essentially relict landforms have met with conflicting results. None has yet proved attributable to formation under periglacial conditions during the Lateglacial, though in view of the importance of debris flows as agents of mass transport in present-day permafrost environments such as Spitsbergen (e.g. Larsson, 1982), it seems likely that some relict cones located outside the limits of the Loch Lomond Readvance began to accumulate under periglacial conditions. Other cones, located within the limits of the readvance, are attributable to formation under paraglacial conditions during the early Holocene. One such cone occurs below Dalness Chasm, in the upper part of Glen Etive in the western Grampians. Here, Brazier *et al.* (1988) found that cone accumulation had occurred between *c.* 10 ka BP and *c.* 4 ka BP, and apparently ceased because supplies of glacigenic and rockfall sediments from upslope were

exhausted. There ensued a long period of stability, which ended abruptly sometime after *c.* 550 yr BP when human interference resulted in destruction of the natural vegetation cover and triggered fluvial incision and the deposition of an alluvial fan of reworked sediment that buried the lower parts of the cone. As such fluvial reworking is characteristic of many upland debris cones, it appears that this sequence of cone evolution may be widespread.

Other cones that have been investigated, however, have proved to be of more recent origin. Debris cones in the Bowland Fells of NW England seem to have accumulated between *c.* 5400 yr BP and *c.* 1900 yr BP, with a later and more widespread phase of aggradation around 900 yr BP. This later episode may reflect anthropogenic interference with the natural vegetation cover, and consequent gully erosion and cone deposition (Harvey & Renwick, 1987). Similarly, a radiocarbon date of 940 ± 95 yr BP obtained on the uppermost part of a soil buried by coalescing debris cones in the Howgill Fells, together with associated palynological evidence, suggests that gullying and consequent debris cone aggradation were triggered in this area by the introduction of grazing sheep by Viking settlers at the beginning of the present millennium (Harvey *et al.*, 1981). The debris cones at the foot of the Black Mountain scarp in south Wales may also be of fairly recent origin. Statham (1976b) calculated that if the rates of debris accumulation that he recorded are representative, then the cones can be no more than 540–700 years old. As his estimate rests on only one year of measurements, however, this conclusion must be regarded as tentative.

Recent cone accumulation is also evident at a site investigated by Brazier & Ballantyne (1989) at the head of Glen Feshie in the western Cairngorms. Here, river erosion of the toe of three coalescing debris cones had exposed a complex sequence of overlapping flow units, several of which are separated by immature soils containing plant roots (Figure 12.16). Radiocarbon dating of these soils and roots indicated that the bulk of the cones (which are up to 10 m thick at the exposed section) had accumulated since the fifteenth century AD, though the lowermost flow unit was found to pre-date *c.* 2000 yr BP. Brazier & Ballantyne found no evidence for vegetation disturbance at this site, and concluded that the episodic nature of cone development was attributable to lateral migration of the River Feshie, with prolonged periods of accumulation when the river occupied the far side of its floodplain alternating with periods of erosion when the river impinged on the cones. They suggested that episodic debris cone accumulation may have occurred much earlier than the radiocarbon dates suggest, but that evidence for earlier cones has subsequently been obscured by lateral erosion by the migrating River Feshie and by burial under the thick deposits laid down by more recent (post fifteenth century AD) flow events. Thus the recency of the Feshie debris cones may be illusory. The extent to which this may be true also of other sites where sections have been exposed by river erosion, for example those in the Howgills, is difficult to judge.

Even though some supposedly 'recent' cones may conceal much earlier debris flow deposits, the growing evidence for debris cone accumulation within the last millennium is impressive; Brazier & Ballantyne (1989), for example, calculated that the three Feshie cones represent the equivalent of an average annual accumulation of 50–60 m³ of sediment over the last 300 years. Even more spectacular evidence for recent acceleration

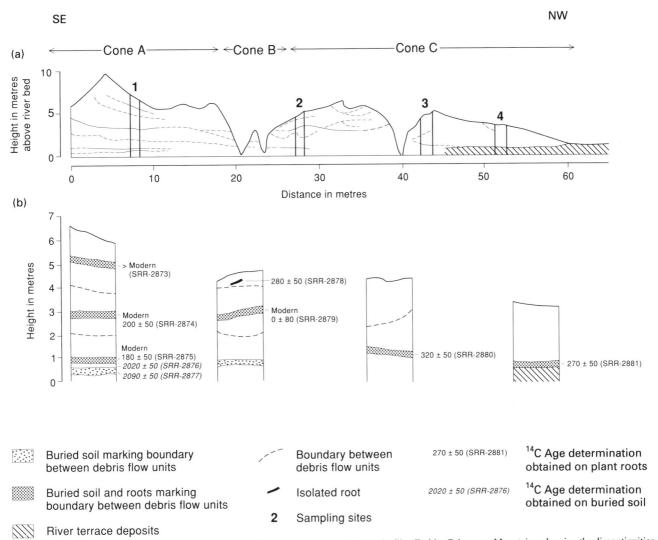

Figure 12.16 (a) Section surveyed across the eroded toes of three coalescing debris flows in Glen Feshie, Cairngorm Mountains, showing the discontinuities between individual debris flow units. (b) Detailed stratigraphy of profiles 1–4 identified on the section. From Brazier & Ballantyne (1989).

of the rate of debris flow activity in upland Britain comes from studies by Innes (1983a) of the age of hillslope debris flows in three areas of the Scottish Highlands, namely An Teallach, the Cairngorms and Glencoe–Glen Etive. On the basis of lichenometric measurements on 780 hillslope flow deposits, Innes concluded that all represented deposition within the last 600 years, and, outside of the Cairngorms, within the last 250 years (Figure 12.17). He discounted progressive weathering and climatic change as possible causes of this dramatic increase in debris flow activity, citing instead land use changes (particularly burning and overgrazing) as possible causes. His findings may be challenged on the grounds that later flows often obscure earlier ones, thus introducing a bias to the lichenometric dating of such features, but it seems indisputable that a drastic increase in debris flow activity has occurred within the last few centuries; had the rates identified by Innes been sustained throughout the Holocene, much greater thicknesses of debris flow deposits would have accumulated than are now evident.

The cause of this intensified activity remains obscure (Ballantyne 1991b). One possibility is that it reflects, directly or

indirectly, a general increase in storminess during the Little Ice Age of the sixteenth to nineteenth centuries AD. For much of this time the oceanic polar front in the north Atlantic region lay much farther south than it does at present, reaching a line between Iceland and the Faroes (Lamb, 1979). The southerly position of this front gave rise to steep thermal gradients, with a resulting increase in the frequency and ferocity of storms tracking across the British Isles (Lamb, 1977, 1979, 1984; Whittington, 1985). Brazier & Ballantyne (1989) have suggested that one or more exceptionally high magnitude rainstorm events with very long return periods may have initiated a general intensification of debris flow activity for many years thereafter, pointing out that slope failure and debris flow erosion resulting from very extreme rainstorm events are liable to have stripped vegetation cover from hillslopes and gullies, thus lowering the threshold for the initiation of subsequent debris flow events. Pending further research, however, this idea remains in the realms of hypothesis. It is fascinating to conjecture, though, that much of the erosion that has scarred and in places devastated the Lateglacial and Holocene taluses of upland Britain

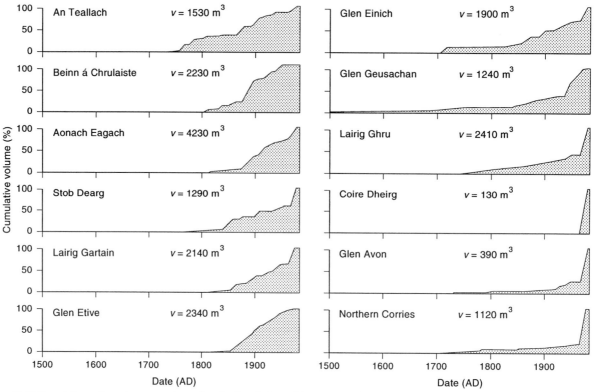

Figure 12.17 Cumulative volume (as a percentage of total volume) of regolith transported by debris flows at 12 sites in the Scottish Highlands during the period AD 1500–1980. The flows were dated using lichenometry. From Innes (1983a).

represents not so much the work of 10 000 years of erosion, but at most that of the last few centuries.

Protalus ramparts

A *protalus rampart* is a ridge or ramp of predominantly coarse detritus, usually located at or near the foot of a talus slope, that formed through the accumulation of debris along the downslope margins of a snowbank or firn field (Figure 12.18). Rampart development has traditionally been attributed to the progressive accumulation of clasts that have fallen from cliffs upslope and rolled, bounced or slid to the foot of the snow, a mechanism first proposed by Drew (1873) and accepted by almost all later writers on the topic (e.g. Daly, 1912, p. 593; Bryan, 1934; White, 1981). Until recently, however, this proposed mode of accumulation remained largely unsubstantiated (Johnson, 1983), as virtually all studies had concerned relict ramparts rather than active examples. Moreover, the lack of data on the characteristics of actively-accumulating ramparts has led to much uncertainty regarding the identification of relict forms, which have been confused with moraines, landslide deposits, avalanche landforms and particularly protalus rock glaciers (e.g. Washburn, 1979, p. 234; Sissons, 1980a; White, 1981; Ballantyne, 1986b). A fundamental constraint on rampart development that has often been ignored is that such forms can develop only within a short distance of the foot of a talus slope. Supposed ramparts that extend more than 50–100 m from the talus foot imply a thickness of 'snow' (actually firn or ice) greater than that required to initiate glacier movement, which would presumably destroy or at least modify any protalus ram-

part that had developed. Several authors have nonetheless described what they have interpreted as protalus ramparts at much greater distances from the slope foot. Some supposed multi-ridged ramparts identified by, for example, Blagbrough & Breed (1967) and Lindner & Marks (1985) possess characteristics more consistent with those of protalus rock glaciers than protalus ramparts.

More recent research has shown that rockfall need not be the only process that contributes to rampart formation. Ono & Watanabe (1986) have demonstrated that the role of rockfall in supplying debris to an active protalus rampart in the Japanese Alps is surpassed by that of frequent debris flows across the snow. Investigations by Harris (1986) on an active rampart in Norway revealed that some of its constituent debris consists not of primary rockfall detritus, but of till transported downslope. Research on two active protalus ramparts on the Lyngen Peninsula in Norway (Figure 12.18) led Ballantyne (1987b) to conclude that although rockfall boulders with sufficient momentum do indeed accumulate at the feet of perennial snowbeds, deposition of avalanche debris containing abundant fine material also contributes substantially to rampart development.

Several early accounts describe protalus rampart formation in upland Britain, though without reference to specific examples (Ballantyne, 1987c). The earliest appears to be that of Ward (1873, p. 426), who identified amongst the various types of moraines in the Lake District a special category consisting of 'mounds of scree material formed at the base of a slope, by the sliding of fragments over an incline of snow lying at the base of crags'. A similar mechanism was envisaged by Gatty (1906) for

Figure 12.18 Active protalus rampart on the Lyngen Peninsula, northern Norway.

the formation of a rampart in Coire na Ciste on Ben Nevis, though this feature was later reinterpreted as an avalanche rampart (Ballantyne, 1989b). Marr (1916, p. 202) considered that 'ridges of loose blocks' that he observed at the foot of taluses in the Lake District had been formed through the sliding of rockfall detritus over snowbanks 'at the end of glacial times'. Only in the last quarter century, however, has the widespread occurrence of protalus ramparts become appreciated. Landforms interpreted as protalus ramparts have been reported in south Wales, mid Wales and Snowdonia (Watson, 1966, 1977a; Lewis, 1970; Unwin, 1975; Watson & Watson, 1977; Gray, 1982), the Lake District (Sissons, 1980a; Oxford, 1985) and west-central Scotland (Rose, 1980). They also occur in various parts of the Scottish Highlands (Sissons, 1977b, 1979a, b) and even on the remote island of St Kilda (Sutherland et al., 1984; Figure 12.19). However, by no means all landforms that have been described as protalus ramparts have been correctly interpreted. For example, that described by Watson (1966, 1977) in Cwm Tinwen in mid Wales is much too remote from the foot of the talus slope to have this origin, and a similar criticism holds for the controversial 'rampart' at the head of Keskadale in the Lake District (Oxford, 1985), which is almost certainly a Loch Lomond Readvance end moraine (cf. Sissons, 1980a). It seems likely that several other putative 'protalus ramparts' that have

been described in upland Britain may in fact represent landforms with a completely different origin.

Perennial snowbeds do not occur below rockwalls in upland Britain at present. Active rampart formation is therefore unlikely, and no examples have been documented, though there are reports of rockfall debris travelling to the foot of steep snowbeds (Vincent & Lee, 1982). There is general consensus that the relict protalus ramparts of upland Britain formed during the Loch Lomond Stadial of c. 11–10 ka BP. This view is based on three lines of argument. First, all recorded examples lie within the limits of the Late Devensian glacial maximum, so that rampart formation must have postdated the decay of the Late Devensian ice sheet. Second, rampart formation requires the development of perennial snowbeds or firn fields, which implies development during a period of renewed cooling following ice-sheet deglaciation, and the Loch Lomond Stadial is the only period of pronounced renewed cooling for which there is evidence throughout upland Britain. Finally, with three questionable exceptions (Unwin, 1975; Sissons, 1979a, p. 72; Oxford, 1985, p. 39) all reported examples lie outside (and often immediately outside) the mapped limits of the Loch Lomond Readvance.

The characteristics of nine protalus ramparts of assumed Lateglacial age in Scotland and the Lake District have been studied by Ballantyne & Kirkbride (1986), though it now seems likely that one of these, a large feature on the floor of the Lairig Ghru in the Cairngorms, may not be a true rampart. Another, a massive arcuate rampart at Baosbheinn in the NW Highlands, is of complex origin and may be more correctly interpreted as a protalus rock glacier (see below). The characteristics of the remainder are summarised in Table 12.3. Ballantyne & Kirkbride found that rampart development has been restricted to sites where there is a marked reduction in gradient at the foot of a steep slope overlooked by a cliff. Such locations include the floors of corries and glacial troughs, but in addition some ramparts are 'perched' on rock steps high on valley sides. All those they examined are situated at the foot of relict talus slopes. Aspect appears to play only a secondary role in rampart location; they noted that of 54 British ramparts, 72% face between north and east, but that this preference may simply reflect the preponderance of rockwalls with such aspects; 13% have southerly aspects. Most ramparts are arcuate in planform, with the extremities of the frontal crest curving upslope at one or both ends, but some are linear, with approximately straight crests. Few ramparts exceed 300 m in length. All of the ramparts studied by Ballantyne & Kirkbride possess a single ramp or ridge crest, which in most cases bounds a shallow linear depression up to 3 m deep (Figure 12.20). On some ramparts, such as the Conachair rampart on St. Kilda (Figure 12.19), this depression is partly infilled with low mounds and discontinuous ridges of debris, which apparently represent chaotic dumping of coarse sediment when the snowbeds at these sites finally melted. The maximum facet angles recorded on the distal slopes of ramparts all fell within the range 34–39°, but proximal slope gradients are generally much gentler (Table 12.3).

An interesting feature of the ramparts investigated by Ballantyne & Kirkbride is that they exhibit a certain amount of morphometric regularity. In particular, there are strong, positive linear relationships between rampart width, rampart thickness and distance from the foot of the talus upslope. The relationship

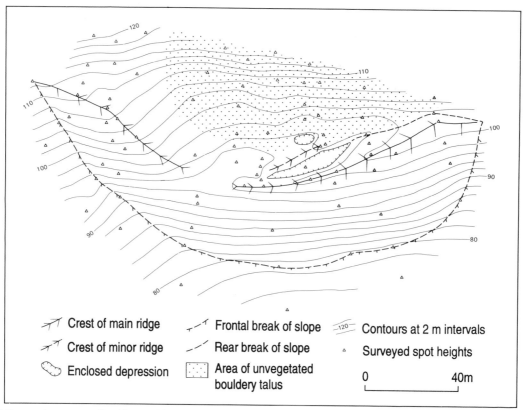

Figure 12.19 Relict protalus rampart of Loch Lomond Stadial age on St Kilda, 60 km west of the Outer Hebrides. After Ballantyne & Kirkbride (1987).

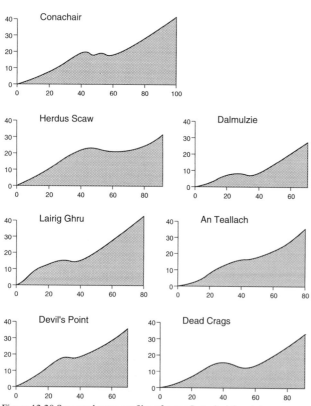

Figure 12.20 Surveyed cross-profiles of seven Lateglacial protalus ramparts in Scotland and the Lake District. After Ballantyne & Kirkbride (1986).

between width and thickness is explicable in terms of the progressive accumulation of coarse cohesionless debris, and effectively reflects the repose angle of such debris. The increase in rampart thickness with increasing crest–talus distance was interpreted as the result of progressive thickening of snowbeds, such that rampart crests moved outwards away from the talus as the ramparts accumulated (Figure 12.21). This model allows for snowbed growth during the Loch Lomond Stadial, and does not require the prolonged maintenance of a delicate balance between snowbed accumulation and snowbed ablation, a balance implicit in the traditional view that protalus ramparts reflect accumulation at the foot of snowbeds of stable dimensions. Although most of the ramparts investigated consist of coarse openwork debris at the surface, abundant fines were observed at some sites. These were attributed by Ballantyne & Kirkbride to granular disintegration of exposed clasts. Although some fines are likely to have originated in this way (exposed clasts tend to be more rounded than subsurface clasts), subsequent research on active ramparts suggests that the bulk of fine material may have been emplaced by avalanches (Ballantyne, 1987b) or debris flows (Ono & Watanabe, 1986). Clasts on rampart surfaces were found to be poorly-sorted, with little variation in size across the surface of any particular rampart, though the size range at any site is strongly conditioned by local lithology. In comparison with clasts of similar lithology contained in nearby end moraines, those on protalus ramparts tend to be generally more angular and 'slabbier' in form, presumably because a proportion of the moraine clasts have experienced modification during transport whereas rampart clasts represent relatively unmodified frost-weathered debris (cf. Boulton, 1978; Ballantyne, 1982a).

Table 12.3: *Characteristics of some Lateglacial protalus ramparts in upland Britain.*

Rampart	Altitude (m)	Lithology	Length (m)	Maximum width (m)	Maximum thickness (m)	Distance to talus (m)	α_{max} (degrees)	β_{max} (degrees)
Lairig Ghru (Cairngorms)	940	Granite	120	37	7	20	39.0	6.5
Devil's Point (Cairngorms)	650	Granite	150	29	6	20	37.5	12.0
Dalmulzie (SE Grampians)	420	Schist	(33)	25	4	16	34.0	13.5
Dead Crags (Lake District)	295	Slate	300	42	10	24	34.0	21.5
Herdus Scaw (Lake District)	195	Basalt	225	48	10	30	34.0	11.0
Conachair (St. Kilda)	90	Granophyre	215	39	8	22	38.0	23.5
An Teallach (NW Highlands)	515	Sandstone	120	34	4	19	36.0	

NB Distance to talus represents the maximum distance from the rampart crest to the foot of the adjacent talus slope; α_{max} is the maximum gradient of the distal slope; β_{max} is the maximum gradient of the proximal slope. The length of the Dalmulzie rampart is truncated, and the An Teallach rampart has no proximal slope.

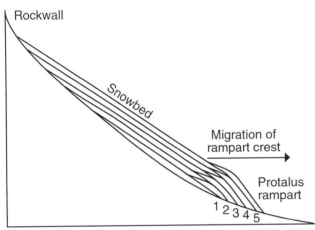

Figure 12.21 Model of protalus rampart development at the foot of a progressively thickening snowbed. The numbers 1–5 represent successive positions of the distal slope of the rampart. After Ballantyne & Kirkbride (1986).

Lateglacial protalus ramparts in upland Britain occur over a wide range of altitudes, from only 40 m in the Outer Hebrides to 950 m in the eastern Cairngorms. Within individual mountain areas, such as the Cairngorms, the Lake District (Sissons, 1980a) or Snowdonia (Gray, 1982) there is considerable variation in rampart altitudes. This appears to reflect local topographic controls, particularly the location of rockwalls overlooking suitable sites for snowbed survival and rampart accumulation. However, Ballantyne & Kirkbride (1986) found that the altitudes of Lateglacial protalus ramparts in the Hebrides and Scottish Highlands also display marked regional trends, with the lowest ramparts in the west and south and the highest in the Cairngorms (Figure 12.22). Allowing for local variations in altitude introduced by topographic controls, this pattern conforms to that of the trend of the equilibrium line altitudes of Loch Lomond Stadial glaciers as reconstructed by Sissons (1979c, 1980b). The general rise in the equilibrium line altitudes of stadial glaciers from the west coast and the Highland Boundary Fault towards the Cairngorms (Figure 12.22) was interpreted by Sissons in terms of decreasing precipitation (particularly snowfall) from 3000–4000 mm y^{-1} in the SW Grampians to 500–600 mm y^{-1} in the Cairngorms. The corresponding trend in rampart altitudes provides support for this interpretation of stadial snowfall patterns, for although the maintenance of perennial snowbeds requires adequate snowfall to offset annual ablation, too much snow (e.g. on high ground in the SW Grampians) would have resulted in rapid snowbed growth and the eventual formation of small glaciers. On a regional scale, therefore, the altitudinal distribution of protalus ramparts provides an indication of former snowfall patterns: assuming limited spatial variation in stadial palaeotemperatures, ramparts at low altitudes indicate formerly abundant snowfall, with both high accumulation and high ablation rates, whereas the restriction of ramparts to high altitudes (as in the Cairngorms) is indicative of light snowfall with correspondingly limited ablation. This interpretation is consistent with ablation season temperature estimates for the Loch Lomond Stadial in Scotland. For the SE Grampians (30–40 km southeast of the Cairngorms) Sissons & Sutherland (1976) calculated that the mean stadial July sea-level temperature was 6 °C, and an identical figure has been calculated by Ballantyne (1989a) for the Isle of Skye on the west coast. Assuming a lapse rate of 0.6 °C/100 m, this implies that mean July temperature at 750 m altitude was only c. 1.5 °C. Under such conditions ablation at high altitudes in the Cairngorms must have been very limited indeed, which in turn implies that snowfall totals must also have been very low in this area under full-stadial conditions.

Another way in which Lateglacial protalus ramparts have been employed to reconstruct palaeoenvironmental conditions is through calculation of former rockfall rates. The volume of debris contained in a rampart is equivalent to that lost from the rockwall upslope during the period of rampart formation. An estimate of the average rockwall retreat implied by a given volume of rampart debris may therefore be calculated by dividing rampart volume (with an appropriate allowance for void space)

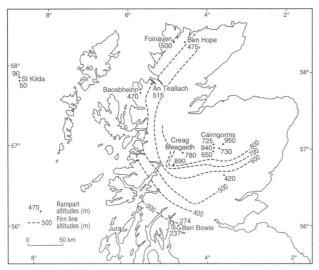

Figure 12.22 Altitudes of the crests of Loch Lomond Stadial protalus ramparts in the Scottish Highlands, and reconstructed regional equilibrium line altitudes of stadial glaciers at their maximum extent. From Ballantyne & Kirkbride (1986).

Table 12.4: *Rockwall retreat rates during the Loch Lomond Stadial, calculated from the volumes of protalus ramparts.*

Rampart	Lithology	Average amount of rockwall retreat (m)	Calculated stadial rockwall retreat rate (mm yr⁻¹) Over 800 yr	Over 400 yr
Lairig Ghru	Granite	1.32	1.65	3.30
Devil's Point	Granite	1.27	1.59	3.18
Dalmulzie	Schist	1.51	1.89	3.78
Dead Crags	Slate	1.33	1.66	3.32
Herdus Scaw	Basalt	1.14	1.43	2.86
Conachair	Granophyre	1.35	1.69	3.38
An Teallach	Sandstone	1.27	1.59	3.18
Mean for all sites		1.31	1.64	3.29

Source: Ballantyne & Kirkbride (1987).

by the area of the rockwall source. Using this method, Ballantyne & Kirkbride (1987) found that for seven Lateglacial ramparts in the Scottish Highlands and the Lake District the average amount of rockwall retreat implied by the rampart volumes was 1.14–1.51 m, with a mean value of 1.31 m (Table 12.4). Conversion of these figures into rockwall retreat *rates* is more difficult, as the exact duration of the period of rampart accumulation is unknown. It can be estimated, however, if it is assumed that the perennial snowbeds required for rampart formation existed for a period of time similar to that of small Loch Lomond Stadial glaciers. Sissons & Sutherland (1976) argued that stadial glacier accumulation in the SE Grampians occurred over *c.* 750 years, but farther south in the Lake District the evidence provided by varves in major lakes suggests that the stadial glaciers existed for only 400–450 years (Pennington, 1978; Sissons, 1980a). On the basis of these figures, Ballantyne & Kirkbride (1987) assumed the minimum period of rampart accumulation to be 400 years and the maximum 800 years, and calculated rockwall retreat rates accordingly (Table 12.4). Although the results must be treated as broad estimates only, they reveal a remarkable consistency from site to site, irrespective of lithology, and suggest that stadial rockfall rates were rather similar to those of present-day high alpine environments (Table 12.2). In part, the high rate of rockfall implied during the Loch Lomond Stadial may reflect the instability of oversteepened rockwalls inherited from ice-sheet glaciation, but, as noted earlier, it seems more likely that it is largely a measure of the effectiveness of freeze–thaw in displacing boulders from cliff faces under the severe periglacial conditions of the stadial.

No account of protalus ramparts in upland Britain would be complete without mention of the massive boulder rampart located at 450 m altitude at Baosbheinn in NW Scotland, at one time believed to be the largest protalus rampart yet described anywhere in the world (Washburn, 1979, p. 234). At this site an arcuate upper ridge of large boulders, up to 55 m high (AB in Figure 12.23) surmounts a lower vegetation-covered ridge (CD in Figure 12.23). Both ridges were interpreted by Sissons

(1976a) as a 'protalus rampart complex' 1 km long. However, Ballantyne (1986b) has shown that the lower ridge is a lateral moraine, and that only the upper boulder ridge may be interpreted as a protalus rampart. Even the interpretation of this upper ridge as a 'conventional' rampart seems questionable, however. Not only is the ridge unusually large for a protalus rampart, but also its volume implies *c.* 14.3 m of rockwall retreat during the Loch Lomond Stadial, an order of magnitude greater than that indicated by contemporaneous protalus ramparts elsewhere (Table 12.4). Ballantyne (1986b) suggested that this anomaly may be accounted for in terms of a major rockslide or series of rockslides over a former snowbed, an interpretation consistent with the distribution of large boulders at the site (Figure 12.23), the absence of steep rock faces above its central and thickest part and the configuration of the source, which resembles a major rockslide scar. Near the centre of the rampart is a depression containing two enclosed hollows that appear to have been produced by the melting of buried ice. This evidence suggests that the Baosbheinn 'protalus rampart' may have experienced forward movement through deformation of internal ice, in which case this landform is more correctly described as a protalus rock glacier, a type of landform considered below.

Protalus rock glaciers

A *rock glacier* is a thick lobate or tongue-like mass of angular debris that has moved slowly downslope as a consequence of the deformation of internal ice or frozen sediments. Two distinct categories may be distinguished. First, *protalus rock glaciers* (also referred to as valley-wall rock glaciers, lobate rock glaciers, talus-derived rock glaciers, talus-foot rock glaciers or simply talus rock glaciers) are those that have developed through the deformation of talus to form step-like or lobate extensions of the lower parts of talus slopes (Figures 12.1 and 12.24). Such features are widely accepted as being permafrost phenomena whose formation is independent of glacier ice. Secondly, *morainic rock glaciers* (also referred to as tongue-shaped rock glaciers, valley-floor rock glaciers, ice-cored rock glaciers or debris rock glaciers) are those that have formed through the burial and subse-

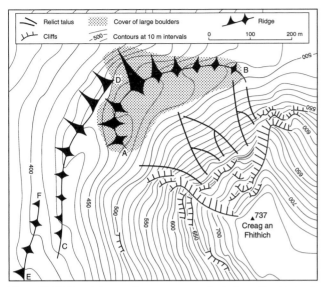

Figure 12.23 Plan of the protalus rampart or rock glacier at Baosbheinn in NW Scotland, showing the upper boulder ridge (AB) and two lower ridges (CD and EF) that represent earlier lateral moraines.

the reality of glacier-cored rock glaciers is simply terminological, and reflects the breadth of definition applied to the term 'rock glacier'. With one or two possible exceptions, all of the relict rock glaciers hitherto identified in upland Britain are protalus rock glaciers.

Active protalus rock glaciers are characterised by steep (up to about 45°) and abrupt frontal and lateral margins, usually 5–40 m high, and tend to be greater in width (across-slope) than length (downslope). They often support a distinctive microrelief that comprises three types of feature: first, meandering and closed depressions, probably attributable to meltout of subsurface ice (Corte, 1976; Haeberli, 1985); second, longitudinal (downslope) ridges and furrows; and third, transverse ridges and depressions that may represent thrusting along subsurface shear planes (White, 1987). Investigations of the structure of active protalus rock glaciers have revealed that they consist of a layer of unfrozen coarse debris, no more than a few metres thick, overlying much finer material with occasional embedded boulders. Excavation and geophysical investigations have shown that in active protalus rock glaciers, this fine debris is perennially frozen and contains a large volume of ice. This may take the form of interstitial ice (filling the voids between particles), matrix ice (surrounding and supporting individual parti-

Figure 12.24 Small active protalus rock glacier in the Green Lakes Valley, Colorado Front Range.

quent deformation of a core of glacier ice under a thick cover of bouldery morainic debris (Barsch, 1987a). Not all authors accept the existence of this second type (e.g. Haeberli, 1985; Barsch, 1987b) but it appears that much of the controversy concerning

cles) or debris-free ice lenses (Fisch, Fisch & Haeberli, 1977; Barsch, 1978; Barsch, Fierz & Haeberli, 1979; Vick, 1981; Wayne, 1981; Calkin, Haworth & Ellis, 1987). It is widely (though not universally) accepted that movement occurs through overburden-induced deformation of ground ice or ice-rich sediments within talus, a mechanism first outlined by Wahrhaftig & Cox (1959) and strongly supported by more recent research (Haeberli, 1985). This interpretation implies that permafrost is a prerequisite for protalus rock glacier development. Reported rates of movement, summarised by Kerschner (1978) and Vitek & Giardino (1987) range from 2 mm yr^{-1} to 3.57 m yr^{-1}.

Only a small number of protalus rock glaciers have been found in upland Britain. The largest concentration is located in the Cairngorms, where there may be as many as eight such forms (Sissons, 1979a; Maclean, 1991). Others occur in the Lake District (Sissons, 1980a) and Snowdonia (J. Rose, personal communication), with isolated examples in NW Scotland (Ballantyne, 1987d) and on the Hebridean island of Jura (Dawson, 1977). All have been ascribed a Loch Lomond Stadial age, on the grounds that they invariably occur outside the limits of the Loch Lomond Readvance and that the Loch Lomond Stadial is the only period during which there is known to have been a return to permafrost conditions following ice-sheet deglaciation. Problems have arisen, however, concerning the correct identification of protalus rock glaciers in Britain. For example, the 'snowbed deposits' identified by Sissons (1979a) in the Cairngorms have subsequently been reinterpreted as rock glaciers (Maclean, 1991), and other features mapped by him as 'snowbed deposits' in the Lake District (Sissons, 1980a) may also represent rock glaciers. Conversely, some of the supposed rock glaciers identified by Sissons in the Lake District appear to have been misinterpreted, as has the dubious 'protalus rock glacier' identified by Oxford (1985) below Grasmoor End. As noted in the context of the Baosbheinn 'protalus rampart', considerable confusion may also exist in the discrimination of protalus rock glaciers from protalus ramparts. The former, however, usually terminate at a much greater distance from the talus foot than do the latter, and in some cases are distinguished by the development of transverse ridges. Such ridges are strongly diagnostic of rock glacier flow, and are particularly well developed on two of the finest rock glaciers in Great Britain, that below Beinn Shiantaidh on the island of Jura (Dawson, 1977) and that at the northern end of Coire Beanaidh in the Cairngorms (Chattopadhyay, 1984).

The dimensions of established examples of protalus rock glaciers in upland Britain vary greatly. That on Jura is 380 m wide (across-slope) and 180 m long; the Coire Beanaidh rock glacier is 480 m wide and 240 m long. The largest protalus rock glacier yet identified, that in Strath Nethy (Cairngorms), is 2400 m wide and 350 m long (Maclean, 1991; Figure 12.25). All, however, are characterised by abrupt bouldery margins standing at angles of 20–37°. Dawson (1977) calculated the volume of the Jura rock glacier as c. 185 000 m³, and on this basis estimated that it represented an average rockwall recession of 2.6–9.2 m. His attempt to translate this into a Loch Lomond Stadial rockwall retreat rate is flawed, however, as the rock glacier contains an unknown amount of pre-stadial talus. Chattopadhay (1984) calculated that the c. 208 000 m³ of debris contained in the Coire Beanaidh rock glacier represents c. 2 m of rockwall retreat during the interval between ice-sheet deglaciation and the end of the Loch Lomond Stadial.

Of the Scottish protalus rock glaciers, it is notable that all documented examples occur on fairly massive lithologies

Figure 12.25 The Strath Nethy rock glacier in the Cairngorm Mountains. This protalus rock glacier extends for 2.4 km along the foot of the slope, and is the largest such feature yet identified in upland Britain. Crescentic failure scars on the granite rockwall above the rock glacier suggest that rockslides provided a large proportion of the rock glacier debris.

(granite, quartzite and Torridon Sandstone) that yield abundant coarse bouldery rockfall debris; none is known to occur on the schistose rocks that underlie much of the Grampians and northern Highlands. This appears consistent with Evin's (1985, 1987) findings that rock glaciers in the French Alps are found mainly below cliffs developed in massive lithologies and rarely on schists. Maclean (1991) has also noted that several Scottish rock glaciers are associated with sites of major rock slope failures. This is true of that at Baosbheinn (Ballantyne, 1986b, 1987d) and also of some in the Cairngorms, where crescentic scars on the cliffs above the rock glaciers are testimony to widespread sliding failure along stress-release joints developed in granite (Figure 12.25). Although this is not a universal feature of protalus rock glaciers in upland Britain, it does suggest that such landforms have developed only at sites that favoured rapid accumulation of talus. The rarity of protalus rock glaciers in Britain may also reflect abundant snowfall over most mountains during the Loch Lomond Stadial. As noted by Thompson (1962), Blagbrough & Farkas (1968) and others, active rock glaciers tend to be located in mountain areas that experience limited snowfall. Haeberli (1985) has pointed out that the absence of rock glaciers from areas of abundant snowfall probably reflects the fact that in such areas the equilibrium line altitudes of glaciers descend below the lower altitudinal limit of permafrost. As a result, the zone of potential rock glacier development is invaded by glacier ice. This may have been the case in Britain during the Loch Lomond Stadial, when many sites potentially favourable for rock glacier formation were occupied by ice glaciers.

The association of protalus rock glaciers with permafrost and limited snowfall implies that they have considerable potential for the reconstruction of palaeoclimatic conditions. This, however, has been exploited only in a very general way. Sissons (1979a) employed the presence of protalus rock glaciers in the Cairngorms to buttress his argument (developed from the small size of contemporaneous glaciers in the area) that the massif was starved of precipitation during the Loch Lomond Stadial. He also inferred from the presence of a relict protalus rock glacier at 300 m in the Lake District that mean annual temperature at this altitude cannot have exceeded c. -1 °C, and

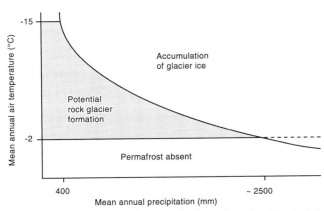

Figure 12.26 Climatic constraints on rock glacier formation. The shaded area represents possible combinations of mean annual temperature and mean annual precipitation that permit the development of protalus rock glaciers. From Haeberli (1985).

hence contemporaneous (Loch Lomond Stadial) mean annual sea-level temperature in NW England cannot have been greater than *c*. +1 °C (Sissons, 1980a). It is now widely recognised that at least sporadic permafrost is required for protalus rock glacier development. Such rock glaciers therefore only form where mean annual temperature is less than -1 °C to -2 °C (White, 1976b; Barsch, 1978). However, Haeberli (1985) has proposed that the climatic constraints on protalus rock glacier development involve simultaneous consideration of both mean annual temperature and mean annual precipitation. According to his model (Figure 12.26), the upper limiting value of mean annual precipitation declines nonlinearly from 2500 mm yr⁻¹ at a mean annual temperature of -2 °C to 400 mm yr⁻¹ at -15 °C. If a particular combination of mean annual temperature and precipitation exceeds this threshold, then glacier ice will form; only below the threshold is protalus rock glacier development possible.

If the relationship proposed by Haeberli is valid, it offers considerable potential for palaeoclimatic reconstruction. For example, Sissons (1979c, 1980b) has proposed from glaciological evidence that mean annual precipitation in the corries of the NW Cairngorms (including Coire Beanaidh, site of a protalus rock glacier) was only 500–600 mm yr⁻¹ during the Loch Lomond Stadial. For rock glacier formation in these corries, Haeberli's curve (Figure 12.26) suggests that the corresponding mean annual temperature cannot have been lower than -11 °C to -12 °C. The Coire Beanaidh rock glacier is 920 m above sea level. Assuming a lapse rate of 0.6°/100 m, the above range implies that mean annual sea-level temperature cannot have been less than -5.5 °C to -6.5 °C. Conversely, even assuming a maximum 'allowable' precipitation value of 2500 mm yr⁻¹, low-level protalus rock glaciers such as that at 350 m on Jura (Dawson, 1977) imply mean annual temperatures of *c*. -2 °C, and corresponding mean annual sea-level temperatures no higher than *c*. 0 °C. The presence of rock glaciers in presently 'wet' areas such as Jura and the Lake District also suggests that stadial precipitation in these areas did not exceed 2500 mm yr⁻¹, rather less than the 3000–4000 mm yr⁻¹ suggested by Sissons (1979c; 1980b) for contemporaneous precipitation in the SW Grampians. Admittedly, the accuracy of such estimates hinges on the validity of Haeberli's model, which has yet to be rigorously tested outside the Alps. Given a fuller record of the locations of protalus rock glaciers in upland Britain, however, their

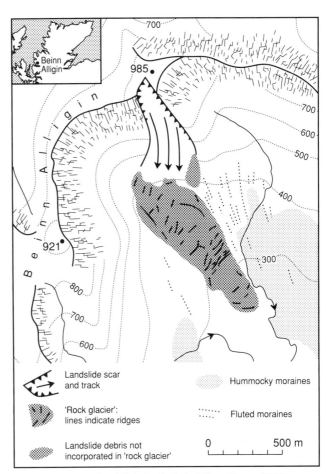

Figure 12.27 The Beinn Alligin 'rock glacier'. From Sissons (1975).

employment in palaeoclimatic reconstructions appears to have much greater potential than has hitherto been realised.

No mention has yet been made of morainic rock glaciers in Britain. Two possible examples have been identified, both (appropriately) inside the limits of the Loch Lomond Readvance. One terminates at 550 m altitude in a valley north of the summit of Beinn an Lochain (912 m) in the SW Grampians. According to Maclean (1991), it is *c*. 500 m long and *c*. 250 m wide, with a mean frontal gradient of *c*. 34°. The constituent debris clearly originated as a major rock slope failure from the cliff at the head of the valley, and it is debatable whether it was transported to its present position on a deforming core of glacier ice, or whether it travelled under its own momentum (Holmes, 1984). A similar controversy surrounds the celebrated 'rock glacier' that lies on the floor of a deep corrie south of the summit of Beinn Alligin in the NW Highlands. This is one of the most impressive landforms in Great Britain. A tongue-shaped accumulation of coarse bouldery debris up to 400 m wide extends downvalley for a distance of 1.2 km (Figure 12.27). The debris rises 12–15 m above the valley floor, and is composed of extremely large sandstone boulders, with no visible interstitial fine material. The deposit rests on a gentle slope that declines from 400 m to 260 m in altitude. Its lateral margins are fairly sharply defined, and the surface relief consists of ridges and depressions. The lowermost 300–400 m of the deposit is, however, much thinner than the remainder, and lacks the steep frontal margin that is the hallmark of all types of rock glacier. The source of the debris was a large rockslide or rock avalanche that left a huge triangular failure scar on the steep cliffs to the north. However, Sissons (1975) argued that the debris accumulation does not represent a simple rockslide deposit, as the debris is not simply banked up against the foot of the cliff, but has travelled 1.2 km downvalley at an angle oblique to that of the failure scar (Figure 12.27). He proposed instead that the debris accumulation is a rock glacier, formed by the reactivation of a remnant of decaying glacier ice at the end of the Loch Lomond Stadial as a result of its burial under rockslide debris. An alternative explanation has been proposed by Whalley (1976), who suggested that the boulder deposit may represent a form of flowslide or 'Sturzstrom' deposit. In reply, Sissons (1976b) outlined various objections to Whalley's explanation, principally that the morphology of the feature is inconsistent with that of a flowslide. A true 'Sturzstrom' would also probably have travelled much farther downvalley. However, the accumulation also lacks many of the features characteristic of a true rock glacier. In particular, rather than terminating in a bold frontal slope, it simply peters out downvalley. In addition, the irregular transverse ridges on its surface are quite unlike those on true rock glaciers. A possible explanation of these anomalies is that though the deposit may represent glacially-deposited rockfall debris, as Sissons suggested, the waning glacier that transported this debris may have been rather larger than he envisaged, so that though the debris was spread downvalley by glacier movement, no 'true' rock glacier was formed (Ballantyne, 1987e). The debate remains unresolved.

Talus and related landforms: summary

The retreat of the last ice sheet exposed an upland landscape of glacially-steepened rockwalls, some of which subsequently remained free of glacier ice whilst others experienced renewed contact with glaciers during the Loch Lomond Stadial of *c.* 11–10 ka BP. Since their final contact with glaciers, such rockwalls have been subject to subaerial weathering and erosion, the products of which have accumulated as talus. Many of the talus slopes that have been investigated in upland Britain bear the hallmarks of unmodified rockfall talus, namely a straight upper slope at *c.* 35–36°, a basal concavity and some degree of fall sorting. Taluses in areas that escaped glaciation during the Loch Lomond Stadial appear to be essentially relict periglacial features of Lateglacial age, and are characterised by a high degree of maturity and often complete or near-complete vegetation cover. During the Holocene, rockfall accumulation has apparently been limited and more than offset by erosion in the form of slope failures, debris flows and gullying. In contrast, talus cones that accumulated during the Holocene in sites deglaciated at the end of the Loch Lomond Stadial tend to be less mature and subject to modest accumulation at present.

Many talus slopes have experienced some modification by processes other than rockfall. Of these, debris flow activity has been the most widespread, in the form of both hillslope and valley-confined flows. The overall effect of debris flows has been to transport debris downslope, thereby lowering the overall talus gradient and increasing the degree of concavity. Where repeated flows have followed the same track, debris cones have accumulated. The history of debris cone deposition is complex: some are likely to be of Lateglacial age, others are Holocene paraglacial landforms and some are relatively recent features that apparently formed in response to anthropogenic interference with vegetation cover. Debris flow events appear to have increased dramatically in frequency over the past 250 years, but the cause of this intensification of activity remains uncertain. Snow avalanches have also been responsible for eroding the upper parts of talus slopes and depositing debris downslope in the form of avalanche boulder tongues, but their present effect is limited to a small number of high-level sites. The influence of avalanche activity in modifying talus slopes during the Late Devensian Lateglacial is unknown.

During the Loch Lomond Stadial, both protalus ramparts and protalus rock glaciers formed at the foot of some taluses. The volume of protalus ramparts indicates rapid rockwall recession during the stadial, due to the efficacy of frost wedging at this time and the instability of glacially-steepened rockwalls. Both types of landform have considerable potential for the reconstruction of stadial palaeoclimate. The altitudinal distribution of protalus ramparts mirrors former precipitation patterns, and the climatic constraints imposed by protalus rock glacier development place limits on both mean annual temperature and mean annual precipitation. In general, talus slopes and their associated landforms are much more environmentally-sensitive features than has commonly been appreciated. Future research on the palaeoenvironmental implications of such landforms may be expected to yield valuable returns in terms of our understanding of past climatic, edaphic and vegetation changes, the role of anthropogenic interference and the consequences for slope form and landscape evolution.

13
Nival, fluvial, aeolian and coastal features

The dominant characteristics of most periglacial landscapes are those produced by ground freezing. Frost action, however, is by no means the only geomorphological process operating within periglacial areas, as the severe climate of such areas introduces a number of ancillary effects that may enhance agents operative in warmer environments. Prolonged snowcover, for example, may intensify processes of weathering and erosion; storage of precipitation as snow for much of the year may result in rapid snowmelt runoff over frozen slopes and within river channels; strong winds blowing over terrain unprotected by vegetation may entrain and transport silt and sand particles; and a combination of frost weathering and wave action may accelerate the erosion of rock coasts. All of these processes may produce distinctive landforms or deposits, and this chapter considers their effects on the landscape of upland Britain.

Nivation and cryoplanation

Snowpatch erosion or *nivation* is not a single geomorphological process but 'a collective term used to designate all aspects of weathering or transport which are accelerated or intensified by the presence of late-lying snow' (Thorn, 1979a, p. 41). Processes traditionally associated with late-lying or perennial snowbeds include intensive freeze–thaw activity (particularly frost weathering of rock), enhanced chemical weathering, slopewash, transport of debris by snowcreep and accelerated solifluction through saturation of regolith downslope from melting snow (Thorn, 1988). Field measurements of nivation processes have confirmed that melt of late-lying snow may enhance slopewash or solifluction (e.g. Rudberg, 1974; Wilkinson & Bunting, 1975; Thorn, 1976a, 1979a; Ballantyne, 1978; Thorn & Hall, 1980), but the role of snowpatches in intensifying rock weathering is more contentious. Thorn (1976a, 1979b, 1988) in particular has questioned the efficacy of frost weathering under or at the margins of late-lying snowbeds, but has championed enhanced chemical weathering as a nivation process. Although considerable shear stresses may be engendered at the base of snowbeds (Mathews & Mackay, 1963; Costin *et al.*, 1973), snowcreep processes appear to be limited to the transport of isolated clasts over smooth bedrock surfaces (Jennings & Costin, 1978).

A combination of some or all of these processes has been held responsible for the development of erosional features ranging in size from miniature hollows (Nichols, 1963) to nivation hollows and benches up to a few tens of metres in length and width (e.g. Cook & Raiche, 1962; St Onge, 1969; Ballantyne, 1978; Figure 13.1) to *cryoplanation terraces* hundreds of metres in extent (Czudek, 1964; Demek, 1968, 1969; Reger & Péwé, 1976). The last-mentioned are amongst the most intrigu-

Figure 13.1 Nivation bench developed on slopes underlain by permafrost near the head of Vendom Fjord, Ellesmere Island, NWT, Canada. Note the marked break of slope where the tread meets the backslope.

ing and least-understood of all periglacial phenomena. They are bedrock steps or terraces cut into hillsides or across hilltops in areas that experience or formerly experienced a severe periglacial climate. According to Priesnitz (1988), they achieve widths of hundreds of metres and lengths (across slope) of several kilometres. The downslope inclination of terrace treads ranges from 1° to 14°. Such treads characteristically meet the backslope at a sharp break of slope that is often covered by a late-lying or perennial snowbed. Treads may cross structural trends and support a veneer of up to 2 m of frost-weathered regolith. Morphologically, the only difference between features labelled 'nivation benches' (Figure 13.1) and supposed 'cryoplanation terraces' would appear to be one of size or maturity of development. Reger & Péwé (1976) have argued that permafrost at shallow depth is a prerequisite of cryoplanation terrace development. The processes responsible for the formation of nivation benches and cryoplanation terraces have generally been assumed rather than assessed (Thorn, 1988). Numerous authors have invoked a combination of headward sapping by frost weathering and transport of comminuted debris across low-gradient treads by wash, solifluction and possibly piping. However, research on the rates at which nivation processes operate (e.g. Thorn, 1976a; Thorn & Hall, 1980) suggests that if large cryoplanation terraces reflect *only* the operation of such processes, an extremely long time would be required for their formation. It is difficult to refute the idea that such terraces may represent inherited features (such as pre-Pleistocene erosion surfaces or structural benches) that have merely been modified by subsequent nivation and frost action processes. Similar strictures apply to the notion that nivation may form or at least initiate large hollows ('nivation cirques') that may be the precursors of true glacial cirques; given what is known of present nivation rates, such an evolutionary sequence seems unlikely (Thorn, 1988).

Relict nivation and cryoplanation landforms in upland Britain

The concept of 'nivation cirques' that represent forms transitional between small-scale nivation hollows and mature glacial cirques has retained a tenacious hold on geomorphological thinking despite accumulating evidence to the contrary. In the Cairngorms, for example, King (1968) mapped as nivation hollows 'any valley possessing a roughly circular or pear-shaped form with no evidence of corrie glaciation...even if evidence of nivation is absent'. The supposed 'nivation hollows' he mapped have a mean diameter of 530 m and a mean depth of 219 m, and are thus an order of magnitude larger than active nivation hollows in arctic areas (e.g. Cook & Raiche, 1962; Ballantyne, 1978; Figure 13.1). Given what is now known of patterns of ice-sheet glaciation in this area (Sugden, 1968), it seems likely that most of the large forms identified by King are simply immature glacial cirques, though the smaller 'sub-plateau hollows' that he described on the fringes of some summits carry late-lying snow even at the present day, and have a more realistic claim to representing nivation (or nivation-modified) hollows. Similarly, on the Isle of Skye, Birks (1973) interpreted 'shallow perched corrie-like features' on the Red Hills and low-level cirques in Trotternish as nivation hollows, but later work has established that most of the former nourished glaciers during the Loch Lomond Stadial, whilst the latter acted as glacier source areas earlier in the Late Devensian (Ballantyne, 1989a, 1990). A glacial origin seems equally likely for the Cove Hill 'nivation hollow' in the Cheviots (Douglas & Harrison, 1985) as this is fully 300 m from crest to lip with a 50 m high backwall.

A closely-argued case for relict large-scale nivation cirques in upland Britain has been made by Watson (1966) with respect to two cirques flanking the Ystwyth Valley in mid-Wales, Cwm Tinwen and Cwm Du. Watson's argument rested on the lack of evidence for glacial erosion in these cirques; instead, he invoked frost-sapping of headwalls, meltwater transport, subnival solifluction and the movement of debris across steep snowpatch surfaces to explain their development. Both cirques are fronted by thick drift accumulations up to 18 m deep. That in the smaller (Cwm Tinwen) takes the form of an arcuate ridge that Watson interpreted as a protalus rampart (Figure 13.2), whilst Cwm Du is fronted by a stepped 'fan' of drift. The drift at both sites comprises a tough diamicton with occasional looser layers deficient in fines, and was interpreted by Watson as a solifluction deposit partly modified by meltwater eluviation. Watson's interpretation has attracted scepticism from those who have preferred to attribute both cirques to glacial erosion (see discussion in Watson (1966)) particularly after the discovery of striated clasts in the associated drifts (Watson, 1977a). A particular difficulty is posed by the thickness of snow or névé required to produce the 'protalus rampart' in Cwm Tinwen. Watson interpreted this as a two stage feature, associated with maximum snowbed gradients of 23° and 25°, and maximum snowpatch thicknesses of 46 m and 33 m respectively (Figure 13.2). Unless densities markedly lower than that of ice are

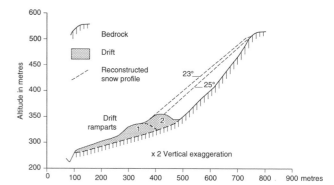

Figure 13.2 Section across Cwm Tinwen, a supposed 'nivation cirque' in the Ystwyth Valley, Wales (after Watson (1966)). The accumulations at the lip of the cirque were interpreted by Watson as protalus ramparts, but the reconstructed snow surfaces shown on the diagram are probably too gentle for downslope movement of debris, and imply implausibly thick depths of stationary névé.

assumed for the accumulated névé, these figures indicate that basal shear stresses must have exceeded 1 bar, so that the 'snowpatch' would have moved as glacier ice, and the 'protalus rampart' probably represents an end moraine. Although the snow surface gradients indicated are maximal, lower gradients would not have permitted movement of debris over the snow surface in the manner envisaged by Watson; indeed, observations on active protalus ramparts suggest that gradients of 23° and 25° are already too low to permit this to occur (cf. Ballantyne, 1987b). Similar considerations apply *a fortiori* to Cwm Du, where Watson's interpretation required former snow thicknesses of up to 82 m. This dilemma nicely illustrates the difficulty of invoking nivation to account for the origin of large cirque-like forms: as a snowbed grows to occupy such forms, basal shear stresses exceed the threshold for glacier movement, and glaciation supplants nivation. Accounts of supposed large-scale 'nivation cirques' in upland Britain and elsewhere therefore deserve scepticism.

More convincing are accounts of proposed cryoplanation terraces (or 'altiplanation terraces') on hills in SW England, particularly as these lie beyond the limits of Pleistocene glaciation and presumably experienced prolonged periglacial conditions during successive glacials. Possible cryoplanation terraces were first observed on Holdsworth Down in north Devon (Guilcher, 1950). These take the form of discontinuous benches up to 120 m wide and 40–200 m long, sloping gently valleyward but bounded both upslope and downslope by distinct breaks of slope. Such benches are apparently cut across bedrock but are not conformable with the bedding of the underlying grits. Guilcher argued that these are analogous to cryoplanation terraces in Alaska, the Urals and Siberia, and inferred that they had been initiated by Pleistocene nivation and progressively lowered by mechanical weathering. Te Punga (1956) took up this theme with an account of cryoplanation terraces on Cox Tor (443 m), which lies near the western margin of the Dartmoor granite but is underlain by the aureole rocks of the intrusion. Cox Tor provides some of the most convincing evidence for cryoplanation in Great Britain (Figure 13.3). According to Gerrard (1988b) the higher terraces are cut in diabase (metadolerite) and the lower ones in hornfels and Culm measures. The treads are 13–65 m wide and up to 800 m long, and have gradients of 4–9°. The risers slope valleywards at

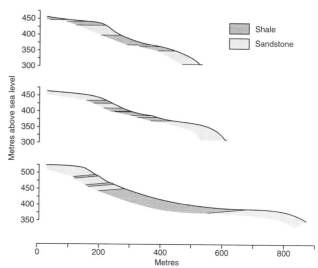

Figure 13.4 Sections across the 'periglacial slope planations' identified by McArthur (1981) in the upper Derwent Basin of the southern Pennines. The lips of the benches tend to overlie relatively resistant sandstone and gritstone strata, whilst the treads are cut across less resistant shales. McArthur interpreted these benches as cryopediments left perched above the valley floor by subsequent stream incision, but they may reflect differential glacial erosion of strata of variable resistance with subsequent modification under periglacial conditions.

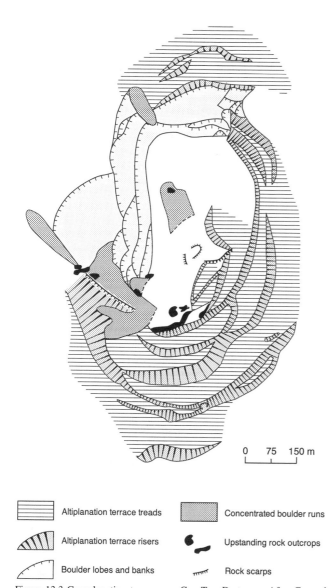

▤ Altiplanation terrace treads	▨	Concentrated boulder runs
◤ Altiplanation terrace risers	●	Upstanding rock outcrops
◠ Boulder lobes and banks	⌐ᴦᴦ	Rock scarps

0   75   150 m

Figure 13.3 Cryoplanation terraces on Cox Tor, Dartmoor. After Gerrard (1988b).

11–20°, and near the summit of the hill become small vertical cliffs cut in diabase. Most terraces also possess a gentle longitudinal slope. The terrace treads are mantled by a thin veneer of regolith, and many are covered by relict patterned ground features. Te Punga suggested that the Cox Tor terraces had been initiated by nivation and extended by a combination of frost weathering and removal of fine debris by wash and possibly deflation, probably when the ground was underlain by permafrost. He observed that terrace development appears to be favoured on horizontal strata of variable resistance, but maintained that cryoplanation terraces may also develop in massive rocks.

Erosional benches resembling cryoplanation terraces appear to be widespread on hills elsewhere in southern England. Waters (1962), for example, identified stepped slopes on nine hills underlain by the aureole rocks that surround the Dartmoor intrusion. These erosional benches are small in comparison with 'active' cryoplanation terraces in Alaska and elsewhere, being only 10–90 m wide with backslopes 2–12 m high, though

they reach lengths of up to 800 m; tread gradients are 3–8°, riser gradients 15–22°. Although these benches are cut across a range of lithologies (altered shales, siltstones, sandstones, dolerite, dyke rocks and lavas), Waters noted that they are restricted to rocks that are well jointed or well bedded. The longitudinal gradient of the terraces is consistent with (but rarely as steep as) that of the underlying bedding, which suggests a measure of structural control. Waters attributed formation of the terraces around Dartmoor to scarp retreat by 'frost sapping', and removal of comminuted debris by solifluction and wash downslope of late-lying snowpatches. Related to such cryoplanation terraces are the small *cryopediments* developed in chalk in parts of southern England. Cryopediments are gently-inclined erosion surfaces that have developed across valley floors as a result of slope retreat under periglacial conditions (Czudek & Demek, 1970b; Priesnitz, 1988). Forms answering this description have been identified by French (1973) at the foot of west- and southwest-facing chalk slopes in the Chilterns, Dorset and Wiltshire. These consist of slightly concave slopes with gradients of 5–9° that extend upslope for 20–100 m and meet the backslope at a clear break of slope. Like cryoplanation terraces, they support a thin regolith mantle (*c.* 1 m thick, but increasing to 2–3 m downslope) consisting of unstratified clay loam with chalk and flint clasts. French (1973) concluded that these pediments developed as a result of headward retreat of backing slopes, with transport of weathered debris towards the valley axes by solifluction (Chapter 8).

A rather similar origin has been inferred for valleyside benches in the upper Derwent basin of the southern Pennines (Figure 13.4). McArthur (1981) termed these benches 'periglacial slope planations' and inferred formation through retreat of sandstone and gritstone scarps and associated removal of rock waste by solifluction. To explain the present bench-like form of these 'periglacial slope planations', McArthur invoked

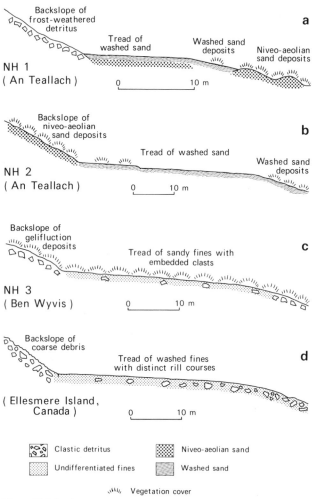

Figure 13.5 Sections surveyed across Holocene nivation hollows cut across niveo-aeolian sand deposits on An Teallach ((a) and (b)), a relict nivation bench on Ben Wyvis (c) and a nivation bench on Ellesmere Island, NWT, Canada (d). After Ballantyne (1985).

nivation processes, the absence of convincing cryoplanation phenomena in areas glaciated during the Late Devensian is not surprising.

It is equally striking that most accounts of relict nivation phenomena in northern England and Scotland pertain to sites that apparently escaped active ice-sheet glaciation during the Late Devensian. One such site is in the basin of the River Esk in NE Yorkshire, a Late Devensian glacier-free enclave. Here Gregory (1965) described many small benches that are most commonly developed on gritstone slopes with northerly aspects at altitudes of 240–330 m. These benches are erosional, but support a thin regolith cover; treads average c. 30 m in width and slope valleyward at gradients of up to 7°, meeting 20° backslopes at a distinct break of slope. Gregory's description of these features suggests strong morphological affinities with small nivation benches in arctic permafrost environments, such as that illustrated in Figure 13.1. He interpreted the Esk benches as the products of 'freeze–thaw processes activated by snow-patch erosion', but pointed out that development of such nivation benches is likely to have been dependent on the pre-existence of flat surfaces with northerly aspects on which snowbeds could have developed. This proposition corroborates with thinking on the restricted geomorphic role of nivation (cf. Thorn, 1976a, 1988), and implies that the Esk benches are 'nivation-modified' landforms rather than features that developed *ab initio* through prolonged occupance by late-lying or perennial snowbeds. Features strikingly similar to nivation benches on arctic permafrost terrain also occur on the higher parts of some Scottish mountains. Fine examples fringe the summit plateau of Ben Wyvis at altitudes above 900 m (Figure 13.5; Ballantyne, 1985). Although late-lying snow still occupies the breaks of slope at the heads of such benches and generates saturation overland flow across the treads, a complete vegetation cover now inhibits erosion and there is no evidence for current nivation activity at such sites. Trimline evidence suggests that the highest parts of Ben Wyvis remained above the last ice sheet (Ballantyne et al., 1987); if this were the case, it seems reasonable to assume that these nivation features are of Devensian age.

Relict nivation features have also been described on the northern slope of Farleton Fell in Cumbria. Here, Vincent & Lee (1982) investigated small, shallow 'snow patch hollows' with rather poorly defined vegetated treads 5–10 m wide backed by frost-weathered detritus. These are developed on limestone, and were interpreted by Vincent & Lee as pre-existing hollows enlarged by nivation processes. Although these hollows appear to be largely relict features, this site was actively glaciated during the Late Devensian, so that nivation activity here must have postdated ice-sheet retreat. Vegetation-covered 'aprons' of reworked Lateglacial loess up to 29 cm deep occupy the lips of the hollows. If such aprons represent downslope washing of loess preferentially deposited on late-lying snowbeds, as implied by Vincent and Lee, then the hollows themselves presumably developed during the Late Devensian Lateglacial.

Relict nivation and cryoplanation landforms: assessment

Relict nivation and cryoplanation landforms rank amongst the most inadequately documented of all periglacial phenomena in Great Britain. Research on their characteristics has been spas-

river incision of 40–50 m following the initiation of scarp retreat. Unlike the sites discussed above, however, the upper Derwent basin lies within the limits of maximum glaciation. As the benches exhibit marked structural control (the lip of each being underlain by resistant sandstone, whilst the treads are cut across shales; Figure 13.4), it seems equally if not more plausible to interpret these benches as the products of differential glacial erosion of strata of variable resistance. In terms of this interpretation, the role of cryoplanation may have been limited to later (i.e. Devensian) scarp retreat and solifluction of regolith over the surfaces of glacially-eroded benches. It is notable in this context that there are no convincing accounts of cryoplanation surfaces in the glaciated uplands of northern Britain. Features interpreted by King (1968) as cryoplanation terraces above 850 m in the Cairngorms do not exceed 15 m in width and are indistinguishable from glacially-scoured rock benches, and a supposed 'altiplanation terrace' on Rhum (Ryder & McCann, 1971) traverses the slope obliquely and apparently reflects underlying structure. Given the extremely long time period implicit in the development of cryoplanation terraces by modic, site-specific and speculative. The uncritical regard with which some researchers have viewed the erosive potential of nivation is evident in the persistent attribution of large 'nivation cirques' to snowpatch erosion alone, even though this requires the accumulation of implausible depths of stationary névé and lacks empirical justification in the form of 'active' analogues. However, examples of actively-developing cryoplanation terraces and nivation hollows or benches have been widely documented in present periglacial environments, and lend credence to the identification of Pleistocene examples in Great Britain. Even so, present information on the erosional potential of nivation processes is woefully inadequate, and the extent to which nivation creates distinctive landforms as opposed to merely modifying pre-existing terraces, benches and hillslope depressions remains debatable; nivation, as Thorn (1988) has observed, is something of a geomorphic chimera.

Despite the paucity of information on relict nivation and cryoplanation phenomena in Great Britain, the above review suggests three issues that deserve further investigation. The first concerns the relationship between documented examples of such phenomena and former glacial limits. On present evidence, convincing cryoplanation phenomena occur only in southern England beyond the limits of maximum glaciation, a distribution consistent with the development of such features over very long timescales throughout the Pleistocene. Nivation benches, such as those of Ben Wyvis and NE Yorkshire, occur in sites that apparently lay within the limits of maximum glaciation, but escaped active glaciation by the last ice sheet and hence may have developed during the Devensian. The only relict nivation features known to postdate ice-sheet retreat are the small snow-patch hollows on Farleton Fell. Assuming the existence of a continuum of nivation forms from small hollows through nivation benches to mature cryoplanation features (Reger & Péwé, 1976), it may be that the regional distribution of such features in Britain reflects the time available for nivation processes to operate without interruption by glacial erosion. This view, however, must be considered speculative pending the identification of relict nivation and cryoplanation phenomena over wider areas, particularly in the light of work (Rapp, 1983; Rapp, Nyberg & Lindh, 1986) which suggests that certain nivation features in Sweden survived ice-sheet glaciation.

A second issue concerns the influence of structure and lithology on the development of nivation benches and cryoplanation features. On this point the literature on British examples is ambiguous. Early authors seeking to stress the periglacial origin of hillslope terraces in southern England (e.g. Te Punga, 1956; Waters, 1962) stressed the lack of conformity between such terraces and underlying structure, but did not provide substantive evidence for this. More telling, perhaps, is the contrast between numerous stepped hillslopes on the metasediments fringing the Dartmoor granite (Waters, 1962) and the apparent lack of equivalent features on the granite itself. This may reflect selective reporting, but suggests that gently-dipping beds of variable resistance constitute more promising sites for cryoplanation terrace development than massive rocks. Such a contrast may, as Gerrard (1988b) has suggested, simply reflect the greater fissility and susceptibility to frost weathering of the former; but the suspicion that some cryoplanation features may reflect strong structural control is difficult to reject.

Finally, no attempt has yet been made to incorporate nivation

phenomena in Quaternary palaeoclimatic reconstructions for Great Britain. According to Reger & Péwé (1976), shallow permafrost underlies all actively-developing cryoplanation terraces in Alaska, and is a prerequisite for terrace formation, providing simultaneously an impermeable substrate and a base level for mass movement and nivation. If true, this implies that permafrost formerly underlay much or all of southern England south of the limits of maximum glaciation. Further, Reger & Péwé inferred that the Alaskan terraces developed under mean annual temperatures no higher than c. -12 °C, and mean July temperatures of only 2–6 °C. Whilst the assumptions underlying these inferences introduce a margin of error, it would appear that the cryoplanation terraces of southern England provide evidence of particularly severe as well as prolonged permafrost conditions. The palaeoclimatic significance of nivation hollows and benches is less well documented. Nivation processes are not constrained by the necessity of permafrost, but it is noticeable that 'true' nivation benches of the type illustrated in Figure 13.1 appear to be restricted to permafrost environments (e.g. Cook & Raiche, 1962; St Onge, 1969; Ballantyne, 1978). It would be premature, however, to conclude that all such features (for example those of NE Yorkshire) developed over permafrost.

Present-day nivation in upland Britain.

There are conflicting opinions regarding the current effectiveness of nivation processes on British mountains. Observations on late-lying snowbeds in the northern Pennines led Tufnell (1971) to speculate that the concentration of moisture downslope from melting snowbeds may contribute to effective needle-ice growth and frost weathering, and may enhance the movement of ploughing boulders and small mass-movement terraces. As evidence of headward erosion by snow lying in hollows and on benches Tufnell cited the development of linear cracks parallel to and immediately upslope of backslopes, together with the collapse of part of one backslope and undermining of a single boulder by meltwater. These observations, however, seem insufficient to support his sweeping claim that 'snow is a major element in the current development of the landscape of the area' (Tufnell, 1971, p. 497). Others have nevertheless embraced this opinion (Vincent & Lee, 1982), but without providing any further convincing evidence. Tufnell (1985) later reiterated his belief in the geomorphic importance of ground saturation downslope of Pennine snowbeds, but conceded that the benches and hollows occupied by late-lying snowbeds have altered little during the Flandrian, and are essentially relict features dating back to the Devensian.

A more detailed assessment of present-day nivation has been carried out on An Teallach in northern Scotland (Ballantyne, 1985). In Coire nan Tota, a series of shallow nivation hollows have formed in thick cohesionless niveo-aeolian sand deposits that have accumulated since the early Holocene. The hollows are characterised by a distinct break of slope at the rear of a gently-sloping wash zone (Figure 13.5), and are enclosed laterally by eroded unvegetated scarps of sand up to 2.5 m high. The treads consist of vegetation-free surfaces crossed by ephemeral networks of shallow braided channels. Snow thickness is estimated to reach 3–7 m. Temperatures measured over two years at a pair of such hollows indicated that the presence of late-lying snow actually reduces the frequency of freeze–thaw cycles, and no evidence was detected for enhanced frost weathering.

Measurements of the conductivity of nival meltwaters indicated lower solute concentrations than those in adjacent non-nival springs, which militates against enhanced chemical activity. Bending of metal rods inserted in snowbed hollows indicated snowcreep, but as the snow contained only low concentrations of sediment, this is unlikely to be geomorphologically important. Solifluction is also unimportant at these sites, as the niveo-aeolian sands on which the hollows have developed are not frost-susceptible, and consequently not affected by downslope movement associated with the growth and melting of ice lenses. Of the processes associated with nivation, only slopewash is unquestionably enhanced by the presence of late-lying snow on An Teallach. Abundant sand provided by active niveo-aeolian deposition onto snowpatch surfaces and the undermining of lateral sand scarps was washed vigorously across the treads and deposited across the lips of the hollows. In places, washed sand deposits have accumulated to form cones several metres long that appear to represent present-day analogues of the 'aprons' of loessic silt described by Vincent & Lee (1982) on Farleton Fell. These observations indicate that the geomorphological role of late-lying snowpatches on An Teallach is restricted to the provision of meltwater that transports fines over restricted distances. Because of the unusual nature of the snowpatch sites, however, this conclusion cannot be extrapolated to most other mountains. Nival meltwater is effective as an erosive agent on An Teallach because it has access to abundant unvegetated, cohesionless fine sediment, which is rarely the case elsewhere, except on other sandstone mountains and on granite plateaux such as those of the Cairngorms and Lochnagar. No such sediment transport was generated by snowmelt wash over the vegetated treads of relict nivation benches on nearby Ben Wyvis. Enhanced wash thus appears to be of local rather than general significance. Conversely, however, other nivation effects that were not evident on An Teallach, such as accelerated solifluction or intensified chemical weathering, may be of greater importance at other upland sites.

The possibility of enhanced chemical weathering under late-lying snow is particularly intriguing. The methodology employed by Ballantyne (1985) to test for enhanced chemical weathering now appears invalid in the light of later research that has emphasised the chemical complexity of the mountain snowpack in the Cairngorms. The ionic composition of snow on this massif has been shown to decrease with altitude (Davies *et al.*, 1988) and shows great spatial variability (Tranter *et al.*, 1987a). Moreover, because of preferential elution of certain ions during the early stages of snowmelt (Tranter *et al.*, 1986, 1987b) meltwater becomes generally more dilute as ablation progresses; this may account for the extremely low levels of dissolved solids measured in meltwaters draining late-lying snowpatches. An alternative approach to the detection of the enhanced weathering by late-lying snow has been attempted by Ballantyne, Black & Finlay (1989), who employed a Schmidt test hammer to compare the aggregate surface hardness of boulders dug from under late-lying snowpatches with that of boulders at adjacent snow-free sites. Results obtained from the Cairngorms were consistent with similar measurements made at sites in Switzerland and Norway, and indicate a significant reduction in aggregate hardness for boulders buried under late-lying snow. Although this test lends support to the concept of enhanced weathering under late-lying snowpatches, it does not indicate the nature of the weathering process involved.

The geomorphic role of running water

The effectiveness of running water as an agent of erosion and transportation in periglacial environments owes much to the distinctive hydrologic regime of such areas (Woo, 1983, 1986, 1990; Chapter 8). A large part of annual precipitation accumulates over the winter as snow, then runs off over a matter of a few days or weeks during a snowmelt flood that recurs annually. During the snowmelt flood, diurnal fluctuations in radiation and temperature are translated into corresponding variations in water discharge, giving a series of flood peaks during which flows of high competence occur. The geomorphic effects of this distinctive runoff regime are less well documented. Estimated sediment yields for arctic rivers are respectable rather than remarkable (e.g. Clark, 1988b, table 16.1). There is widespread agreement, though, that the critical control on denudation rate is sediment availability; denudation may be an order of magnitude greater in recently-deglaciated terrain where there is an abundance of entrainable sediment (Church & Ryder, 1972). Moreover, sediment transport alone is an incomplete indicator of the geomorphic role of periglacial rivers. Arctic uplands are often incised by deep canyons and steep-sided river valleys, whilst many arctic lowlands comprise broad pediments and extensive periglacial sandar. Such landforms are eloquent testimony to the long-term effect of fluvial activity in cold environments.

Surface wash in periglacial environments is favoured by protracted snowmelt over seasonally-frozen ground or permafrost, giving saturation overland flow. Widespread rillwash downslope of late lying snowbeds is a conspicuous feature of some permafrost areas (e.g. Ballantyne, 1978; Lewkowicz & French, 1982; Woo & Steer, 1982, 1983; Lewkowicz & Young, 1990). The geomorphic efficacy of slopewash is, however, strongly conditioned by sediment availability, and hence by vegetation cover (Lewkowicz, 1983). Although concentrations of suspended sediments may reach high levels (up to 1700 ppm according to Wilkinson & Bunting (1975)), such values appear exceptional. Moreover, suspended sediment transport is rarely sustained throughout the snowmelt period. As a result, estimates of associated denudation rates are low (Lewkowicz, 1988). However, Wilkinson & Bunting (1975, p. 105) noted that valley floors on Cornwallis Island are covered by 'thick accumulations of niveo-alluvial sediments', which suggests that slopewash in some arctic environments may be of greater importance than is indicated by the rather limited data on sediment discharge.

Relict fluvial landforms in the British uplands

In those parts of lowland Britain that remained beyond the margins of the last ice sheet, evidence for fluvial activity under periglacial conditions relates mainly to the Dimlington Stadial and earlier (Chapter 8). In contrast, evidence for periglacial river action in upland Britain postdates the recession of the last ice sheet, and relates mainly to the Loch Lomond Stadial of *c.* 11–10 ka BP. Stratigraphic evidence for enhanced fluvial transport at this time is evident in cores taken from the floors of lakes that lacked Loch Lomond Readvance glaciers in their catchments, and hence were fed by 'periglacial' rivers rather than glacial runoff. In such lakes in NW Scotland, deposition during the stadial is represented by dominantly minerogenic sediments containing a high representation of inwashed *Artemisia* pollen.

Figure 13.6 Dissected alluvial fans of Lateglacial age in Glen Roy, Lochaber, Scotland. Note the abandoned lake shorelines in the background. Photograph by Dr J.E. Gordon.

Such sediments have been interpreted by Pennington (1977b) as indicative of climatic severity, breakup of continuous vegetation cover, frost disturbance of soils and enhanced delivery of sediment to rivers. At Cam Loch in Sutherland, for example, the stadial is represented by 0.19 m of faintly banded clay sandwiched between similar or greater thicknesses of fine detrital mud; the period of predominantly minerogenic sedimentation at this site occurred between *c.* 11 ka BP and *c.* 10.4 ka BP.

Morphological evidence for contemporaneous erosion by nonglacial rivers is scant, though Sissons (1969) observed that the lower parts of valleys excavated in sandy raised beach deposits in Scotland are infilled with early Flandrian marine deposits. From this he inferred that such valleys were excavated during the Loch Lomond Stadial, when subsurface drainage through the sand was prevented by permafrost. Contemporaneous fluvial deposits, however, are more extensively documented. Sissons (1976c) argued that many of the relict alluvial fans of upland Scotland were deposited under periglacial conditions during the stadial, pointing out that sediment supply would have been enhanced by sparse vegetation, an absence of peat cover, feeble soil development and concentrated slope-wash over frozen subsoil. Large, relict alluvial fans at the foot of the Ochil Hills in central Scotland, for example, can be shown to date at least in part to the Late Devensian Lateglacial through their relationship to raised beaches in the same area. Similarly, Windermere Interstadial lake deposits at Costorphine

near Edinburgh are extensively covered by a broad fan of sand and gravel. This fan is in turn buried beneath Flandrian lake deposits, and thus appears to be of Loch Lomond Stadial age. Valley-mouth fans of angular debris in Shropshire also overlie Windermere Interstadial deposits, and are thus also attributable to deposition by floods under periglacial conditions during the Loch Lomond Stadial (Rowlands & Shotton, 1971). A similar interpretation has been placed on relict alluvial fans in Swaledale, north Yorkshire, (Pounder, 1985), though in this case it rests on the coarseness of constituent gravels (which include boulders up to 0.5 m long) and evidence for subsequent fan dissection rather than stratigraphic evidence. The Sandbeds fan in the north-eastern Lake District is more securely dated to the Loch Lomond Stadial, as the boundary between two gravel units exhibits involutions indicative of deposition under periglacial conditions and soliflucted till overlies the fan gravels. Boardman (1985b) calculated that the dimensions of this fan imply an average catchment lowering of 0.58 m during the stadial; if stadial conditions are assumed to have persisted for *c.* 1000 years, this in turn implies an extremely high denudation rate, of the order of 0.6 mm yr^{-1}.

In Scotland, particular attention has been focused on the dissected alluvial fans that debouch into Glen Roy (Figure 13.6). The slopes of Glen Roy are fringed by three spectacular 'parallel roads' that represent shorelines cut by a sequence of lakes dammed up in the glen by glacier ice during the Loch Lomond

Stadial. Sissons & Cornish (1983) interpreted these fans as the products of subaqueous deposition by torrential floods during the stadial. However, Peacock (1986) noted that interbedded lacustrine deposits do not occur within the fan sediments, which are dominated by clast-supported subhorizontally bedded gravels. Peacock interpreted these sediments as representing subaerial deposition of reworked glacigenic deposits by high-velocity sheet floods. He inferred that the fans had been deposited after ice-sheet retreat but *before* the damming of lakes by glacier ice during the Loch Lomond Stadial. Peacock's work in Glen Roy highlights one of the difficulties inherent in interpreting relict alluvial fans as periglacial landforms of Loch Lomond Stadial age. As he noted, the deposition of such fans in the interval between ice-sheet retreat and the culmination of the Loch Lomond Readvance may reflect paraglacial reworking of unstable sediment on recently-deglaciated terrain early in the Late Devensian Lateglacial rather than vigorous fluvial activity during the Loch Lomond Stadial. Such paraglacial alluviation has been widely recognised elsewhere (e.g. Ryder, 1971; Church & Ryder, 1972; Jackson *et al.*, 1982), and may equally apply to many relict alluvial fans in British uplands. Brazier (1987), for example, found that alluvial fans outside the limits of the Loch Lomond Readvance in the Grampian Highlands tend to be significantly larger than those that have developed during the Flandrian within these limits, but was unable to discount the possibility that these represent redeposition of sediment immediately after ice-sheet retreat rather than aggradation under periglacial conditions during the Loch Lomond Stadial. The same caveat applies to river terraces that have been ascribed to stadial alluviation, though in Glen Feshie (western Cairngorms), Robertson-Rintoul (1986) has demonstrated that a pronounced terrace attributable to Loch Lomond Readvance deglaciation at *c.* 10 ka BP may be distinguished from a higher kettled terrace deposited during ice-sheet deglaciation (*c.* 13 ka BP) on soil-stratigraphic grounds, with no evidence for intervening paraglacial terrace development.

Slopewash during the Late Devensian Lateglacial

Stratigraphic evidence for accelerated slopewash during the Loch Lomond Stadial is widespread throughout the British uplands. Numerous cores sunk in enclosed peat-filled basins outside the limits of the Loch Lomond Readvance have revealed a distinctive tripartite sequence of Late Devensian Lateglacial deposits buried under organic sediments of Flandrian age. Basal sediments, characteristically minerogenic and non-polleniferous, are overlain by organic mud or gyttja containing pollen assemblages typical of the Windermere Interstadial. These in turn are overlain by a layer of predominantly minerogenic sediments containing pollen assemblages indicative of a return to stadial conditions. This minerogenic layer has been widely interpreted as the product of slopewash during the Loch Lomond Stadial, and numerous radiocarbon age determinations support this view. In an infilled meltwater channel north-west of the Cairngorms, for example, slopewash during the Loch Lomond Stadial is represented by a layer of silty sand 0.45 m thick, dated by radiocarbon assay to between 11 115 ± 220 yr BP and 9740 ± 170 yr BP (Birks & Mathewes, 1978). Similarly, in a dead-ice hollow at Traeth Mawr, 600 km farther south near the Brecon Beacons (south Wales), a layer of fine silt and clay 0.8 m thick was deposited between 10 620 ± 100 yr BP and 9970 ± 115 yr BP, and was attributed by Walker (1980) to climatic deterioration at *c.* 10.6 ka BP, break up of the interstadial vegetation cover and consequent minerogenic inwash. At some sites, the texture of inwashed sediment reflects environmental changes during the Loch Lomond Stadial. An interesting example is provided by Macpherson (1980), who investigated the nature of changes in stadial sedimentation within a kettle hole in the upper Spey Valley in the Grampians. A strong representation of chionophilous taxa in the early part of the stadial pollen record suggests the initial development of extensive snowbeds. This stage was accompanied by a progressive coarsening of inwashed mineral sediment. The most extreme conditions of the stadial, however, appear to have been marked by increased continentality, with a reduction in the size of snowbeds and a concomitant decline in sedimentation rate and decrease in the size of mineral particles. Oceanic conditions appear to have returned at the end of the stadial, with renewed snowbed growth and an increase in the rate of sediment inwash.

The operation of slopewash during the Late Devensian Lateglacial is also manifest in stratified slope deposits that exhibit rhythmic layering distinguished by particle-size variations. British authors have frequently adopted the term *grèzes litées* to describe such deposits, but as DeWolf (1988) has pointed out, the original use of this term referred to alternating bands of granules and fines (Guillien, 1951); in Great Britain only the 'bedded growan' recorded in exposures on Dartmoor (Gerrard, 1988b) appears to satisfy this description. Coarser stratified beds of the type described elsewhere are more accurately described as *éboulis ordonnées*. The general term 'stratified slope deposits' is employed here. Such deposits are particularly widespread on the lower slopes of upland valleys in central Wales (Watson, 1965b; Ball, 1966; Potts, 1971; Figure 13.7). Such deposits form slopes of 8–30°, commonly have a southerly or westerly aspect, and are restricted in distribution to areas underlain by fissile rocks such as slate, shale, greywacke and schist. Both Watson (1965b) and Potts (1971) reported a recurring tripartite stratigraphy associated with stratified slope deposits in Wales (Figure 13.8). The basal zone consists of openwork angular platy debris dominated by particles 2–50 mm long. The overlying stratified deposits are composed of alternating layers of coarse angular fragments (0.5–35 mm long) and fine debris containing a high proportion of silt. The uppermost zone again comprises openwork angular platy debris. In all zones the coarse debris displays marked downslope orientation. Structures interpreted as ice-wedge casts penetrate the stratified deposits in several locations.

Both Watson (1965b) and Potts (1971) have argued that the tripartite stratigraphy of these deposits reflects environmental changes during the Lateglacial. Watson pointed out that a tripartite stratigraphy also exists in the slope deposits of the Nant Iago valley near Cader Idris, where Lateglacial gravels are sandwiched between head deposits (Watson, 1969a), and suggested that the three layers represent respectively the cold conditions of the Older Dryas, climatic amelioration during the Allerød, then renewed cooling during the Younger Dryas (Loch Lomond Stadial). He inferred from the Nant Iago stratigraphy that the first and last stages were dominated by solifluction, whilst the intervening stage was dominated by slopewash. This explanation raises a number of difficulties. The openwork units overlying and underlying stratified deposits *sensu stricto* are

Figure 13.7 Stratified slope deposits near Dolgellau, mid-Wales.

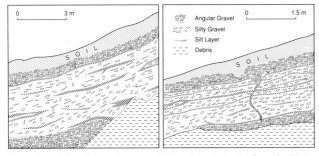

Figure 13.8 Tripartite stratigraphy of slope deposits in the Chwefru Valley (left) and Tywi Valley, Wales (after Potts (1971)). The upper and lower gravels were attributed by Potts to solifluction, the intervening stratified deposits to slopewash.

certainly not characteristic of solifluction; the status of an 'Older Dryas' climatic revertance in upland Britain is uncertain (Walker, 1984); the observed changes in stratigraphy could represent changes in dominant process within a single period of climatic deterioration; and the presence of ice-wedge casts within the deposits seems to indicate accumulation and stabilization *before* the development of continuous permafrost. Indeed, it seems plausible that the stratified slope deposits of Wales simply represent rapid paraglacial accumulation of loose sediment under periglacial conditions at the end of the Dimlington Stadial.

Stratified slope deposits have also been described by Boardman (1977, 1978, 1985b) at sites in the northern Lake District. At these sites small angular mudstone clasts display downslope orientation and strong imbrication, but true rhythmic bedding is absent from most exposures. At the principal site at Latrigg, north of Keswick, borehole data proved up to 12.5 m of such deposits near the slope foot, and Boardman distinguished two units. The upper, *c.* 1 m thick, is bedded and rich in fines; the lower is dominated by platy imbricate gravels. The deposits overlie till deposited by the last ice sheet, and on this basis Boardman inferred deposition during the Loch Lomond Stadial, though again it is possible that the deposits relate to periglacial conditions at the close of the Dimlington Stadial. Remarkably, there are few reports of stratified slope deposits in Scotland, though Geikie (1869) reported bedded shales up to 8 m thick on mountains in the Southern Uplands. Kotarba (1984) observed bedding within scree on the island of Rhum, and Tivy (1962) noted that gravel in the uppermost part of the C horizon in soils on the Lowther Hills is deficient in fines and exhibits sorting and rude stratification.

The stratified slope deposits of upland Britain have been widely attributed to slopewash over vegetation-free ground during the Late Devensian Lateglacial. Tivy (1962) and Watson (1965b) invoked snowmelt sheetflow under permafrost conditions; Potts (1971) attributed beds with abundant fines to either redistribution by nival meltwater or concentration of fines after initial deposition; and the imbricate nature of the stratified slope deposits in the northern Lake District led Boardman (1977, 1978, 1985b) to conclude that these represent redistribution of scree by snowmelt-generated sheetwash. This apparent unanimity of opinion is, however, somewhat misleading. A small galaxy of eminent periglacialists who visited some Lake Distict sites in 1985 were unconvinced by the prevailing 'slopewash' explanations yet unable to generate plausible alternatives. Such problems of interpretation appear to arise largely through lack of information on present-day analogues (French, 1976, p. 235; van Steijn, van Brederode & Goedheer, 1984) and partly through confusion of the relatively coarse stratified deposits of upland Britain with true *grèzes litées*. Slopewash appears a plausible explanation for the latter, which are characteristically fine-grained and gently bedded, but the coarseness of the stratified slope deposits of upland Britain appears to require the operation of some much more competent agent of transportation. Gelifluction has been widely invoked, with subsequent eluviation of fines responsible for the formation of openwork beds (DeWolf, 1988; Francou, 1989, 1990), but this explanation seems to require implausibly rapid movement of thin sheets of gelifluctate. The alternative view that openwork layers may accumulate by the sliding of clasts across snowbeds or frozen ground (e.g. van Steijn *et al.*, 1984) appears even less likely. The apparent inadequacy of such interpretations led van Steijn (1986, 1988) to advocate debris flow as the transportational mechanism. Though his laboratory experiments lend credence to this explanation, hillslope debris flows tend to follow linear tracks and their deposits are characteristically massive and unstratified (Chapter 12); the sort of sheet-like debris flows envisaged have not been reported on steep or moderate slopes underlain by frozen ground.

In sum, little can be said with certainty about the origin of the coarse stratified slope deposits in upland Britain. It seems reasonable to infer that transport of debris was accomplished by some form of highly competent sheet-like flow (slopewash, debris flow, slushflow or possibly gelifluction), and that this occurred over a frozen substrate. It also seems reasonable to suggest that wash was important as an agent of eluviation (removing fines from openwork layers and/or depositing fine beds), and that late-lying or perennial snow provided a major source of water. A Lateglacial periglacial origin thus seems likely, though paraglacial reworking of frost-weathered debris under cold conditions also appears possible, particularly in areas where ice-wedge casts imply permafrost development after stabilisation of the deposit. It is also notable that such deposits appear to be restricted to areas underlain by fissile rocks that have yielded small, platy clasts in response to macrogelivation (Potts, 1971); this probably accounts for the absence of such deposits over wide areas of upland Britain, where frost-weathered debris is too coarse to have undergone transport by slopewash or flow.

Present runoff and sediment transport in mountain catchments

Information on the hydrology of mountain streams in Great Britain is rather fragmentary, but clearly indicates that snowmelt-generated runoff is much less important than in continental periglacial environments. The influence of snowmelt is nonetheless evident in secondary spring runoff maxima in some Scottish rivers, and diurnal discharge fluctuations caused by variations in the melt of mountain snows are also detectable in the runoff records for upland catchments. That for the River Feshie, which drains a 106 km² catchment in the western Cairngorms, provides a good example. Streamflow in the Feshie shows persistent diurnal oscillations between March and June, and during this period runoff augmented by snowmelt greatly exceeds precipitation (Ferguson, 1984; Figure 13.9). Peak snowmelt flows, however, tend to be much less than those engendered by rainstorms. There are several reasons for this. First, because thaw may occur several times in the course of any winter, the snowpack on British mountains rarely accumulates to great thickness. Second, snow on the lower parts of mountain catchments often melts whilst that at higher altitudes is still frozen. This 'dissipation' of snowmelt is accentuated by a slow and sporadic increase in spring temperatures. Finally, the contribution of snowmelt is often obscured by melt taking place during periods of rainfall. At such times severe flooding may result (Archer, 1981), and Johnson (1975) has found that the highest recorded discharges at 27% of all gauging stations in Scotland contained a snowmelt component. In many upland catchments, more than 70% of rainfall contributes to stream runoff, and the highest runoff coefficients may exceed 90% (Reynolds, 1969). In winter, upland rivers exhibit a flashy response to the passage of cold fronts with rainfall intensities exceeding 6 mm hr⁻¹, but the most damaging floods are often caused by brief, intensive convectional rainstorms during the summer months (Newson, 1981). Spectacular examples have been documented for Lochaber (Common, 1954), the Cairngorms (Baird & Lewis, 1957; McEwen & Werritty, 1988), northern England (Carling, 1986; Harvey, 1986) and central Wales (Newson, 1980). In Wester Ross, a slow-moving occlusion caused 140 mm of rainfall in 24 hr in September 1981, and generated specific discharges of over 4.5 m³ s⁻¹ km⁻² in the catchment of Ardessie Burn, which drains a 13.3 km² catchment west of An Teallach (Acreman, 1983).

Despite the violence of some upland floods, rates of sediment transport in mountain streams are unremarkable. Studies of the entrapment of non-dissolved sediment in reservoirs fed by upland streams in northern England and southern Scotland suggest average denudation rates of 25–54 t km⁻² yr⁻¹ (Ledger, Lovell & McDonald, 1974), and even this relatively modest range may reflect accelerated sediment yield due to land use practices. In a fascinating study of recent variations in the rate of sediment accumulation in Llyn Peris, in Snowdonia, Dearing, Elner & Happey-Wood (1981) demonstrated that sediment influx has increased from *c.* 5 t km² yr⁻¹ in the mid-eighteenth century to *c.* 42 t km⁻² yr⁻¹ in 1966–76. The rapid increase in sediment transport in the present century they attributed to overgrazing, trampling pressure and construction works in the catchment. Afforestation of upland catchments also results in substantial increases in sediment delivery (e.g. Stott *et al.,* 1986). The dominant mode of nondissolved sediment transport also varies greatly from catchment to catchment, and depends on the coarseness of bed material and the availability of fine-grained sediment. Lewin, Cryer & Harrison (1974) cited a 4:1 ratio of bedload to suspended load for upland rivers in mid-Wales, but Acreman (1983) suggested that for the coarse-bedded Ardessie Burn in Wester Ross suspended load may constitute almost the entire nondissolved fraction of normal average sediment output. If so, sediment transport from this basin would

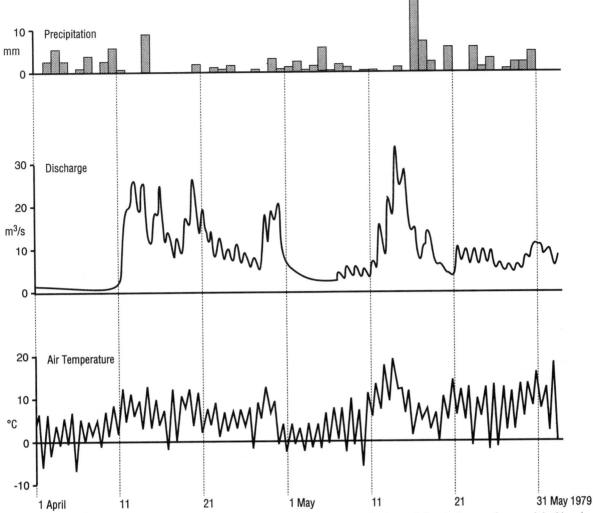

Figure 13.9 Diurnal oscillations in the discharge of the River Feshie, western Cairngorms, reflecting variations in the rate of snowmelt in this upland basin. After Ferguson (1984). Compare with Figure 8.1.

appear to be low, as the average yield of suspended sediment is only *c.* 9.3 t km^{-2} yr^{-1}. Conversely, however, Richards & McCaig (1985) calculated that minimum bedload yield in Allt a'Mhuillin, which drains a 6.2 km^2 catchment north of Ben Nevis, is *c.* 26 t km^{-2} yr^{-1}; they estimated a total solids yield for this catchment of 30–50 t km^{-2} yr^{-1}, a figure consistent with those inferred from sediment accumulation in upland reservoirs. Rather higher bedload yields (29–77 t km^{-2} yr^{-1}) have been calculated for three steep catchments in the English Lake District (Newson & Leeks, 1985).

The importance of dissolved load is more difficult to gauge from the data available. For small catchments this may be the dominant component of load. Measurements made by Reynolds (1986) on the solute composition of streamwater draining a small upland tributary of the River Wye indicated a solute output of 37.6 t km^{-2} yr^{-1}, compared with a nondissolved output of 10.8 t km^{-2} yr^{-1}. Similarly, Ballantyne (1981) calculated a solute output of *c.* 50 t km^{-2} yr^{-1} for the catchment of Ardessie Burn in Wester Ross, compared with an average suspended sediment yield of only 9.3 t km^{-2} yr^{-1}. These figures, however, exaggerate the role of solution in upland denudation as they take no

account of the solute input of precipitation and dry deposition. For the Glendye catchment in north-east Scotland, Reid *et al.* (1981) showed that the solute output of nine major elements and ions totals *c.* 29.5 t km^{-2} yr^{-1}, but that this includes a precipitation input of *c.* 10.5 t km^{-2} yr^{-1} and a dry deposition component of over *c.* 6.5 t km^{-2} yr^{-1}, figures which suggest that less than half of solute output represents denudation within the catchment. A further consideration that confounds estimates of current sediment yield from upland catchments is that river load is invariably assessed at fairly low altitudes, and may not accurately reflect denudation of high ground. In particular, the middle courses of most upland rivers flow through thick glacial deposits, and erosion of these often supplies much of the nondissolved sediment; the load carried by tributaries draining upper slopes may be much less than the figures cited above suggest.

The action of wind

Perhaps the most notable feature of upland climate in Great Britain is the strength of the wind. On Great Dun Fell (847 m) in

Figure 13.10 Deflation surface at an altitude of 760 m on the northern plateau of An Teallach in NW Scotland, showing the gravel lag deposit characteristic of such surfaces.

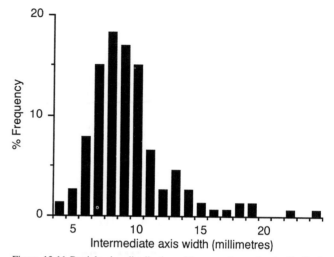

Figure 13.11 Particle size distribution of lag gravels on the An Teallach deflation surface (Figure 13.10). Truncation of the left-hand side of the distribution indicates that most particles smaller than 6–8 mm have been removed by the wind.

the Pennines, gusts exceeding 180 km h^{-1} occur every few years, whilst the strongest gusts recorded annually on Cairngorm (1245 m) range from 177 to 275 km h^{-1}. The strongest winds tend to occur in winter, when deep depressions track regularly over northern Britain. Such strong winds have given rise to a number of distinct landforms and deposits on exposed upper slopes and plateaux. These include deflation surfaces, niveo-aeolian deposits, wind-patterned ground and turf-banked terraces. All of these are actively developing under present-day conditions, and all reflect some degree of lithological control, being primarily (but not exclusively) associated with cohesionless sand-rich regolith such as that developed on sandstone or granitic rocks.

Deflation surfaces

On exposed high plateaux underlain by such lithologies, strong winds have often stripped virtually all vegetation cover, and have winnowed away loose sand and silt to create extensive deflation surfaces armoured by flat-lying boulders and carpeted by a lag deposit of fine gravel. The archetypal examples occur on the sweeping granite plateaux of Lochnagar and the Cairngorms in the eastern Highlands of Scotland, and on the Torridon Sandstone mountains of NW Scotland (Figure 13.10). Deflation surfaces are, however, also developed at altitudes as low as 400 m on windswept islands such as Rhum, Orkney and Shetland (Ball & Goodier, 1974; Goodier & Ball, 1975). Although most widespread on sandstone and granite hills, examples occasionally occur on summits and plateaux underlain by other lithologies, for example on volcanic rocks in Snowdonia and on Cambrian Quartzite near the summits of Arkle (787 m) and Foinaven (908 m) in Sutherland. The summit plateaux of Moruisg (928 m) and Maoile Lunndaidh (1007 m) in NW Scotland provide superb if rare examples of deflation surfaces developed on Moine schists. Elsewhere, schistose regolith tends to support a complete vegetation mat that inhibits deflation. The effectiveness of deflation on such wind-scoured surfaces may be judged by the coarseness of the gravel lag from which all finer particles have been swept away. Analysis of the size of exposed grains on the An Teallach deflation surface (Figure 13.10) reveals that most grains smaller than 6–8 mm

had been removed by the wind (Figure 13.11). In comparison, Thorn & Darmody (1985) found that aeolian transport on the Niwot Ridge in Colorado rarely involves material coarser than 4 mm. This highlights the extraordinary effectiveness of Scottish mountain winds in scavenging fines from exposed, unvegetated plateaux.

An intriguing feature of some deflation surfaces is that they appear to have developed fairly recently as a result of the erosion of overlying soil or of a cover of windblown sand deposits. Deflation surfaces on Ronas Hill in Shetland and Ward Hill in Orkney, for example, are surrounded by eroding scarps that fringe otherwise vegetated aeolian sand deposits, and remnant 'islands' of vegetated sand deposits with eroded margins occur amid expanses of wind-scoured detritus (Ball & Goodier, 1974; Goodier & Ball, 1975). Similarly, the impressive deflation surface that occupies much of the northern plateau of An Teallach (Figure 13.10) is fringed by eroding sand deposits, and interrupted by remnant islands of vegetated sand up to about 1.2 m thick (Figure 13.12). Islands of felspathic soil supporting a complete vegetation mat but surrounded by eroding margins also occur on deflation surfaces on the Cairngorms. Such eroded remnants suggest that areas now occupied by broad deflation surfaces were once partly or completely covered by a stable, vegetated blanket of soil or windblown sand. However, comparison of aerial photographs with field observations made 20–30 years later has indicated no discernible change in the position of sand scarp margins on Ronas Hill and An Teallach, which suggests that despite the eroded appearance of such covers, current rates of scarp retreat are low. This is confirmed by erosion pin measurements on An Teallach, which indicate an average scarp retreat of only 28 mm yr^{-1} (Ballantyne & Whittington, 1987).

Scarp retreat at the margins of a pre-existing stable soil or sand cover has been investigated in Iceland by Troll (1973) and Schunke (1975), who have shown that it is largely due to loosening of exposed sediment by frost heave and needle ice growth, and subsequent removal of loose grains by the wind. King (1968, 1971a) has demonstrated the importance of this

Figure 13.12 Eroded 'island' of sand on the An Teallach deflation surface, a remnant of a formerly more extensive sand cover.

process on the high plateaux of the Cairngorms. On An Teallach, however, scarp retreat at the margins of the sand cover is more rapid in summer than winter. This may in part reflect protection of unvegetated sand scarps under snowbeds in winter, but also sheep sheltering behind and rubbing against scarps dislodge sand that is subsequently blown or washed away (Ballantyne & Whittington, 1987). Irrespective of the mode of scarp retreat, the widespread erosion evident on and around some deflation surfaces implies extensive disruption of a previously-stable vegetated cover of soil or windblown sand. Goodier & Ball (1975) suggested that a stable sand cover on Ronas Hill may have accumulated under favourable conditions early in postglacial times, and that opening up of vegetation cover during a period of general climatic deterioration in the sixteenth to eighteenth centuries AD may have precipitated rapid scarp erosion under the combined action of frost and strong winds. Increased storminess across the British Isles at this time is now well established (Lamb, 1977, 1979, 1984; Whittington, 1985). An alternative possibility is that opening up of vegetation cover on exposed plateaux and consequent erosion of soil or sand covers resulted from the use of high ground as pasturage for sheep from the late eighteenth century onwards (Ballantyne & Whittington, 1987).

Windblown sand deposits

On some Scottish mountains the products of deflation have accumulated in the form of windblown sand or silt deposits on sheltered cols or lee slopes. Such deposits were first described by Peach et al. (1913, p. 112), who attributed windblown sands on mountains in NW Scotland to disintegration and aeolian reworking of the underlying Torridon Sandstone, and described on An Teallach 'dunes comparable to those on the sea shore'. Godard (1965) suggested that these 'dunes' were formed by drifting sand under conditions colder and drier than those of the present, and maintained that recent activity had been limited to erosion of 'dune' margins, a view endorsed by Sissons (1967, p. 225; 1976c, p. 109). Similar 'hill dunes' were described by Ball & Goodier (1974; Goodier & Ball, 1975) on the granite regolith of Ronas Hill in Shetland and the sandstone detritus of Ward Hill in Orkney. At both sites they interpreted vegetated 'dunes' as eroded remnants of a formerly more extensive cover of sand, and buried soil horizons within the 'dunes' were

inferred to represent evidence of cyclic deposition and stability. Aeolian deposits on Scottish mountains have also been described by Birse (1980), who noted that though these are composed of sand, some contain humic matter derived from erosion of nearby organic soils. Not all high-level aeolian deposits are predominantly sand-sized, however; an extensive blanket of windblown fines derived from Cambrian Quartzite bedrock on Ben Arkle in Sutherland contains 54% silt, with 40% fine sand, 6% medium sand and less than 1% clay (Pye & Paine, 1983).

The accumulation of loose windblown sand and silt deposits on the higher parts of some of the most wet and exposed summits in Great Britain poses a conundrum. On Torridon Sandstone hills and those of Orkney and Shetland, such deposits cling to slopes of 20–28°, and it is difficult to envisage how drifting sand could have accumulated on such slopes without being washed into adjacent valleys during storms. A solution to this problem has been proposed by Ballantyne & Whittington (1987), who carried out a detailed study of the windblown sands on the northern plateau of An Teallach, which supports some of the most extensive high-level aeolian deposits in upland Britain (Figure 13.13). The An Teallach deposits form vegetation-covered sheets of sand that often terminate laterally at vegetation-free eroded scarps (Figure 13.14). Scarps like these have previously been described by numerous authors as 'dunes', but lack the sedimentary structures characteristic of true dunes. On windward slopes the An Teallach sands nowhere exceed 1.2 m in depth, but at the heads of lee valleys they reach thicknesses of up to 4 m. Thin remnant islands of sand occur on shallow cols on the plateau surface, and on slopes of up to 28° (Figures 13.12 - 13.14).

Present accumulation on lee slopes is manifest as a layer of sand on the melting snowpack in May and June. This sand derives partly from weathering of Torridon Sandstone outcrops and clasts on exposed plateau surfaces, and partly from erosion of the margins of remnant sand deposits upwind. During snowmelt, sand blown onto the snowpack throughout the winter becomes concentrated on the snow surface. When the snow finally disappears, this layer of sand is lowered onto the underlying vegetation, which traps the sand deposits and thus stabilises them. The presence of roots and root casts in sections cut in the sand deposits indicates that vegetation growth keeps pace with sand deposition, thus providing a protective cover for the accumulating deposit (cf. Pissart, Vincent & Edlund, 1977). The An Teallach sands are thus niveo-aeolian in origin, and represent a form of sediment accumulation now widely recognised in arctic, subarctic and montane environments (Pissart, 1966; Rochette & Cailleux, 1971; Cailleux, 1972, 1976; Thorn & Darmody, 1980; Koster, 1988; Koster & Dijkmans, 1988). These observations indicate that vegetation cover is likely to have persisted throughout the period of sand accumulation, which explains the somewhat paradoxical survival of thick deposits of cohesionless sand on steep slopes in an environment of extreme wetness. The niveo-aeolian deposits on An Teallach are massive, poorly-sorted and coarser than most aeolian deposits, comprising mainly medium sand (0.2–0.6 mm) with a substantial proportion of coarse sand (Figure 13.15). The coarseness of the deposits reflects the size of grains weathered from the parent Torridon Sandstone, the strength of the winter winds and short transport paths. The poor sorting and absence

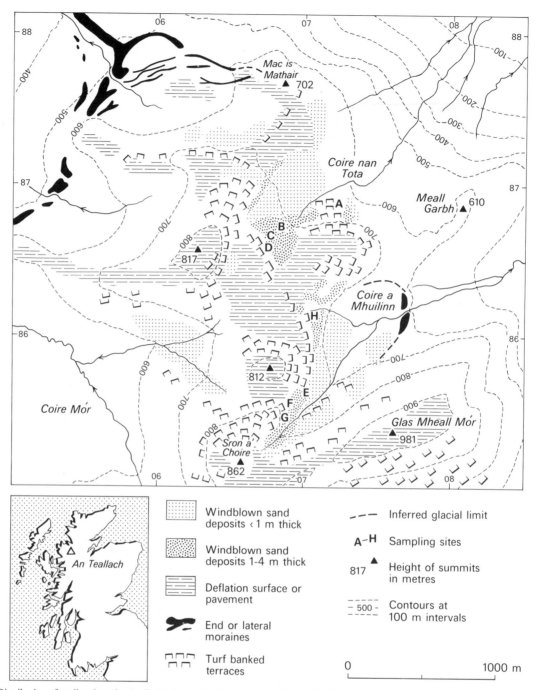

Figure 13.13 Distribution of aeolian deposits, An Teallach, NW Scotland. From Ballantyne & Whittington (1987).

of grain-size stratification are due to admixture of different grades of sand during deposition from snowcover (cf. Rochette & Cailleux, 1971; Jahn, 1972; Thorn & Darmody, 1980). By measuring the concentration of sand covering late-lying snowbeds, Ballantyne & Whittington (1987) calculated that *c.* 9.3 t of sand had been blown into a lee valley on An Teallach during a single winter. Roughly one third of this sand appears to have been reworked from the eroding margins of remnant deposits upwind, such as that illustrated in Figure 13.12. The remaining two-thirds were apparently supplied by granular disaggregation and deflation of particles derived

from exposed clasts and rock outcrops on the plateau surface upwind.

Perhaps the most interesting feature of the An Teallach sands is their history of deposition. Eight sections excavated in deep (>2.5 m) sand deposits on lee (east-facing) slopes all revealed two distinct stratigraphic units, sometimes separated by an unconformity (Figures 13.16 and 13.17). The upper unit comprises fresh, unweathered reddish-brown or brown sand, interrupted only by occasional faint bands of manganese staining. In contrast, the lower unit is heavily weathered and sometimes contains organic-rich humic horizons with occasional black

Figure 13.14 Dissected sand sheets bounded by unvegetated scarps 1.0–1.5 m high at 750 m in Coire nan Tota, a shallow valley in the lee of the An Teallach plateau.

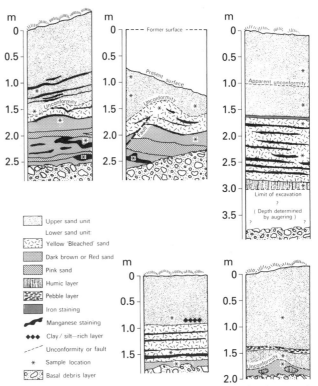

Upper sand unit:
Lower sand unit:
Yellow 'Bleached' sand
Dark brown or Red sand
Pink sand
Humic layer
Pebble layer
Iron staining
Manganese staining
Clay / silt—rich layer
Unconformity or fault
Sample location
Basal debris layer

Figure 13.16 Sections excavated in deep sand deposits on lee slopes on An Teallach, showing the contrast between the fresh upper sand unit and the weathered lower sand unit. From Ballantyne & Whittington (1987).

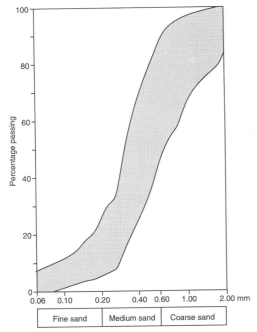

Figure 13.15 Grain-size distribution envelope (stippled) defining the grain-size limits of 57 samples of niveo-aeolian sand from the deposits on An Teallach.

Figure 13.17 Section cut in a deep lee-slope sand deposit on An Teallach. An unconformity separating the fresh upper sand unit from the weathered lower sand unit runs from top right to bottom left. The lower unit is down-faulted parallel to the unconformity on the left of the photograph.

peaty layers (Figure 13.18). Pollen analysis and radiocarbon dating have shown that the lower unit sands began to accumulate in the early Flandrian with the establishment of a heathland flora on slopes flanking the An Teallach plateau. By c. 7900 yr BP the sand had reached a thickness of c. 0.5 m at the head of lee valleys, and progressive accumulation at such sites appears to have continued until the deposits were up to 2.2 m thick. There followed a long period of minimal accumulation on lee slopes, which Ballantyne & Whittington (1987) attributed to establishment of a stable unbroken vegetation cover over much of the plateau source area. The subsequent accumulation of the fresh upper sand unit seems to imply that this phase was abruptly and relatively recently terminated by the breaking of the vegetation mat on plateau sand deposits, and consequent redeposition of eroded plateau sands on sheltered lee slopes. As

noted in the previous section, this recent stripping of wind-blown sands from exposed cols (and the consequent emergence of underlying deflation surfaces) may have been triggered by increased storminess during the Little Ice Age of the seventeenth and eighteenth centuries AD; alternatively, destruction of vegetation cover on col deposits may have resulted from overgrazing following the introduction of sheep to the mountain in the late eighteenth or nineteenth centuries. Although the investigations by Ballantyne & Whittington (1987) were confined to a single massif, they noted that niveo-aeolian sands on other Torridonian Sandstone mountains appear to be struc-

Figure 13.18 Organic-rich humic horizons containing occasional black peaty layers in the lower weathered sand unit on An Teallach.

turally and sedimentologically similar to those on An Teallach, and inferred that their findings have wide applicability for the interpretation of high-level windblown sands elsewhere in upland Britain. Further work on the history of sand accumulation, particularly in Orkney and Shetland, may yield rich rewards in terms of our understanding of the history of Holocene environmental changes on British uplands.

Wind-patterned ground

The term *wind-patterned ground* is here employed to refer to surface patterns formed by the alternation of vegetated and vegetation-free regolith on exposed ground where high winds are the dominant agent of patterning. At least three types of pattern have been created by wind action: *deflation scars*, which are patches of bare ground on otherwise vegetated terrain; *wind stripes*, which are straight or wavy lines of vegetation that alternate regularly with unvegetated (or thinly vegetated) soil; and *wind crescents*, which are arcuate patches of vegetation scattered across otherwise vegetation-free deflation surfaces. Deflation scars roughly 1 m wide and 2–4 m long have been described by King (1971a) on plateaux in the Cairngorms that are otherwise covered by a complete vegetation mat, and similar phenomena were observed on other Scottish mountains by Kelletat (1970a). These patches of unvegetated ground are floored by a granular lag deposit similar to that found on more widespread deflation surfaces, and are often bordered by what King termed 'turf scarps' and 'sand ripples'. The former are miniature scarps of vegetation-capped soil up to 0.3 m high, and apparently result from undermining of the vegetation cover by a combination of needle ice erosion and deflation; the latter are spreads of loose sand carried by the wind onto surrounding vegetation. On particularly exposed cols and spur crests, deflation scars tend to be elongated in the direction of dominant winds (Figure 13.19). It is not known whether disruption by frost heave, needle ice and strong winds is responsible for initiating as well as extending deflation scars, or whether this requires some other agency, such as grazing animals.

Of greater interest are wind stripes, which consist of alternating bands of vegetated and vegetation-free terrain that are aligned at right angles to the direction of prevailing winds (Figure 13.20). Striking examples occur on the summit plateaux

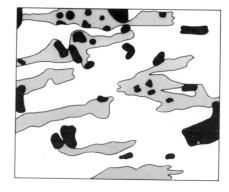

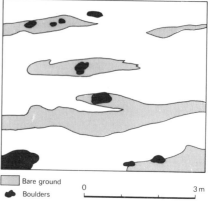

Bare ground

Boulders

0   3 m

Figure 13.19 Deflation scars cut across vegetation cover at 600 m (top) and 900 m on sandstone regolith, An Teallach. The prevailing westerly winds blow from left to right.

of Ronas Hill on Shetland and Ward Hill on Hoy (Ball & Goodier, 1974; Goodier & Ball, 1975). The vegetated stripes on these hills are 0.2–1.7 m wide, and are separated by unvegetated zones with a similar range of widths. In cross-section the vegetated stripes are characterised by an undercut windward edge and a gentle leeward slope. Ball & Goodier suggested that destruction of the windward edges by frost action and deflation is counterbalanced by plant colonisation at the relatively protected leeward margins. A state of dynamic equilibrium is thus maintained, with vegetated stripes migrating slowly downwind. Wind crescents 0.5–1.2 m wide and 0.6–3.0 m long also occur on more exposed parts of Ronas Hill and have a similar cross-sectional form, though these tend to be less regularly spaced.

Our understanding of these fascinating microforms owes much to ecological studies of the dynamics of montane vegetation communities. Crampton (1911) noted that on the higher parts of quartzite hills in Caithness the growth habit of *Calluna vulgaris* tends to form a series of parallel ridges and troughs at right angles to the direction of prevailing winds. The troughs, he noted, 'are occupied by naked older twigs, with much lichen, and the younger shoots and leafy growth are strongest on the lee side of every wave...it is evident that the young shoots of the heather grow best in the sheltered side of the wave, and it follows that the waves must really slowly advance by growth in the same direction as the wind blows' (p. 38). His observations were further developed in a series of classic papers devoted to the ecology of the Cairngorms (Watt, 1947; Watt & Jones, 1948; Metcalfe, 1950; Burges, 1951; Ingram, 1958). On exposed plateaux at altitudes of 700–1000 m, the vegetation of

Figure 13.20 Wind stripes developed normal to the prevailing winds on a high-level plateau in the Cairngorms. Photograph by Dr N.G. Bayfield.

these mountains is dominated by *Calluna vulgaris*, which is frequently arranged in wave-like wind stripes with intervening bare ground (Figure 13.20). The vegetated areas consist of *Calluna* plants aligned side by side across the direction of the dominant winds. Watt (1947) and Metcalfe (1950) have described how each vegetated stripe forms a regeneration complex: the apices of the *Calluna* plants colonise the sheltered zone leeward of each stripe, whilst the older exposed stems die away to windward. Frequently the lichen *Cladonia sylvatica* colonises the old *Calluna* stems to windward. When the *Cladonia* mat eventually disintegrates, bare soil is exposed between the vegetated areas and this in turn may eventually be colonised by a further wave of advancing *Calluna*. At higher altitudes, similar wind stripes may be developed in other vegetation assemblages. Sometimes *Empetrum hermaphroditum* forms the colonising wave, and is progressively overgrown in a downwind direction by *Rhacomitrium lanuginosum* and *Vaccinium* species. Watt (1947) estimated that such assemblages are advancing downwind at *c.* 20 mm yr[-1]. On the highest and most windswept ground *Juncus triffidus* occasionally forms ribbon-like wind stripes, but more often forms a patchwork of wind crescents sheltering around the lee sides of prominent boulders (Ingram, 1958).

The dynamics of heather (*Calluna*) stripes in the Cairngorms have been studied by Bayfield (1984), who found that these are currently advancing downwind at an average rate of 9 mm yr[-1]. There is little net loss or gain in the vegetated area, which suggests that the stripe patterns are in dynamic equilibrium with the ambient environment. Bayfield showed that individual vegetated stripes have the profile of miniature sand dunes, with slightly convex bare surfaces on the windward side and concave surfaces on the leeward vegetated side. The wavelength of the stripe patterns he examined was 0.85–1.38 m, and their amplitude was 5–10 cm. The origin of these features is enigmatic. Bayfield pointed out that there are two possibilities, namely erosion of an intact heather mat or selective colonisation of bare ground. He rejected the former as unlikely to produce stripe patterns. Instead, he suggested that the stripes formed through preferential colonisation of *Calluna* on the lee sides of a pre-existing pattern of unvegetated sand ripples or miniature dunes, possibly after climatic deterioration during the Little Ice Age

had resulted in completely bare deflation surfaces. Although the suggestion that such patterns developed only during subsequent climatic amelioration must be considered speculative, this hypothesis is attractive in view of the dune-like cross-section of vegetated stripes. Equally, though, the miniature dune form of vegetated stripes may have developed in *response* to a pre-existing patterned vegetation cover. This alternative was envisaged by Barrow, Costin & Lake (1968), who studied the causes of equal spacing amongst clumps of *Epacris petrophila* on Mount Koskiusko in Australia. They observed that *Eparcis* initially colonised bare surfaces in the lee of obstacles, then migrated downwind. From these observations they inferred that as migration occurred the uneven distribution of plants modified the wind profile and was itself modified by the wind to produce a fairly regular spacing of vegetation that was thereafter maintained. Whether this feedback mechanism could produce regularly-spaced vegetated stripes (as opposed to clumps or crescents of vegetation) is uncertain, however, and on balance Bayfield's explanation appears more satisfactory for the origin of wind stripes on British mountains.

Turf-banked terraces

Closely related to wind stripes are *turf-banked terraces*, which are step-like landforms with steep vegetated risers and gently-sloping, sparsely-vegetated treads (Figure 13.21). Such terraces are usually elongate across-slope, giving a 'staircase' appearance, and are most common on windswept upper slopes. Examples were first reported in the northern Highlands of Scotland (Peach *et al.*, 1912, 1913) and on Scaraben in Caithness (Crampton & Curruthers, 1914). Turf-banked terraces on mountains in the Lake District were vividly described by Hollingworth (1934, p. 177), who observed that 'bare areas tend to form elongated gently-inclined terraces held up by rather steeper banks of turf and connected to lower terraces by comparatively broad belts of debris that usually run obliquely down the slope' (Figure 13.22). Hollingworth envisaged that such terraces represent retardation of mass movement. 'The stony debris', he wrote, 'is a sea of moving material sweeping around islands of more or less stationary turf'. He observed fine examples on Bowscale Fell (703 m), noting that on steep gradients the terraces often dip obliquely across-slope.

Turf-banked terraces are amongst the most widespread of all periglacial landforms on British mountains. Like other wind-related features, they tend to be best developed on sandy, cohesionless regolith, but have nevertheless developed on a wide range of lithologies. They have been observed on many uplands, including the hills of Orkney and Shetland (Ball & Goodier, 1974; Goodier & Ball, 1975; Veyret & Coque-Delhuille, 1989), the mountains of northern Scotland (Galloway, 1958, 1961a; Godard, 1965; Mottershead & White, 1969; Kelletat, 1970a; Ballantyne, 1977, 1981, 1987f), the Cairngorms (Watt & Jones, 1948; Metcalfe, 1950; King, 1971a), the Lake District (Hollingworth, 1934; Hay, 1937), the northern Pennines (Tufnell, 1969) and Snowdonia (Ball & Goodier, 1970), and have been variously termed steps or denuded steps, parallel terraces, terracettes and solifluction terraces. Across northern Scotland the lower limit of terrace development rises eastwards from 450 m on Rhum to 600 m on An Teallach and 700 m on Ben Wyvis. On Shetland and Orkney

Figure 13.21 Turf-banked terraces on lee slopes on An Teallach, NW
Scotland.

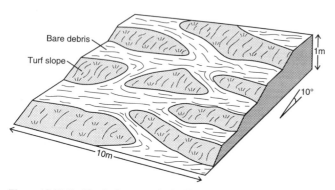

Figure 13.22 Turf-banked terraces in the English Lake District, as represented by Hollingworth (1934), showing the oblique ramps that connect the
treads of horizontal terraces. Hollingworth believed that debris movement
is concentrated along such ramps.

they descend to 300 m. These variations in altitude appear to be
related to exposure to strong winds. Turf-banked terraces of the
type described here have received little attention outside Great
Britain, and are not mentioned in any of the standard texts on
periglacial geomorphology. Similar features have, however,
been reported from other areas, including Mount Washington
(Antevs, 1932), Jan Mayen (Wilson, 1952), the mountains of
Sweden (Lundqvist, 1962), the Faroes (Lewis & Lass, 1965)
and the Colorado Front Range (Benedict, 1970).

On Ronas Hill in Shetland and Ward Hill in Orkney, Ball &
Goodier (1974; Goodier & Ball, 1975) recognised two categories of terrace, horizontal and oblique. A third intermediate
type, interconnecting, was identified by Ballantyne (1981) on
An Teallach. The relationship between these types is interesting. The treads of horizontal terraces often end in oblique ramps
that connect treads at different levels (Figures 13.21 and 13.22).
On some slopes these ramps approach or exceed the horizontal
treads in size, giving rise to interconnecting (horizontal and
oblique) terraces. On other slopes, turf-banked ramps dipping
steeply across-slope are continuous for many metres, producing
impressive oblique terraces. Even on the most fully-developed
oblique terraces, however, horizontal elements are present in
the form of shallow turf-banked steps. There is therefore a continuum from horizontal terraces through interconnecting terraces to oblique terraces. Terrace distribution is strongly influ-

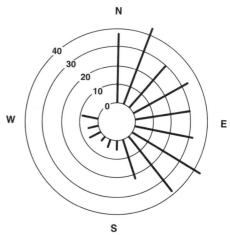

Figure 13.23 Aspects of a random sample of 333 horizontal turf-banked terraces on five mountains in the northern Highlands of Scotland.

enced by the direction of prevailing winds (Wilson, 1952;
Benedict, 1970). In the northern Highlands of Scotland, where
the dominant winds blow from the west and south-west, 90% of
turf-banked terraces occupy lee slopes with aspects concentrated within a semicircular range from 350° through 90° to
170° (Figure 13.23). The exceptions almost always occupy cols
and grade into elongate deflation scars downslope. The predominance of easterly aspects reflects vegetation colonisation of
sheltered terrace risers, which are also sites of persistent snow-
lie. Different types of terrace also show systematic relationships
with aspect. Horizontal terraces generally occupy slopes facing
east or north-east; oblique terraces invariably dip upwind (i.e. to
the west or south-west) and occupy both north-facing and
south-facing slopes; and interconnecting terraces occupy transitional aspects (Figure 13.24). Similar relationships between terrace distribution and dominant wind direction have been
observed on Shetland (Ball & Goodier, 1974).

Turf-banked terraces vary widely in dimensions. On Ronas
Hill in Shetland, the bare treads of horizontal terraces are
0.8–4.2 m wide and up to 23 m long, and have downslope gradients of 4–12°; vegetated risers are 0.6–3.0 m wide and have gradients of 13–32° (Ball & Goodier, 1974; Veyret & Coque-
Delhuille, 1989). Most horizontal terraces in the northern
Highlands have tread widths of 0.5–4.0 m and lengths of
4.0–25.0 m, with riser heights ranging from a few centimetres
to a maximum of about 1.4 m. Such terraces occupy slopes of
6–31° and dip only gently across-slope at angles of less than
12° (Ballantyne, 1981). Larger specimens occur on frost-susceptible schist regolith in this area, with treads up to 12 m wide
and 87 m long, but such features appear to be transitional
between turf-banked terraces *sensu stricto* and solifluction
sheets of the type described in Chapter 11. Oblique terraces
tend to be larger than their horizontal counterparts. On Ward
Hill in Orkney, oblique treads are 2–3 m wide but risers are 3–5
m wide. (Goodier & Ball, 1975). On An Teallach, oblique terrace treads are 1–7 m wide and 5–45 m long; risers are 0.8–3.4
m high, and the across-slope dip is 15–31°. The factors controlling terrace morphology have been analysed by Ballantyne
(1981) for a sample of over 330 horizontal terraces on five
mountains in the northern Highlands. Gradient exercises a

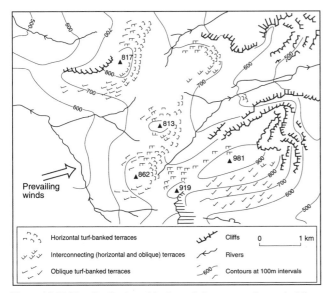

Figure 13.24 The distribution of horizontal, interconnecting and oblique turf-banked terraces on the northern plateau of An Teallach, N.W. Scotland. Horizontal terraces occupy sheltered slopes with easterly aspects; oblique terraces occupy northerly and southerly slopes, and invariably dip upwind (westwards); intersecting terraces occupy intermediate aspects.

major control. Increases in overall slope tend to be reflected in a decrease in tread width, accompanied by increases in riser height, tread gradient and across-slope obliquity (Figure 13.25). Aspect is also important: with departures from westerly (windward) aspects, risers tend to increase in both width and height but to decrease in slope, so that overall terrace sizes tend to be greatest on sheltered lee slopes. Finally, terrace size is strongly related to the coarseness of constituent debris: terrace length, width and height are all positively related to the mean size of clasts underlying terrace risers (Figure 13.26).

Excavation of horizontal terraces has revealed clasts concentrated on terrace treads above a relatively clast-free zone 0.2–0.7 m deep, a structure that points to the operation of vertical frost sorting and/or deflation of surficial fines. Clasts also tend to be concentrated under the risers. The largest clasts often underlie the riser crests, whilst those within terraces tend to be much smaller (Figure 13.27). Exposed clasts tend to be edge-rounded by microgelivation, buried clasts to be angular, a contrast that suggests limited recent exchange of surface and subsurface debris. Matrix granulometry is variable, but generally dominated by sand; terraces occur on both frost-susceptible and nonfrost-susceptible soils. Organic-rich horizons sometimes occur at the base of terraces. These may be overridden soils, but may equally represent downwash of organic material through the overlying regolith.

There have been two schools of thought concerning the nature of terrace activity. Following Wilson (1952), researchers

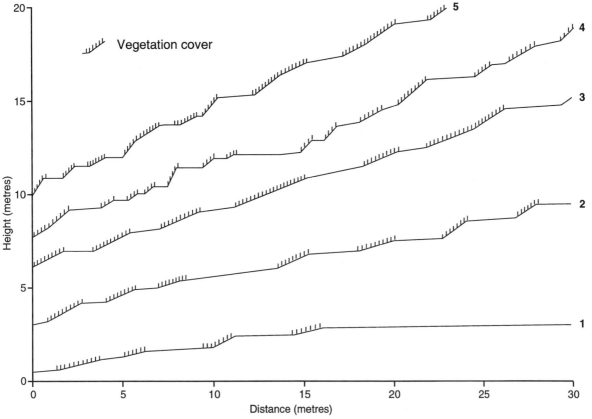

Figure 13.25 A transect surveyed upslope across a flight of horizontal terraces on An Teallach, showing the changes in terrace morphology with decreasing gradient (the transect continues from each line to the next).

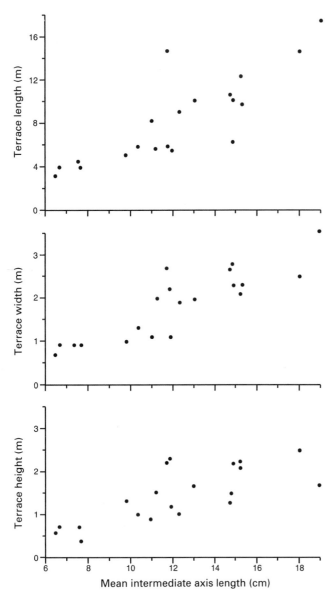

Figure 13.26 Lengths, widths and heights of 20 horizontal turf-banked terraces on An Teallach, plotted against the mean intermediate axis lengths of 50 clasts sampled from the risers of each terrace, showing how terrace size tends to increase with increasing coarseness of contituent debris.

working on Ronas Hill have described how progressive encroachment of sediment onto vegetation at riser crests is apparently counterbalanced by downslope extension of vegetation cover across terrace treads, so that the entire terrace system migrates downslope (Ball & Goodier, 1974; Veyret & Coque-Delhuille, 1989). This conception is clearly related to the notion of wind stripe migration , as outlined by Bayfield (1984) and others. It is difficult, however, to envisage how downslope migration of vegetation cover (if it occurs) could cause *en masse* movement of the terrace system. Other sites, moreover, display no evidence for *en masse* movement of either terraces or vegetation. The alternative view, introduced by Hollingworth (1934), is that the vegetated risers act as immobile 'dams', and that downslope movement of debris is concentrated along the oblique ramps that connect horizontal terraces. There are also differences of opinion regarding the process

responsible for movement of debris on terrace treads. Hollingworth advocated frost creep, but Mottershead & White (1969) regarded this as unlikely in view of the coarseness of debris at the sites they investigated. Metcalfe (1950) and Ball & Goodier (1974) favoured slopewash. 'Solifluction' has also been invoked, but the implications of this term have not been made explicit by its various advocates (Hay, 1937; Galloway, 1961a; Ryder & McCann, 1971).

On An Teallach, the average downslope movement of clasts across terrace treads was 3.9–4.8 mm yr^{-1} over a three-year measurement period (Ballantyne, 1981). Displacement during the summer months was negligible. Though the soils underlying terrace treads comprise mainly medium or coarse sand, repeated episodes of frost heave were recorded on terraces in winter and ice needles up to 30 mm long developed under surface particles. In spring, individual clasts appeared displaced downtread from enclosing 'sockets' of fines. These observations strongly suggest that frost creep is the dominant mechanism of sediment transfer across the An Teallach terraces. This is consistent with the displacement profiles of markers inserted vertically into terrace treads, as these displayed after three years an approximately exponential decline with depth, a characteristic of soils moving downslope as a result of repeated heave and resettlement (Figure 13.28). Displacement was confined to the uppermost 75–140 mm of the terraces. These findings, however, do not eliminate the possibility of other forms of sediment transport on terraces elsewhere. In particular, terraces developed on frost-susceptible soils may move through a combination of superficial frost creep and gelifluction; as noted earlier, turf-banked terraces developed on frost-susceptible soils are often transitional to solifluction sheets of the type described in chapter 11.

Although it is widely agreed that the development of turf-banked terraces reflects both movement of unvegetated debris and the influence of dominant winds on patterns of vegetation colonisation, the manner in which terraces originate is uncertain. There seem to be three possibilities: (1) modification by mass movement of deflation scars on otherwise vegetated slopes; (2) selective vegetation colonisation on bare slopes, causing retardation of creeping debris; and (3) vegetation colonisation in the lee of pre-existing terrace risers. The first explanation was advocated by Ball & Goodier (1974), who proposed that turf-banked terraces originate through partial destruction of the vegetation mat, with movement of debris across the resulting deflation scars producing low-gradient treads and steep vegetated risers. Though it is difficult to envisage the formation of deflation scars aligned normally or obliquely to prevailing winds, this explanation accounts for the formation of horizontal terraces on windswept cols or spurs aligned athwart the dominant wind direction. Such sites often support elongate deflation scars (Figure 13.19), and terraces in such locations may therefore simply be deflation scars that have adopted a step-like form as a result of downslope creep of debris across the scar surface. Ballantyne (1987a) referred to such terraces as *deflation terraces*, to distinguish them from those aligned normal to dominant wind direction on lee slopes. Such deflation terraces tend to have straight-fronted risers, flat bare treads and few interconnecting ramps (Figure 13.29).

Horizontal terraces on lee slopes may result from initial colonisation of bare slopes by vegetation stripes aligned normal

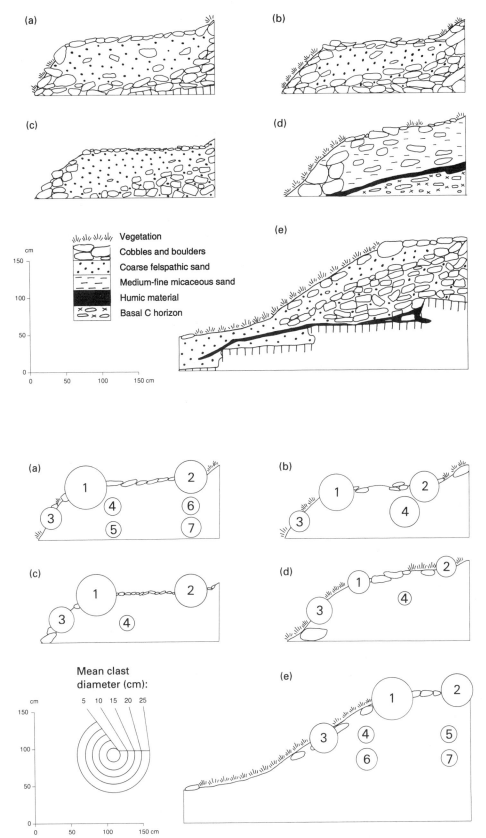

Figure 13.27 Top: the structure of five horizontal turf-banked terraces on mountains in the northern Highlands of Scotland. Bottom: variations in mean clast diameter within the same five terraces.

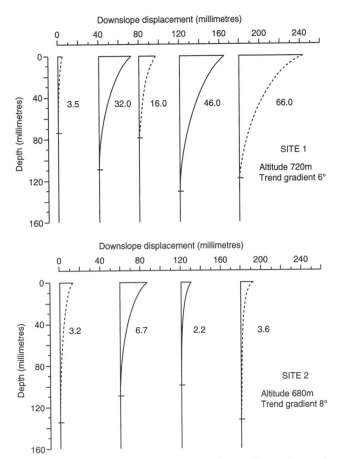

Figure 13.28 Displacement profiles of rubber tubes (continuous lines) and columns of glass beads (dashed lines) inserted vertically into the treads of two horizontal terraces on An Teallach and excavated after three years. The horizontal dash indicates the maximum depth of displacement.

Figure 13.29 Deflation terraces at 600 m on the Ruinsival ridge on the Island of Rhum. These represent modification of deflation scars by the downslope creep of debris.

to the wind (*cf.* Bayfield, 1984). Once established, such vegetation stripes would effectively retard the downslope movement of creeping debris, which would then accumulate upslope of each stripe, forming an incipient riser. This in turn would form a favourable location for further vegetation colonisation by providing shelter from the wind. Ultimately, the process of terrace growth would cease only when tread gradients became so low that the supply of debris to the growing riser was negligible. This mode of formation was anticipated by King (1971a), but evidence for its validity is lacking. Oblique terraces may develop in a similar fashion, if it is assumed that vegetation stripes are capable of forming oblique patterns that dip upwind on slopes with northerly and southerly aspects. Alternatively, on steep slopes with such aspects it is possible that vegetation colonisation in the lee of obstacles locally arrests the downslope creep of debris. As this debris accumulates, it may provide sufficient shelter for the zone of colonisation to extend obliquely upslope, thus allowing incremental growth of an oblique terrace that dips upwind (Figure 13.30).

All of the putative origins considered above assume that turf-banked terraces develop as a *result* of moving debris being influenced by (and in turn influencing) the pattern of vegetation cover. A final possibility is that the terraces develop first, as a result of frost creep and possibly solifluction, and thus provide sheltered sites for vegetation colonisation in the lee of terrace risers. By impeding creep, this vegetation would then contribute to terrace growth. There is some evidence in favour of this on An Teallach, where poorly-developed unvegetated terraces occupy windswept west-facing slopes. The formation of these debris terraces appears to have been independent of vegetation colonisation. On lee slopes, similar features may have evolved into mature turf-banked terraces by providing sheltered sites for vegetation growth, which in turn caused modification of the original terraces by arresting downslope movement of debris, causing riser growth and flattening of treads.

These competing explanations are difficult to evaluate. Deflation terraces certainly occur, but represent only a minority of all turf-banked terraces. Interpretation of both horizontal and oblique terraces as vegetation stripes modified by the accumulation of creeping debris offers the simplest general explanation, but is difficult to substantiate. The final hypothesis, that turf-banked terraces represent modified debris terraces, appears to be supported by the structural characteristics of such landforms, in particular by the evidence for frost sorting and by the concentration of the largest clasts under terrace risers. Until the genesis of turf-banked terraces is established, their significance must remain uncertain. We do not know whether turf-banked terraces represent a decrease in regolith stability as a result of partial destruction of vegetation cover, or an increase in stability resulting from partial vegetation colonisation. It is not certain whether terraces are essentially static, or whether they move *en masse* in response to downslope migration of vegetation cover. Turf-banked terraces occur within the limits of the Loch Lomond Readvance, hence presumably formed during the Holocene; otherwise we know nothing of their age, though on both Ronas Hill and An Teallach there is some rather inconclusive evidence that some terraces *may* have developed since the stripping of an overlying cover of niveo-aeolian sand within the last few centuries (Ball & Goodier, 1974; Ballantyne, 1981). The uncertainties concerning the significance of turf-banked terraces are symptomatic of our incomplete understanding of aeolian phenomena in upland Britain. Features such as niveo-aeolian sands, deflation surfaces, scars and terraces, wind stripes and turf-banked terraces indicate the vulnerability of upland environments (and particularly upland plant communities) to disruption by frost, strong winds and grazing animals,

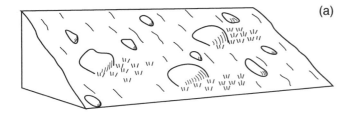

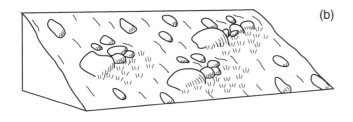

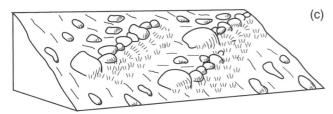

Figure 13.30 Schematic representation of the development of oblique terraces through vegetation colonisation in the lee of boulders at sites where the dominant winds blow across-slope.

yet we lack the understanding necessary to interpret such forms with confidence in terms of past environmental changes. Herein lies a challenge for geomorphologists and ecologists alike.

Shoreline erosion under periglacial conditions

Although the effects of frost weathering in Britain are most prominent on mountain summits (Chapter 9), various researchers have suggested that the breakup of rock under periglacial conditions has also been of great importance in the development of the coastal landscape, particularly in the fjords and sounds of western Scotland. The evidence for coastal erosion under periglacial conditions forms the final topic of this chapter. The idea that a combination of frost weathering and wave action has been responsible for cutting extensive shore platforms in resistant bedrock is not new, however, but was proposed nearly a century ago by the great Norwegian explorer and humanitarian Fridtjof Nansen (1904, 1922) to account for the development of the 'strandflat' of western Norway and certain other coastal areas. The strandflat is an uneven, partly submerged rock platform that extends seawards from the coastal mountains, and takes the form of numberless skerries and low islands offshore, and a broad coastal platform that abuts steep cliffs. Nansen concluded that the strandflat resulted from frost-shattering of sea cliffs under periglacial conditions, and removal of the resultant debris by wave action and sea ice. His view was echoed by several of his contemporaries (Rekstad,

1915; Vogt, 1918; Grønlie, 1924) and still commands support today (Larsen & Holtedahl, 1985).

Wider interest in the possibility of rapid development of rock-cut shore platforms under periglacial conditions is a much more recent development, and one that has been stimulated, at least in part, by debate concerning the origin of relict rock platforms along the western seaboard of Scotland. Several authors have stated that erosion of rock along high latitude coasts is ineffective (e.g. Zenkovich, 1967; French, 1976; Davies, 1980), and others have been sceptical of the efficacy of frost weathering as an agent of rapid platform erosion (Andrews, 1981, p. 113; Trenhaile, 1983, p. 79). This view is understandable when seen in the context of high arctic coasts such as those of the Canadian arctic archipelago. In such areas fetch and intertidal range are often limited, and for much of the summer the littoral zone is protected from erosion by an *ice foot*, an accumulation of fast ice pushed ashore by movement of pack ice in winter (McCann & Carlisle, 1972; Evenson & Cohn, 1979). In some arctic areas the development of shore platforms may also have been inhibited by rapid glacio-isostatic uplift that has prevented the establishment of a stable relative sea level over a period sufficient for platform development.

Whatever the cause of the absence of rock platforms around many arctic coasts, a growing body of evidence seems to suggest that under propitious circumstances the combined effects of frost action (as a weathering agent) and wave action or sea ice (as transportation agents) may indeed cause rapid retreat of coastal cliffs and the formation of an associated intertidal rock platform (Fairbridge, 1977). Convincing evidence for this has been found in a range of coastal environments, from Québec (Cailleux & Hamelin, 1967; Guilcher, 1981; Trenhaile & Rudakas, 1981; Dionne, 1988; Dionne & Brodeur, 1988) to Hudson Bay (Allard & Tremblay, 1983) and Spitsbergen (Moign, 1974, 1976). Moreover, some of these studies have suggested that frost weathering not only causes recession of sea cliffs, but also results in breakup of rock within the intertidal zone, though possibly only below the mid-tide level where the bedrock attains critical levels of saturation (Trenhaile & Mercan, 1984). Empirical estimates of rates of sea cliff retreat under periglacial conditions range from 6–8 mm yr[1] for the South Shetland Islands (Hansom, 1983) to 25–50 mm yr[1] for Spitsbergen (Jahn, 1961), and thus support the notion of rapid platform development at the foot of receding cliffs. Observations on lake shores also indicate the importance of frost action on shoreline development in areas of severe winter freezing. Of particular interest in this respect is a study by Matthews, Dawson & Shakesby (1986; also Dawson *et al.*, 1987) on the processes of erosion that operated on the shores of a small glacier-dammed lake in Jotunheimen, Norway. Matthews and his co-researchers found that miniature shore platforms up to 5.3 m wide had been cut in metamorphic rock over a probable period of 75–125 yr, at an average rate of 26–44 mm yr[1]. They proposed that these platforms had developed as a result of the deep penetration of the annual freezing cycle at the lake margin, movement of lakewater towards the freezing plane and consequent development of segregation ice in bedrock fissures. This interpretation therefore stresses headward erosion of the rock platforms at the lake margin as the prime agent of platform formation. In larger lakes, processes of lake ice push and pull may also be important in removing blocks detached by such frost action (Dionne, 1979).

Figure 13.31 The Main Rock Platform in the Firth of Lorn. Photograph by Dr J.M. Gray.

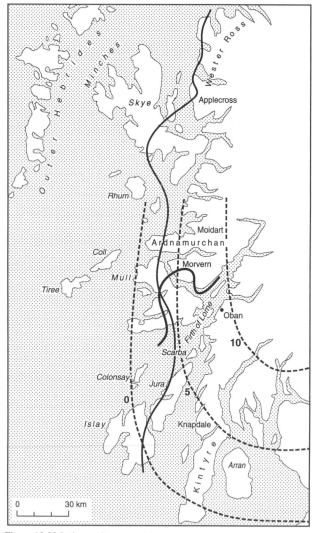

Figure 13.32 Isobases (in metres OD) for the Main Rock Platform between Ardnamurchan and Kintyre on the west coast of Scotland (dashed lines). The short line drawn through Mull delimits the area around the Firth of Lorn within which the Main Rock Platform is extensively developed. The longer line drawn between Wester Ross and Mull represents the eastern limit of high rock platform fragments.

The Main Rock Platform in western Scotland

Along much of the west coast of Scotland between Kintyre and the Isle of Skye are fragments of a conspicuous raised rock platform, generally known as the Main Rock Platform (Figure 13.31). This is best developed around the relatively sheltered shores of the Firth of Lorn and eastern Mull, where it is cut across a variety of rock types, including limestone, slate, conglomerate and basalt. In this area it is typically 20–30 m broad, but occasionally achieves widths exceeding 150 m, and is often backed by a steep rock cliff. Numerous caves occur at the rear of the platform, which also supports stacks and natural arches. The altitude of the platform reaches a maximum of *c.* 10–11 m in the Oban area, and declines westwards and south-westwards, passing below present sea level in western Mull, northern Islay and southern Kintyre (Figure 13.32). Like the Norwegian strandflat, the platform is often well developed in sheltered bays. Early commentators (e.g. Wright, 1914) considered that the Main Rock Platform and its associated cliffline had developed in postglacial times, but this explanation appears inconsistent with the negligible amount of rock erosion evident on the present shore. An interglacial origin was then advocated by numerous researchers, who saw the platform as correlative with a low-level till-covered platform in Ireland (e.g. McCann, 1966; Synge & Stephens, 1966; Gray, 1974). This view, however, was forcefully disputed by Sissons (1974a), who pointed out the equivocal nature of the evidence for glaciation of the platform and the implausibility of delicate features such as stacks and arches surviving the ravages of glacial erosion. Sissons also observed that the tilt of the shoreline in the Firth of Lorn area (0.16 m km^{-2} according to Gray (1974)) is similar to that of a buried marine planation surface of Lateglacial age in SE Scotland (Sissons, 1969). From this he inferred that the Main Rock Platform is contemporaneous with the buried planation surface, and that both form part of a single isostatically-tilted shoreline that he termed the Main Lateglacial Shoreline. This inference is supported also by the very restricted development of the Main Rock Platform within the limits of the Loch Lomond Readvance. Sissons' interpretation of the age of the Main Rock Platform implies extremely rapid platform development within a limited period of stable relative sea level. He therefore advanced the hypothesis that the platform had developed not as a consequence of prolonged marine abrasion under temperate conditions, but through a combination of frost weathering and removal of debris by wave action under the severe periglacial conditions of the Loch Lomond Stadial. 'Erosion of bedrock', he suggested (1974a, p. 46), 'would be greatly assisted by semi-diurnal freezing and thawing in the intertidal zone', and he noted that this process 'may account for the strong development of the erosional features in many sheltered localities'.

This revolutionary hypothesis met some initial opposition (Peacock, 1975; Peacock, Graham & Wilkinson, 1978) but has subsequently been supported, clarified and extended by others, notably Gray (1978) and Dawson (1979, 1980, 1988). Gray traced the Main Rock Platform south to Kintyre, where it passes below sea level. He showed that in the area of the SW Highlands there is no evidence for glaciation of the platform, and that it is again often anomalously well developed in sheltered localities exposed to limited fetch. Dawson (1979) widened the argument to consider the morphological differences between 'polar' (periglacial) rock platforms and those developed under 'nonpolar' conditions, and subsequently applied some of these criteria to the Main Rock Platform on the islands of Jura, Scarba and Islay. Here he found that it is again well developed on coasts exposed to limited fetch, is associated with numerous stacks, arches, geos and caves, and exhibits no evidence of having been glaciated (Dawson, 1980). He also made an interesting comparison between the appearance of low-level platform fragments of interglacial age and that of the Main Rock Platform. The former are often ice-moulded, striated and potholed, but the latter is irregular and fretted in appearance. He noted that the regional gradient of the Main Rock Platform across Jura, Scarba and Islay (0.13 m km^{-2}) is intermediate between that of the highest Lateglacial shorelines in this area (0.61 m km^{-2}) and that of the Main Postglacial Shoreline of mid-Flandrian age (0.05 m km^{-2}). As all of these gradients appear to relate to the same centre of glacio-isostatic uplift, it follows that the Main Rock Platform formed in the interval between ice-sheet deglaciation and the mid Flandrian, an interpretation consistent with a Lateglacial origin. Finally, on the assumption that the Main Rock Platform was cut entirely during the Loch Lomond Stadial, Dawson calculated that the associated cliff retreat rate must have been c. 70 mm yr^{-1}. This represents a phenomenal rate of erosion, surpassing even the 25–50 mm yr^{-1} suggested by Jahn (1961) for coastal erosion on Spitsbergen and much greater than the rates inferred by Hansom (1983) for the South Shetland Islands. It is comparable, however, with an estimate by Rasmussen (1981) of 40 mm yr^{-1} for erosion of a rock platform in northern Norway during the Younger Dryas (Loch Lomond) Stadial, particularly as the quartzite bedrock across which the Main Rock Platform is cut on Jura, Scarba and Islay is well jointed and hence prone to frost wedging. Interestingly, Dawson's calculated rate of cliff recession is at least one order of magnitude greater than estimates of mountain rockwall retreat rates during the stadial (Table 12.4). This difference, if valid, probably reflects undermining of coastal cliffs and removal of debris by sea ice and wave action.

Although there is now widespread acceptance that the Main Rock Platform probably represents the product of marine erosion under periglacial conditions during the Loch Lomond Stadial, some problems of interpretation remain. A speleothem from a limestone cave associated with the platform on the island of Lismore yielded a uranium-series date of 103 ka BP, which if valid indicates that the cave and hence the platform at this locality were in existence long before the Late Devensian Lateglacial (Gray & Ivanovitch, 1988). One solution to this anomaly is that the present platform is a polycyclic feature, and represents retrimming of a shoreline of much greater antiquity (Peacock et al., 1978). If this is the case, of course, it does not invalidate the notion of periglacial coastal erosion during the Loch Lomond Stadial (Dawson, 1988), but it does imply that such erosion may not have been as spectacularly rapid as hitherto proposed. Another problem that has not yet been satisfactorily explained is the relatively poor development of the Main Rock Platform north and west of a line drawn across the Island of Mull (Figure 13.32) and also in parts of the SW Highlands. Various ingenious and possibly ingenuous explanations have been proposed to account for the fragmentary nature of the platform in these areas. These include differential rates of glacio-

isostatic recovery (Gray, 1978), restricted development of a putative ancestral platform because of the former existence of an ice mass over Ardnamurchan and Moidart, the possible influence of Lateglacial neotectonic activity (Dawson, 1988, 1989) and geological control (Gray, 1989). It is difficult to evaluate the validity of these competing explanations. We may reasonably conclude, however, that although the *present* configuration and distribution of the Main Rock Platform represent marine erosion under the severe periglacial conditions of the Loch Lomond Stadial, its history may have been very much more complex indeed.

Other British shore platforms

Growing awareness of the possible effectiveness of shore erosion under periglacial conditions has also stimulated reinterpretation of other rock-cut shore platforms in Great Britain. Particular attention has been focused on a series of high-level platforms at altitudes of 18–51 m OD that are well developed west of a line drawn through Wester Ross, Skye, Ardnamurchan, Mull, Jura and Islay (Figure 13.32), but which are almost unrepresented east of this line. Many of these platform fragments achieve widths of 100–200 m, and some are up to 1 km wide. Significantly, some are striated, ice-moulded or overlain by till, and such evidence led to their initial interpretation as interglacial or even preglacial landforms. Sissons (1981a, 1982), however, noted that this explanation encounters three difficulties. First, the required periods of prolonged interglacial sea-level stability are not recorded from oxygen isotope curves based on evidence from ocean cores. Second, because other British shore platforms of inferred interglacial age occur at altitudes close to present sea level, the presence of much higher 'interglacial' platforms in the Hebrides implies substantial tectonic uplift for which there is no independent evidence. Finally, the interglacial hypothesis fails to explain the restriction of high-level platforms to the area west of the line alluded to above. In Wester Ross, moreover, Sissons & Dawson (1981) found that some high and apparently unglaciated rock platform fragments form part of a Lateglacial shoreline that is elsewhere in this area represented by raised beaches and shingle ridges, and hence manifestly not of interglacial origin. To reconcile these inconsistencies, Sissons proposed that the high platform fragments of western Scotland represent a series of shorelines cut and retrimmed by periglacial shore erosion during successive glacials, and that the altitude of these features reflects the influence of glacio-isostatic uplift. The glaciated fragments, he proposed, may be attributed to formation during ice-sheet advance or readvance. From the westerly distribution of high rock platforms Sissons inferred that throughout much of the later Devensian the western limit of the last Scottish ice sheet lay amongst the Inner Hebrides, so that platforms were formed or enlarged under periglacial conditions beyond this limit, but not within it.

Sissons (1981a) concluded his arguments concerning the origin of the rock platforms of western Scotland by proposing a general model of platform age and development that involves three distinct groups of platform fragments (Figure 13.33). The first group consists of near-horizontal interglacial fragments that occur around present sea level. The second group comprises the high rock platforms of Wester Ross, Ardnamurchan

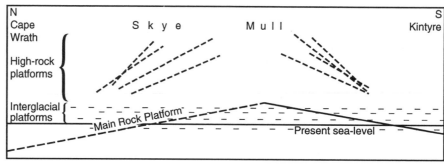

Figure 13.33 Theoretical relationships of rock platforms along a north–south transect in western Scotland. After Sissons (1981a).

and the Inner Hebrides. These Sissons interpreted as periglacial features formed by a combination of frost weathering and wave action outside the margins of successive ice sheets, and lifted to their present altitudes by differential glacio-isostatic uplift. The third 'group' contains only a single feature, the Main Rock Platform or Main Lateglacial Shoreline, interpreted as having been cut under periglacial conditions during the Loch Lomond Stadial (see above). Though bold in conception, elements of this scheme remain conjectural. In particular, the notion that the westerly distribution of the high rock platforms reflects a stable position of the last Scottish ice sheet is so far unsupported by independent evidence for a land-based ice margin in the Hebrides, except possibly on Islay (Dawson, 1982). Moreover, the attribution of a Lateglacial age to the high-level platforms in Wester Ross is perhaps more tenuous than Sissons & Dawson (1981) have suggested, as the rock-cut fragments of this shoreline may represent exhumation of an ancient platform rather than rapid platform development *ab initio* during the retreat of the last ice sheet. Whilst Sissons' general model possesses a seductive elegance and simplicity, its explicit recognition of the existence of interglacial platform fragments and different generations of rock platforms suggests that the true ancestry of these enigmatic landforms may yet prove to be rather more complex than the model suggests.

In southern Britain, intertidal and raised rock platforms have traditionally been interpreted as having developed during prolonged periods of interglacial sea-level stability (Kidson, 1977). The question as to whether such rock platforms may owe their origin to coastal erosion under periglacial conditions has been addressed by Dawson (1986). A major problem with this interpretation is that, unlike western Scotland, such areas were not subject to glacio-isostatic depression during the early stages of ice-sheet buildup. It follows that sea levels during glacial episodes seem likely to have been lower than at present because of glacio-eustatic lowering of the oceans. Dawson suggested that this problem may be overcome by invoking rapid southwards movement of the North Atlantic oceanic polar front at the onset of glacial periods, so that cold water surrounded the British Isles and severe periglacial conditions prevailed *before* sea-level lowering associated with ice-sheet buildup became effective. In support of this conjecture he cited instances of raised beaches associated with head deposits and the presence of ice-rafted erratics on platform surfaces. Such evidence, however, does not demonstrate *contemporaneity* of the shorelines and the overlying head deposits or erratics. Pending more conclusive evidence for periglacial shoreline development in southern Britain, Dawson's proposal must be regarded as interesting but essentially speculative.

Periglacial erosion of lake shorelines

Although the idea of accelerated erosion of the British coastline under periglacial conditions has attracted considerable interest, much less attention has been focused on the possible effects of frost action on lake shorelines. An outstanding exception is provided by the work of Sissons (1978) on the formation of what are perhaps the most famous landforms in Great Britain, the 'parallel roads' of Glen Roy and adjacent valleys in the western Highlands of Scotland. These 'parallel roads' represent the shorelines of lakes that were impounded in Glen Roy, Glen Gloy and Glen Spean by the up-valley advance of glacier ice during the Loch Lomond Stadial (Figure 13.6). The altitudes of the three main shorelines in Glen Roy (260 m, 325 m and 350 m) correspond with those of outlet cols that became successively blocked during ice advance and open during ice retreat. The origin of the Glen Roy shorelines and the related sequence of lake drainage was first worked out over a century ago by Jamieson (1863, 1892), but Sissons (1978) was the first to consider in detail the processes of shoreline formation. He found that the 'roads' range in width from 1.6 m to 63.6 m, slope valleywards at an average gradient of 10.2°, and were formed by a combination of erosion at the rear and deposition at the front. He also found that in many areas the steepened slopes backing the shorelines are cut in rock, and that in some exposures the platform is revealed to be cut entirely across bedrock (Figure 13.34). The mechanism of erosion he investigated by testing for relationships between the volume of rock or drift removed by shore erosion and a number of possible controlling variables. Though he detected a strong positive correlation between volume eroded and the length of fetch from the southwest, Sissons argued that the development of some rock-cut shorelines in areas of very limited fetch indicates that wave action alone would have been inadequate to account for shoreline erosion during the limited life of the lakes. Instead, he cited the occurrence of large slabs detached from the backslopes of shorelines as evidence of powerful frost action during shoreline formation, and invoked incorporation of coarse debris in lake ice to account for removal of such slabs. This inference is supported by the presence of angular ice-rafted dropstones embedded in lacustrine silts on the lower parts of the glen.

During the Loch Lomond Stadial a number of other lakes were dammed up in Highland glens by glaciers, for example

Figure 13.34 Shorelines of the former ice-dammed lake that occupied Glen Roy during the Loch Lomond Stadial. The nearest shoreline bench is cut across Moine schists. Photograph by Dr J.M. Gray.

near Loch Tulla (Ballantyne, 1979b) and in Glen Doe (Sissons, 1977a). Although such lakes were short-lived, their existence is reflected in the occurrence of shorelines high above the valley floors. It is likely that these, like the shorelines in Glen Roy, owe their formation to rapid excavation of drift and bedrock by a combination of frost action, wave erosion and removal of debris by floating ice. The rapid development of shorelines at the margins of former ice-dammed lakes also has much wider implications, for it suggests that shoreline development at the margins of many *existing* lakes may also have been accelerated under periglacial conditions. This proposition has been explored by Firth (1984), who mapped a low-level relict rock platform along the shores of Loch Ness. This platform is cut across schists, backed by a cliff 3–5 m high and achieves a width of up to 20 m in places. Firth's research demonstrated that the platform must have developed during the Loch Lomond Stadial, when Loch Ness was a proglacial lake, and he invoked frost wedging of bedrock, wave action and entrainment of

debris within ice floes to account for its development during the brief period when Loch Ness was raised above its present level. His work suggests that analogous platforms may exist at or just below the levels of many other lakes in upland Britain in areas that lay outside the limits of the Loch Lomond Readvance.

Assessment

Research on shorelines subject to severe winter freezing strongly suggests that frost action and removal of debris by floating ice may under propitious conditions contribute to accelerated shore erosion. Whether the effectiveness of these processes is as great as some advocates have claimed is more open to question. In the context of upland Britain, it may be concluded that 'periglacial' shore erosion during the Loch Lomond Stadial almost certainly *contributed* to the development of the Main Rock Platform of western Scotland, and probably also to the formation of shorelines at the margins of former ice-dammed or proglacial lakes. Its wider significance in explaining the high platform fragments in the Inner Hebrides, low-level platforms elsewhere and lacustrine rock-cut shorelines is less certain. The problem in determining the role of shoreline erosion under periglacial conditions is one of equifinality: similar erosional landforms may result from wave action alone, and hence the influence of 'periglacial' shore erosion has often been argued largely on the basis of distribution, inferred stadial age and rapidity of formation rather than morphology or associated deposits. This difficulty impedes accurate evaluation of the contribution of 'periglacial' shore erosion to the evolution of coastal landforms and lake shorelines in upland Britain. Despite this, it is indisputable that the *hypothesis* of accelerated shoreline erosion under periglacial conditions has provided exciting new insights concerning the development of littoral and lacustrine rock platforms, and has radically influenced recent thinking on the nature of erosional landforms around the British coast.

PART 4
Periglacial environments

The previous ten chapters of this book have constituted a systematic survey of the periglacial phenomena of Great Britain. The final chapter brings together many of these phenomena by outlining the components and climatic implications of periglaciation during three contrasting time periods. The first of these, the Dimlington Stadial of *c*. 26–13 ka BP, represents an interval of prolonged and very severe cold conditions. During this period the last ice sheet advanced to a line between south Wales and the Vale of York, covering some two-thirds of Great Britain, then gradually withdrew to expose almost all of the country to the influence of periglaciation. The second reconstruction is of periglacial conditions during the Loch Lomond Stadial of *c*. 11–10 ka BP, a much briefer and slightly less severe cold period during which glacier ice remained largely confined to upland Britain. The third environment to be considered is that of British mountains at present. In contrast to the previous two examples, this is an extremely marginal maritime periglacial environment, dominated not so much by extreme cold or ground freezing as by exposure to high winds and rain. As a corollary, the final part of Chapter 14 outlines some of the major themes and prospects for future research on periglacial phenomena in Great Britain, in particular considering how such features may be employed to gain deeper insights into the environmental conditions that pertained at different times during the Quaternary era.

14

Past and present periglacial environments

A recurrent theme of many of the chapters in this book has concerned the environmental constraints that affect the development of different types of periglacial phenomena. Because particular periglacial features develop within a specific range of environmental circumstances, relict features provide useful information concerning the conditions that pertained at the time of their formation, and thus constitute valuable proxy indicators of former environments. Moreover, as the climatic contraints on particular features differ, recognition of coeval *assemblages* of relict forms may provide us with considerable insight into the nature and severity of climate at the time of their formation. In this chapter we explore the usefulness of such periglacial assemblages for the reconstruction of environmental conditions during the two well-established stadial episodes of the Late Devensian, the Dimlington Stadial of *c.* 26–13 ka BP and the Loch Lomond Stadial of *c.* 13–10 ka BP, and compare these with the 'interglacial' assemblage represented by presently-active periglacial phenomena on British mountains.

Our grounds for selecting these three time periods to exemplify changing periglacial environments in Britain are partly pragmatic and partly heuristic. Much of the evidence for earlier periglaciation in Britain has been destroyed by subsequent erosion or obliterated by burial under later drift deposits, and in consequence such evidence is often fragmentary. In contrast, Late Devensian periglacial phenomena are widespread throughout Great Britain, and thus provide a much more extensive data base than is available for earlier Pleistocene cold periods. Happily, the three periods chosen for investigation encompass a particularly wide range of environmental conditions. The Dimlington Stadial is representative of protracted pleniglacial conditions during which the last ice sheet extended across two-thirds of the land surface. Those areas that remained ice-free experienced prolonged periglaciation throughout this time, whilst areas that were glaciated during the Dimlington Stadial were subject to periglacial modification during both the waxing and waning of the ice sheet. The Loch Lomond Stadial, on the other hand, was a relatively short-lived cold episode during which glacier ice was largely confined to mountain areas; contemporaneous periglaciation was neither as protracted nor as severe as during the Dimlington Stadial. Present-day climatic conditions on British mountains lie between the Holocene extremes represented by the Climatic Optimum and the Little Ice Age, and may thus be considered fairly representative of interglacial periglaciation, though there is some evidence that certain forms of periglacial activity have been unwittingly enhanced by human activity (Ballantyne, 1991b,c).

Palaeoenvironmental reconstruction

There are two basic approaches that may be employed in reconstructions of former environmental conditions on the basis of the evidence provided by contemporaneous assemblages of relict periglacial phenomena. The first involves the search for a representative analogue from the range of present-day periglacial environments. The rationale underlying this procedure is that if a particular relict assemblage corresponds closely in composition with one in a present-day periglacial environment, such as Alaska, northern Scandinavia or Spitsbergen, it is assumed that the relict assemblage implies fairly similar climatic conditions to those that currently prevail in the selected analogue. It is now recognised, however, that this assumption is fraught with pitfalls. Not only is the insolation budget of Great Britain (and other mid-latitude areas) substantially different from that of arctic and subarctic environments, but it also appears likely that Pleistocene stadials were characterised by radically different global atmospheric circulation from that of the present. Such considerations led Williams (1975) to conclude that appropriate present-day analogues do not exist for past periglacial climates in Britain, a view later echoed by French (1987).

The second approach to palaeoenvironmental reconstruction is synthetic. The climatic implications of particular relict phenomena are assessed on the basis of what is known of their present climatic range. As different features exhibit differences in climatic tolerance, their mutual occurrence in relict form implies that the climate at the time of their development must have been such as to permit their simultaneous development. This approach is preferable to that outlined above in that it makes no assumptions about the suitability of particular analogues, but relies on what is known about the present climatic range of individual features. When employed in conjunction with the evidence provided by other proxy indicators of palaeoenvironmental conditions, such as contemporaneous pollen and faunal assemblages, the use of relict periglacial phenomena in this way has enormous potential in constraining the possible range of former climates.

A number of particular problems arise, however, from the use of periglacial phenomena as palaeoenvironmental indicators. One concerns the dating of such features. Of the great variety of relict features discussed in this book only loess and loessic gelifluctate have been directly dated (using thermoluminescence techniques), and determination of the age of other features relies largely on their stratigraphy or on radiocarbon dating of subjacent or supradjacent organic material. A particular problem arises in assessing the ages of relict periglacial landforms and near-surface soil structures such as supraformational ice-wedge casts, fragipans and cryoturbations. In parts of Britain that escaped glaciation during the Dimlington Stadial, such features may be of Late Devensian age, but may equally relate to some earlier cold phase. In areas occupied by the last

Table 14.1: *Environmental implications of relict periglacial phenomena.*

NB Evidence for continuous permafrost implies MAAT less than *c.* -6 °C to -8 °C; evidence for discontinuous permafrost implies MAAT in the range *c.* -1 °C to -8 °C.

Feature	Environmental implications	Critical references
1. Near-surface cryogenic structures in unconsolidated sediments		
Ice-wedge casts	Severe winter ground cooling. In fine sediments: continuous or discontinuous (?) permafrost; MAAT < -3 °C to -4 °C. In sand and gravel: continuous permafrost; MAAT < -6 °C.	Péwé (1966a,b) Washburn (1979, 1980) Romanovskij (1985) Harry & Gozdzik (1988) Burn (1990)
Tundra polygons	As above.	As above.
Sand wedges	As above. Abundant supply of aeolian sand may indicate regional aridity.	As above; also: Black (1973) Péwé (1974)
Soil wedges	Severe winter cooling of seasonally frozen ground. In fine sediments: MAAT < +1 °C. In sand and gravel: MAAT < -1 °C.	Romanovskij (1985)
Pingo scars and related depressions	Shallow discontinuous permafrost. MAAT probably <-4 °C to -5 °C, but not markedly lower.	Lagerbäch & Rodhe (1985) Mackay (1988b)
Thermokarst depressions	Continuous (?) ice-rich permafrost	
Large-scale nonsorted patterns	Probably indicative of permafrost	Williams (1964, 1965) Mackay (1979b, 1980) Van Vliet-Lanoë (1988a,b)
Large-scale sorted patterns	Strong but not conclusive indicator of permafrost: MAAT < 0 °C to -2 °C. May indicate active layer depth.	Goldthwait (1976) Washburn (1979, 1980)
Cryoturbation structures	Depth of annual freeze-thaw; some indicate depth of active layer above permafrost.	Van Vliet-Lanoë (1988a) Vandenberghe (1988)
Fragipans	Indicate permafrost and depth of former active layer.	FitzPatrick (1956b) Van Vliet-Lanoë & Langhor (1981)
Soil microstructures	Depth of annual freeze-thaw of frost-susceptible soil. May indicate depth of active layer.	Harris (1985) Van Vliet-Lanoë (1985)
2. Mass-wasting landforms and structures		
Solifluction sheets and lobes	Seasonal freeze-thaw of soil.	Harris (1987b) Lewkowicz (1988)
Active layer detachment slides	Permafrost. May indicate depth of former active layer.	As above.
Ground-ice slumps	Rapid thaw of ice-rich permafrost.	As above.
Granular head deposits	Seasonal freeze-thaw of soil.	As above. Also Hutchinson (1991)
Clayey head deposits	Active layer sliding (*qv*): permafrost.	As above. Also Spink (1991)
Cambering and valley bulging	Thaw of ice-rich permafrost.	Hutchinson (1991) Parks (1991)
3. Aeolian and niveo-aeolian features		
Loess deposits	Dominant wind direction.	Pye (1984) Catt (1985a,b)
Coversands	As above.	Buckland (1982)
Sand dunes	As above.	Matthews (1970)

Table 14.1: *Continued.*

Feature	Environmental implications	Critical references
4. Frost-weathered detritus and related features		
Mountain-top detritus	*In situ* detritus indicates upper limit of former glaciers, and may indicate permafrost and depth of former active layer.	Strömqvist (1973) Thorp (1981, 1986)
Cryoplanation	Continuous permafrost (?).	Reger & Péwé (1976)
Tors	Frost-shattered tors may indicate areas that escaped glacial erosion during the Late Devensian.	Linton (1955) Goudie & Piggott (1981)
Periglacial trimlines	Indicate upper limits of active glacial erosion.	Thorp (1981, 1986) Nesje *et al.* (1987, 1988) Ballantyne (1990)
5. Talus-related landforms		
Protalus ramparts	Former regional snowfall patterns.	Ballantyne & Kirkbride (1986)
Protalus rock glaciers	Discontinuous permafrost: MAAT < -1 °C to -2 °C; former snowfall patterns; mean annual precipitation.	Haeberli (1985)

ice sheet, there is often ambiguity as to whether periglacial landforms and near-surface soil structures formed during the retreat of the last ice sheet or reflect renewed thermal decline during the Loch Lomond Stadial. Even if associated stratigraphic, morphostratigraphic or dating evidence places a particular feature firmly within a particular stadial episode, in some cases it is not clear whether formation took place under full-stadial conditions or under a less severe periglacial climate during the preceding period of climatic deterioration or the succeeding phase of climatic amelioration. Such considerations may explain the apparent co-existence of features indicative of continuous permafrost with others indicative of discontinous permafrost. The variable severity of stadial climate makes the interpretation of features indicative of former active layer depths (fragipans and cryoturbations) particularly uncertain.

Another source of difficulty arises from problems of interpretation. True ice-wedge casts are mimicked by wedge-shaped structures of completely different origin; involutions may be produced by a variety of mechanisms, not all of which relate to ground freezing and thawing; ramparted ground-ice depressions are similarly polygenetic; and the literature on upland periglaciation in Britain contains several unfortunate examples of mistaken identity. Even where particular landforms and structures have been correctly interpreted, their implications in terms of former climatic parameters may be ambiguous. The interpretation of ice-wedge casts nicely illustrates this problem. Ice wedges develop through cracking of permafrost in response to rapid and severe winter ground cooling (to perhaps -20 °C or lower). The degree and rate of cooling required to induce cracking are also influenced by site-specific factors such as snow-cover, the rheology of the host sediment, ground ice content and ground temperature gradient. The distribution of active ice wedges is thus only indirectly linked to mean annual air temperature (MAAT), which itself reflects the summer as well as winter temperature regime and is insensitive to both secular climatic change and the occurrence of extreme winter freezing events (Burn, 1990). The tenuous physical relationship between

MAAT and the mechanics of permafrost cracking has even prompted Harry & Gozdzik (1988) to conclude that 'although ice-wedge pseudomorphs provide unambiguous evidence of former permafrost conditions, it remains difficult to define their palaeoenvironmental significance with any greater precision'.

This appears an unnecessarily pessimistic standpoint, though it rightly emphasises the need for caution in translating the occurrence of ice-wedge casts into former MAAT. The much-cited work of Péwé (1966a,b) demonstrated that in Alaska the distribution of active ice-wedges corresponds to a MAAT no higher than -6 °C to -8 °C, and Washburn (1979, 1980) advocated adoption of a more conservative 'warm-side' limiting MAAT of -5 °C on the basis of a wider range of published evidence. More recent research has suggested that even this figure may be too low, citing MAATs of -4 °C for permafrost cracking in fine-grained glaciolacustrine sediments (Burn, 1990), -2.5 °C in frozen loam (Romanovskij, 1985) and -3.5 °C in peaty sediments (Hamilton *et al.*, 1983). From such findings, it seems reasonable to infer that wedge casts in fine-grained sediments normally imply MAATs no higher than -3 °C to -4 °C, with at least occasional severe winter freezing. Wedge casts in sands and gravels would appear to indicate that former MAAT was no higher than about -6 °C, with severe winter chilling of continuous permafrost.

Similar caveats apply to the interpretation of other ground freezing phenomena, such as pingo scars, ground-ice depressions, sand wedges, cryoturbations and large-scale patterned ground. The known climatic constraints on the development of such features have been considered in previous chapters, and are summarised in Table 14.1. In all cases, reconstruction of former climatic parameters on the basis of the evidence provided by relict periglacial phenomena involves a degree of uncertainty and imprecision, but in this respect periglacial evidence is no different from that provided by other proxy indicators of palaeoclimate such as pollen assemblages, insect, molluscan or mammalian fauna, or the dimensions of former glaciers. Clearly, though, greater confidence may be placed on palaeoenvironmental interpretations based on converging evi-

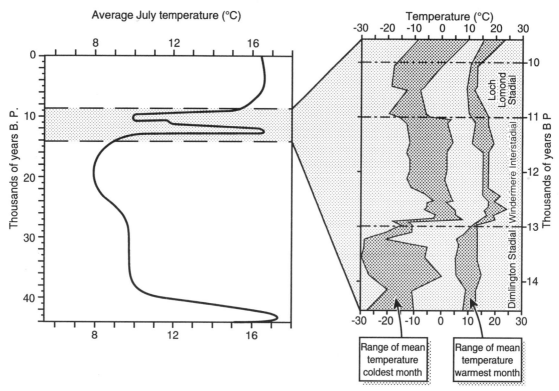

Figure 14.1 Left: Variations in mean July temperature since 44 ka BP as inferred from fossil coleopteran assemblages (after Coope, 1977b). Right: Variations in mean monthly temperatures for the coldest and warmest months during the period 14–10 ka BP, as inferred from fossil coleopteran assemblages. Based on Atkinson *et al.* (1987).

dence from various sources than on those arising from the interpretation of periglacial features in isolation.

Periglacial environments during the Dimlington Stadial

Following the Upton Warren Interstadial of the Middle Devensian (an event conventionally ascribed to *c.* 43–40 ka BP), the climate of Great Britain appears to have reverted to conditions of arctic severity. Coleopteran assemblages indicate that throughout the remainder of the Middle Devensian the English Midlands experienced a cold, dry and markedly continental climate with mean July temperatures of around 10 °C, mean January temperatures as low as -20 °C to -30 °C, and mean annual temperatures within the range -8 °C to -12 °C (Coope *et al.*, 1971; Coope, 1977b). Such conditions imply the re-establishment of continuous permafrost and the widespread development of ice-wedge polygons, though expansion of glacier ice at this time was apparently inhibited by aridity (Lowe & Walker, 1984). The changes that brought about the subsequent growth of the Late Devensian ice sheet during the Dimlington Stadial (26–13 ka BP) were probably heralded mainly by increases in precipitation rather than temperature decline. Significantly, a July palaeotemperature curve constructed by Coope (1975) on the basis of the changing composition of coleopteran assemblages in the English Midlands (Figure 14.1) suggests that though mean July temperatures may have declined slightly at around 25 ka BP, this summer cooling was also accompanied by an influx of species associated with more humid conditions and thus a more oceanic regime.

The transition to more a more oceanic (though still very cold) climate at the beginning of the Dimlington Stadial was probably associated with changes in the efficacy of certain periglacial processes, though details of these are difficult to reconstruct due to a lack of adequately dated sedimentary sequences. A notable exception, however, occurs near the village of Beckford in Worcestershire. At this location the terrace gravels of the Current Valley are overlain by cross-bedded sands containing organic silts that have yielded a radiocarbon age of 27 650 ± 250 yr BP (Briggs *et al.*, 1975). The terrace gravels appear to have been transported to the valley bottom by solifluction and subsequently reworked by the river during nival flood events. They contain numerous interformational ice-wedge casts indicative of continuous permafrost. At around 27 ka BP, however, aeolian activity resulted in an influx of sands that were also subjected to fluvial reworking. Mineralogical analysis has linked these sands with the nearby Cheltenham Sands, and has established a generally westerly provenance of windblown sediment. The associated coleopteran assemblage suggests a less markedly continental climate than during the preceding episode of gravel accumulation, though the presence of interformational ice-wedge casts within the sands confirms the continued presence of permafrost. The cessation of sand accumulation was followed by a period when ice-wedge polygons developed on a stable terrace surface and the river began to incise its course below the general terrace level. The final phase of periglacial activity represented at this site took the form of the accumulation of up to a metre of loessic and solifluction deposits on the terrace surface, where they were subjected to severe cryoturbation. Correlation with loess accumulation elsewhere in southern Britain suggests that this final phase of activity may have continued until near the end of the Dimlington Stadial.

The palaeoclimatic changes implied by this depositional sequence may be interpreted as follows. The earliest phase represented indicates a dry continental climate with continuous permafrost and mean annual temperatures below -5 °C to -6 °C, a hydrologic regime dominated by spring nival floods, and pronounced solifluction activity on valley sides. After *c.* 27 ka BP a strengthening of oceanic influence is evident with stronger westerly winds, though the climate remained sufficiently cold (MAAT <-5 °C to -6 °C) for the persistence of continuous permafrost and ice-wedge formation, and sufficiently dry (at least for part of the year) to permit deflation of sand. Diminution of sand accumulation and the associated onset of river incision may indicate a return to more continental conditions with diminished westerly airflow. The final phase of loess accumulation, solifluction and cryoturbation suggests sustained cold continental conditions, though the absence of ice-wedge casts in the uppermost sediments may indicate that winter temperatures were insufficiently severe to promote thermal contraction cracking of permafrost.

The sequence of events represented at Beckford has been summarised in detail to illustrate the potential of sites with adequate dating control for the sequential reconstruction of palaeoenvironments on the basis of periglacial structures and deposits. An alternative and no less valuable strategy is to consider the distribution of periglacial phenomena of Dimlington Stadial age across Britain as a whole with a view to reconstructing variations in regional palaeoclimate, a methodology pioneered by Williams (1969, 1975). A slightly different approach was adopted in a synthesis by Watson (1977b), who sought to employ the relationships between periglacial phenomena and climate in arctic environments to reconstruct the changing periglacial environment of Britain during the Devensian. Williams concentrated his analysis on the known distribution of ice-wedge casts and polygonal crop marks, the reconstruction of former active layer depths as recorded by cryoturbation structures, and the evidence of former wind directions and precipitation patterns afforded by the distribution and characteristics of loess, coversands and dunes. Watson based his reconstruction primarily on the palaeoclimatic significance of ice-wedge casts and pingo scars. Both assumed that discontinuous permafrost may develop where MAAT is lower than −1 °C and that continuous permafrost implies MAATs lower than -6 °C to -8 °C. Williams also inferred that ice-wedge casts in England probably imply a former MAAT no higher than -8 °C to -10 °C, primarily on the grounds that many supraformational casts are developed in sands and gravels, which require more extreme winter temperatures for contraction cracking than fine-grained sediments. Both authors noted that severe winter temperatures lower than -15 °C to -20 °C are necessary for cracking of permafrost, and observed the necessity for limited snowcover. Their conclusions are evaluated below.

Mean annual temperatures during the Dimlington Stadial

The widespread distribution of supraformational ice-wedge casts of inferred Dimlington Stadial age outside the limits of the Late Devensian ice sheet implies extensive continuous permafrost at this time, with mean annual temperatures certainly below -6 °C to -8 °C and possibly lower than the -8 °C to -10 °C range advocated by Williams (1969, 1975). Such figures, however, provide only a 'warm-side' estimate of former MAAT. One way of estimating the absolute value of MAAT at this time is to identify the contemporaneous continuous/discontinuous permafrost boundary, which may be inferred to represent the approximate position of the former -6 °C to -8 °C MAAT isotherm. In this respect, Williams' (1969) reconstruction is of particular interest. Williams noted that although supraformational wedge casts and associated polygonal crop marks are abundant in East Anglia and the English Midlands, documented wedge casts are rare in southern England and the southwest peninsula. From this he inferred that permafrost had been less extensive in the latter areas, even during the coldest period of the Dimlington Stadial. He also suggested that the paucity of wedge casts recorded in SW England was a reflection of less severe winter freezing, with mean annual temperatures possibly exceeding -6 °C to -8 °C. Although he recognised that greater snowcover in SW England may have inhibited permafrost cracking during the stadial, he discounted this explanation in view of the absence of evidence for contemporaneous glaciation on Dartmoor. Although Williams did not specifically attempt to delimit the former boundary between continuous and discontinuous permafrost, his map (Figure 14.2(a)) implies that this lay along an irregular line between the Bristol Channel and the Isle of Wight. If correct, this implies that the -6 ° to -8 °C stadial MAAT isotherm was similarly located during the coldest part of the stadial.

The validity of this proposal may be interrogated in the light of subsequent research on the nature and distribution of relict periglacial phenomena in southern and south-western England, and indeed on the neighbouring European mainland. In general, such research has vindicated William's conclusion that wedge structures are much rarer in south-western England than in central England and East Anglia. In the latter areas numerous further occurrences have been located, but very few have been discovered in the south-west peninsula, even though parts of this area have been subject to detailed scrutiny (Scourse, 1985, 1987; Miller, 1990; Gallop, 1991). On the northern margins of the south-west peninsula, however, relict ice-wedge polygons exposed on the shores of the Severn Estuary and on the coast of south Wales imply former continuous permafrost (Allen, 1987; Harris, 1989). Similarly, ice-wedge casts occur at Highcliffe near Bournemouth on the south coast (Lewin, 1966; Barton, 1984). These are developed in gravels and both mantled by and partly infilled by a loessic loam that has yielded a thermoluminescence date of *c.* 17 ka BP (Parks & Rendell, 1992). If it is assumed that the ice wedges at Highcliffe formed immediately prior to the accumulation of the overlying loess, then they demonstrate that the continuous permafrost zone during the Dimlington Stadial extended farther west than William's reconstruction indicated (Figure 14.2(a)). This inference is supported by research in northern France. A reconstruction by Van Vliet-Lanoë (1988c) suggests that although Brittany was underlain by discontinuous permafrost during the Dimlington Stadial, the continuous/discontinuous permafrost transition lay west of the Cotentin Peninsula (Figure 14.2(b)). If this boundary is extrapolated northwards, it suggests that all but the extreme south-west of England experienced continuous permafrost during the Dimlington Stadial.

Features that have been interpreted as ice-wedge casts occur in this area, though in many instances both the age and signifi-

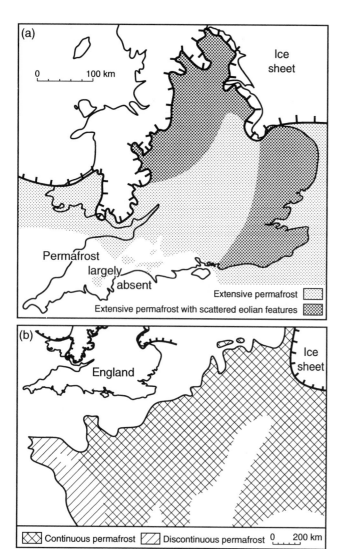

Figure 14.2 (a) Conjectural distribution of permafrost in England at the time of the Late Devensian glacial maximum, as proposed by Williams (1969). (b) Distribution of continuous and discontinuous permafrost in northern France, Belgium and the Netherlands at the time of the last glacial maximum, as proposed by Van Vliet-Lanoë (1988c).

cance of these features are uncertain. Most of the wedge structures observed on the north coast of Devon and Cornwall by Stephens (1961, 1970) occur near the top of head deposits and are less than a metre in length, and consequently may represent frost cracks that developed in seasonally-frozen ground rather than ice wedges formed by the cracking of continuous permafrost. The same interpretation is possible for the wedge forms described by Waters (1961) in east Devon. Although these are 1–2 m long and bedding in the host gravels has been contorted upwards, the casts are narrow and crack-like. Likewise, contraction crack structures near Tregunna on the Camel Estuary may represent former active layer soil wedges rather than ice-wedge casts (Scourse, 1987). Other periglacial phenomena in the south-west peninsula also yield only equivocal information regarding the status of former permafrost during the Dimlington Stadial. Large-scale sorted patterns on Dartmoor provide some additional evidence favouring former permafrost in this area, but their distribution is restricted to

higher ground and such features are not necessarily diagnostic of *continuous* permafrost. Disturbed Liassic shales at Lyme Regis in Dorset have been interpreted by Hutchinson & Hight (1987) as indicating that permafrost extended to a minimum depth of 7 m at some time prior to the Windermere Interstadial, but the full depth and age of former permafrost at this site remain uncertain. Similarly, the blocky to platy structure that occurs *c.* 2 m below the surface of head deposits at Lannacombe in south Devon appears to reflect ice segregation in former permafrost, but neither the age nor thickness of this permafrost horizon have been determined. The cryoturbation structures that occur in many coastal exposures in SW England indicate thawing to a depth of at least 2 m (Figure 6.25), but may equally have developed above thin discontinuous permafrost or a much thicker layer of continuous permafrost.

In sum, although the location of the continuous/discontinuous boundary during the Dimlington Stadial remains indeterminate, it appears to have lain considerably farther to the west than was implied by Williams' (1969) reconstruction (Figure 14.2(a)). Northwards extrapolation of Van Vliet-Lanoë's mapping (Figure 14.2(b)) suggests that only parts of Devon and Cornwall lay outside the continuous permafrost zone. However, it appears equally possible on present evidence that most if not all of the south-west peninsula was underlain by continuous permafrost during the coldest part of the stadial. If such were the case, the lack of convincing ice-wedge casts in this area may be attributable to the insulating effect of greater snowcover rather than less severe winter temperatures. Both effects would be expected in view of the peninsula's exposure to Atlantic weather systems, though the influence of these may have been more marked in summer than in winter (Kutzbach & Wright, 1985; COHMAP, 1988). A possible complicating factor is the relatively high geothermal heat flux associated with granite batholiths in this area (Hutchinson & Thomas-Betts, 1990). This may have reduced the depth of permafrost under Dartmoor and other areas of intrusive rock, but seems unlikely to have affected neighbouring areas underlain by sedimentary rocks. At best, we may conclude that the available periglacial evidence is not inconsistent with the notion that the continuous/discontinuous permafrost boundary (and thus the -6 °C to -8 °C MAAT isotherm) lay in the extreme south-west of England during the coldest part of the Dimlington Stadial, though the evidence on which this conclusion is based remains debatable.

The marked increase in the frequency of reported ice-wedge casts and tundra polygons to the north and east of the south-west peninsula has been generally interpreted in terms of increased continentality and lower temperatures (particularly winter temperatures) during the Dimlington Stadial (Williams, 1969, 1975; West, 1977). Watson (1977b, p. 195) noted that such widespread permafrost cracking probably implies a former MAAT no higher than *c.* -10 °C, but argued that at the maximum of Late Devensian glaciation MAAT was actually several degrees lower. He based his argument on the grounds that 'the southern limit of ice-wedge formation in the last glaciation lay some 500 km south of the southern English coast', and by analogy with present conditions in Alaska contended that this implied that 'full polar desert conditions' probably existed in the English Midlands, with mean annual temperatures of around -15 °C during the pleniglacial episode of *c.* 18 ka BP. This argument appears debatable in view of Britain's much greater exposure to maritime influences and his estimate conflicts with the evidence provided by coleopteran assemblages. This suggests that though summer temperatures were depressed several degrees below 10 °C at this time (Coope, 1977b) the winters were actually markedly warmer during the period of maximum ice-sheet expansion than either before or afterwards, a conclusion consistent with reduced continentality and increased precipitation during the pleniglacial (Atkinson et al., 1987). It is also difficult to reconcile Watson's argument with the notion that continuous permafrost failed to become established on the south-west peninsula. Had MAAT plunged to -15 °C in central England during the Dimlington Stadial, it seems unlikely that it could have exceeded -8 °C only 200 km to the south-west. If it is accepted that the continuous/discontinuous permafrost boundary was located in the extreme south-west of England during the thermal nadir of the stadial, a mean annual temperature of around -10 °C appears a more reasonable estimate for central England at this time, and is consistent with the coleopteran evidence. This estimate relates only to the coldest part(s) of the stadial and is not necessarily representative of pleniglacial conditions.

Summer temperatures and the annual temperature regime

In general, periglacial landforms and structures are much poorer indicators of former summer temperatures than the evidence provided by pollen analyses and studies of faunal assemblages. The former demonstrate that trees were absent from the British landscape during the Dimlington Stadial, and thus suggest that mean July temperatures cannot have exceeded 10 °C. Studies of coleopteran assemblages offer potentially greater precision, but only for the last two millennia of the stadial is the evidence adequate for summer temperatures to be estimated with confidence (Atkinson et al., 1987). Reconstructions by Coope and his co-researchers (Coope, 1975, 1977a,b; Coope et al., 1971; Figure 14.1) suggest that July temperatures in central England averaged around 10 °C during the period 35-25 ka BP, then declined to around 7 °C during the Late Devensian pleniglacial of 20–18 ka BP. That by Atkinson et al. (1987) indicates with somewhat greater confidence that cool summers with mean July temperatures not exceeding 10 °C persisted until the astonishingly rapid warming that terminated the stadial at around 13 ka BP. It should be noted, though, that the pleniglacial estimate of c. 7 °C rests largely on a single well-dated but impoverished coleopteran assemblage in silts underlying Devensian till at the type site at Dimlington, and could be several degrees in error (Coope et al., 1971, p. 93).

An interesting attempt to use periglacial features as indicators of summer warmth has been made by Williams (1975) on the basis of the depth of the former active layer as determined by the amplitude of cryoturbations of inferred Dimlington Stadial age. Williams showed a keen awareness of the potential pitfalls of this approach. Although active layer depths in present permafrost environments are generally related to summer warming, they are also affected by various local factors such as the nature of the substrate, vegetation cover, snowcover, soil moisture regime and aspect. Moreover, not all involutions represent freezing and thawing within a former active layer, and even those that do may have been beheaded by erosion or truncated at sedimentary horizons at depths less than that of the former active layer. Finally, it

is usually impossible to tell whether the amplitude of cryoturbations reflects former active layer depth under pleniglacial conditions, or during earlier or later periods when summer temperatures were rather higher. The lack of consistency in the amplitudes of cryoturbation structures of inferred Dimlington Stadial age (Table 6.1) is a measure of these uncertainties.

Williams noted that cryoturbations in sandy substrates in East Anglia and the Midlands have modal amplitudes in the range 2.0–2.3 m, and by comparison with published active layer depths for Alaska and Siberia inferred that these are consistent with a thawing index (accumulated temperatures above zero) of 900–1500 degree-days (°C). Williams assumed that the supraformational (near-surface) cryoturbations in these areas post-date the Upton Warren Interstadial, and argued that as some of those in the Midlands developed after the Late Devensian ice-sheet maximum, they are of Dimlington Stadial age. His evidence, however, is insufficient to pinpoint the timing of formation more accurately. By assuming a thawing index of 900 degree-days and a mean July temperature of 10 °C, he constructed a conjectural mean monthly temperature curve for 'the coldest part of the last glaciation' (Figure 14.3(a)). This predicts a MAAT of c. -8 °C and a mean January air temperature of c. -25 °C, and fairly closely resembles a curve produced by Coope et al. (1971) on the basis of coleopteran evidence for 'a cold, continental, phase of the last glaciation' (Figure 14.3b). This apparent convergence of evidence prompted Williams (1975, p. 109) to conclude that 'it suggests that involutions provide a reasonable picture of the climate despite the many assumptions that have to be made about them'.

Though Williams' approach certainly does not lack ingenuity, scrutiny of his evidence and methodology suggests that such enthusiastic endorsement of its success may be misplaced. The representativeness of the modal depths selected (2.0–2.3 m) is questionable (Table 6.1); the appropriateness of his analogues is uncertain; the cryoturbations on which the analysis are based are probably asynchronous features and of uncertain age; and the coldest (pleniglacial) summer temperatures of the Dimlington Stadial may have been several degrees lower than the 10 °C assumed. Williams gave no indication of why he selected the minimum thaw index (900 degree-days) implied by his arctic analogues, though it is notable that assumption of 1500 degree-days under the constraint of a mean July temperature of 10 °C generates a low-amplitude mean monthly temperature curve with MAAT barely below 0 °C, and thus is inconsistent with the much lower MAAT implied by the (assumed) contemporaneous formation of ice wedges. A particular difficulty that has emerged since Williams presented his reconstruction of mean monthly temperatures is that the coolest summer temperatures during the pleniglacial probably coincided with more humid, less continental conditions during which winter temperatures (and possibly MAAT) were higher than during the preceding and succeeding periods of cold, dry continental conditions (Atkinson et al., 1987). Because it assumes a 10° mean July temperature, Williams' reconstruction appears to be more representative of the 'continental' regime of the stadial, rather than the more maritime 'pleniglacial' regime with its lower summer temperatures.

The above considerations suggest that it is inappropriate to attempt to represent the Dimlington Stadial in terms of a single temperature curve. An alternative is to generate a family of such curves, each of which reflects particular constraints, to repre-

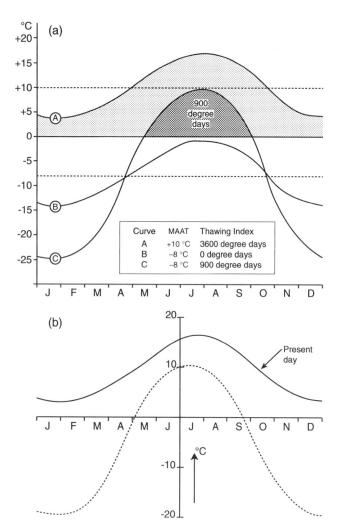

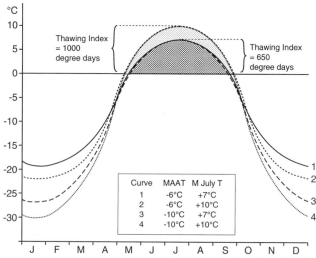

Figure 14.4 The likely range of mean monthly temperatures for central England during the Dimlington Stadial, based on assumption of MAATs of -6 °C and -10 °C and mean July temperatures of +7 °C and +10 °C.

Figure 14.3 (a) Present mean monthly temperatures in central England (curve A) and conjectural values for the coldest part of the last glaciation based on the assumption of MAAT = -8 °C and a thawing index of 900 degree-days (curve C). Curve B is based on MAAT = -8 °C with the same range of mean monthly temperatures as at present. After Williams (1975). (b) Present mean monthly temperatures for the English Midlands compared with those inferred from fossil coleopteran assemblages for a cold continental phase of the last glaciation. After Coope *et al.* (1971).

sent stadial palaeotemperatures in central England. The four curves depicted in Figure 14.4 are constrained by two sets of assumptions. The first is the value of mean July temperature: the maximum value is set at 10°, as suggested by coleopteran assemblages and the absence of trees, and the minimum at 7 °C, the pleniglacial figure suggested by coleopteran evidence, though this figure is acknowledgedly uncertain. The second set of constraints is generated by MAATs of -6 °C and -10 °C. The former is based on the assumption that -6 °C represents the maximum MAAT under which ice-wedges could have formed in coarse sediments, the latter on the assumption that the continuous/discontinuous permafrost boundary lay in the extreme south-west of England. It is worth noting, however, that there appears to be no firm evidence for ice-wedge formation under the pleniglacial conditions of c. 20–18 ka BP. Though wedge casts are developed in Late Devensian tills at sites close to the glacial limit, these could have developed with the return of

more continental conditions after the last ice-sheet achieved its maximum extent, and are not necessarily indicative of permafrost cracking during the pleniglacial. This raises the possibility that even the lowest-amplitude curve depicted in Figure 14.4 (curve 1) depicts a more extreme range of temperatures than prevailed during the pleniglacial: both MAAT and mean January temperature may have been slightly higher, and mean July temperature slightly lower.

If it is assumed that MAAT during the pleniglacial did not exceed -6 °C, the family of curves depicted in Figure 14.4 indicates approximately the range of thermal regimes of the stadial. Curves 2 and 4 represent the probable range of temperatures for the 'continental' regime, and curves 1 and 3 depict the likely range of 'pleniglacial' temperatures. A notable feature of the latter two curves is that they imply a thawing index of only about 650 degree-days, whereas the 'continental' curves imply a thawing index of about 1000 degree-days. The difference reflects the sensitivity of this index to the assumed value of mean July temperature, and reinforces the conclusion reached above that William's (1975) reconstruction relates not to the pleniglacial, but to the more continental regime that preceded and succeeded it. It will be apparent from the various caveats introduced in this discussion, however, that even the fairly wide range of values depicted by these curves may be no more than a first approximation of the full range of Dimlington Stadial palaeotemperatures.

Precipitation during the Dimlington Stadial

Because snow is an effective insulator, deep snowcover is incompatible with thermal contraction cracking of permafrost and the growth of ice wedges. Williams (1975) amassed evidence which suggests that a cover of 250 mm of snow (c. 100 mm water equivalent) effectively impedes ice-wedge formation, and on this basis concluded that winter precipitation during the Dimlington Stadial was much lighter than at present. He noted that as snowfall presently constitutes only about 40% of the total precipitation in arctic areas, mean annual precipitation in areas of England with ice-wedge casts could not have exceeded *c.* 250 mm. Whilst the precision of this estimate rests on the validity of the analogue, it seems indisputable that at the time of ice-wedge development annual precipitation was much less than the *c.* 600–750 mm that falls on lowland England at present. Various other lines of evidence appear to confirm that lowland England experienced a rather dry climate during at least part of the Dimlington Stadial. Halophyte elements in macrofossil floras indicate saline soils characteristic of dry summers with high rates of evaporation (e.g. Bell, 1969), and the widespread if thin loess deposits of southern and eastern England (Figure 8.21) indicate at least periodic summer aridity.

The low precipitation values suggested by Williams (1975) are, however, difficult to reconcile with the extension of the Late Devensian ice sheet across two-thirds of Great Britain during the Dimlington Stadial. The buildup of the ice sheet suggests that precipitation totals must have been substantially higher during at least part of the first half of the stadial and particularly during the pleniglacial of 20–18 ka BP. Of interest in this context are the estimates by Lockwood (1979) of lowland precipitation and evaporation in lowland Britain during the Dimlington Stadial. Lockwood based his reconstructions partly on the assumption that the Asiatic tundra offers a reasonable analogue for conditions in lowland England during the stadial, and on this basis provided two contrasting water balance estimates that he believed to represent a realistic range of possibilities. His lower estimate of mean annual precipitation (260 mm yr^{-1}, with 80–120 mm falling as snow) is very similar to that of Williams (1975), and implies very dry summer conditions with extremely limited (10 mm yr^{-1}) runoff and low soil moisture conditions, the kind of climatic environment that might favour the development of saline soils and the aeolian entrainment of silt particles. His higher estimate (367 mm yr^{-1}, again with 80–120 mm falling as snow) implies moist, cloudy summers, with much greater runoff (57 mm yr^{-1}) and much higher soil water contents, conditions inimical to the formation of saline soils or aeolian silt entrainment. Lockwood suggested that this second estimate may approximate pleniglacial conditions around 18 ka BP, citing in support global atmospheric circulation models that suggest increased wetness in Great Britain at this time, whilst the lower estimate represents more continental conditions earlier in the stadial. However, he noted that even his higher estimate appears inadequate to permit the extension of the Late Devensian ice sheet, unless there were very steep precipitation gradients between the maritime margins of the ice sheet and the unglaciated lowlands of England. Alternatively, of course, his higher estimate may undervalue snowfall during the pleniglacial and the preceding period of ice-sheet buildup.

Some light is thrown on this conundrum by the palaeohydrological reconstructions of Cheetham (1976, 1980). Cheetham calculated that the discharge of the mean annual (nival) flood on the River Kennet during the deposition of the Late Devensian Beenham Grange Terrace gravels lay within the range 153–418 m^3 s^{-1}, with a best estimate of 253 m^3 s^{-1}. As the catchment area of the Kennet is 1037.7 km^2, these palaeodischarges translate into daily peak runoff values of 12.7–34.8 mm and 21.1 mm respectively. If we reduce Cheetham's figures by 50% to give an estimate of average diurnal runoff during the nival flood (minimum estimate 6.3 mm per day, maximum estimate 17.4 mm per day, best estimate 10.5 mm per day), Lockwood's suggestion of a total annual runoff of only 57 mm yr^{-1} during the

pleniglacial implies a very abbreviated annual nival flood (minimum 3 days, maximum 9 days, best estimate 5–6 days). This seems unrealistically attenuated for a basin 1037.7 km^2 in area, and suggests that Lockwood's snowfall estimates are too low. This may imply either that his estimated mean annual precipitation (367 mm) under-represents conditions during the wettest part of the stadial, or alternatively that snowfall represented a larger proportion of mean annual precipitation than he allowed in his calculations. There are, of course, many uncertainties and assumptions in all of the analyses discussed above. At best we may infer that though a mean annual precipitation of *c.* 250 mm for periods of 'continental' climate during the stadial accords with the evidence implied by ice-wedge casts, saline soils and loess deposition, the increase of precipitation that accompanied and probably triggered the expansion of the last ice sheet is much less certain. It seems likely, however, that under the relatively maritime conditions of the pleniglacial, mean annual precipitation in lowland England exceeded the 367 mm yr^{-1} suggested by Lockwood, possibly by a considerable margin.

Loess deposits and former wind directions

Recognition of systematic regional trends in the thickness, granulometry and mineralogy of loess and coversands offers the potential for reconstruction of the dominant winds at the time of deposition. In Great Britain, most areas of coversand reflect deposition during the Loch Lomond Stadial, and hence information on wind direction during the Dimlington Stadial must be derived largely from the loessic sediments deposited at that time. There seems to be general consensus that most if not all surficial loess deposits in Britain are of Dimlington Stadial age, a proposition first advanced by Catt (1977, 1978) on stratigraphic, mineralogical and archaeological grounds and subsequently vindicated by thermoluminescence dating of loess and loess-enriched gelifluctate at a variety of sites in southern and eastern England (Wintle, 1981; Wintle & Catt, 1985a,b). Parks & Rendell (1988, 1992) have, however, demonstrated the survival of isolated pockets of Early Devensian and pre-Ipswichian loess at a small number of sites, and have shown that at one site (Reculver, on the Thames Estuary) loess deposition was probably renewed during the Loch Lomond Stadial. The loess deposits of Britain tend to be much thinner and more discontinuous than those of continental Europe, exceeding 1 m in depth only in the extreme south-east (Figure 8.21). In part, the discontinuity and shallowness of the loess mantle reflects reworking by solifluction or subsequent erosion (Catt, 1977), but the principal cause of the relative dearth of loess in Britain appears to have been the relative moistness of the stadial climate compared with that of the continent, and possibly a more restricted sediment supply.

The limitation of most of the surficial loess deposits of Britain to the south and east of England was interpreted by Williams (1975) in terms of an eastwards increase in aridity and decrease in vegetation cover, from which he inferred that the moisture-bearing winds came from the west. He acknowledged, however, that the prevailing circulation during the stadial was probably dominated by a stable blocking anticyclone across Europe, so that the prevailing winds were easterlies. Subsequent studies of the loess of southern and eastern England also indicate a dominant easterly airflow, though opinions differ as to the source of the loess. Lill & Smalley (1978) argued that most of the quartz component of loessic deposits in these areas was derived from continental Europe, but Catt (1977, 1978, 1985a,b) envisaged the main source of loess as outwash sediments in the then exposed North Sea Basin. Eden (1980) agreed that loess deposits in eastern England were derived from the North Sea Basin, but proposed transportation by north to northeasterly winds, a proposition supported by West's (1991) interpretation of certain Fenland depressions as the sites of former oriented thaw lakes (Chapter 5). In complete contrast to these findings, the coarser and mineralogically-distinct loesses of Cornwall and the Scilly Isles (Catt & Staines, 1982) and those that mantle limestone hills in north Wales and NW England (Lee & Vincent, 1981; Vincent & Lee, 1981) appear to be derived from glacigenic deposits in the then exposed Irish Sea Basin, and thus imply transportation by westerly winds. The aeolian Cheltenham Sands, which apparently date to the onset of the Dimlington Stadial, have also been interpreted as having been transported by westerly winds (Briggs et al., 1975).

The conflicting evidence outlined above offers no simple interpretation of dominant wind directions during the Dimlington Stadial. One possible explanation for the different wind directions suggested by loess deposits is that the latter are not contemporaneous, and represent deposition at different times during the stadial. It is conceivable, for example, that loesses indicative of transport from easterly source areas were emplaced during periods of more continental climate with enhanced easterly airflow, whilst those indicative of transportation from westerly sources reflect enhanced maritime influence. This possibility seems unlikely, however, in view of the wide scatter of Late Devensian thermoluminescence ages obtained from loesses of apparent eastern provenance in southern and SE England (Parks & Rendell, 1992). An alternative explanation is that the climate of the Dimlington Stadial was characterised by seasonal changes in wind direction. This interpretation is supported by climatic modelling that indicates a predominantly southerly surface airflow in winter around a large anticyclone centred over NE Europe, and a contrasting northerly surface airflow in summer as a result of high pressure to the north and west of Britain, possibly reinforced by katabatic effects (Kutzbach & Wright, 1985). If this model is valid, it suggests that deflation of the exposed North Sea Basin occurred mainly in summer, whilst dust from continental sources may have been carried into SE England during the winter. The eastwards transportation of loess from the exposed Irish Sea Basin may reflect occasional frontal incursions in the summer months, accompanied by strong westerly winds.

A final possibility is that both the age and provenance of aeolian silt in the loess deposits of Britain owe at least as much to changing sources of sediment supply as they do to surface circulation patterns. The extensive series of thermoluminesence determinations made by Parks & Rendell (1992) on the loessic deposits of southern and SE England reveals that though many are of Late Devensian (26–10 ka BP) age and some have survived since the Early Devensian, there is virtually no evidence for loess deposits of Middle Devensian (50–26 ka BP) age, even though most of the Middle Devensian appears to have been characterised by a very cold, arid, continental climate (Lowe & Walker, 1984). A solution to this enigma is offered by consideration of the sources of loess. Apart from a possible continental

component (Lill & Smalley, 1978), it is generally agreed that much of the British loess derived from silts originally deposited as outwash on the exposed floors of the North Sea Basin and Irish Sea Basin. These areas were exposed to deflation only during periods of exceptionally low sea level. Such conditions certainly pertained during the Late Devensian and possibly during the Early Devensian, but a marked rise in global sea level is known to have occurred around 60 ka BP (Shackleton, 1987), and it seems likely that throughout the Middle Devensian these two offshore basins were extensively inundated, and hence cannot have constituted major sources of windblown silt at this time. In contrast, the build up of the last mid-latitude ice sheets after c. 26 ka BP resulted in a dramatic lowering of global sea level to about -120 m to -130 m below present, exposing the offshore basins of the North Sea and Irish Sea to deflation. At the same time, rivers draining the last ice sheet deposited the extensive spreads of outwash that apparently formed the source of much Late Devensian loess. It therefore appears likely that periods of extensive loess deposition in Britain may have been jointly constrained by sea-level change and by the availability of fine-grained outwash sediments on the exposed floor of the surrounding continental shelf. This explanation helps to account for the lack of loessic deposits in those parts of northern Britain that experienced marked glacio-isostatic depression during the Dimlington Stadial. In such areas the retreat of the last ice sheet was accompanied by relative sea levels higher than those of the present day, and consequently potential offshore sources of fine-grained sediment lay under water as the land surface emerged from the ice.

The periglacial environment during ice-sheet retreat

There is abundant evidence that the progressive contraction of the Late Devensian ice sheet between c. 18 ka BP and c. 13 ka BP took place under sustained cold conditions. According to Ruddiman & McIntyre (1981), the North Atlantic oceanic polar front remained at the latitude of Portugal or northern Spain throughout this period, implying that Britain was surrounded by cold polar waters and possibly pack ice. Direct evidence for the persistence of cold water in the North Sea is provided by the arctic faunas of the Errol Beds, estuarine clays deposited in eastern Scotland following the withdrawal of the last ice sheet inland (Peacock, 1981). Coleopteran assemblages relating to the final millennium of the Dimlington Stadial (14–13 ka BP) indicate that Britain continued to experience full stadial conditions, with mean July temperatures in central England no higher than c. 10 °C and January temperatures of approximately -20 °C to -25 °C (Coope & Brophy, 1972; Atkinson et al., 1987; Figure 14.1). The persistence of such a cold continental regime at the very end of the stadial has been widely interpreted as indicating that retreat of the last ice sheet was caused mainly by snowfall starvation rather than any marked increase in ablation season temperatures.

This interpretation is strongly supported by periglacial evidence. In particular, ice-wedge casts and tundra polygons are of widespread occurrence in outwash deposits and tills inside the limits of Late Devensian glaciation (Figure 4.25) and provide conclusive evidence for the survival or re-establishment of permafrost in the wake of the retreating ice sheet (Worsley, 1987), as well as indicating general aridity and winter chilling sufficiently severe to induce permafrost cracking. The relict polygons of the west Midlands, a short distance inside the glacial limit, have been interpreted by Morgan (1971a,b, 1973) as indicating a single phase of permafrost following ice-sheet retreat. Morgan also suggested that the modest depth of associated ice-wedge casts may signify a climatic regime close to the threshold for active thermal contraction cracking. If valid, this suggests that these features probably developed when MAAT stood within the range -6 °C to -10 °C. Of particular interest are the numerous wedge casts and relict tundra polygons that have been found in central and eastern Scotland. Although some of these have been attributed to the Loch Lomond Stadial (see below), a Dimlington Stadial age seems to be more likely for the majority (Armstrong & Patterson, 1985). Given that central and eastern Scotland were deglaciated between 16 ka BP and 14 ka BP (Sutherland, 1991b), the development of ice wedges in these areas implies re-establishment of continuous permafrost as late as the final millennium of the stadial. The deposition of loess deposits as far north as the Vale of Eden (Figure 8.21) provides some confirmatory evidence of persistent aridity during ice-sheet retreat.

If conditions in central England were marginal for ice-wedge development following the retreat of the ice sheet, as suggested by Morgan, it is possible that the northwards withdrawal of the ice sheet was accompanied by gradual replacement of continuous permafrost by discontinuous permafrost in more southerly parts of Britain. Degradation of permafrost before the very end of the stadial is suggested by the occurrence in East Anglia of ice-wedge casts mantled by loessic sediment that must have been deposted after wedge casting had occurred. It is possible, too, that the replacement of continuous permafrost by thin discontinuous permafrost in the latter part of the Dimlington Stadial resulted in the localised formation of pingos and related ground-ice mounds. There is reasonable stratigraphic evidence to suggest that pingo scars in the Wey Valley in Surrey (Carpenter & Woodcock, 1981) and the 'older generation' of ramparted ground ice depressions in East Anglia (Sparks et al., 1972) are of Dimlington Stadial age. As these areas appear to have been underlain by continuous permafrost during the coldest part of the stadial, the development of pingos or related ground-ice mounds implies a transition from continuous permafrost to discontinuous permafrost, with an associated MAAT of around -5 °C (Chapter 5). Watson (1977b) has developed this idea further. Noting that the formation of mature pingos in arctic areas appears to require several centuries or millennia, he considered it unlikely that pingo development in England and Wales could have been confined to the Loch Lomond Stadial. He therefore suggested that pingos in Britain began to form with a change from continuous to discontinuous permafrost near the end of the Dimlington Stadial, and continued to develop under conditions of discontinuous permafrost during both the Windermere Interstadial and the Loch Lomond Stadial. This suggestion, however, is at variance with climatic reconstructions for the Windermere Interstadial, which suggest that by c. 12.5 ka BP the climate of England and Wales was as warm as at present (Atkinson et al., 1987). It also conflicts with the evidence provided by sections through ground-ice depressions in East Anglia, which indicate that an older (Dimlington Stadial) generation of features is stratigraphically distinct from a younger (Loch Lomond Stadial) generation of similar forms, with no evidence for continuous development (Sparks et al., 1972).

Although the notion that pingos and related ground-ice forms developed in response to thinning of permafrost in the the closing stages of the Dimlington Stadial appears reasonable, it conflicts with the evidence of coleopteran assemblages, which suggest that the climate of central England during the final millennium of the stadial remained very severe (with mean January temperatures of c. -20 °C to -25 °C) until interrupted by the very rapid warming that heralded the onset of the Windermere Interstadial (Atkinson et al., 1987; Figure 14.1). If this was indeed the case, continuous permafrost may have survived across much of southern Britain until c. 13 ka BP, when it is likely to have undergone rapid degradation in response to the very much warmer climate of the interstadial. Such considerations suggest that the pingos and ground-ice mounds of East Anglia and elsewhere may have formed rather earlier in the stadial. One interesting possibility is that such features developed in mid-stadial, in response to permafrost thinning under the rather more maritime conditions of the pleniglacial (20–18 ka BP), when winters in Britain appear to have been warmer than either before or afterwards, and were on average perhaps 10 °C warmer than during the period 14.5–13 ka BP (Atkinson et al., 1987). The development of pingos and related ground-ice mounds implies a MAAT of around -5 °C, which is consistent with the (admittedly limited) evidence provided by pleniglacial coleopteran assemblages.

In sum, the development of ice-wedges in the wake of the retreating Late Devensian ice sheet strongly supports other climatic evidence in suggesting that contraction of the ice sheet took place under cold, arid, continental conditions. Moreover, the existence of wedge casts and tundra polygons of inferred Dimlington Stadial age in central and eastern Scotland implies that the ice sheet had retreated as far as the central and western Highlands of Scotland before the onset of the rapid warming that marked the beginning of the Windermere Interstadial at c. 13 ka BP. The evidence for progressive thinning of permafrost in southern Britain is equivocal. Although pingos and other ground-ice mounds may have developed in response to a gradual change from continuous to discontinuous permafrost during the period of ice-sheet retreat, other evidence indicates that cold, arid conditions persisted until the very end of the stadial, and implies that permafrost degradation was swift. If this were the case, a further implication is that pingos and related ground-ice mounds of inferred Dimlington Stadial age formed rather earlier, possibly under the relatively maritime conditions of the Late Devensian pleniglacial.

The periglacial environment of the Dimlington Stadial: discussion.

One weakness of previous attempts to reconstruct the periglacial environment of the Dimlington Stadial is that they have implicitly tended to assume uniformity of climate (Williams, 1969, 1975), with the most severe temperature regime coinciding with the maximum extension of the last ice sheet (Watson, 1977b). Subsequent research on coleopteran assemblages relating to different parts of the stadial has suggested, however, that this was a period of marked climatic variability (Coope et al., 1971; Briggs et al., 1975; Coope, 1977b; Lowe & Walker, 1984; Atkinson et al., 1987). As a first approximation, three broad climatic episodes may be recognised. First, before c. 27–25 ka BP the climate of central England appears to have been dominated by cold, arid continental conditions, with MAAT in the range -8 °C to -12 °C, mean July temperatures of c. 10 °C and mean January temperatures within the range -20 °C to -30 °C. Second, the pleniglacial episode of c. 20–18 ka BP appears to have been characterised by rather more maritime conditions, with higher MAAT and less severe winter temperatures, lower summer temperatures and a general increase in humidity. Both the growth of the last ice sheet to its maximum extent at this time and coleopteran evidence suggest that the transition to more maritime conditions may have occurred well before the pleniglacial, possibly as early as c. 25 ka BP. Finally, it appears that cold, dry continental conditions similar to those of the Middle Devensian–Dimlington Stadial transition had become re-established by c. 14.5 ka BP, and lasted until very rapid warming heralded the onset of the Windermere Interstadial at c. 13 ka BP. The retreat of the last ice sheet from its maximum southern extent after 18 ka BP suggests that such continental conditions persisted for much of the period 18–13 ka BP.

The contribution of relict periglacial phenomena to our understanding of climate and climatic change during the stadial is hamstrung by poor dating control. The evidence provided by ice-wedge casts and tundra polygons confirms that during the coldest episode(s) of continental climate, continuous permafrost apparently underlay all unglacierised parts of Britain except possibly the extreme south-west of England. These wedge casts imply MAAT <-6 °C, severe winter chilling to temperatures of less than - 20 °C and probably precipitation totals no higher than about 250 mm yr^{-1}. If it is accepted that the continuous/discontinuous permafrost boundary and hence the -6 °C to -8 °C MAAT isotherm lay in the extreme south-west of England under the cold continental conditions of the early and late stadial, then an associated MAAT of c. -10 °C seems reasonable for central England at these times, an estimate that agrees well with that based on coleopteran assemblages. The occurrence of wedge casts and polygon networks of inferred Dimlington Stadial age in central and eastern Scotland confirms that very cold continental conditions persisted until the very end of the stadial, by which time the snowfall-starved ice sheet had retreated to its main source areas in the Highlands of Scotland. The question as to whether the temperature regime of the pleniglacial was sufficiently severe to permit ice-wedge development remains open, though Worsley (1987) has provided evidence which suggests that the last ice sheet advanced over continuous permafrost in Cheshire. The development of wedge casts in Late Devensian tills and outwash deposits cannot be taken as evidence of permafrost cracking during the pleniglacial, as such features may have developed during the subsequent return to more continental conditions.

The evidence provided by pingo scars and related ground-ice depressions is temporally ambiguous. The conventional view is that as such features occur in areas for which there is evidence of continuous permafrost, their development reflects the thinning of permafrost (with MAAT of around -5 °C) near the end of the stadial. This view, however, is at variance with the evidence favouring persistence of cold continental conditions until the very end of the stadial, though if correct would appear to imply rapid development of these features. An alternative solution is that pingos and other ground-ice mounds developed

under the more maritime conditions of the pleniglacial, but this is equally unsubstantiable on present evidence. Climatic inferences based on the thickness of cryoturbated layers (Williams, 1975) are also potentially undermined by uncertainties concerning the age of such features, though it appears that cryoturbations with an amplitude of c. 2.0 m probably developed under relatively continental conditions (with a mean July temperature of c. 10 °C) rather than during the cooler summers of the pleniglacial.

Although the precipitation estimate of c. 250 mm yr^{-1} reached independently by Williams (1975) and Lockwood (1979) appears a reasonable figure for the relatively continental conditions that characterised the early and later parts of the stadial in central England, the growth of the Late Devensian ice sheet during the period prior to c. 18 ka BP implies much higher precipitation. The palaeohydrological data of Cheetham (1976, 1980) suggest that Lockwood's upper estimate of 367 mm yr^{-1} (with 80–120 mm falling as snow) is probably too low to represent pleniglacial conditions, unless steep precipitation gradients are assumed. Uncertainties also attend interpretation of the significance of Dimlington Stadial loess deposits. Dating evidence (Parks & Rendell, 1992) suggests that widespread loess deposition required exposure of extensive areas of the surrounding shelves by falling sea level. The easterly provenance of much of the loess in Britain is consistent with a prevailing easterly airflow dictated by a stable high pressure cell over continental Europe, with possible seasonal variations in wind direction. Equally, however, the presence of loess of westerly provenance in Wales and SW England implies at least intermittent incursions of maritime air. Unfortunately, the present dating evidence is insufficient to allow periods of increased westerly airflow to be identified, though an obvious possibility is that these relate primarily to the period of ice-sheet build up (c. 25–18 ka BP), whilst easterlies prevailed during the preceding and succeeding periods of more continental conditions.

The periglacial environment of the Loch Lomond Stadial

The Loch Lomond Stadial, conventionally assigned to the period 11–10 ka BP, is the equivalent of the Younger Dryas chronozone of continental Europe and represents the final episode of severe periglaciation to affect Great Britain. The return to stadial conditions after the Windermere Interstadial was ushered in by renewed southwards movement of the North Atlantic oceanic polar front, which according to Ruddiman et al. (1977) reached its southernmost position off the coast of SW Ireland around 10.2 ka BP. The associated cooling resulted in a recrudescence of glacier ice in mountain areas (Figure 2.1), though for the most part glaciers were confined to upland valleys and many summits and plateaux remained ice-free. Virtually all of lowland Britain was exposed to periglaciation throughout the stadial, though the effects of such exposure are generally much less evident than for the Dimlington Stadial. In part this difference may reflect less severe climatic conditions during the Loch Lomond Stadial, though the relative brevity of this later cold period offers a more likely explanation for the limited impact of periglacial processes on the lowland landscape at this time. In contrast, most if not all relict periglacial phenomena in upland Britain developed or were last active during the Loch Lomond Stadial.

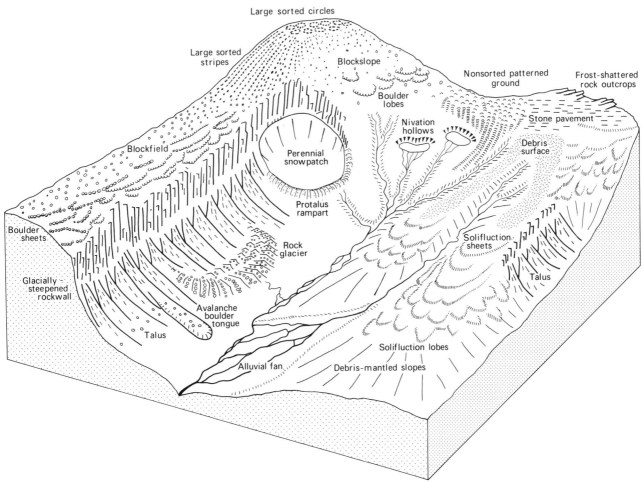

Figure 14.5 Schematic representation of the range of Late Devensian periglacial features on Scottish mountains, most if not all of which were last active during the Loch Lomond Stadial. After Ballantyne (1984).

Periglacial features of Loch Lomond Stadial age

The range of periglacial phenomena known to have developed in upland Scotland following the downwastage of the Late Devensian ice sheet is illustrated in Figure 14.5. A similar assemblage of relict features is also found in other mountainous parts of Britain, such as Snowdonia and the Lake District. Although only a limited number of these features can be conclusively attributed to the Loch Lomond Stadial as opposed to the period of ice-sheet decay, there is widespread belief that most relict periglacial phenomena on high ground were last active during the stadial, a belief encouraged by the spatial complementarity of such phenomena and the limits of the Loch Lomond Readvance (Ballantyne, 1984). Many features, however, seem likely to have formed initially during ice-sheet downwastage in the latter part of the Dimlington Stadial. This was certainly true of the cover of frost-weathered regolith that mantles many hills, though in some areas such regolith appears to be of much greater antiquity (Chapter 9). It seems likely that many Lateglacial taluses also began to accumulate immediately after the withdrawal of the last ice sheet, though reconstructed rockwall retreat rates indicate a substantial further addition of rockfall debris during the Loch Lomond Stadial (Ballantyne &

Kirkbride, 1987). Most relict upland patterned ground features appear to have either formed or been reactivated during the Loch Lomond Stadial, and the same is true of relict upland solifluction features such as boulder lobes, sorted and nonsorted solifluction lobes, and the solifluction terraces that occupy the floors of some upland valleys (Ballantyne, 1984; Douglas & Harrison, 1987). The protalus ramparts and rock glaciers of upland Britain can be attributed to the Loch Lomond Stadial on the grounds that both types of landform imply renewed climatic deterioration after the retreat of the last ice sheet from high upland valleys and corries, and the Loch Lomond Stadial represents the only subsequent period of sufficiently severe climate for the development of these forms (Dawson, 1977; Chattopadhyay, 1984; Ballantyne & Kirkbride, 1986; Maclean, 1991).

Whilst the characteristics of upland periglaciation during the Loch Lomond Stadial have been reasonably well established, the range of lowland periglacial features attributable to this period is rather less certain. Such uncertainty stems mainly from the difficulties of differentiating lowland landforms and near-surface soil structures of Loch Lomond Stadial age from those that developed during the latter part of the Dimlington Stadial in the wake of the retreating ice sheet. This difficulty is particularly pronounced with respect to ice-wedge casts and tundra polygons. Although various authors have argued for the widespread formation of such features during the Loch Lomond Stadial (e.g. Gemmell & Ralston, 1984; Maizels, 1986), the stratigraphic evidence for permafrost cracking and wedge formation at this time is often ambiguous and for the most part poorly-documented. A survey by Galloway (1961c) of ice-wedge casts in Scotland produced no unequivocal evidence that any of the numerous features he inspected are of Loch Lomond Stadial age. Sissons (1976c, p.111) reported wedge casts just inside the limit of the Loch Lomond Readvance on Mull and in the NW Highlands, but descriptions of the Mull 'wedge casts' have never been published and the feature he interpreted as a wedge cast in the NW Highlands is almost certainly of other origin. Similarly, Coope (1977b) briefly mentioned ice-wedge casts in sediments of Loch Lomond Stadial age at Glen Ballyre on the Isle of Man, but subsequent inspection of these features revealed that they are of low amplitude and could be interpreted as the product of cracking of seasonally-frozen ground. The same interpretation probably applies to small wedge-like structures that occur in Loch Lomond Stadial sediments at Glanllynau in north Wales (Harris & McCarroll, 1990). Finally, structures in a Lateglacial beach ridge near Glasgow have been interpreted by Rose (1975) as ice-wedge casts of Loch Lomond Stadial age, but the origin of these features is debatable (Worsley, 1987) and even if the structures Rose described do represent ice-wedge casts, the possibility that they represent permafrost cracking during the final stages of ice-sheet retreat cannot be discounted. In sum, the evidence for ice-wedge casts and tundra polygons of Loch Lomond Stadial age remains tantalisingly inconclusive. One intriguing feature of this debate is that all the putative wedge casts mentioned above occur in the west of the Britain, which is more exposed to moderating maritime influences and which during the Loch Lomond Stadial appears to have experienced a wet and stormy climate. Sissons (1979c, 1980b) estimated that mean annual precipitation in the SW Highlands of Scotland may have been as high as 3000–4000 mm, of which perhaps 75% fell as snow. Although this may be regarded as a liberal estimate, it does suggest that the areas in which the supposed ice-wedge casts have been reported are actually the least likely to have undergone permafrost cracking during the stadial.

More certain evidence for at least discontinuous permafrost on low ground during the Loch Lomond Stadial is provided by pingo scars and related ground-ice depressions. Sparks et al. (1972) demonstrated that the 'younger generation' of such features in East Anglia certainly formed at this time, though the dating evidence from other sites is more ambiguous. Radiocarbon dating and pollen analyses indicate that accumulation of organic sediment within pingo basins at various sites in Wales began around 10 ka BP (Handa & Moore, 1976; Watson, 1977b), and the apparent absence of organic deposits of Windermere Interstadial age in these basins suggests that they represent pingo growth during the Loch Lomond Stadial. Such absence of evidence is inconclusive, however, as Windermere Interstadial deposits may have been buried under inwashed minerogenic sediments during the stadial, and such sediments are notoriously difficult to penetrate using conventional coring techniques. Some pingo scars, such as those at Llangurig in east Wales, Ballaugh on the Isle of Man (Watson, 1971) and the

Whicham Valley in Cumbria (Bryant et al., 1985) lie well within the limits of the last ice sheet, and hence represent pingo formation either in the later stages of the Dimlington Stadial or during the Loch Lomond Stadial.

There is fairly widespread evidence for the development of cryoturbations up to 2 m in amplitude during the Loch Lomond Stadial (Table 6.1). Cryoturbated palaeosols of Windermere Interstadial age occur as far south as Kent, Buckinghamshire and Berkshire (Kerney, 1965; Evans, 1966; Paterson, 1971), but are not necessarily indicative of former permafrost. Two generations of cryoturbations occur at Glanllynau in north Wales: the older features are of Dimlington Stadial age, but the younger cryoturbations occur in sands and gravels that postdate the Windermere Interstadial (Harris & McCarroll, 1990). This latter group is associated with wedge structures that appear to represent former frost cracks indicative of thermal contraction cracking of seasonally-frozen ground, but again do not necessarily imply former permafrost. In Leicestershire, however, cryoturbated coversands of inferred Loch Lomond Stadial age have been interpreted by Douglas (1982) as indicating the presence of at least discontinuous permafrost during the stadial. Although the characteristics of these features are consistent with their formation above a former permafrost table, they cannot be considered conclusive evidence of permafrost during the Loch Lomond Stadial, particularly as the coversands in which they occur are dated only by analogy with securely-dated deposits in Lincolnshire. At most, therefore, cryoturbation structures of certain Loch Lomond Stadial age indicate deep (up to 2 m) seasonal ground freezing and thawing, but the question as to whether some of these developed above a former permafrost table remains open. Less equivocal in this respect are fragipans or indurated soil horizons, which have been widely interpreted as indicators of former permafrost and active layer thickness (FitzPatrick, 1956b; Van Vliet-Lanoë & Langhor, 1981), and are known to occur within the area occupied by the last ice sheet in Britain (e.g. Crampton, 1965; Harris, 1991). Unfortunately, such structures as have been described are chronologically ambiguous, in that they may relate either to the later stages of the Dimlington Stadial or to the Loch Lomond Stadial, and hence again do not provide conclusive evidence for permafrost during the latter.

The widespread occurrence of solifluction in lowland areas during the Loch Lomond Stadial is demonstrated by sites where organic material of Windermere Interstadial age is overlain by solifluction deposits or soliflucted till, for example at Springburn and Robroyston near Glasgow (Dickson et al., 1976), Woodhead in NE Scotland (Hall, 1984), Eston Beacon in the Tees Valley and Brinziehill in Kincardineshire (Harkness & Wilson, 1979). Such sites, however, demonstrate only seasonal freezing and thawing of the ground, as solifluction is known to occur in permafrost-free environments. More critical in this respect are sites where clay-rich head deposits overlie organic deposits of Windermere Interstadial age, as at Sevenoaks in Kent (Weeks, 1969; Skempton & Weeks, 1976), the Gwash Valley in Northamptonshire (Chandler, 1976) and Horsecombe Vale in the Cotswolds (Chandler et al., 1976). Shear planes at the base of such head deposits indicate emplacement by sliding rather than flow, and in many respects such slides resemble the active layer detachment slides of arctic environments. If this analogy is valid, it indicates that failure took place over a former permafrost table (Chapter 7), which in turn implies the re-establishment of at least discontinuous permafrost in parts of central and southern England during the Loch Lomond Stadial.

The stadial also appears to have been a time of vigorous fluvial activity and enhanced coastal erosion. Floods during the stadial appear to have been responsible for the deposition of alluvial fans in Shropshire (Rowlands & Shotton, 1971), Swaledale in Yorkshire (Pounder, 1985) and the Lake District (Boardman, 1985b), and Sissons (1976c) has argued that many of the relict alluvial fans of upland Scotland were deposited at this time. Increased flood discharges also affected lowland catchments, such as the Gipping Valley in Suffolk, where Rose et al. (1980) found evidence of incision of Windermere Interstadial sediments to a depth of at least 9 m followed by infilling of the incised channel under cold-climate conditions. The efficacy of slopewash at this time is manifest in numerous kettle holes and other enclosed depressions, in which organic sediments of Windermere Interstadial age are overlain by stadial slopewash deposits. Similarly, widespread slopewash is recorded in the chalklands of southern England where on lower valley slopes a layer of colluvium frequently overlies a Windermere Interstadial palaeosol (Kerney et al., 1964; Evans, 1966). Stratified slope deposits in Wales and the Lake District have also been attributed to the operation of slopewash during the Loch Lomond Stadial (Watson, 1965b; Boardman, 1977, 1978), though a late Dimlington Stadial age for such deposits cannot be discounted (Chapter 13). By far the best evidence for enhanced littoral erosion during the Loch Lomond Stadial is the Main Rock Platform of western Scotland, and its probable correlative, the buried marine planation surface of SE Scotland (Sissons, 1974; Gray, 1978; Dawson, 1979, 1980, 1988). Although there is some evidence that the Main Rock Platform may represent retrimming of a much older shoreline (Gray & Ivanovitch, 1988), this does not invalidate the notion of greatly enhanced coastal erosion during the stadial. Accelerated littoral erosion at this time is supported by research by Sissons (1978), who attributed the rapid erosion of lake shorelines in Glen Roy to the disruption of bedrock by frost action and the removal of debris by wave action and lake ice.

A distinctive feature of the periglacial environment of lowland Britain during the Loch Lomond Stadial was the localised deposition of coversands in areas such as the lower Trent Valley, Lincolnshire and Lancashire. Only in the Vale of York, however, are contemporaneous sand dunes preserved in any number (Matthews, 1970). Radiocarbon dating and pollen analyses of organic material underlying and overlying aeolian sand deposits at a number of sites have confirmed that most coversands in lowland England were deposited at this time (Matthews, 1970; Gaunt et al., 1971; Buckland, 1982; Wilson et al., 1981), though those in the Breckland of East Anglia appear to be much older (West et al., 1974). Mineralogical and textural studies of the Loch Lomond Stadial coversands in Lancashire and the Vale of York indicate that they were derived locally as a result of deflation of nearby fluvioglacial or glaciolacustrine deposits. This interpretation is supported by the observations of Buckland (1982), who found that the Lincolnshire coversands pass laterally into shallow channel structures infilled with waterlain sand, and thus probably represent localised reworking of adjacent fine-grained fluvial deposits.

Permafrost and its palaeoclimatic implications

There can be no doubt that the Loch Lomond Stadial witnessed the widespread re-establishment of permafrost in Great Britain. In upland areas, the formation of protalus rock glaciers at this time implies the development of at least discontinuous permafrost with an associated MAAT of less than about -2 °C (Haeberli, 1985). Regional occurrences of relict large-scale sorted patterns such as occur on British mountains are also widely believed to indicate former permafrost (Williams, 1975; Goldthwait, 1976; Washburn, 1979, 1980), though both the age and the palaeoenvironmental significance of these features are less certain. In lowland areas, the dubious status of proposed ice-wedge casts of Loch Lomond Stadial age (see above) means that these cannot be employed with any confidence as indicators of permafrost at this time. Other diagnostic features consequently assume greater importance. Of these, probably the most useful are the securely-dated ground-ice depressions or pingo scars investigated by Sparks et al. (1972) in East Anglia, which imply the development of discontinuous permafrost in this area during the stadial. The pingo scars of Wales have similar palaeoenvironmental implications, but for these an earlier (Dimlington Stadial) age cannot be discounted on present evidence. The extension of discontinuous permafrost into other parts of England during the Loch Lomond Stadial is, however, strongly suggested by the occurrence of what appear to be active layer detachment slides of Loch Lomond Stadial age in the Cotswolds (Chandler et al., 1976), east Midlands (Chandler, 1976) and even as far south as Kent (Weeks, 1969, Skempton & Weeks, 1976). Such southerly extension of permafrost during the stadial is consonant not only with the development of cryoturbations in Windermere Interstadial palaeosols in southern England (Kerney, 1965; Evans, 1966; Paterson, 1971), but also with the presence of arctic beetle assemblages in southern England at this time (Coope, 1977b).

Although discontinuous permafrost may exist under favourable circumstances in locations where MAAT is as high as -1 °C, the formation of open-system pingos or related ground-ice mounds appears to require a rather more severe regime, with a sustained MAAT no higher than about -4 °C to -5 °C (Mackay, 1988b; Chapter 5). On the other hand, MAATs cannot have been much lower than this, or the formation of continuous permafrost would have inhibited the development of such features. It follows that the best estimate of stadial MAAT for East Anglia and central England on the basis of periglacial evidence is c. -5 °C. This figure is identical to an earlier estimate by Watson (1977b) and also to that proposed for the coldest part of the stadial by Atkinson et al. (1987) on the basis of evidence provided by beetle assemblages. Assumption of a MAAT of -5 °C for central England during the coldest part of the stadial suggests that continuous permafrost probably existed at sea level in the more northerly parts of Britain. Present MAAT in NE Scotland, for example, is about 2 °C lower than that in East Anglia and central England, and it is probable that latitudinal temperature gradients during the Loch Lomond Stadial were steeper than at present due to the greater influence in northern Britain of such factors as increased distance from warm oceanic waters, probable winter freezing of the North Sea and proximity to the Loch Lomond Readvance icefield. Such considerations suggest that during the coldest part of the stadial,

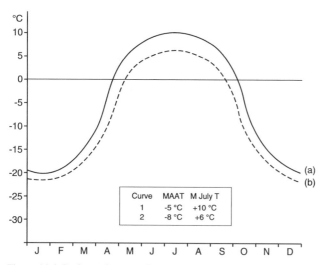

Figure 14.6 Conjectural range of mean monthly temperatures in central England (upper curve) and NE Scotland (lower curve) during the coldest part of the Loch Lomond Stadial. The upper curve is based on an assumed MAAT of -5 °C and a mean July temperature of +10 °C, the lower on an assumed MAAT of -8 °C and a mean July temperature of +6 °C.

sea-level MAAT in NE Scotland was no higher than -7 °C, and possibly a degree or two lower, which in turn implies the return of continuous permafrost to at least the more northerly and easterly parts of Scotland during the stadial. The status of permafrost in western Scotland is less certain, however, as the heavy snowfall that fed the glaciers of the western Grampians may have insulated the ground from the most severe winter temperatures, thereby widening the difference between MAAT and mean annual temperature at the ground surface. The development of a protalus rock glacier at 350 m on the the Hebridean island of Jura, however, confirms that at least discontinuous permafrost existed on low ground in western Scotland during the stadial.

Studies of Loch Lomond Stadial beetle assemblages (Coope, 1977b; Atkinson et al., 1987) place the mean July air temperature of central England during the coldest part of the stadial at or fractionally below +10 °C. Estimates based on the dimensions of contemporaneous glaciers suggest a mean July sea-level temperature of 7.5 °C for the English Lake District (Sissons, 1980a,b), 7.0 °C for the SW Grampians (Sissons, 1979c, 1980b), and 6.0 °C for both the eastern Grampians (Sissons & Sutherland, 1976) and the Isle of Skye (Ballantyne, 1989a). Given that the present difference in mean July temperature between central England and Skye or the eastern Grampians is c. 2.5 °C and that a steeper northwards temperature gradient probably existed during the stadial, such estimates appear reasonably consistent. Together with the MAAT estimates outlined in the previous paragraph, they permit reconstruction of the approximate mean monthly temperature regime of the coldest part of the stadial for central England (assuming a MAAT of -5 °C, and a mean July air temperature of c. 10 °C) and NE Scotland (assuming a MAAT of -8 °C, and mean July temperature of c. 6 °C). The results (Figure 14.6) suggest that mean January air temperatures in the former area were of the order of -20 °C during the coldest part of the stadial, whilst those in NE Scotland were about 2 °C lower. It is of interest to note that though the seasonal temperature regime of central

England as reconstructed here appears marginal for permafrost cracking and ice-wedge development, even in fine-textured sediments (Table 14.1), that of NE Scotland was probably sufficiently severe for ice wedges to have formed, particularly as this latter area experienced marked aridity (see below). It may well be that at least some of the numerous supraformation ice-wedge casts and tundra polygons in northern and eastern Scotland developed during the Loch Lomond Stadial, though confirmatory evidence for this remains to be established.

Dominant winds and precipitation patterns during the Loch Lomond Stadial

An attempt has been made by Sissons (1979c, 1980b) to reconstruct prevailing winds and precipitation patterns during the Loch Lomond Stadial on the basis of the distribution and dimensions of former glaciers. By far the largest ice mass of the time existed in the Western Highlands (Figure 2.1), which suggests that snowfall during the stadial was primarily associated with airstreams from the west or south-west. Sissons noted, however, that on a more local scale the largest glaciers in many areas tended to have southerly aspects, and that equilibrium line altitudes in the eastern Grampians rose northwards from the Highland boundary. From these more detailed patterns he inferred that most snow-bearing winds blew from the south. He suggested that this implies a synoptic situation in which warm or occluded fronts approached from the south-west, and that major snowfalls were often associated with southerly or south-easterly winds in advance of the passage of such fronts. Frequent repetition of this situation, he argued, implies that many depressions followed more southerly tracks than at present, and this he related to the southerly position of the North Atlantic oceanic polar front off SW Ireland (Ruddiman et al., 1977). Sissons also deduced from the contrasting dimensions of Loch Lomond Readvance glaciers in western and eastern Scotland that snowfall must have declined markedly eastwards across the Scottish Highlands during the stadial. He noted that although the low temperatures of the stadial would have tended to depress precipitation totals below present-day levels, this effect may have been at least partly offset by stormier conditions resulting from the vigorous interaction of airmasses due to the southerly position of the oceanic (and hence the atmospheric) polar front. On the somewhat speculative assumption that these two contrary influences may have cancelled each other out, he suggested that stadial precipitation over the mountains of the SW Grampians was fairly similar to that of the present (3000–4000 mm yr^{-1}). Conversely, the diminutive size of the glaciers that developed in the high corries of the NW Cairngorms led him to suggest that mean annual precipitation at 1000 m in this massif may have been as low as 500–600 mm (25–30% of present values) with perhaps only 200–300 mm falling annually in the adjacent Spey Valley. The dominance of the xerophyte *Artemisia* in Loch Lomond Stadial pollen assemblages from NE Scotland also suggests that this area experienced regional aridity during the stadial (Birks & Mathewes, 1978; Tipping, 1985).

The evidence provided by various periglacial phenomena helps us to assess the validity of the climatic patterns suggested by Sissons. Of particular value in evaluating the predominant airflow of the stadial are the coversands of England. Although some coversands apparently reflect reworking of fluvioglacial sands by winds blowing from a variety of directions (Wilson *et al.*, 1981), mineralogical studies of the provenance of coversands in eastern England indicate deposition by winds blowing from a direction slightly south of west (Buckland, 1982), and thus provide support for Sisson's advocacy of a predominantly westerly airflow. The presence of blown sand on the eastern side of the Trent Valley in Nottinghamshire (Edwards, 1967) also suggests transportation of grains from the valley floor by west to south-west winds. Further evidence for the dominance of westerlies and a steep west–east precipitation gradient at this time is provided by the altitudes of protalus ramparts of inferred Loch Lomond Stadial age. Across the Scottish Highlands and Hebrides, these rise eastwards from under 100 m on St Kilda and Harris to around 500 m on the western mainland and to over 900 m in the Cairngorms (Ballantyne & Kirkbride, 1986; Figure 12.22). The development of protalus ramparts implies the prolonged survival of perennial firn fields with fairly constant dimensions, and hence maintenance of a delicate balance between winter snowfall and summer ablation. As ablation must have been much greater on low ground in the Hebrides than at 900 m in the Cairngorms, it follows that snowfall in the Hebrides must also have been much greater than in the Cairngorms, an interpretation consistent with Sissons' notion of snowfall associated with prevailing westerlies.

The steep south-west to north-east precipitation gradient proposed by Sissons is also open to investigation using periglacial evidence, in the form of protalus rock glaciers. Two of the finest examples of these features occur at 920 m in Coire Beanaidh in the NW Cairngorms (Chattopadhyay, 1984) and some 170 km to the south-west at 350 m on the Hebridean island of Jura (Dawson, 1977). It was argued above that sea-level MAAT in NE Scotland during the Loch Lomond Stadial was not higher than -7 °C and possibly a degree or two lower. If we assume a possible MAAT range of -7 °C to -9 °C for this area and also a decline in MAAT with altitude within the range 0.5 °C/100 m to 0.8 °C/100 m, the predicted MAAT for 920 m in the Cairngorms (the altitude of the Coire Beanaidh rock glacier) falls within the range -11.6 °C to -16.4 °C. In terms of the constraints proposed by Haeberli (1985) for protalus rock glacier formation in the Alps (Figure 12.26), this temperature range implies that mean annual precipitation in Coire Beanaidh cannot have exceeded 375–550 mm. This estimate supports Sissons' suggestion that the Cairngorms (and thus by implication all of NE Scotland) experienced marked aridity during the Loch Lomond Stadial, and suggests that if anything his estimate of 500–600 mm yr[-1] for 1000 m altitude in the Cairngorms is slightly generous. On Jura, it is likely that MAAT was slightly higher than in NE Scotland, in view of its more southerly latitude and greater exposure to maritime influences. If we assume a MAAT range of -5 °C to -7 °C and a similar range of altitudinal temperature gradients to that employed above, the predicted MAAT for 350 m (the altitude of the Beinn Shiantaidh rock glacier) falls within the range -6.7 °C to -9.8 °C. In terms of Haeberli's model, this implies that mean annual precipitation at this altitude on Jura did not exceed 800–1250 mm.

Despite the uncertainties involved in the above calculations, they appear to confirm in a general way that there was a very steep south-west to north-east precipitation gradient across the Scottish Highlands during the Loch Lomond Stadial. Though the differences in altitude between the two rock glacier sites make direct comparison of precipitation values invalid, it is notable that the Jura estimates imply that stadial precipitation at 350 m on the island was approximately 50–78% of the present precipitation, whereas the Coire Beanaidh estimates imply only 19–27% of present precipitation. Thus whilst the calculations support the general pattern of precipitation decline advocated by Sissons (1979c, 1980b), they also suggest that he may have overestimated stadial precipitation, particularly in assuming that in the mountains of the SW Grampians it was similar to that of the present day. It would be wrong, however, to regard either set of estimates as sacrosanct, not only because of the assumptions made in their calculation, but also because stadial precipitation is likely to have fluctuated in response to changes in the average position of the atmospheric polar front. Given that the dominant airflow was from the west or south-west, the most obvious cause of the very steep eastwards or north-eastwards decline in precipitation during the stadial was the scavenging effect of the extensive icefield that occupied the Western Highlands at this time. Cooling of moist air that was forced to rise over the icefield must have caused heavy precipitation in the Western Highlands, whilst locations in the lee of the icefield experienced marked aridity. An analogous situation exists at present in south-central Norway, where so much snow is lost to the Jostedals Ice Cap (Jostedalsbreen) and its satellites that the higher Jotunheim massif to the east supports only relatively small glaciers, and low ground in the lee of Jotunheimen in places receives less than 400 mm precipitation yearly.

The periglacial environment of the Loch Lomond Stadial: discussion

As with the Dimlington Stadial, it would be misleading to present the climate of the Loch Lomond Stadial as uniform. In general, the palaeoclimatic information provided by most of the periglacial phenomena that developed in Britain during the stadial may be assumed to relate to conditions at or around the time that the glaciers of the Loch Lomond Readvance reached their maximum extent. This appears to have occurred sometime after (possibly shortly after) 10.5 ka BP (Rose *et al.*, 1988; Peacock *et al.*, 1989). It is now known, however, that the initial stages of glacier retreat were marked by numerous glacier stillstands and minor readvances. This initial retreat phase appears to have been prolonged, and has been interpreted as representing a decline in snowfall during the latter part of the stadial rather than any marked thermal improvement (Benn *et al.*, 1992). Northwards movement of the oceanic polar front resulted in a return of warm waters to the shores of Britain after *c.* 10.1 ka BP (Peacock & Harkness, 1990), an extremely rapid increase in air temperatures (Atkinson *et al.*, 1987) and consequently uninterrupted glacier retreat with localised instances of ice stagnation (Benn *et al.*, 1992). Inferences regarding the periglacial climate of the stadial therefore probably relate to the period of most severe conditions, which lasted from some time before 10.5 ka BP until around 10.1 ka BP.

The Loch Lomond Stadial appears to have been dominated during the period of glacier expansion by a vigorous westerly atmospheric circulation. This is evident not only from the marked concentration of glacier ice in the Western Highlands of Scotland, but also from the eastwards transport of windblown sand in eastern England. The eastwards and north-eastwards rise in the average altitudes of glacier equilibrium lines and pro-talus ramparts across the Scottish Highlands and Hebrides also provides evidence that most precipitation was carried across northern Britain on westerly or south-westerly airstreams, though the heaviest snowfalls were probably associated with southerly or south-easterly winds in advance of eastwards-tracking warm or occluded fronts. There is some evidence that precipitation in western Scotland may have been slightly lower than at present, and much stronger evidence that NE Scotland, which lay in the lee of the Grampian icefield, experienced marked aridity. In the absence of conclusive evidence for ice-wedge development during the stadial, palaeotemperature esti-mates based on contemporaneous periglacial phenomena rest mainly on the evidence provided by dated ground-ice depres-sions in East Anglia and probable active layer detachment slides in central and southern England. These suggest that MAAT was of the order of -5 °C in central England during the coldest part of the stadial, an estimate identical to that reached on the basis of beetle assemblages. A corresponding MAAT of -8 °C is inferred for NE Scotland. However, owing to the uncer-tainties and poor resolution of radiocarbon dating, we cannot at present be certain that such temperatures necessarily coincided with the strong westerly airflow adduced from other glacial and periglacial evidence. It is possible that the thermal minimum of the stadial was achieved *after* the main period of glacier expan-sion, possibly in response to a brief re-establishment of high pressure over NW Europe. This would in turn have led to a greater dominance of easterly winds, drier conditions and ini-tially slow glacier retreat as a result of diminished snowfall in the final centuries of the stadial. Pending further evidence, how-ever, this proposal must be regarded as conjectural.

The present-day periglacial environment of upland Britain

The rapid increase in temperatures at the beginning of the Flandrian (present) Interglacial effectively terminated periglacial activity in lowland Britain, but on the higher moun-tains of Britain a restricted range of periglacial processes con-tinued to operate. Unfortunately, our knowledge of the nature of geomorphic activity on British mountains during much of the Flandrian is extremely scant, and restricted to a small number of upland sites where there is dated evidence for solifluction activ-ity (Sugden, 1971; Mottershead, 1978), nive-oaeolian deposi-tion (Ballantyne & Whittington, 1987), colluviation (Innes, 1983c) and accumulation of debris cones (Harvey *et al.*, 1981; Harvey & Renwick, 1987; Brazier *et al.*, 1988). A number of studies, however, have suggested that erosion of upland areas has accelerated within the last few centuries. Such erosion has taken the form of stripping of soil cover from high plateaux by strong winds (Ball & Goodier, 1974; Pye & Paine, 1983; Ballantyne & Whittington, 1987), accelerated solifluction (Ballantyne, 1986d) and a massive increase in debris flow activ-ity (Innes, 1983a; Brazier & Ballantyne, 1989). The causes of such accelerated activity are open to debate. Some have sug-gested that climatic deterioration and the increased frequency and severity of storms during the Little Ice Age of the sixteenth to nineteenth centuries AD may have been responsible. Others, however, have placed the blame on human interference, partic-ularly burning of vegetation cover and overgrazing by sheep

introduced to upland pastures (Ballantyne, 1991b,c). Irrespective of cause, one consequence of such activity has been the exposure of wide tracts of upland regolith to superficial frost action, and the gullying and consequent destabilisation of the debris cover on some steep mountain slopes. It follows that though the range of periglacial processes active at present in upland Britain is probably representative of those that have operated throughout the Flandrian, they may be rather more effective and more widespread than has been the case during most of this period.

The nature of the present climate of upland Britain has been outlined in Chapter 2, and need only be summarised here. In general, it is characterised by extreme wetness and strong winds rather than extreme cold. Even on the highest summits, MAATs are slightly above 0°C and temperatures below -10 °C are infre-quent. Permafrost is consequently absent, though annual ground freezing on high ground occasionally reaches depths of 0.5 m or more. Above 600 m, all but the most easterly summits experience a mean annual precipitation of over 2000 mm, with some westerly summits receiving over 4000 mm. Average snow-lie (>50% cover) generally exceeds 100 days at 600 m in the Scottish Highlands, but is less in the mountains of England and Wales. The highest parts of all British uplands experience frequent strong winds, with gusts exceeding 100 km h[-1] and occasionally (on the very highest ground) 200 km h[-1]. This wet and windy *maritime periglacial* regime is very different from those of the Loch Lomond Stadial or the Dimlington Stadial, so it is not surprising that the range of periglacial features now active in upland Britain (Figure 14.7) has little in common with those that were active on the same mountains during the Late Devensian (Figure 14.5). Three sets of periglacial processes are currently operative: those related to frost action and nivation, those resulting from aeolian activity and those affecting talus and other steep debris-mantled slopes.

Frost action and nivation processes

The operation of current freeze–thaw activity on the upper slopes of British mountains is manifest in three ways: rock breakdown, the development of frost-sorted patterned ground and downslope movement of debris by solifluction. All three, however, operate on a much reduced scale to that of similar effects during the Late Devensian. The absence of *in situ* frost-weathered detritus inside the limits of the Loch Lomond Readvance indicates that macrogelivation has been largely inoperative since the termination of severe periglacial condi-tions at the end of the Loch Lomond Stadial. Some researchers, such as Godard (1959, 1965) and Hills (1969) have suggested that frost splitting may operate on certain lithologies under pre-sent conditions, but there is only localised evidence for this. Plateau frost debris often supports a cover of moss, soil or peat that betrays its relict status. Granular disintegration and flaking of rock are certainly active under present conditions, but their effects are highly discriminatory with respect to lithology. In terms of roundness of exposed surfaces and abundance of recent granular detritus, some sandstones and granites appear most susceptible to microgelivation (King, 1968; Ballantyne, 1981). Current granular disintegration of such rocks is also evi-dent in the form of thin accumulations of windblown sand on the surfaces of ablating snowbeds located downwind from

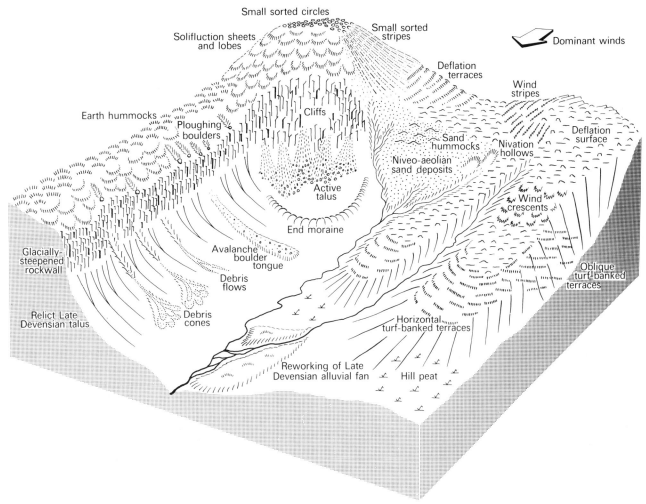

Figure 14.7 Schematic representation of the range of active periglacial phenomena on British mountains. After Ballantyne (1987a).

unvegetated plateaux (Ballantyne & Whittington, 1987). In contrast, fine-grained igneous and metamorphic rocks such as basalt and quartzite exhibit limited rounding and are apparently resistant to granular disaggregation under the present relatively mild climatic regime. Freshly exposed surfaces on rocks with strong foliation or cleavage (such as slate and mica-schist) suggest that these are currently subject to occasional flaking. The marked contrast that exists between exposed and buried rock and clast surfaces indicates that the current effects of granular disaggregation are restricted to exposed rock. This implies that the current effects of frost weathering are negligible on high ground mantled by a cover of soil or peat.

Small but active frost-sorted patterned ground occurs on unvegetated regolith in virtually all parts of upland Britain, including Snowdonia (Tallis & Kershaw, 1959; Ball & Goodier, 1970), the Lake District (Hollingworth, 1934; Hay, 1936, 1943; Caine, 1963b, 1972; Warburton, 1987), the Pennines (Tufnell, 1969), southern Scotland (Miller *et al.*, 1954), the Hebrides (Godard, 1958, 1965; Ryder & McCann, 1971; Ballantyne, 1991a) and various parts of the Scottish Highlands (Simpson, 1932; Kelletat, 1970a). Several common characteristics are evident from the accounts cited above. Active sorted patterns rarely exceed 0.6 m in width and sorting

is usually confined to the uppermost 0.2 m of the soil. The maximum size of sorted clasts is invariably reported as being 15 cm or less. These data suggest that shallow ground freezing and thawing limits the maximum depth of sorting, which in turn dictates the dimensions of the resultant patterns and the maximum size of sorted clasts. If this interpretation is correct, it follows that clast size imposes a severe constraint on the distribution of active sorted patterns in upland Britain. Such patterns also appear to be restricted to areas of frost-susceptible soil, and consequently are common only on lithologies (such as lavas, shales and mica-schists) that have yielded an abundance of fine-grained soil and small clasts in response to past weathering. A final crucial limitation on pattern occurrence is vegetation cover. Active patterning occurs only on bare ground, and on many mountains this constraint restricts its distribution to areas where vegetation cover has been stripped by deflation. Remarkably, climate seems to impose little constraint on altitude of occurrence: sorted stripes occur on till bluffs at 230 m on the Isle of Skye (Ballantyne, 1991a) and sorted nets have been observed on unvegetated ground near sea level in Shetland (Spence, 1957). Pattern formation, moreover, may occur rapidly over a few winters (Miller *et al.*, 1954), though more fragile forms may experience periodic destruction by wind, rain and surface wash (Tallis & Kershaw, 1959). On slopes, surface clasts associated with sorted stripes are known to move rapidly downslope at rates averaging up to 300 mm yr^{-1} (Caine, 1963a; Ballantyne, 1987a) as a result of the frequent growth and collapse of needle ice crystals, though subsurface movement is much slower. The status of possibly active nonsorted patterns in the form of earth hummocks or thufur is much less certain. Researchers working in northern England have proposed that earth hummocks may be actively developing there, on the basis that they occur on ground that has probably experienced recent disturbance (Tufnell, 1975; Pemberton, 1980), but a periglacial origin for the forms described by these authors has not been convincingly established. Hummocks and associated relief stripes on some Scottish mountains are underlain by mature podzols, and hence are certainly relict features, probably of Loch Lomond Stadial age (Ballantyne, 1986c).

Although many of the solifluction sheets and lobes on British mountains are relicts of formerly more severe periglacial conditions, active solifluction forms are widespread above 550 m in the Scottish Highlands and at rather higher altitudes on more southerly mountains (Tufnell, 1969; Ball & Goodier, 1970; Kelletat, 1970a). Such active forms differ from their relict counterparts in four ways. First, they are generally smaller, with risers only 0.2–1.0 m high. Secondly, they occasionally occur in areas that were occupied by glacier ice during the Loch Lomond Stadial. Thirdly, whereas the risers of relict features are usually degraded, those of active lobes are steep (generally >45 °) and often bulge downslope. Finally, active lobes often overlie organic soil horizons or peat fragments containing Flandrian pollen grains (White & Mottershead, 1972). Radiocarbon dating of organic matter buried by lobe advance indicates at least intermittent downslope movement during the past 5500 years (Sugden, 1971; Mottershead, 1978; Ballantyne, 1986d), but recorded rates of current movement are rather low. On the Drumochter Hills, Chattopadhyay (1982) recorded 50 mm of surface displacement over three winters, declining to zero at 20 cm depth, and on Ben Wyvis the maximum recorded surface displacement on a steep slope was 18 mm yr^{-1} (Ballantyne, 1981). Such rates of surface movement are modest in comparison with those recorded in other periglacial environments (cf. Benedict, 1976; Harris, 1987b), and may be attributable to the operation of near-surface frost creep alone. Commonly associated with active solifluction sheets and lobes are active ploughing boulders. These occur down to altitudes of 450 m on many British uplands (Kelletat, 1970a; Tufnell, 1972), and represent perhaps the most widespread indicator of current periglacial activity on British hills, though they appear to be restricted to frost-susceptible regolith. Recent boulder movement is indicated by the presence of a deep niche a few centimetres wide that separates such boulders from the furrows upslope. Average rates of current boulder movement range from 0.4 to 64 mm yr^{-1} on the Pennines (Tufnell, 1976), 0.3 to 8.7 mm yr^{-1} on Lochnagar (Shaw, 1977), 6 to 34.5 mm yr^{-1} on the Fannich Mountains (Ballantyne, 1981) and 1.5 to 7.0 mm yr^{-1} on the Drumochter Hills (Chattopadhyay, 1986).

The current effectiveness of nivation processes on British mountains is debatable. Researchers working on high ground in northern England have viewed nivation as an important element of current landscape evolution (Tufnell, 1971; Vincent & Lee, 1982), but this conclusion remains unsubstantiated by detailed measurements, and the nivation benches and hollows identified in this area may well be Devensian relicts that have experienced no more than superficial modification during the Flandrian (Tufnell, 1985). Observations made at snowpatch sites in the northern Highlands of Scotland (mainly nivation hollows formed in niveo-aeolian sand deposits) have indicated that no increase in freeze–thaw activity, chemical weathering or mass movement is associated with the presence of late-lying snow, and showed that though snowcreep occurs it is of negligible importance as a transportation agent (Ballantyne, 1985). Snowpatch-related activity at the sites studied appears to be limited to the localised redistribution of sandy sediments by surface wash across unvegetated terrain. Subsequent research in the Cairngorms has nevertheless suggested that the presence of late-lying snow has enhanced clast weathering in this area (Ballantyne et al., 1989), though the nature of such enhanced weathering remains uncertain.

Wind action on British mountains

On some exposed plateaux, particularly those mantled by coarse sandy regolith, erosion by strong winds has created extensive deflation surfaces. The most outstanding examples occupy the high granite plateaux of the Cairngorms and Lochnagar, and the Torridon Sandstone ridges of NW Scotland, but deflation surfaces occur on almost all lithologies and are developed as low as 400 m on the windswept hills of Orkney and Shetland. Many deflation surfaces appear to have developed through destruction of a pre-existing vegetation cover and winnowing away of the underlying soil. On some granite and sandstone mountains, 'islands' of vegetated sand deposits stand proud of wind-scoured plateaux, demonstrating that sand cover was once more extensive (Ball & Goodier, 1974; Goodier & Ball, 1975; Ballantyne & Whittington, 1987). Similarly, isolated remnants of vegetated soil with eroding margins occur on deflation surfaces elsewhere (Kelletat, 1970a; Birse, 1980) and indicate that such surfaces were once mantled by a stable vegetated soil cover. The erosion of mountain-top soils or sand deposits is attributable to loosening of exposed fine sediment by needle ice and frost heave, and removal of loose grains by strong winds (King, 1971a). Surface wash and trampling by sheep play subsidiary roles (Ballantyne, 1985). Several studies suggest that the stripping of vegetated covers of soil or sand from exposed plateaux is a fairly recent phenomenon (Ball & Goodier, 1974; Pye & Paine, 1983). On An Teallach in NW Scotland, for example, up to 2.3 m of fresh, unweathered sand has been deposited on top of much older niveo-aeolian deposits on lee slopes, indicating massive recent erosion of plateau surface sand upwind, and consequent progressive exposure of the present deflation surface (Ballantyne & Whittington, 1987). It is widely acknowledged that such accelerated recent erosion results from the breaking of a stable protective vegetation cover; what remains in doubt is the timing and cause of such vegetation destruction. Ball & Goodier (1974) proposed that the opening up of vegetation cover on Ronas Hill in Shetland occurred under the relatively severe conditions of the Little Ice Age of the sixteenth to nineteenth centuries A.D., but others have suggested that overgrazing by sheep introduced to mountain areas in the eighteenth or nineteenth centuries may equally be held responsible (Pye & Paine, 1983; Ballantyne & Whittington, 1987).

On some mountains, soil eroded from cols or plateaux has been redeposited as vegetated sheets of windblown sand on lee slopes. Research on the sandstone massif of An Teallach has shown that these are currently accumulating through sand blown from plateaux upwind becoming trapped in the winter snowpack at sheltered lee sites. As the snow melts, the sand is lowered onto the underlying vegetation, which grows through the accumulating deposit and stabilises it. Thufur-like 'sand hummocks' are sometimes developed on the surface of these niveo-aeolian deposits as a result of concentrated slopewash between vegetation tussocks (Ballantyne, 1986c). Radiocarbon dating and pollen analysis of organic-rich horizons within the deposits on An Teallach have demonstrated that sand has been accumulating in this way since the early Flandrian, reaching thicknesses of up to 4 m, though as noted above there has been a massive increase in recent accumulation due to erosion of pre-existing deposits on the plateau surface upwind. Present accumulation declines with distance downwind from the plateau source area and ranges from 10 to 300 g m^{-2} yr^{-1}, but is balanced to some extent by retreat of the unvegetated margins of the sand sheets at rates averaging 25–30 mm yr^{-1} (Ballantyne & Whittington, 1987).

Commonly associated with deflation surfaces are three forms of wind-patterned ground: deflation scars (patches of bare ground on vegetated terrain), wind stripes (straight or wavy lines of vegetation that alternate regularly with unvegetated soil) and wind crescents (arcuate patterns of vegetation scattered across otherwise unvegetated deflation surfaces). Deflation scars result from selective opening up of vegetation cover, and tend to migrate downwind (King, 1971a). Wind stripes are aligned normal to dominant wind direction, and also migrate downwind as vegetation colonises the sheltered lee side of each stripe whilst older stems die away on the windward side (Watt, 1947; Metcalfe, 1950), though the origin of the regular patterning remains uncertain (Bayfield, 1984). Equally enigmatic is the origin of the related slope forms known as turf-banked terraces, which have steep vegetated risers and sparsely vegetated treads. Such forms are common throughout upland Britain (e.g. Tufnell, 1969; Ball & Goodier, 1970, 1974; Kelletat, 1970a; Goodier & Ball, 1975; Veyret & Coque-Delhuille, 1989). The attitude and form of turf-banked terraces is clearly a response to dominant wind direction. Three types may be differentiated: horizontal deflation terraces aligned parallel to dominant wind direction and apparently formed by creep operating on the treads of deflation scars; other horizontal terraces on lee slopes, possibly formed by the accumulation of creeping debris behind 'dams' of vegetation (Hollingworth, 1934); and oblique terraces that dip steeply upwind on otherwise vegetation-free slopes and which appear to reflect the progressive anchoring of creeping debris by vegetation growing in the lee of large boulders. Current movement of debris on terrace treads and around the relatively immobile risers seems to be dominated by frost creep, which averages a few millimetres per year at the surface and declines rapidly to zero at shallow depth (Ballantyne, 1981). Gelifluction may also be partly responsible for terrace movement on frost-susceptible regolith, and it is notable that in some locations large turf-banked terraces merge imperceptibly into solifluction sheets.

Rockfall, talus, debris flows and avalanche activity

Most talus slopes in upland Britain, particularly those located outside the limits of the Loch Lomond Readvance, are largely or entirely vegetated and have been extensively modified by gullying, debris flows and shallow landslips. These features suggest that such taluses are essentially relict Lateglacial landforms on which erosion has replaced accumulation as the dominant mode of geomorphic activity. This conclusion is supported by calculations indicating that rockwall retreat rates during the Loch Lomond Stadial were two orders of magnitude greater than those associated with relict talus at present (Ballantyne & Kirkbride, 1987). At upland sites in Scotland and Snowdonia the latter average only c. 0.015 mm yr^{-1} (Ballantyne & Eckford, 1984; Stuart, 1984). Inside the limits of the Loch Lomond Readvance, however, talus cones that must have developed entirely during the Flandrian often present a fresher appearance. Such cones are commonly overlooked by high cliffs, and fresh debris is often strewn over their sparsely-vegetated slopes, indicating that such localities currently experience much more frequent rockfall than do relict taluses outside the readvance limit. This conclusion is supported by the observations of Ballantyne & Eckford (1984), who recorded an average frequency of two rockfalls per three hours of observation in a corrie on An Teallach. Although these observations were of limited duration and gave no information on the magnitude of individual falls, they suggest that talus accumulation may still be appreciable at favourable sites within the limits of the Loch Lomond Readvance.

The role of recent snow avalanches in modifying talus slopes has been very limited. Even in Scotland, where mountain snows are most persistent, the great majority of avalanches are rather small, with runout lengths of less than 250 m. Most are cornice falls, sluffs or shallow slab avalanches; full-depth avalanches and airborne avalanches are rare (Ward, 1980, 1984). Irrespective of magnitude, few avalanches contain significant amounts of debris, so that even in areas where avalanches occur frequently, they are currently of little importance as agents of talus modification and debris transport (Ward, 1985; Davison & Davison, 1987). At a small number of high level sites, however, the geomorphic effect of recent avalanches is more prominent. In the Lairig Ghru (Cairngorms) a number of active avalanche boulder tongues extend across the valley. In Coire na Ciste, on Ben Nevis, avalanches have lowered the gradient of talus slopes and deposited an apron of fresh debris across the corrie floor (Ballantyne, 1989b).

Of much more widespread importance as agents of current mass transport are debris flows. Hillslope flows occur on steep (>30°) slopes throughout upland Britain, but tend to be most abundant on debris with a coarse sandy matrix. Conversely, they are rare in areas of glacial scouring where debris flow activity is limited by the lack of sediment on steep slopes (Innes, 1983a; Ballantyne, 1986a). Debris flows in Britain are generally triggered by intense rainstorms (Baird & Lewis, 1957), and though the frequency of flow events varies greatly, it may under favourable circumstances be remarkably high. Innes (1983a), for example, has demonstrated that over 70 hillslope flows were mobilised in the Lairig Ghru between 1970 and 1980. The great majority of such flows carry less than 30 m^3 of sediment (Innes, 1985), but individual valley-confined flows may transport much greater volumes of debris (Brazier & Ballantyne, 1989). Debris cones that have accumulated at the foot of gullies or steep tributary valleys as a result of repeated debris flow activity are common along the flanks of glacial troughs. Most are relict forms that accumulated during the Lateglacial or as a result of paraglacial sedimentation in the early Flandrian (Brazier et al., 1988), but some were deposited much more recently or are currently subject to renewed deposition or reworking. Debris cones in NW England, for example, are known to have experienced renewed accumulation as a result of vegetation changes induced by human activity at the time of Viking settlement expansion in the tenth century AD (Harvey et al., 1981; Harvey & Renwick, 1987). Others in upper Glen Feshie appear to have been deposited mainly within the last 400 years, and represent an average annual accumulation of 50–60 m³ of sediment during the last three centuries (Brazier & Ballantyne, 1989). Even more convincing evidence for a recent intensification of debris flow activity in the Scottish Highlands emerges from a study by Innes (1983a), who demonstrated on the basis of lichenometric measurements that the great majority of flow tracks represent deposition within the last 250 years. Although his findings are open to dispute on the grounds that later flow tracks often obscure earlier ones, it appears incontestable that a drastic increase in debris flow activity has occurred within the past few centuries; had the rates identified by Innes been sustained throughout the Flandrian, much greater thicknesses of debris flow deposits would have accumulated than are now evident.

The maritime periglacial environment of upland Britain

The range of periglacial processes and landforms active on the mountains of Great Britain is strongly determined by the nature of the present climatic regime, which is dominated by strong winds and extreme wetness rather than severe cold or deep ground freezing. The comparative mildness of this regime limits mechanical weathering to superficial disaggregation and flaking of exposed rock and clast surfaces. Because of limited frost penetration, active sorted patterns are small and superficial, and solifluction is shallow and generally intermittent. Nivation processes, talus accumulation and transport of debris by snow avalanches are restricted to particularly favourable sites. Conversely, present conditions favour widespread ploughing boulder movement and debris flow activity, and the strong winds that sweep across British mountains have promoted the development of a rich variety of wind-related features, including deflation surfaces, niveo-aeolian sand deposits, and several types of wind-patterned ground and turf-banked terrace.

In view of the absence of permafrost in upland Britain and the relatively limited role of frost-action processes, some geomorphologists have expressed doubts regarding both the current effectiveness of periglacial activity on British mountains and the 'periglacial' status of the present upland environment. Worsley (1977, p. 217), for example, has asserted that 'despite some devotees who emphasise the amount of current periglaciation, there is no way of escaping the fact that approximately 10 000 yr BP truly periglacial environments within the British Isles were abruptly terminated and have not since returned'. The validity of this standpoint depends, of course, on what is

regarded as a 'truly periglacial environment', a consideration that in the past has engendered much heated and ultimately sterile debate. The view adopted here is that the present environment of the higher parts of British mountains satisfies the definition of 'periglacial' advocated by Harris et al. (1988) and cited on the first page of Chapter 1. What is indisputable, however, is that even the highest and coldest parts of Great Britain are climatically on the very margins of the wider periglacial realm. Consequently, the range of periglacial phenomena presently active in upland Britain has little in common with those of high arctic lowlands or mid-latitude alpine environments, but constitutes a distinct maritime type.

Despite the relative mildness of the present periglacial environment on British mountains, certain mass-transport processes appear to operate at least as effectively as in mountain areas that experience a more severe climatic regime. Some attempt has been made in the account above to present representative figures for various types of mass transport, but the measurements cited are given in different units, relate to different areas and hence do not permit direct comparison of the effectiveness of different processes. In Britain, this has been attempted for only one mountain, the Torridon Sandstone massif of An Teallach (1062 m) in NW Scotland. For comparative purposes, measured sediment transport rates on An Teallach were converted by Ballantyne (1981) to a common unit, watts per square kilometre. This unit represents potential energy loss associated with the downslope transfer of sediment over a uniform time interval and standard area of 1 km² (Caine, 1976). The results (Table 14.2) must be treated with some caution, as An Teallach is a fairly high, steep mountain exposed to high precipitation (c. 3600 mm yr⁻¹ at 670 m) and very strong winds that have stripped upper slopes of all but a patchy vegetation cover. The rates cited are therefore likely to be greater than for most British mountains. Comparison of the data highlights three features: (1) the greater effectiveness of rapid mass-transport processes (rockfall and debris flow) compared with frost creep and gelifluction, though admittedly the last-mentioned is of little importance on An Teallach because of the coarseness of the regolith matrix; (2) the effectiveness of aeolian transport, though this certainly represents an optimum case in view of the abundance of readily-entrainable loose sand on exposed plateau surfaces; and (3) the importance of solute transport, though the figure cited includes atmospheric inputs and hence is certainly an overestimate.

Of perhaps greater interest is the fact that for all process sets except avalanches, the An Teallach rates are of the same order of magnitude as (and sometimes exceed) those measured by Rapp (1960a) in Karkevagge, a high mountain area in northern Sweden, and also those cited by Caine (1976) for the San Juan Mountains in Colorado. It is also notable that the relative effectiveness of different agents (i.e. solute transport > rapid mass transport > slow mass movement) is similar for An Teallach and Karkevagge. In general the figures cited indicate that, admittedly under favourable circumstances, rates of mass-transport under the maritime periglacial conditions of upland Britain are not dissimilar to (and in some cases exceed) those on mountains subject to a more extreme periglacial climate. Many more data are required, however, to substantiate the relative importance of of different forms of upland mass-transport activity on a country-wide basis.

Table 14.2: *Comparative rates of mass-transport activity in three mountain areas[a].*

Process set	An Teallach, NW Scotland (Ballantyne, 1981)	Karkevagge, N Sweden (Rapp, 1960a)	San Juan Mts, Colorado (Caine, 1976)
Rockfall	1.689	0.405	0.133
Avalanches	slight	0.453	0.001
Landslides and debris flows	0.560	1.997	0.362
Frost creep and gelifluction	0.175	0.166	0.109
Surface wash	slight	slight	0.147
Wind action	0.394		
Solute transport[b]	9.017	2.828	0.104

[a] All values are expressed in watts per square kilometre.

[b] Probably a considerable overestimate, as it includes atmospheric inputs.

Themes and prospects

Throughout this book an attempt has been made not only to consolidate our understanding of the periglacial phenomena of Great Britain, but also to highlight areas of uncertainty and potentially rewarding themes for further investigation. It is appropriate that we conclude by considering some of the prospects and challenges for future periglacial research in Britain. These are subsumed under two general aims that underpin most previous work and seem likely to dominate research during the foreseeable future. The first is increased understanding of periglacial phenomena, particularly their characteristics, mode of formation and environmental or palaeoenvironmental significance. This aim extends equally to active phenomena and those now represented only in relict form. The second aim is employment of periglacial phenomena as an aid to reconstruction of Quaternary stratigraphies and palaeoenvironments, and is to some extent dependent on our success in satisfying the first.

One route to a fuller understanding of the formation and significance of certain periglacial phenomena is increased use of the mountains of Britain as a 'laboratory' for process studies. Perhaps in response to some atavistic colonial instinct, British periglacialists have often travelled to far-flung periglacial environments to investigate periglacial phenomena that are currently active on high ground in Britain. Such long-range investigations, however, are frequently undermined by a lack of observations outside of the summer field season. Use of high-level sites in Britain for process experiments offers a much more accessible alternative, though one that has barely been explored. The most promising topics for investigation in this way are small-scale frost sorting and patterned ground formation, solifluction of seasonally-frozen ground, ploughing boulder movement, nivation processes (particularly the influence of late-lying snow on clast weathering), solution rates and processes and wind erosion. Even such basic information as rates of current mass transport has been measured at only a few upland sites to date, and hence further studies would be of value in obtaining a representative picture of present activity. It is also worth noting that the genesis of certain upland phenomena, both relict and active, is very imperfectly understood. Examples include boulder lobes, nonsorted patterned ground (earth hummocks and nonsorted relief stripes), wind-patterned ground and turf-banked terraces; elucidatation of the origins and implica-

tions of these features offers an interesting opportunity for geomorphologists who are not daunted by the rigours of the maritime periglacial climate. Moreover, the possibilities of achieving a fuller understanding of periglacial processes are not limited to the upland zone. Studies of the sedimentology and geotechnical properties of head deposits, for example, have already thrown much useful light on the ways in which former movement occurred, and a further challenge exists in the differentiation of head from paraglacial sediment-gravity flow deposits that accumulated during retreat of the last (and earlier) ice sheets. Similarly, studies of cryoturbation structures in Britain have so far been restricted largely to the palaeoenvironmental interpretation of such forms. The abundance of these in Devensian drifts offers an opportunity for a much more detailed analysis of the mode of formation of different types.

The wide range of rock types that underlie both the uplands and lowlands of Great Britain presents an unrivalled but underexploited opportunity for examining the effects of lithology in constraining the development of various periglacial phenomena. Though the nature of lithological control has been outlined in general terms (e.g. Ballantyne, 1984, 1987a; Harris 1987a), the links between lithology, regolith characteristics, periglacial processes and resulting landforms have not been rigorously established. In particular, analysis of the different ways in which rocks have responded to Late Devensian periglacial weathering could usefully be used to complement laboratory studies of rock breakdown, and laboratory experiments such as that pioneered by Gallop (1991) appear to have great potential for establishing under controlled conditions the links that exist between regolith type and the distribution of particular periglacial phenomena.

The major challenge that confronts periglacial research in Britain, however, is unquestionably the use of periglacial phenomena for the reconstruction of Quaternary stratigraphies and palaeoenvironments. Although some very useful work has already been carried out in this respect (e.g. Williams, 1975; Rose *et al.,* 1985b; Worsley, 1987), it is notable that in many site-specific studies only the most obvious indicators of former periglacial conditions (such as ice-wedge casts and cryoturbations) are employed as diagnostic criteria. Less obvious pedological indicators of frozen ground such as fragipans, platy or lenticular soil structures, cutans and vesicles have been reported in only a small minority of studies. This appears to be an area that could be fruitfully developed by Quaternary stratigraphers.

A number of particular research problems await resolution: we may identify, for instance, establishment of the cryostratigraphy of the interval between the Hoxnian and Ipswichian Interglacials and that of the Early and Middle Devensian as particular priorities, though of course progress in such areas is likely to be at least partly serendipitous in that it is dependent on the discovery of representative stratotypes and other key sites at which periglacial structures are well preserved.

Reconstruction of Late Devensian palaeoenvironments on the basis of periglacial evidence also poses a number of challenges. By far the most urgent of these is probably establishment of better dating control. It is now apparent that climatic variation during the Dimlington Stadial and Loch Lomond Stadial may have been considerable, so that periglacial phenomena once believed to be essentially coeval may have developed at very different times (and under different climatic circumstances) within a single cold period. Again, however, despite improved dating techniques (particularly thermoluminescence dating and ^{14}C dating of small samples of organic material using accelerator mass spectrometry), resolution of the ages of climatically-diagnostic periglacial phenomena within individual stadial episodes will ultimately depend on the fortuitous discovery of appropriate sites. This does not mean that positive steps cannot be taken to force the pace of discovery. Uncertainties concerning the age of former pingos in Wales, for example, may be resolved by trenching entire pingo scars using a mechanical excavator. Similarly, those concerning the status and distribution of permafrost during the Loch Lomond Stadial may be resolvable by searching for fragipans in contemporaneous glacial and fluvioglacial sediments, or through systematic monitoring and detailed reinvestigation of critical sites (such as Glanllynau in north Wales or Glen Ballyre on the Isle of Man) where stadial deposits are exposed in section above those of Windermere Interstadial age. The gradual development of tephrochronology in Britain may also eventually offer a solution to the vexed problem of differentiating supraformational ice-wedge casts of Loch Lomond Stadial age from those that developed in the later stages of the Dimlington Stadial.

Several upland features may also prove to have underexploited potential as palaeoenvironmental indicators. Earlier in this chapter we saw how protalus rock glaciers may be used to constrain palaeoprecipitation estimates for the Loch Lomond Stadial. This approach may be fruitfully extended to incorporate all known examples of such features, and by integrating the findings with data on the equilibrium line altitudes of Loch Lomond Readvance glaciers. Potentially great rewards may also accrue from the detailed mapping of high-level periglacial trimlines in mountain areas peripheral to the main accumulation zones of the last ice sheet. If such trimlines delimit nunataks that remained above the upper surface of the ice, it follows that they offer unique evidence for constraining or testing proposed ice-sheet models. A further possible priority for research in upland Britain concerns the implications of periglacial phenomena for establishing climatic variations during the Holocene. In Scandinavia and particularly the Swiss Alps, detailed chronologies identifying episodes of Holocene climatic deterioration have been established on the basis of radiocarbon-dated evidence for accelerated solifluction, avalanche deposition and debris-flow activity. There is great potential for the establishment of a similar record in upland Britain, but the data hitherto collected are fragmentary and consequently of limited value.

The observations and suggestions outlined above represent a number of research strategies that appear likely to yield particularly useful results in terms of our understanding of the periglacial phenomena of Great Britain and the environments in which these developed. Our list is by no means exhaustive, and other researchers will doubtless dispute our priorities and deprecate our omissions. If this volume has stimulated such disagreement, so much the better; in periglacial research, as in all fields of scientific endeavour, dissent is the natural mother of discovery.

References

Aartolahti, T. (1970). Fossil ice-wedges, tundra polygons and recent frost cracks in southern Finland. *Annales Academiae Scientiarum Fennicae, Series A, III. Geologica-Geographica*, 107, 5–26.

Ackerman, K.J. & Cave, R. (1967). Superficial deposits and structures, including landslips, in the Stroud district, Gloucestershire. *Proceedings of the Geologists' Association*, 78, 567–586.

Ackroyd, P. (1986). Debris transport by avalanche, Torlesse Range, New Zealand. *Zeitschrift für Geomorphologie*, 30, 1–14.

Acreman, M.C. (1983). The significance of the flood of September 1981 on the Ardessie Burn, Wester Ross. *Scottish Geographical Magazine*, 99, 150–160.

Addison, K. (1987). Debris flow during intense rainfall in North Wales: a preliminary survey. *Earth Surface Processes and Landforms*, 12, 561–566.

Akagawa, S. & Fukuda, M. (1991). Frost heave mechanism in welded tuff. *Permafrost and Periglacial Processes*, 2, 301–309.

Åkerman, H.J. (1984). Notes on talus slope morphology and processes in Spitsbergen. *Geografiska Annaler*, 66A, 267–284.

Åkerman, H.J. & Malmström, B. (1986). Permafrost mounds in the Abisko area, northern Sweden. *Geografiska Annaler*, 68A, 155–165.

Albers, G. (1930). Notes on tors and the clitter of Dartmoor. *Reports of the Devonshire Association for the Advancement of Science*, 62, 373–378.

Alexander, C.S. & Price, L.W. (1980). Radiocarbon dating and the rate of movement of two solifluction lobes in the Ruby Range, Yukon Territory. *Quaternary Research*, 13, 365–379.

Allard, M. & Tremblay, G. (1983). Les processus d'érosion littorale périglaciaire de la région de Port-de-la-Baleine et des îles Manitoumik sur la côte est de la mer d'Hudson, Canada. *Zeitschrift für Geomorphologie, Supplementband*, 47, 27–60.

Allen, J.R.L. (1984). Truncated fossil thermal contraction polygons (?Devensian) in the Mercia Mudstone Formation (Trias), Oldbury upon Severn, Gloucestershire. *Proceedings of the Geologists' Association*, 95, 263–273

Allen, J.R.L. (1987). Dimlington Stadial (Late Devensian) ice-wedge casts and involutions in the Severn Estuary, southwest Britain. *Geological Journal*, 22, 109–118

Allen, P. (1983). *Middle Pleistocene Stratigraphy and Landform Development of South-East Suffolk*. Unpublished PhD thesis, University of London.

Anderson, F.W. & Dunham, K.W. (1966). The Geology of Northern Skye. *Memoir of the Geological Survey of the United Kingdom*. Edinburgh: HMSO.

Andersson, J.G. (1906). Solifluction, a component of subaerial denudation. *Journal of Geology*, 14, 91–112.

André, M-F. (1985). Lichénométrie et vitesses d'évolution des versants arctiques pendant l'Holocene (Région de la Baie du Roi, Spitsberg, 79° N). *Révue de Géomorphologie Dynamique*, 1985 (2), 49–72.

André, M-F. (1986). Dating slope deposits and estimating rates of rockwall retreat in northwest Spitsbergen by lichenometry. *Geografiska Annaler*, 68A, 65–75.

Andrew, R. & West, R.G. (1977). Appendix: pollen analysis from Four Ashes, Worcs. *Philosophical Transactions of the Royal Society, London*, B 280, 242–6.

Andrews, J.T. (1961). The development of scree slopes in the English Lake District and central Quebec-Labrador. *Cahiers de Géographie de Québec*, 10, 219–230.

Andrews, J.T. (1981). Book review. *Arctic and Alpine Research*, 13, 113.

Anketell, J.M., Cegla, J. & Dzulynski, S. (1970). On the deformational structures in systems with reversed density gradients. *Rocznik Polskiego Towarzystwa Geologicznego*, 40L, 3–29.

Antevs, E. (1932). *Alpine Zone of Mt Washington Range*. Maine: Auburn Press.

Archer, D.R. (1981). Severe snowmelt runoff and its implications. *Proceedings of the Institution of Civil Engineers*, Part 2, 71, 1047–1060.

Arkell, W.J. (1947). *The Geology of Oxford*. Oxford: Clarendon Press.

Armstrong, M. & Patterson, I.B. (1985). Some recent discoveries of ice-wedge cast networks in north-east Scotland – a comment. *Scottish Journal of Geology*, 21, 107–108.

Arnborg, L., Walker, H.J. & Peippo, J. (1966). Water discharge of the Colville River, 1962. *Geografiska Annaler*, 48, 195–210.

Arnborg, L., Walker, H.J. & Peippo, J. (1967). Suspended load in the Colville River, Alaska, 1962. *Geografiska Annaler*, 49A, 131–144.

Atkinson, T.C., Briffa, K.R. & Coope, G.R. (1987). Seasonal temperatures in Britain during the past 22,000 years, reconstructed using beetle remains. *Nature*, 325, 587–592.

Atkinson, T.C., Briffa, K.R., Coope, G.R., Joachim, M.J. & Perry, D.W. (1986). Climatic calibration of coleopteran data. In *Handbook of Holocene Palaeoecology and Palaeohydrology*, ed. B.E. Berglund, pp. 851–858. Chichester: Wiley.

Austen, R.A.C. (1851). Superficial accumulations on the coasts of the English Channel. *Quarterly Journal of the Geological Society*, 7, 118–136.

Avebury, Lord (1902). *The Scenery of England*. London: Macmillan.

Babinski, Z. (1982). Pingo degradation in the Bayan–Nuurin–Khotnor Basin, Khangai Mountains, Mongolia. *Boreas*, 11, 291–300.

Bailey, E.B. & Maufe, H.B. (1916). The geology of Ben Nevis and Glencoe. *Memoirs of the Geological Survey: Scotland*. Edinburgh: HMSO.

Baird, P.D. & Lewis, W.V. (1957). The Cairngorm floods, 1956: summer solifluction and distributory formation. *Scottish Geographical Magazine*, 73, 91–100.

Baker, C.A. (1976). Late Devensian periglacial phenomena in the upper Cam Valley, north Essex. *Proceedings of the Geologists' Association*, 87, 285–306.

Baker, C.A., & Jones, D.K.C. (1980). Glaciation of the London Basin and its influence on the drainage pattern: a review and appraisal. In *The Shaping of Southern England*, Institute of British Geographers Special Publication, 11, pp. 131–175. London: Academic Press.

Bakkahøi, S. & Bandis, C. (1988). Meteorological conditions' influence on the permafrost ground in Sveagruva, Spitsbergen. In *Proceedings of the 5th International Conference on Permafrost, Trondheim*, pp. 39–43. Trondheim: Tapir Press.

Ball, D.F. (1960). *The Soils and Land Use of the District around Rhyl and Denbigh*. London: HMSO.

Ball, D.F. (1966). Late-glacial scree in Wales. *Biuletyn Peryglacjalny*, 15, 151–163.

Ball, D.F. (1976). Close-packed patterned arrangement of stones and shells on shore-line platforms. *Biuletyn Peryglacjalny*, 25, 5–7.

Ball, D.F. & Goodier, R. (1968). Large sorted stone-stripes in the Rhinog Mountains, North Wales. *Geografiska Annaler*, 50A, 54–59.

Ball, D.F. & Goodier, R. (1970). Morphology and distribution of features resulting from frost action in Snowdonia. *Field Studies*, 3, 193–217.

Ball, D.F. & Goodier, R. (1974). Ronas Hill, Shetland: a preliminary account of its ground pattern features resulting from the action of frost and wind. In *The Natural Environment of Shetland*, ed. R. Goodier, pp. 89–106. Edinburgh: Nature Conservancy Council.

Ballantyne, C.K. (1977). An Teallach. In *The Scottish Highlands: Excursion Guide for the 10th INQUA Congress*, ed. J.B. Sissons, pp. 35–40. Norwich: Geobooks.

Ballantyne, C.K. (1978). The hydrologic significance of nivation features in permafrost areas. *Geografiska Annaler*, 60A, 51–54.

Ballantyne, C.K. (1979a). Patterned ground on an active medial moraine, Jotunheimen, Norway. *Journal of Glaciology*, 22, 396–401.

Ballantyne, C.K. (1979b). A sequence of Lateglacial ice-dammed lakes in East Argyll. *Scottish Journal of Geology*, 15, 153–160.

Ballantyne, C.K. (1981). *Periglacial Landforms and Environments on Mountains in the Northern Highlands of Scotland*. Unpublished Ph.D. Thesis, University of Edinburgh, 2 volumes.

Ballantyne, C.K. (1982a). Aggregate clast form characteristics of deposits at the margins of four glaciers in the Jotunheimen Massif, Norway. *Norsk Geografisk Tidsskrift*, 36, 103–113.

Ballantyne, C.K. (1982b). Depths of open joints and the limits of former glaciers. *Scottish Journal of Geology*, 18, 250–252.

Ballantyne, C.K. (1983). Precipitation gradients in Wester Ross, North-West Scotland. *Weather*, 38, 379–387.

Ballantyne, C.K. (1984). The Late Devensian periglaciation of upland Scotland. *Quaternary Science Reviews*, 3, 311–343.

Ballantyne, C.K. (1985). Nivation landforms and snowpatch erosion on two massifs in the Northern Highlands of Scotland. *Scottish Geographical Magazine*, 101, 40–49.

Ballantyne, C.K. (1986a). Landslides and slope failures in Scotland: a review. *Scottish Geographical Magazine*, 102, 134–150.

Ballantyne, C.K. (1986b). Protalus rampart development and the limits of former glaciers in the vicinity of Baosbheinn, Wester Ross. *Scottish Journal of Geology*, 22, 13–25.

Ballantyne, C.K. (1986c). Nonsorted patterned ground on mountains in the Northern Highlands of Scotland. *Biuletyn Peryglacjalny*, 30, 15–34.

Ballantyne, C.K. (1986d). Late Flandrian solifluction on the Fannich Mountains, Ross-shire. *Scottish Journal of Geology*, 22, 395–406.

Ballantyne, C.K. (1987a). The present-day periglaciation of upland Britain. In *Periglacial Processes and Landforms in Britain and Ireland*, ed. J. Boardman, pp. 113–126. Cambridge: Cambridge University Press.

Ballantyne, C.K. (1987b). Some observations on the morphology and sedimentology of two active protalus ramparts, Lyngen, northern Norway. *Arctic and Alpine Research*, 19, 167–174.

Ballantyne, C.K. (1987c). Winter-talus ridges, nivation ridges, and protalus ramparts. *Journal of Glaciology*, 33, 246–247.

Ballantyne, C.K. (1987d). The Baosbheinn protalus rampart. In *Wester Ross: Field Guide*, ed. C.K. Ballantyne & D.G. Sutherland, pp. 167–171. Cambridge: Quaternary Research Association.

Ballantyne, C.K. (1987e). The Beinn Alligin 'rock glacier'. In *Wester Ross: Field Guide*, eds. C.K. Ballantyne & D.G. Sutherland, pp. 134–137. Cambridge: Quaternary Research Association.

Ballantyne, C.K. (1987f). An Teallach. In *Wester Ross: Field Guide*, eds. C.K. Ballantyne & D.G. Sutherland, pp. 71–92. Cambridge: Quaternary Research Association.

Ballantyne, C.K. (1989a). The Loch Lomond Readvance on the Isle of Skye, Scotland: glacier reconstruction and palaeoclimatic implications. *Journal of Quaternary Science*, 4, 95–108.

Ballantyne, C.K. (1989b). Avalanche impact landforms on Ben Nevis, Scotland. *Scottish Geographical Magazine*, 105, 38–42.

Ballantyne, C.K. (1990). The Late Quaternary glacial history of the Trotternish Escarpment, Isle of Skye, Scotland, and its implications for ice-sheet reconstruction. *Proceedings of the Geologists' Association*, 101, 171–186.

Ballantyne, C.K. (1991a). Periglacial features on the mountains of Skye. In *The Quaternary of the Isle of Skye*, eds. C.K. Ballantyne, D.I. Benn, J.J. Lowe & M.J.C. Walker, pp. 68–81. Cambridge: Quaternary Research Association.

Ballantyne, C.K. (1991b). Late Holocene erosion in upland Britain: climatic deterioration or human influence? *The Holocene*, 1, 81–85.

Ballantyne, C.K. (1991c). Holocene geomorphic activity in the Scottish Highlands. *Scottish Geographical Magazine*, 107, 84–98.

Ballantyne, C.K. (1992). Rock slope failure and debris flow, Gleann na Guiserein, Knoydart. *Scottish Journal of Geology*, 28, 77–80.

Ballantyne, C.K., Black, N.M. & Finlay, D.P. (1989). Enhanced boulder weathering under late-lying snowpatches. *Earth Surface Processes and Landforms*, 14, 745–750.

Ballantyne, C.K. & Eckford, J.D. (1984). Characteristics and evolution of two relict talus slopes in Scotland. *Scottish Geographical Magazine*, 100, 20–33.

Ballantyne, C.K. & Gray, J.M. (1984). The Quaternary geomorphology of Scotland: the research contribution of J.B. Sissons. *Quaternary Science Reviews*, 3, 259–289.

Ballantyne, C.K. & Kirkbride, M.P. (1986). The characteristics and signifi-cance of some Lateglacial protalus ramparts in upland Britain. *Earth Surface Processes and Landforms*, 11, 659–671.

Ballantyne, C.K. & Kirkbride, M.P. (1987). Rockfall activity in upland Britain during the Loch Lomond Stadial. *Geographical Journal*, 153, 86–92.

Ballantyne, C.K. & Matthews, J.A. (1982). The development of sorted circles on recently-deglaciated terrain, Jotunheimen, Norway. *Arctic and Alpine Research*, 14, 341–354.

Ballantyne, C.K. & Matthews, J.A. (1983). Desiccation cracking and sorted polygon development, Jotunheimen, Norway. *Arctic and Alpine Research*, 15, 339–349.

Ballantyne, C.K., Sutherland, D.G. & Reed, W.J. (1987). Introduction. In *Wester Ross: Field Guide*, eds. C.K. Ballantyne & D.G. Sutherland, pp. 1–63. Cambridge: Quaternary Research Association.

Ballantyne, C.K. & Wain-Hobson, T. (1980). The Loch Lomond Advance on the Island of Rhum. *Scottish Journal of Geology*, 16, 1–10.

Ballantyne, C.K. & Whittington, G.W. (1987). Niveo-aeolian sand deposits on An Teallach, Wester Ross, Scotland. *Transactions of the Royal Society of Edinburgh: Earth Sciences*, 78, 51–63.

Banham, P.H. (1971). Pleistocene beds at Corton, Suffolk. *Geological Magazine*, 108, 281–285.

Bard, E., Arnold, M., Maurice, P., Duprat, J., Moyes, J. & Duplessy, J-C. (1987). Retreat velocity of the North Atlantic polar front during the last deglaciation determined by ^{14}C mass spectrometry. *Nature*, 328, 791–794.

Barrow, M.D., Costin, A.B. & Lake, P. (1968). Cyclical changes in an Australian Fjaeldmark community. *Journal of Ecology*, 56, 89–96.

Barry, R.G. (1992). *Mountain Weather and Climate*. 2nd Edition. London: Routledge.

Barsch, D. (1977). Eine Abschälung von Schuttproduktion und Schutt-transport im Bereich aktiver Blockgletscher der Schweizer Alpen. *Zeitschrift für Geomorphologie Supplementband*, 28, 148–160.

Barsch, D. (1978). Rock glaciers as indicators for discontinuous alpine permafrost: an example from the Swiss Alps. *Proceedings of the 3rd International Conference on Permafrost, Edmonton, Alberta, Canada*, Volume 1, pp. 349–352. Ottawa: National Research Council of Canada.

Barsch, D. (1987a). Rock glaciers: an approach to their systematics. In *Rock Glaciers*, eds. J.R. Giardino, J.F. Shroder & J.D. Vitek, pp. 41–44. London: Allen and Unwin.

Barsch, D. (1987b). The problem of the ice-cored rock glacier. In *Rock Glaciers*, eds. J.R. Giardino, J.F. Shroder & J.D. Vitek, pp. 45–53. London: Allen and Unwin.

Barsch, D., Fierz, H. & Haeberli, W. (1979). Shallow core drilling and bore-hole measurements in the permafrost of an active rock glacier near the Grubengletscher, Wallis, Swiss Alps. *Arctic and Alpine Research*, 11, 215–228.

Barton, M.E. (1984). Periglacial features exposed in the coastal cliffs at Naish Farm, near Highcliffe. *Proceedings of the Hampshire Field Club and Archaeological Society*, 40, 5–20.

Bate, C.S. (1871). On the clitter of the tors of Dartmoor. *Reports of the Devonshire Association for the Advancement of Science*, 4, 517–519.

Bayfield, N.G. (1984). The dynamics of heather (*Calluna vulgaris*) stripes in the Cairngorm Mountains, Scotland. *Journal of Ecology*, 72, 515–527.

Bell, F.G. (1969). The occurrence of southern, steppe and halophyte elements in Weichselian (Last-Glacial) floras from southern Britain. *New Phytologist*, 68, 913–922.

Bell, F.G., Coope, G.R., Rice, R.J. & Riley, T.H. (1972). Mid-Weichselian fossil-bearing deposits at Syston, Leicestershire. *Proceedings of the Geologists' Association*, 83, 197–211.

Benedict, J.B. (1970). Downslope soil movement in a Colorado Alpine region: rates, processes and climatic significance. *Arctic and Alpine Research*, 2, 165–226.

Benedict, J.B. (1976). Frost creep and gelifluction features: a review. *Quaternary Research*, 6, 55–76.

Benn, D.I. (1989). Debris transport by Loch Lomond Readvance glaciers in Northern Scotland: basin form and the within-valley asymmetry of lateral moraines. *Journal of Quaternary Science*, 4, 243–254.

Benn, D.I. (1990). *Scottish Lateglacial Moraines: Debris Supply, Genesis and Significance*. Unpublished PhD Thesis, University of St Andrews.

Benn, D.I. (1991). Glacial landforms and sediments on Skye. In *The Quaternary of the Isle of Skye: Field Guide*. eds. C.K. Ballantyne, D.I. Benn, J.J. Lowe & M.J.C. Walker, pp. 35–67. Cambridge: Quaternary Research Association.

Benn, D.I., Lowe, J.J. & Walker, M.J.C. (1992). Glacier response to climatic change during the Loch Lomond Stadial and early Flandrian: geomorphological and palynological evidence from the Isle of Skye, Scotland. *Journal of Quaternary Science*, 7, 125–144.

Bennett, L.P. & French, H.M. (1988). Observations on near-surface creep in permafrost, Eastern Melville Island, Arctic Canada. *Proceedings of the 5th International Conference on Permafrost, Trondheim, Norway*, pp. 682–688. Trondheim: Tapir Press.

Bennett, L.P. & French, H.M. (1990). *In situ* permafrost creep, Melville Island, and implications for global change. *Proceedings of the 5th Canadian Permafrost Conference*, pp. 119–123. Québec: University of Laval.

Bennett, M. (1990). The deglaciation of Glen Croulin, Knoydart. *Scottish Journal of Geology*, 26, 41–46.

Berry, F.G. (1979). Late Quaternary scour hollows and related features in central London. *Quarterly Journal of Engineering Geology*, 12, 9–29.

Beskow, G. (1935). Tjäbildningen och tjällyftningen med särskild hänsyn till vägar och järnägar. *Sveriges Geologiska Undersökning*, Arsbok 26, Ser. C, No 375.

Beven K., Lawson, A. & McDonald, A. (1978). A landslip/debris flow in Bilsdale, North York Moors, September 1976. *Earth Surface Processes*, 3, 407–419.

Biczysko, S.J. (1981). Relic landslip in west Northamptonshire. *Quarterly Journal of Engineering Geology*, 14, 169–174.

Bik, M.J.J. (1969). The origin and age of the prairie mounds of southern Alberta, Canada. *Biuletyn Periglacjalny*, 19, 85–130.

Billings, W.D. (1974). Arctic and alpine vegetation: plant adaptations to cold summer climates. In *Arctic and Alpine Environments*, eds. Ives, J.D. & Barry, R.G., pp. 403–444. London: Methuen.

Birks, H.H. & Mathewes, R.W. (1978). Studies in the vegetation history of Scotland. V. Late Devensian and early Flandrian pollen and macrofossil stratigraphy at Abernethy Forest, Inverness-shire. *New Phytologist*, 80, 455–484.

Birks, H.J.B. (1973). *The Past and Present Vegetation of the Isle of Skye: a Palaeoecological Study*. Cambridge: Cambridge University Press.

Birks, H.J.B. (1977). The Flandrian forest history of Scotland: a preliminary synthesis. In *British Quaternary Studies: Recent Advances*, ed. F.W. Shotton, pp. 119–135. Oxford: Clarendon.

Birse, E.L. (1980). Suggested amendments to the world soil classification to accommodate Scottish aeolian and mountain soils. *Journal of Soil Science*, 31, 117–124.

Birse, E.L. & Robertson, L. (1970). *Assessment of Climatic Conditions in Scotland, 2: Based on Exposure and Accumulated Frost*. Aberdeen: Macauley Institute.

Bishop, W.W. & Coope, G.R. (1977). Stratigraphical and faunal evidence for Lateglacial and Flandrian environments in south-west Scotland. In *Studies in the Scottish Lateglacial*, eds. J.M. Gray & J.J. Lowe, pp. 61–88. Oxford: Pergamon.

Black, R.F. (1973). Growth of patterned ground in Victoria Land, Antarctica. *North American Contribution to the 2nd International Conference on Permafrost, Yakutsk*, pp. 193–203. Washington DC: National Academy Press.

Black, R.F. (1976). Periglacial features indicative of permafrost: ice and soil wedges. *Quaternary Research*, 6, 3–26.

Black, R.F. (1978). Fabric of ice-wedges in central Alaska, USA. *Proceedings of the 4th International Conference on Permafrost, Fairbanks Alaska*, pp. 68–73. Washington DC: National Academy Press.

Blagbrough, J.W. & Breed, W.J. (1967). Protalus ramparts on Navajo Mountain, Utah. *American Journal of Science*, 265, 759–773.

Blagbrough, J.W. & Farkas, S.E. (1968). Rock glaciers in the San Mateo Mountains, south-central New Mexico. *American Journal of Science*, 266, 812–823.

Bleasdale, A. & Chan, Y.K. (1972). Orographic influences on the distribution of precipitation. In *Distribution of Precipitation in Mountainous Areas*, volume 2, pp. 161–170. Geneva: World Meteorological Organisation.

Boardman, J. (1977). Stratified screes in the northern Lake District. *Proceedings of the Cumberland Geological Society*, 3, 233–237.

Boardman, J. (1978). Grèzes litées near Keswick, Cumbria. *Biuletyn Peryglacjany*, 27, 23–34.

Boardman, J. (1985a). Editor. *Field Guide to the Periglacial Landforms of Northern England*. Cambridge: Quaternary Research Association.

Boardman, J. (1985b). The northeastern Lake District: periglacial slope deposits. In *Field Guide to the Periglacial Landforms of Northern England*, ed. J. Boardman, pp. 23–37. Cambridge: Quaternary Research Association.

Boardman, J. (1987). Editor. *Periglacial Processes and Landforms in Britain and Ireland*. Cambridge: Cambridge University Press.

Bones, J.G. (1973). Process and sediment-size arrangement on high arctic talus, southwest Devon Island, NWT, Canada. *Arctic and Alpine Research*, 5, 29–40.

Borlase, W. (1758). *The Natural History of Cornwall*. Oxford.

Bosikov, N.P. (1988). General moistening of the area and intensity of cryogenic processes. *Proceedings of the 5th International Conference on Permafrost, Trondheim, Norway*. pp. 695–699. Trondheim: Tapir Press.

Boulton, G.S. (1978). Boulder shapes and grain-size distribution of debris as indicators of transport paths through a glacier and till genesis. *Sedimentology*, 25, 773–799.

Boulton, G.S. (1979). Processes of glacial erosion on different substrata. *Journal of Glaciology*, 23, 15–37.

Boulton, G.S., Jones, A.S., Clayton, K.M. & Kenning, M.J. (1977). A British ice-sheet model and patterns of glacial erosion and deposition in Britain. In *British Quaternary Studies*, ed. F.W. Shotton, pp. 231–246. Oxford: Clarendon Press.

Boulton, G.S., Peacock, J.D. & Sutherland, D.G. (1991). Quaternary. In *Geology of Scotland*, ed. G.Y. Craig, pp. 503–543. London: The Geological Society.

Boulton, G.S., Smith, G.D., Jones, A.S. & Newsome, J. (1985). Glacial geology and glaciology of the last mid-latitude ice sheets. *Journal of the Geological Society, London*, 142, 447–474.

Bout, P. & Godard, A. (1973). Aspects du modelé périglaciaire en Scandinavie du nord; problèmes de genèse, comparaisons. *Biuletyn Peryglacjalny*, 22, 49–79.

Bowen, D.Q. (1973a). The Pleistocene history of Wales and the borderland. *Geological Journal*, 8, 207–224.

Bowen, D.Q. (1973b). The Pleistocene succession of the Irish Sea. *Proceedings of the Geologists' Association, London*, 84, 249–272.

Bowen, D.Q. (1974). The Quaternary of Wales. In *The Upper Palaeozoic and Post-Palaeozoic Rocks of Wales*, ed. T.R. Owen, pp. 373–426. Cardiff: University of Wales Press.

Bowen, D.Q. (1991). Time and space in the glacial sediment systems of the British Isles. In *Glacial Deposits in Great Britain and Ireland*, eds. J. Ehlers, P.L. Gibbard & J. Rose, pp. 3–11. Rotterdam: Balkema.

Bowen, D.Q., Hughes, S., Sykes, G.A. & Miller, G.H. (1989). Land-sea correlations in the Pleistocene based on isoleucine epimerization in non-marine molluscs. *Nature*, 340, 49–51.

Bowen, D.Q., Rose, J., McCabe, A.M. & Sutherland, D.G. (1986). Correlation of Quaternary glaciations in England, Ireland, Scotland and Wales. *Quaternary Science Reviews*, 5, 299–340.

Bowen, D.Q. & Sykes, G.A. (1988). Correlation of marine events and glaciations on the northeast Atlantic margin. *Philosophical Transactions of the Royal Society of London*, B 318, 619–635.

Boyer, S. & Pheasant, D. (1974). Delimitation of weathering zones in the fiord area of eastern Baffin Island, Canada. *Geological Society of America, Bulletin*, 85, 805–810.

Bradshaw, R. & Smith, D.I. (1963). Permafrost structures on Sully Island, Glamorgan. *Geological Magazine*, 100, 556–564.

Brandon, A. & Sumbler, M.G. (1991) The Balderton Sand and Gravel: pre-Ipswichian cold stage fluvial deposits near Lincoln, England. *Journal of Quaternary Science*, 6, 117–129.

Brazier, V. (1987). *Late Quaternary Alluvial Fans, Debris Cones and Talus Cones in the Grampian Highlands, Scotland*. Unpublished PhD. Thesis, University of St Andrews.

Brazier, V. & Ballantyne, C.K. (1989). Late Holocene debris cone evolution in Glen Feshie, western Cairngorm Mountains, Scotland. *Transactions of the Royal Society of Edinburgh: Earth Sciences*, 80, 17–24.

Brazier, V., Whittington, G.W. & Ballantyne, C.K. (1988). Holocene debris cone evolution in Glen Etive, Western Grampian Highlands, Scotland. *Earth Surface Processes and Landforms*, 13, 525–531.

Brewer, R. & Pawluk, S. (1975). Investigations of some soils developed in hummocks of the Canadian sub-Arctic and southern Arctic regions: I, morphology and micromorphology. *Canadian Journal of Soil Science*, 55, 301–319.

Bridge, M.C., Haggart, B.A. & Lowe, J.J. (1990). The history and palaeo-climatic significance of subfossil remains of *Pinus silvestris* in blanket peats from Scotland. *Journal of Ecology*, 78, 77–99.

Bridgland, D.R. (1988). Report of Geologists' Association field meeting in northeast Essex, May 22nd–24th, 1987. *Proceedings of the Geologists' Association*, 99, 315–333.

Bridgeman, P.W. (1912). Water, in the liquid and five solid forms, under pressure. *American Academy of Arts and Sciences, Proceedings,* 47, 441–558.

Briggs, D.J., Coope, G.R. & Gilbertson, D.D. (1975). Late Pleistocene terrace deposits at Beckford, Worcestershire, England. *Geological Journal,* 10, 1–16.

Briggs, D.J., Coope, G.R. & Gilbertson, D.D. (1985). The chronology and environmental framework of early man in the upper Thames Valley. *British Archaeological Series,* 137.

Briggs, D.J. & Courtney, F.M. (1972). Ridge-and-trough topography in the north Cotswolds. *Proceedings of the Cotswold Naturalists and Field Club,* 37, 94–103.

Briggs, D.J. & Gilbertson, D.D. (1980). Quaternary processes and environments in the upper Thames Valley. *Transactions of the Institute of British Geographers,* New Series, 5, 53–65.

Bristow, C.R. & Bazley, R.A. (1972). *Geology of the Country around Royal Tunbridge Wells.* British Geological Survey Memoir. London: HMSO.

Broeker, W.S., Peteet, D.M. & Rind, U. (1985). Does the ocean-atmosphere system have more than one stable mode of operation? *Nature,* 315, 21–25.

Brown, A.P. (1977). Late Devensian and Flandrian vegetational history of Bodmin Moor, Cornwall. *Philosophical Transactions of the Royal Society,* B 276, 251–320.

Brown, R.J.E. (1960). The distribution of permafrost and its relation to air temperature in Canada and the USSR. *Arctic,* 13, 167–177.

Brown, R.J.E. (1966). The relation between mean annual air and ground temperatures in the permafrost regions of Canada. In *Proceedings, 1st International Permafrost Conference,* pp. 241–246. Washington DC: National Academy of Science.

Brown, R.J.E. (1970). *Permafrost in Canada.* Toronto: University of Toronto Press.

Brown, R.J.E & Kupsch, W.O. (1974). Permafrost terminology. *National Research Council, Canada, Associated Committee on Geotechnical Research, Technical Memorandum,* 111, 62 pp.

Brown, R.J.E., Johnston, G.H., Mackay, J.R., Morgenstern, N.R., & Smith, W.W. (1981). Permafrost distribution and terrain characteristics. In *Permafrost Engineering,* ed. G.H. Johnson, pp. 31–72. Toronto: Wiley.

Brown, R.J.E. & Péwé, T.L. (1973). Distribution of permafrost in North America and its relationship to the environment, 1963–73, a review. In *North American Contribution, 2nd International Conference on Permafrost, Yakutsk,* pp. 71–100. Washington DC: National Academy of Sciences.

Bryan, K. (1934). Geomorphic processes at high altitudes. *Geographical Review,* 24, 655–656.

Bryant, I.D. (1982). *Periglacial River Systems: Ancient and Modern.* Unpublished PhD. Thesis, University of Reading.

Bryant, I.D. (1983a). Facies sequences associated with some braided river deposits of late Pleistocene age from Southern Britain. In *Modern and Ancient Fluvial Systems: Sedimentology and Process,* eds. J.D. Collinson & J. Lewin, pp. 267–275. International Association of Sedimentologists Special Publication 6. Oxford: Blackwell.

Bryant, I.D. (1983b). The utilization of arctic river analogue studies in the interpretation of periglacial river systems in southern Britain. In *Background to Palaeohydrology: A Perspective,* ed. K. Gregory, pp. 413–431. London: John Wiley and Sons Ltd.

Bryant, I.D., Holyoak, D.T. & Moseley, K.A. (1983). Late Pleistocene deposits at Brimpton, Berkshire, England. *Proceedings of the Geologists' Association,* 94, 321–343.

Bryant, R.H. & Carpenter, C.P. (1987). Ramparted ground ice depressions in Britain and Ireland. In *Periglacial Processes and Landforms in Britain and Ireland,* ed. J. Boardman, pp. 183–190. Cambridge: Cambridge University Press.

Bryant, R.H., Carpenter C.P. & Ridge, T.S. (1985). Pingo scars and related features in the Whicham Valley, Cumbria. In *Field Guide to the Periglacial Landforms of Northern England,* ed. J. Boardman, pp. 47–53. Cambridge: Quaternary Research Association.

Brunsden D. (1979). Mass movements. In *Process in Geomorphology,* eds. C. Embleton & J. Thornes, pp. 130–186. London: Edward Arnold.

Buckland, P.C. (1982). The coversands of north Lincolnshire and the Vale of York. In *Papers in Earth Science: Lovatt Lectures, Worcester,* eds. B.H. Adlam, C.R. Fenn & L. Morris, pp. 143–178. Norwich: Geo Books.

Bull, A.J. (1936). Studies in the geomorphology of the South Downs. *Proceedings of the Geologists' Association,* 47, 99–129.

Bull, A.J. (1940). Cold conditions and landforms in the South Downs. *Proceedings of the Geologists' Association,* 51, 63–70.

Burchell, J.P.T. (1933). The Northfleet 50-foot submergence later than the Coombe Rock of post-early Mousterian times. *Archaeologia,* 83, 67–92.

Burek, C. (1991). Quaternary history and glacial deposits in the Peak District. In *Glacial Deposits in Great Britain and Ireland,* ed. J. Ehlers, P.L. Gibbard & J. Rose, pp. 193–201. Rotterdam: A.A. Balkema.

Burges, A. (1951). The ecology of the Cairngorms. III. The *Empetrum-Vaccinium* zone. *Journal of Ecology,* 39, 271–284.

Burn, C.R. (1990). Implications for palaeoenvironmental reconstruction of recent ice-wedge development at Mayo, Yukon Territory. *Permafrost and Periglacial Processes,* 1, 3–14.

Burn, C.R. & Smith, M.W. (1988). Thermokarst lakes at Mayo, Yukon Territory, Canada. *Proceedings of the 5th International Conference on Permafrost, Trondheim, Norway,* pp. 700–711. Trondheim: Tapir Press.

Burrin, P.J. (1981) Loess in the Weald. *Proceedings of the Geologists' Association,* 92, 87–92.

Burt, T.P. (1980). A study of rainfall in the southern Pennines. *Department of Geography, Huddersfield Polytechnic, Occasional Paper,* 8, 95 pp.

Burt, T.P. & Williams, P.J. (1976). Hydraulic conductivity in frozen soils. *Earth Surface Processes and Landforms,* 1, 349–360.

Burton, R.G.O. (1987). The role of thermokarst in landscape development in eastern England. In *Periglacial Processes and Landforms in Britain and Ireland,* ed. J. Boardman, pp. 203–208. Cambridge: Cambridge University Press.

Cailleux, A. (1961). Mares et lacs ronds et loups de glace du sol. *Biuletyn Peryglacjalny,* 10, 35–41.

Cailleux, A. (1968). Periglacial of McMurdo Strait (Antarctica). *Biuletyn Peryglacjalny,* 17, 57–90.

Cailleux, A. (1972). Les formes et dépôts nivéo-eolien actuels en Antarctique et au Nouveau-Québec. *Cahiers de Géographie de Québec,* 16/39, 377–409.

Cailleux, A. (1976). Formes et dépôts nivéo-éoliens sur le pied du glace à Porte-de-la-Baleine, Québec subarctique. *Révue de Géographie, Montréal,* 30, 213–219.

Cailleux, A. (1978). Niveo-eolian deposits. In *Encyclopedia of Sedimentology,* eds. R.W. Fairbridge & J. Bourgeois, pp. 501–503. New York: Van Nostrand Reinhold.

Cailleux, A. & Calkin, P.E. (1963). Orientation of hollows in cavernously weathered boulders. *Biuletyn Peryglacjalny,* 12, 147–150.

Cailleux, A. & Hamelin, L.E. (1967). Périglaciaire actuel sur le littoral du Bic (Bas-Saint Laurent). *Cahiers de Géographie de Québec,* 23, 361–378.

Caine, N. (1963a). Movement of low angle scree slopes in the Lake District, Northern England. *Revue de Géomorphologie Dynamique,* 14, 171–177.

Caine, N. (1963b). The origin of sorted stripes in the Lake District, northern England. *Geografiska Annaler,* 45, 172–179.

Caine, N. (1967). The tors of Ben Lomond, Tasmania. *Zeitschrift für Geomorphologie,* 11, 418–429.

Caine, N. (1968). The blockfields of northeastern Tasmania. *Australian National University Department of Geography Publication,* G6, 127 pp.

Caine, N. (1972). The distribution of sorted patterned ground in the English Lake District. *Revue de Géomorphologie Dynamique,* 21, 49–56.

Caine, N. (1974). The geomorphic processes of the alpine environment. In *Arctic and Alpine Environments,* eds. J.D. Ives & R.G. Barry, pp. 721–748. London: Methuen.

Caine, N. (1976). A uniform measure of subaerial erosion. *Bulletin of the Geological Society of America,* 87, 137–140.

Caine, N. (1980). The rainfall intensity-duration control of shallow landslides and debris flows. *Geografiska Annaler,* 62A, 23–27.

Calkin, P.E., Haworth, L.A. & Ellis, J.M. (1987). Rock glaciers of the central Brooks Range, Alaska. In *Rock Glaciers,* eds. J.R. Giardino, J.F. Shroder & J.D. Vitek. London: Allen and Unwin, 65–82.

Cameron, T.D.J., Stoker, M.S. & Long, D. (1987). The history of Quaternary sedimentation in the U.K. sector of the North Sea Basin. *Journal of the Geological Society of London,* 144, 43–58.

Campbell R.H. (1974). Debris flows originating from soil slips during rainstorms in southern California. *Quarterly Journal of Engineering Geology,* 7, 339–349.

Carling, P.A. (1986). The Noon Hill flash floods: July 17th 1983. Hydrological and geomorphological aspects of a major formative event in an upland landscape. *Transactions of the Institute of British Geographers,* 11, 105–118.

Carling, P.A. (1987). A terminal debris-flow lobe in the northern Pennines, United Kingdom. *Transactions of the Royal Society of Edinburgh: Earth Sciences,* 78, 169–176.

Carlston, C.W. (1963). Drainage density and streamflow. *US Geological Survey Professional Paper*, 422-C, 8pp.

Carpenter, C.P. & Woodcock, M.P. (1981). A detailed investigation of a pingo remnant in Western Surrey. *Quaternary Studies*, 1, 1–26.

Carruthers, R.G. & Anderson, W. (1941). Early post-glacial conditions in England. *Nature*, 147, 28.

Carson, M.A. (1977). Angles of repose, angles of shearing resistance and angles of talus slopes. *Earth Surface Processes*, 2, 363–380.

Carter, L.D. (1988). Loess and deep thermokarst basins in arctic Alaska. *Proceedings of the 5th International Conference on Permafrost, Trondheim*, pp. 706–711. Trondheim: Tapir Press.

Carter, L.D. & Galloway, J.P. (1981). Earthflows along Henry Creek, northern Alaska. *Arctic*, 34, 325–328.

Castleden, R. (1977). Periglacial pediments in central southern England. *Catena*, 4, 111–121.

Castleden, R. (1980). Fluvioperiglacial sedimentation: a general theory of fluvial valley development in cool temperate lands, illustrated from western and central Europe. *Catena*, 7, 135–152.

Catt, J.A. (1977). Loess and coversands. In *British Quaternary Studies: Recent Advances*, ed. F.W. Shotton, pp. 221–229. Oxford: Clarendon Press.

Catt, J.A. (1978). The contribution of loess to soils in lowland Britain. In *The Effect of Man on the Landscape: the Lowland Zone*, eds. S. Limbrey & J.G. Evans, pp. 12–20. Research Report No 21, Council for British Archaeology.

Catt, J.A. (1979). Soils and Quaternary geology in Britain. *Journal of Soil Science*, 30, 607–642.

Catt, J.A. (1985a). Particle size distribution and mineralogy as indicators of pedogenic and geomorphic history: examples from soils of England and Wales. In *Geomorphology and Soils*, eds. K.S. Richards, R.R. Arnett & S. Ellis, pp. 202–218. London: George Allen and Unwin.

Catt, J.A. (1985b). Soils and Quaternary stratigraphy in the United Kingdom. In *Soils and Quaternary Landscape Evolution*, ed. J. Boardman, pp. 161–178. Chichester: John Wiley & Sons.

Catt, J.A. (1991a). Late Devensian glacial deposits and glaciations in eastern England and the adjoining offshore regions. In *Glacial Deposits in Great Britain and Ireland*, eds. J. Ehlers, P.L. Gibbard & J. Rose, pp. 61–68. Rotterdam: Balkema.

Catt, J.A. (1991b). The Quaternary history and glacial deposits of East Yorkshire. In *Glacial Deposits in Great Britain and Ireland*, eds. J. Ehlers, P.L. Gibbard & J. Rose, pp. 185–193. Rotterdam: Balkema.

Catt, J.A. & Penny, L.F. (1966). The Pleistocene deposits of Holderness, East Yorkshire. *Proceedings of the Yorkshire Geological Society*, 35, 375–420.

Catt, J.A. & Staines, S.J. (1982). Loess in Cornwall. *Proceedings of the Ussher Society*, 5, 368–375.

Catt, J.A., Weir, A.H. & Madgett, P.A. (1974). The loess of eastern Yorkshire and Lancashire. *Proceedings of the Yorkshire Geological Society*, 40, 23–39.

Chambers, M.J.G. (1966). Investigations of patterned ground at Signy Island, South Orkney Islands: I, Interpretation of mechanical analyses. *British Antarctic Survey Bulletin*, 9, 21–40.

Chambers, M.J.G. (1967). Investigation of patterned ground at Signy Island, South Orkney Islands: III, Miniature patterns, frost heaving, and general conclusions. *British Antarctic Survey Bulletin*, 12, 1–22.

Chandler, R.H. (1909). On some dry chalk valley features. *Geological Magazine*, 5, 538.

Chandler, R.J. (1970a). The degradation of Lias Clay slopes in an area of the East Midlands. *Quarterly Journal of Engineering Geology*, 2, 161–181.

Chandler, R.J. (1970b). A shallow slab slide in the Lias Clay near Uppingham, Rutland. *Geotechnique*, 20, 253–260.

Chandler, R.J. (1972a). Periglacial mudslides in Vestspitsbergen and their bearing on the origin of fossil 'solifluction' shears in low angled clay slopes. *Quarterly Journal of Engineering Geology*, 5, 223–241.

Chandler, R.J. (1972b). Lias clay: weathering processes and their effect on shear strength. *Géotechnique*, 22, 403–431.

Chandler, R.J. (1973). The inclination of talus, arctic talus terraces, and other slopes composed of granular materials. *Journal of Geology*, 81, 1–14.

Chandler, R.J. (1976). The history and stability of two Lias clay slopes in the upper Gwash Valley, Rutland. *Philosophical Transactions of the Royal Society, London*, A283, 463–491.

Chandler, R.J., Kellaway, G.A., Skempton, A.W. & Wyatt, R.J. (1976).

Valley slope sections near Bath, Somerset. *Philosophical Transactions of the Royal Society*, A 283, 527–556.

Chattopadhyay, G.P. (1982). *Periglacial Geomorphology of parts of the Grampian Highlands of Scotland*. Unpublished PhD Thesis, University of Edinburgh.

Chattopadhyay, G.P. (1983). Ploughing blocks on the Drumochter Hills in the Grampian Highlands, Scotland: a quantitative report. *Geographical Journal*, 149, 211–15.

Chattopadhyay, G.P. (1984). A fossil valley-wall rock glacier in the Cairngorm Mountains. *Scottish Journal of Geology*, 20, 121–125.

Chattopadhyay, G.P. (1986). Ploughing block movement on the Drumochter hills in the Grampian Highlands, Scotland. *Biuletyn Peryglacjalny*, 30, 57–60.

Cheetham, G.H. (1975). *Late Quaternary Palaeohydrology, with Reference to the Kennet Valley*. Unpublished PhD thesis, University of Reading.

Cheetham, G.H. (1976). Palaeohydrological investigations of river terrace gravels. In *Geo-archaeology: Earth Science and the Past*, eds. D. A. Davidson & M. Shakle, pp. 335–344. London: Duckworth.

Cheetham, G.H. (1980). Late Quaternary palaeohydrology: the Kennet Valley case-study. In *The Shaping of Southern England*, ed. D.K.C. Jones, pp. 203–223. London: Academic Press.

Chmal, H., Klementowski, J. & Migala, K. (1988). Thermal currents of active layer in Hornsund area. In *Proceedings of the 5th International Conference on Permafrost, Trondheim*, pp. 44–49 Trondheim: Tapir Press.

Chorley, R.J., Stoddart, D.R., Haggett, P. & Slaymaker, H.O. (1966). Regional and local components in the aerial distribution of surface sand facies in the Breckland, Eastern England. *Journal of Sedimentary Petrology*, 36, 209–220.

Church, M. (1972). Baffin Island sandurs: a study of arctic fluvial processes. *Geological Survey of Canada, Bulletin*, 216, 208 pp.

Church, M. (1974). Hydrology and permafrost with reference to northern North America. In *Permafrost Hydrology*, pp. 7–20. Ottawa: Environment Canada.

Church, M. & Ryder, J.M. (1972). Paraglacial sedimentation; a consideration of fluvial processes conditioned by glaciation. *Bulletin of the Geological Society of America*, 83, 3059–3067.

Church, M., Stock R.F. & Ryder, J.M. (1979). Contemporary sedimentary environments on Baffin Island, NWT, Canada: debris slope accumulations. *Arctic and Alpine Research*, 11, 371–402.

Clapperton, C.M. (1970). The evidence for a Cheviot ice cap. *Transactions of the Institute of British Geographers*, 50, 115–127.

Clapperton, C.M. (1975). Further observations on the stone runs of the Falkland Islands. *Biuletyn Peryglacjalny*, 24, 211–217.

Clark, B.B. (1973). The Camel Estuary section west of Tregunna House. *Proceedings of the Ussher Society*, 2, 551–553.

Clark, M.J. (1988a) Editor. *Advances in Periglacial Geomorphology*. Chichester: Wiley.

Clark, M.J. (1988b). Periglacial hydrology. In *Advances in Periglacial Geomorphology*, ed. M.J. Clark, pp. 415–462. Chichester: Wiley.

Clark, M.J., Gurnell, A.M. & Threlfall, J.L. (1988). Suspended sediment transport in arctic rivers. *Proceedings of the 5th International Conference on Permafrost, Trondheim*, pp. 558–565. Trondheim: Tapir Press.

Clark, M.R. & Dixon, A.J. (1981). The Pleistocene braided river deposits in the Blackwater Valley area of Berkshire and Hampshire, England. *Proceedings of the Geologists' Association*, 92, 139–157.

Clark, R. (1971). Periglacial landforms and landscapes in Northumberland. *Proceedings of the Cumberland Geological Society*, 3, 5–20.

Clayton, K. (1979). The Midlands and southern Pennines. In *The Geomorphology of the British Isles: Eastern and Central England*, by A. Straw & K. Clayton, pp. 142–240. London: Methuen.

Cogley, J.G. (1972). Processes of solution in an arctic limestone terrain. *Institute of British Geographers Special Publication*, 4, 201–211.

Cogley, J.G. & McCann, S.B. (1976). An exceptional storm and its effects in the Canadian high arctic. *Arctic and Alpine Research*, 8, 105–110.

COHMAP Members (1988). Climatic changes of the last 18 000 years: observations and model simulations. *Science*, 241, 1043–1052.

Common, R. (1954). A report on the Lochaber, Appin and Benderloch floods. *Scottish Geographical Magazine*, 70, 6–20.

Common, R. & Galloway, R.W. (1958). Ice wedges in Midlothian – a note. *Scottish Geographical Magazine*, 74, 44–46.

Connell, E.R., Edwards, K.J. & Hall, A.M. (1982). Evidence for two pre-Flandrian palaeosols in Buchan, Scotland. *Nature*, 297, 570–572.

Connell, E.R. & Hall, A.M. (1987). The periglacial history of Buchan, north-east Scotland. In *Periglacial Processes and Landforms in Britain and Ireland*, ed. J. Boardman, pp. 277–285. Cambridge: Cambridge University Press.

Conrad, V. (1946). Polygon nets and their physical development. *American Journal of Science*, 244, 274–296.

Conway, B.W. (1974). The Black Venn landslip, Charmouth, Dorset. *Institute of Geological Sciences Report*, 74/73. London: HMSO.

Conway, B.W. & Devoy, R.J. (1977). Cooling Marshes, Rochester. In *South East England and the Thames Valley: Excursion Guide for the 10th INQUA Congress*, eds. E.R. Shephard-Thorn & J.J. Wymer, pp. 46–50. Norwich: Geo Abstracts.

Cook, F.A. (1967). Fluvial processes in the high arctic. *Geographical Bulletin*, 9, 263–268.

Cook, F.A. & Raiche, V.G. (1962). Simple transverse nivation hollows at Resolute, NWT. *Geographical Bulletin*, 18, 79–85.

Cook, J.D. (1989). *Active and Relict Sorted Circles, Jotunheimen, Norway: a Study of the Altitudinal Zonation of Periglacial Processes*, Unpublished PhD thesis, University of Wales, 2 volumes.

Cooke, R.U. & Warren, A. (1973). *Geomorphology in Deserts*. London: Batsford.

Coombe, D.E., Frost, L.C., Le Bas, M. & Watters, W. (1956). The nature and origin of the soils over the Cornish serpentine. *Journal of Ecology*, 44, 605–615.

Coope, G.R. (1959). A late Pleistocene insect fauna from Chelford, Cheshire. *Proceedings of the Royal Society, London*, B 151, 70–86.

Coope, G.R. (1975). Climatic fluctuations in north-west Europe since the last interglacial, indicated by fossil assemblages of Coleoptera. In *Ice Ages: Ancient and Modern*, eds. A.E. Wright & F. Mosely, pp. 153- 168. Liverpool: Seel House Press.

Coope, G.R. (1977a). Quaternary coleoptera as aids in the interpretation of environmental history. In *British Quaternary Studies: Recent Advances*, ed. F.W. Shotton, pp. 55–68. Oxford: Clarendon Press.

Coope, G.R. (1977b). Fossil coleopteran assemblages as sensitive indicators of climatic changes during the Devensian (last) cold stage. *Philosophical Transactions of the Royal Society of London*, B 280, 313–340.

Coope, G.R. & Angus, R.B. (1975) An ecological study of a temperate interlude in the middle of the Last Glaciation, based on fossil Coleoptera from Isleworth, Middlesex. *Journal of Animal Ecology*, 44, 365–391.

Coope, G.R. & Brophy, J.A. (1972). Lateglacial environmental changes indicated by a coleopteran succession from North Wales. *Boreas*, 1, 97–142.

Coope, G.R., Morgan, A. & Osborne, P.J. (1971). Fossil coleoptera as indicators of climatic fluctuations during the last glaciation in Britain. *Palaeogeography, Palaeoclimatology, Palaeoecology*, 10, 87–101.

Coope, G.R., Shotton, F.W. & Strachan, I. (1961). A late Pleistocene fauna and flora from Upton Warren, Worcestershire. *Philosophical Transactions of the Royal Society of London*, B 244, 379–421.

Cooper, R.G. (1980). A sequence of landsliding mechanisms in the Hambleton Hills, northern England, illustrated by features at Peak Scar, Hawnby. *Geografiska Annaler*, 62A, 149–156.

Corner, G.D. (1980). Avalanche impact landforms in Troms, north Norway. *Geografiska Annaler*, 62A, 1–10.

Cornish, R. (1981). Glaciers of the Loch Lomond Stadial in the western Southern Uplands of Scotland. *Proceedings of the Geologists' Association*, 92, 105–114.

Corte, A.E. (1961). The frost behaviour of soils: laboratory and field data for a new concept. I: vertical sorting. *US Army Corps of Engineers, Cold Regions Research and Engineering Laboratory*, Research Report 85 (1), 22 pp.

Corte, A.E. (1962a). Vertical migration of particles in front of a moving freezing plane. *Journal of Geophysical Research*, 67 (3), 1085–1090.

Corte, A.E. (1962b). The frost behaviour of soils: laboratory and field data for a new concept. II: horizontal sorting. *US Army Corps of Engineers, Cold Regions Research and Engineering Laboratory*, Research Report, 85 (2), 20pp.

Corte, A.E. (1963). Relationships between four ground patterns, structure of the active layer, and type and distribution of ice in permafrost. *Biuletyn Peryglacjalny*, 12, 7–90.

Corte, A.E. (1966). Particle sorting by repeated freezing and thawing. *Biuletyn Peryglacjalny*, 15, 176–240.

Corte, A.E. (1971). Laboratory formation of extrusion features by multicyclic freeze–thaw in soils. *Bulletin, Centre de Géomorphologie du CNRS*, 13–15, 117–131.

Corte, A.E. (1976). Rock glaciers. *Biuletyn Peryglacjalny*, 26, 175–197.

Costa, J.E. (1984). Physical geomorphology of debris flows. In *Developments and Applications of Geomorphology*, ed. J.E. Costa & P.J. Fleisher, pp. 268–317. Berlin: Springer-Verlag.

Costin, A.B., Jennings, J.N., Bautovitch, B.C. & Wimbush, D.J. (1973). Forces developed by snowpatch action, Mount Twynam, Snowy Mountains, Australia. *Arctic and Alpine Research*, 5, 121–126.

Cotton, D.E. (1968). *An Investigation of the Structure, Morphology and Ecology of Some Pennine Soil Microtopographic Features*. Unpublished PhD thesis, University of Lancaster, 203 pp.

Coutard, J-P., Gabert, P. & Ozouf, J.C. (1988). Étude du processus de cryoreptation en divers sites de la Haute-Ubaye (Alps du Sud). *Bulletin, Centre de Géomorphologie du CNRS*, 34, 9–28.

Coxon, P. (1978). The first record of a fossil naled in Britain. *Quaternary Newsletter*, 24, 9–11.

Coxon, P. & O'Callaghan, P. (1987). The distribution and age of pingo remnants in Ireland. In *Periglacial Processes and Landforms in Britain and Ireland*, ed. J. Boardman, pp. 183–190. Cambridge: Cambridge University Press.

Crampton, C.B. (1911). *The Vegetation of Caithness Considered in Relation to the Geology*. Edinburgh, 132 pp.

Crampton, C.B. (1965). An indurated horizon in soils of South Wales. *Journal of Soil Science*, 16, 230–241.

Crampton, C.B. & Carruthers, R.G. (1914). The geology of Caithness. *Memoirs of the Geological Survey, Scotland*. Edinburgh.

Crampton, C.B. & Taylor, J.A. (1967). Solifluction terraces in south Wales. *Biuletyn Peryglacjalny*, 16, 15–36.

Cresswell, D. (1983). Deformation of weathered profiles below head at Constantine Bay, north Cornwall. *Proceedings of the Ussher Society*, 5, 487.

Cunningham, F.F. (1964). A detail of process on scarp edges of millstone grit. *East Midland Geographer*, 3, 322–325.

Cunningham, F.F. (1965). Tor theories in the light of south Pennine evidence. *East Midland Geographer*, 3, 424–433.

Cunningham, F.F. (1969). The Crow Tors, Laramie Mountains, Wyoming. *Zeitschrift für Geomorphologie*, 13, 56–74.

Curry, R.R. (1966). Observations of alpine mudflows in the Tenmile Range, central Colorado. *Bulletin of the Geological Society of America*, 77, 771–776.

Curtis, L.F. & James, J.H. (1959). Frost heaved soils of Barrow, Rutland. *Proceedings of the Geologists' Association*, 70, 310–314.

Czeppe, A. (1964). Exfoliation in a periglacial climate. *Geografica Polonica*, 2, 5–10.

Czudek, T. (1964). Periglacial slope development in areas of the Bohemian Massif, in northern Moldavia. *Biuletyn Peryglacjalny*, 14, 169–193.

Czudek, T. & Demek, J. (1970a). Thermokarst in Siberia and its influence on the development of lowland relief. *Quaternary Research*, 1, 103–120.

Czudek, T. & Demek, J. (1970b). Pleistocene cryopedimentation in Czechoslovakia. *Acta Geographica Lodziensia*, 24, 101–108.

Czudek, T. & Demek, J. (1971). Pleistocene cryoplanation of the Ceska Vysocina highlands, Czechoslovakia. *Transactions of the Institute of British Geographers*, 52, 95–112.

Dackombe, R.V. & Thomas, G.S.P. (1985). Editors. *Field Guide to the Quaternary of the Isle of Man*. Cambridge: Quaternary Research Association.

Dahl, R. (1966). Block fields, weathering pits and tor-like forms in the Narvik Mountains, Nordland, Norway. *Geografiska Annaler*, 48A, 55–85.

Daly, R.A. (1912). Geology of the North American Cordillera at the forty-ninth parallel. *Canadian Geological Survey Memoirs*, 38, 857 pp.

Davidson, C. (1889). On the creeping of the soil cap through the action of frost. *Geological Magazine*, 6, 255–261.

Davies, J.D. (1980). *Geographical Variation in Coastal Development*. 2nd Edition. London: Longman.

Davies, T.D., Brimblecombe, P., Blackwood, I.L., Tranter, M. & Abrahams, P.W. (1988). Chemical composition of snow in the remote Scottish Highlands. In *Acid Deposition at High Elevation Sites*, eds. M.H. Unsworth & D. Fowler, pp. 517–539. Amsterdam: Kluwer Scientific Publishers.

Davis, G.H. (1962). Erosional features of snow avalanches, Middle Fork, King's River, California. *United States Geological Survey Professional Paper*, 450-D, 122–125.

Davison R.W. & Davison S.K. (1987). Characteristics of two full-depth slab avalanches on Meall Uaine, Glen Shee, Scotland. *Journal of Glaciology*, 33, 51–54.

Dawson, A.G. (1977). A fossil lobate rock glacier in Jura. *Scottish Journal of Geology*, 13, 37–42.

Dawson, A.G. (1979). Polar and non-polar shore platform development. *Bedford College, University of London, Papers in Geography*, 6, 28 pp.

Dawson, A.G. (1980). Shore erosion by frost: an example from the Scottish Lateglacial. In *Studies in the Lateglacial of North-West Europe*, eds. J.J. Lowe, J.M. Gray & J.E. Robinson, pp. 45–53. Oxford: Pergamon Press.

Dawson, A.G. (1982). Lateglacial sea-level changes and ice-limits in Islay, Jura and Scarba, Scottish Inner Hebrides. *Scottish Journal of Geology*, 18, 253–265.

Dawson, A.G. (1986). Quaternary shore platforms and sea-levels in southern Britain: a speculative hypothesis. In *Essays for Professor R.E.H. Mellor*, ed. W. Ritchie, J.C. Stone & A.S. Mather, pp. 377–382. Aberdeen: University of Aberdeen.

Dawson, A.G. (1988). The Main Rock Platform (Main Lateglacial Shoreline) in Ardnamurchan and Moidart, western Scotland. *Scottish Journal of Geology*, 24, 163–174.

Dawson, A.G. (1989). Distribution and development of the main rock platform: reply. *Scottish Journal of Geology*, 25, 233–238.

Dawson, A.G., Matthews, J.A. & Shakesby, R.A. (1987). Rock platform erosion on periglacial shores: a modern analogue for Pleistocene rock platforms in Britain. In *Periglacial Processes and Landforms in Britain and Ireland*, ed. J. Boardman, pp. 173–182. Cambridge: Cambridge University Press.

Dawson, M. (1985). Environmental reconstructions of a Late Devensian terrace sequence. Some preliminary findings. *Earth Surface Processes and Landforms*, 10, 237–246.

Dawson, M. (1987). Sedimentological aspects of periglacial terrace aggradations: a case study from the English Midlands. In *Periglacial Processes and Landforms in Britain and Ireland*, ed. J. Boardman, pp. 265–275. Cambridge: Cambridge University Press.

Dawson, M. & Bryant, I.D. (1987). Three dimensional facies geometry in Pleistocene outwash sediments, Worcestershire, UK. In *Recent Developments in Fluvial Sedimentology*, ed. F.G. Etheridge, Society of Economic Palaeontologists & Mineralogists, Special Publication, 39, 191–196.

De Gans, W. (1981). *The Drentsche Aa Valley System: a Study in Quaternary Geology*. Thesis, Vruje Universiteit te Amsterdam, 132pp.

De Gans, W. (1983). Lithology, stratigraphy, palynology and radiocarbon dating of the Middle and Late Weichselian deposits in the Drensche Aa valley system. *Geologie en Mijnbouw*, 62, 285–295.

De Gans, W. (1988). Pingo scars and their identification. In *Advances in Periglacial Geomorphology*, ed. M.J. Clark, pp. 300–324. Chichester: Wiley.

De la Beche, H.T. (1839). Report on the geology of Cornwall, Devon and west Somerset. *Memoirs of the Geological Survey*. London.

Dearing, J.A., Elner, J.K. & Happey-Wood, C.M. (1981). Recent sediment flux and erosional processes in a Welsh upland lake catchment based on magnetic susceptibility measurements. *Quaternary Research*, 16, 356–372.

Demek, J. (1964a). Castle koppies and tors in the Bohemian Highland (Czechoslovakia). *Biuletyn Peryglacjalny*, 14, 195–216.

Demek, J. (1964b). Slope development in granite areas of Bohemian Massif (Czechoslovakia). *Zeitschrift für Geomorphologie*, 5, 83–106.

Demek, J. (1968). Cryoplanation terraces in Yakutia. *Biuletyn Peryglacjalny*, 17, 91–116.

Demek, J. (1969). Cryogene processes and the development of cryoplanation terraces. *Biuletyn Peryglacjalny*, 18, 115–125.

DeWolf, Y. (1988). Stratified slope deposits. In *Advances in Periglacial Geomorphology*, ed. M.J. Clark, pp. 91–110. Chichester: Wiley.

Dickson, J.H., Jardine, W.G. & Price, R.J. (1976). Three late-Devensian sites in west central Scotland. *Nature*, 262, 43–44.

Dijkmans, J.W.A. (1988). Seasonal frost mounds in an eolian sand sheet near Søndre Strømfjord, W. Greenland. *Proceedings of the 5th International Conference on Permafrost, Trondheim*, pp. 728–733. Trondheim: Tapir Press.

Dimbleby, G.W. (1952). Pleistocene ice wedges in north-east Yorkshire. *Journal of Soil Science*, 3, 1–19.

Dineley, D.L. (1963). Contortions in the Bovey Beds (Oligocene), SW England. *Biuletyn Peryglacjalny*, 12, 151–160.

Dines, H.G., Hollingworth, S.E., Edwards, H., Buchan, S. & Welch, F.B.A. (1940). The mapping of head deposits. *Geological Magazine*, 77, 198–226.

Dines, H.G., Holmes, S.C.A. & Robbie, J.A. (1954). Geology of the country around Chatham. *Memoir of the Geological Survey, UK*. London: HMSO.

Dionne, J.-C. (1978). Formes et phénomènes périglaciaires en Jamésie, Québec subarctique. *Géographie Physique et Quaternaire*, 32, 187–247.

Dionne, J.-C. (1979). Ice action in the lacustrine environment. A review with particular reference to subarctic Quebec, Canada. *Earth Science Reviews*, 15, 185–212.

Dionne, J.-C. (1988). Frost weathering and ice action in shore platform development with particular reference to Quebec, Canada. *Zeitschrift für Geomorphologie, Supplementband*, 71, 117–130.

Dionne, J.-C. & Brodeur, D. (1988). Érosion des plates-formes rocheuses littorales par affouillement glaciel. *Zeitschrift für Geomorphologie*, 32, 101–115.

Dixon, J.C., Thorn, C.E. & Darmody, R.G. (1984). Chemical weathering processes on the Vantage Peak nunatak, Juneau Icefield, Southern Alaska. *Physical Geography*, 5, 111–131.

Doornkamp, J.C. (1974). Tropical weathering and the ultra-microscopic characteristics of regolith quartz on Dartmoor. *Geografiska Annaler*, 56, 73–82.

Douglas, T.D. (1982). Periglacial involutions and the evidence for coversands in the English Midlands. *Proceedings of the Yorkshire Geological Society*, 44, 131–143.

Douglas, T.D. & Harrison, S. (1984). Solifluction sheets – a review and case study from the Cheviot Hills. *Newcastle upon Tyne Polytechnic Occasional Series in Geography*, 7, 35 pp.

Douglas, T.D. & Harrison, S. (1985). Periglacial landforms and sediments in the Cheviots. In *Field Guide to the Periglacial Landforms of Northern England*, ed. J. Boardman, pp. 68–75. Cambridge: Quaternary Research Association.

Douglas, T.D. & Harrison, S. (1987). Late Devensian periglacial slope deposits in the Cheviot Hills. In *Periglacial Processes and Landforms in Britain and Ireland*, ed. J. Boardman, pp. 237–244. Cambridge: Cambridge University Press.

Drew, F. (1873). Alluvial and lacustrine deposits and glacial records of the Upper-Indus Basin. *Quarterly Journal of the Geological Society of London*, 29, 441–471.

Dubois, A.D. & Ferguson, D.K. (1985). The climatic history of pine in the Cairngorms based on radiocarbon dates and stable isotope analyses, with an account of the events leading up to its colonization. *Review of Palaeobotany and Palynology*, 46, 55–80.

Dunbar, M.J. (1968). *Ecological Development in Polar Regions*. New York: Prentice Hall, 127pp.

Dunn, J.R. & Hudec, P.P. (1966). Water, clay and rock soundness. *Ohio Journal of Science*, 66, 153–168.

Dunn, J.R. & Hudec, P.P.(1972). Frost and sorption effects in argillaceous rocks. *Highway Research Record*, 393, 65–78.

Dzulynski, S. (1963). Polygonal structures in experiments and their bearing upon some 'periglacial phenomena'. *Academie Polonaise des Sciences, Bulletin. Géologie et Géographie*, 11, 145–150.

Eden, D.N. (1980). The loess of north-east Essex, England. *Boreas*, 9, 165–177.

Eden, M.J. & Green, C.P. (1971). Some aspects of granite weathering and tor formation, Dartmoor, England. *Geografiska Annaler*, 53A, 92–99.

Edmunds, F.H. (1930). The coombe rock of the Hampshire and Sussex coast. *Memoirs of the Geological Survey: Summary of Progress for 1929*, 2, 63–68.

Edwards, W. (1967) Geology of the country around Ollerton. *Memoir of the Geological Survey of the United Kingdom*.

Egginton, P.A. & French, H.M. (1985) Solifluction and related processes, eastern Banks Island, NWT. *Canadian Journal of Earth Sciences*, 22, 1671–1678.

Ehlen, J. (1992). Analysis of spatial relationships among geomorphic, petrographic and structural characteristics of the Dartmoor tors. *Earth Surface Processes and Landforms*, 17, 53–67.

Ehlers, J. & Gibbard, P.L. (1991). Anglian glacial deposits in Britain and the adjoining offshore regions. In *Glacial Deposits in Great Britain and Ireland*, eds. J. Ehlers, P.L. Gibbard & J. Rose, pp. 17–24. Rotterdam: Balkema.

Ehlers, J., Gibbard, P.L. & Rose, J. (1991a). Glacial deposits of Britain and Ireland: general overview. In *Glacial Deposits in Great Britain and Ireland*, eds. J. Ehlers, P.L. Gibbard & J. Rose, pp. 493–502. Rotterdam: Balkema.

Ehlers, J., Gibbard, P.L. & Rose, J. (1991b). Editors. *Glacial Deposits in Great Britain and Ireland*. Rotterdam: Balkema.

Ellis, S. (1979). The identification of some Norwegian mountain soil types. *Norsk Geografisk Tidsskrift*, 33, 205–212.

Ellis, S. (1981). Patterned ground at Wharram Percy, North Yorkshire: its origin and palaeoenvironmental implications. In *The Quaternary in Britain*, eds. J. Neale & J. Flenley, pp 98–107. Oxford: Pergamon Press.

Evans, J.G. (1966). Late-glacial and Post-glacial subaerial deposits at Pitstone, Buckinghamshire. *Proceedings of the Geologists' Association*, 77, 347–364.

Evans, J.G. (1968). Periglacial deposits in the Chalk of Wiltshire. *Wiltshire Archaeological and Natural History Magazine*, 63, 12–27.

Evans, R. (1976). Observations on a stripe pattern. *Biuletyn Peryglacjalny*, 25, 9–22.

Evenson, E.B. & Cohn, B.P. (1979). The ice-foot complex: its morphology, formation and role in sediment transport and shoreline protection. *Zeitschrift für Geomorphologie*, 23, 58–75.

Evin, M. (1985). Contribution a l'étude de la macrofissuration dans les parvis situées a l'amont des glaciers rocheux Alpes du Sud (France, Italie). *Revue de Géomorphologie Dynamique*, 34, 17–30.

Evin, M. (1987). Lithology and fracturing control of rock glaciers in southwestern Alps of France and Italy. In *Rock Glaciers*, eds. J.R.Giardino, J.F Shroder & J.D. Vitek. pp. 83–106. London: Allen and Unwin.

Eyles, N. (1983). Modern Icelandic glaciers as depositional models for 'hummocky moraine' in the Scottish Highlands. In *Tills and Related Deposits*, eds. E.B. Evenson, C. Schlüchter & J. Rabassa, pp. 47–59. Rotterdam: Balkema.

Fagg, C.C. (1923). The recession of the Chalk escarpment and the development of Chalk valleys. *Proceedings and Transactions of the Croydon Natural History and Scientific Society*, 9, 93–112.

Fahey, B.D. (1983). Frost action and hydration as rock weathering mechanisms on schist: a laboratory study. *Earth Surface Processes and Landforms*, 8, 535–545.

Fahey, B.D. & Dagesse, D.F. (1984). An experimental study of the effect of humidity and temperature variations on the granular disintegration of argillaceous carbonate rocks in cold climates. *Arctic and Alpine Research*, 16, 291–298.

Fahnestock, R.K. (1969) Morphology of the Slims River. In *Icefield Ranges Research Project*, ed. V.C. Bushnell, pp. 161–172. Washington: American Geographical Society.

Fairbridge, R. (1977). Rates of sea-ice erosion of Quaternary littoral platforms. *Studia Geologica Polonica*, 52, 135–142.

Ferguson, R.I. (1984). Magnitude and modelling of snowmelt runoff in the Cairngorm Mountains. *Hydrological Sciences Journal*, 29, 49–62.

Firth, C.R. (1984). *Raised Shorelines and Ice Limits in the Inner Moray Firth and Loch Ness areas.* Unpublished PhD thesis, Coventry Polytechnic.

Fisch, W., Fisch, W. & Haeberli, W. (1977). Electrical DC resistivity soundings with long profiles on rock glaciers and moraines in the Alps of Switzerland. *Zeitschrift für Gletscherkunde und Glazialgeologie*, 13, 239–260.

Fitzharris, B.B. & Owens, I.F. (1984). Avalanche tarns. *Journal of Glaciology*, 30, 308–312.

FitzPatrick, E.A. (1956a). Progress report on the observations of periglacial phenomena in the British Isles. *Biuletyn Peryglacjalny*, 4, 99–115.

FitzPatrick, E.A. (1956b). An indurated soil horizon formed by permafrost. *Journal of Soil Science*, 7, 248–254.

FitzPatrick, E.A. (1958). An introduction to the periglacial morphology of Scotland. *Scottish Geographical Magazine*, 74, 28–36.

FitzPatrick, E.A. (1969). Some aspects of soil evolution in north-east Scotland. *Soil Science*, 107, 403–408.

FitzPatrick. E.A. (1987). Periglacial features in the soils of North East Scotland. In *Periglacial Processes and Landforms in Britain and Ireland*, ed. J. Boardman, pp. 153–162. Cambridge: Cambridge University Press.

Flemal, R.C. (1972). Ice injection origin of the DeKalb mounds, north central Illinois, USA. In *Proceedings of the 24th International Geological Congress, Montreal*, Section 12, 130–135. Montreal: University of Montreal.

Flemal, R.C. (1976). Pingos and pingo scars: their characteristics, distribution and utility in reconstructing former permafrost environments. *Quaternary Research*, 6, 37–53.

Ford, T.D. (1967). Deep weathering, glaciation and tor formation in Charnwood Forest, Leics. *Mercian Geologist*, 2, 3–14.

Ford, T.D. (1969). Dolomite tors and sandfilled sinkholes in the Carboniferous Limestone of Derbyshire, England. In *The Periglacial Environment*, ed. T.L. Péwé, pp. 387–397. Montreal: McGill University Press.

Foster, H.D. (1970). Establishing the age and geomorphological significance of sorted stone-stripes in the Rhinog Mountains, North Wales. *Geografiska Annaler*, 52A, 96–102.

Francis, E.A. (1970). Quaternary. *Transactions of the Natural History Society of Northumberland*, 41, 134–152.

Francou, B. (1988). *Talus Formation in High Mountain Environments: Alps and tropical Andes.* Thèse de Doctorat d'Etat, Université de Paris XVII.

Francou, B. (1989). La stratogenèse dans les formations de pente soumises à l'action du gel. *Bulletin de l'Association Française pour l'Étude du Quaternaire*, 1989–4, 185–199.

Francou, B. (1990). Stratification mechanisms in slope deposits in high subequatorial mountains. *Permafrost and Periglacial Processes*, 1, 249–263.

French, H.M. (1971). Slope asymmetry of the Beaufort Plain, northwest Banks Island, NWT, Canada. *Canadian Journal of Earth Sciences*, 8, 717–731.

French, H.M. (1972). Asymmetrical slope development in the Chiltern Hills. *Biuletyn Peryglacjalny*, 21, 51–73.

French, H.M. (1973). Cryopediments on the chalk of southern England. *Biuletyn Peryglacjalny*, 22, 149–156.

French, H.M. (1976). *The Periglacial Environment*. London: Longman.

French, H.M. (1986). Periglacial involutions and mass displacement structures, Banks Island, Canada. *Geografiska Annaler*, 68A, 167–174.

French, H.M. (1987). Periglacial processes and landforms in the Western Canadian Arctic. In *Periglacial Processes and Landforms in Britain and Ireland*, ed. J. Boardman, pp. 27–43. Cambridge: Cambridge University Press.

French, H.M. (1988). Active layer processes. In *Advances in Periglacial Geomorphology*, ed. M.J. Clark, pp. 151–177. Chichester: Wiley.

French, H.M. & Dutkiewicz, L. (1976). Pingos and pingo-like forms, Banks Island, Western Canadian Arctic. *Biuletyn Peryglacjalny*, 26, 211–222.

French, H.M. & Egginton, P. (1973). Thermokarst development, Banks Island, Western Canadian Arctic. In *Permafrost: North American Contribution to the 2nd International Conference on Permafrost, Yakutsk, USSR.* pp. 203–212, Washington: National Academy of Sciences Publication 2115.

French, H.M. & Harry, D.G. (1988). Nature and origin of ground ice, Sandhills Moraine, southwest Banks Island, western Canadian Arctic. *Journal of Quaternary Science*, 3, 19–30.

French, H.M. & Harry, D,G. (1990). Observations on buried glacier ice and massive segregated ice, western arctic coast, Canada. *Permafrost and Periglacial Processes*, 1, 31–43.

French, H.M. & Hegginbottom, J.A. (1983). Editors. *Northern Yukon Territory and Mackenzie Delta, Canada.* Guidebook 3, Fourth International Conference on Permafrost, Fairbanks, Alaska. Division of Geological & Geophysical Surveys, Department of Natural Resources, Fairbanks, Alaska, 185pp.

French, H.M. & Lewkowicz, A.G. (1981). Periglacial slopewash investigation, Banks Island, Western Arctic. *Biuletyn Peryglacjalny*, 28, 33–45.

Frich, P. & Brandt, E. (1985). Holocene talus accumulation rates, and their influence on rock glacier growth. A case study from Igpik, Disko, West Greenland. *Geografisk Tidsskrift*, 85, 32–43.

Fukuda, M. & Matsuoka, T. (1982). Pore water pressure profile in freezing porous rocks. In *Low Temperature Science. Series A: Physical Sciences.* pp. 217–224. Hokkaido University, Sapporo, Japan.

Furrer, G. (1965). Die subnivale Höhenstufe und ihre Untergrenze in den Bündner und Walliser Alpen. *Geographica Helvetica*, 20, 185–192.

Gailey, R.A. (1961). Fossil ice wedge at Poltalloch. *Scottish Geographical Magazine*, 77, 88.

Gallois, R.W. (1965). *The Wealden District*. London: Institute of Geological Sciences, HMSO.

Gallop, M. (1991). *Quaternary Periglacial Slope Processes in South West England: Field Data and Laboratory Simulation.* Unpublished PhD. Thesis, University of Wales.

Galloway, R.W. (1958). *Periglacial Phenomena in Scotland.* Unpublished PhD. Thesis, University of Edinburgh.

Galloway, R.W. (1961a). Solifluction in Scotland. *Scottish Geographical Magazine*, 77, 75–87.

Galloway, R.W. (1961b). Periglacial phenomena in Scotland. *Geografiska Annaler*, 43, 348–353.

Galloway, R.W. (1961c). Ice wedges and involutions in Scotland. *Biuletyn Peryglacjalny*, 10, 169–193.

Gamper, M. (1981). Heutige solifluktionsbeträge von Erdströmen und klimamorphologische Interpretation fossiler Böden. *Ergebnisse der Wissenschaftlichen Untersuchungen im Sweizerischen Nationalpark*, 15, 443pp.

Gamper, M. (1983). Controls and rates of movement of solifluction lobes in the eastern Swiss Alps. *Proceedings of the 4th International Conference on Permafrost, Fairbanks, Alaska*, pp. 328–333. Washington: National Academy Press.

Gamper, M. & Suter, J. (1982). Postglaziale Klimageschichte der Schweitzer Alpen. *Geographica Helvetica*, 37, 105–114.

Gangloff, P. (1983). Les fondements géomorphologiques de la théorie des paléonunataks: le cas des Monts Torngats. *Zeitschrift für Geomorphologie, Supplementband*, 47, 109–136.

Gangloff, P. & Pissart, A. (1983). Évolution géomorphologique et palses minérales dans la région de Kuujjuak (Fort-Chimo, Québec). *Bulletin de la Société Géomorphologique de Liège*, 19, 119–132.

Gardner, J. (1970). Geomorphic significance of avalanches in the Lake Louise area, Alberta, Canada. *Arctic and Alpine Research*, 2, 135–144.

Gardner, J. (1983). Rockfall frequency and distribution in the Highwood Pass area, Canadian Rocky Mountains. *Zeitschrift für Geomorphologie*, 27, 311–324.

Gardner, K. & West, R.G. (1975). Fossil ice-wedge polygons at Corton, Suffolk. *Bulletin of the Geological Society of Norfolk*, 27, 47–53.

Gatty, V.H. (1906). The glacial aspect of Ben Nevis. *Geographical Journal*, 27, 487–492.

Gaunt, G.D. (1976a). The Devensian maximum ice limit in the Vale of York. *Proceedings of the Yorkshire Geological Society*, 40, 631–637.

Gaunt, G.D. (1976b). *The Quaternary Geology of the Southern Part of the Vale of York*. Unpublished PhD. Thesis, University of Leeds.

Gaunt, G.D., Jarvis, R.A. & Matthews, B. (1971). The Late Weichselian sequence in the Vale of York. *Proceedings of the Yorkshire Geological Society*, 38, 281–284.

Geikie, A. (1865). *The Scenery of Scotland*. 1st Edition. London and Cambridge: Macmillan

Geikie, A. (1869). Surface-wash. *Memoirs of the Geological Survey: Scotland. Explanation of Sheet 24*, 21.

Geikie, J. (1877). *The Great Ice Age*. 2nd Edition. London: Daldy, Ibister and Company.

Geikie, J. (1894). *The Great Ice Age*. 3rd Edition. London: Daldy, Ibister and Company

Gemmell, A.M.D. & Ralston, I.B.M. (1984). Some recent discoveries of ice-wedge cast networks in north-east Scotland. *Scottish Journal of Geology*, 20, 115-118.

Gerrard, A.J. (1974). The geomorphological importance of jointing on Dartmoor granite. *Institute of British Geographers, Special Publication*, 7, 39–51.

Gerrard, A.J. (1978). Tors and granite landforms of Dartmoor and eastern Bodmin Moor. *Proceedings of the Ussher Society*, 4, 204–210.

Gerrard, A.J. (1984). Multiple working hypotheses and equifinality in geomorphology: comments on a recent article by Haines-Young and Petch. *Transactions of the Institute of British Geographers*, 9, 364–366.

Gerrard, A.J. (1988a). *Rocks and Landforms*. London: Unwin Hyman.

Gerrard, A.J. (1988b). Periglacial modification of the Cox Tor–Staples Tors area of western Dartmoor, England. *Physical Geography*, 9, 280–300.

Gibbard, P.L. (1977). Pleistocene history of the Vale of St. Albans. *Philosophical Transactions of the Royal Society of London*, B 280, 445–483.

Gibbard, P.L. (1979). Middle Pleistocene drainage in the Thames Valley. *Geological Magazine*, 116, 35–44.

Gibbard, P.L. (1985). *The Pleistocene History of the Middle Thames Valley*. Cambridge: Cambridge University Press.

Gibbard, P.L. (1988). The history of the great northwest European rivers during the past three million years. *Philosophical Transactions of the Royal Society of London*, B 318, 559–602.

Gibbard, P.L. *et al.* (1991). Early and early Middle Pleistocene correlations in the southern North Sea Basin. *Quaternary Science Reviews*, 10, 23–52.

Gibbard, P.L., Wintle, A.G. & Catt, J.A. (1987). Age and origin of clayey silt 'brickearth' in west London, England. *Journal of Quaternary Science*, 2, 3–10.

Girling, M.A. (1974). Evidence from Lincolnshire of the age and intensity of the mid-Devensian temperate episode. *Nature*, 250, 270.

Gladfelter, B.G. (1975). Middle Pleistocene sedimentary sequences in East Anglia (United Kingdom). In *After the Australopithecines: Stratigraphy, Ecology and Cultural Change in the Middle Pleistocene*, eds. K.E. Butzer & G.L. Isaac, pp. 225–258. The Hague: Mouton.

Gleason, K.J., Kranz, W.B., Caine, N., George, J.H. & Gunn, R.D. (1986). Geometrical aspects of sorted patterned ground in recurrently frozen soil. *Science*, 232, 216–230.

Glentworth, R. (1944). Studies on the soils developed on basic igneous rocks in central Aberdeenshire. *Transactions of the Royal Society of Edinburgh*, 61, 149–170.

Glentworth, R. (1954). The soils of the country around Banff, Huntly and Turiff. *Memoirs of the Soil Survey of Great Britain, Scotland*. Edinburgh: HMSO.

Glentworth, R. & Dion, H.G. (1949). The association or hydraulic sequence in certain soils of the podzolic zone of north-east Scotland. *Journal of Soil Science*, 1, 35–49.

Godard, A. (1958). Quelques observations sur le modelé des regions volcaniques du nord-ouest de l'Écosse. *Scottish Geographical Magazine*, 74, 37–43.

Godard, A. (1959). Contemporary periglacial phenomena in western Scotland. *Scottish Geographical Magazine*, 75, 55.

Godard, A. (1965). *Recherches de Géomorphologie en Écosse du Nord-Ouest*. Paris: Masson.

Godwin, H. (1956). *The History of the British Flora*. Cambridge: Cambridge University Press.

Godwin, H. (1959). Studies of the post-glacial history of British vegetation. XIV. Late glacial deposits at Moss Lake, Liverpool. *Philosophical Transactions of the Royal Society*, B 242, 127–149.

Goldthwait, R.P. (1976). Frost-sorted patterned ground: a review. *Quaternary Research*, 6, 27–35.

Good, T.R. & Bryant, I.D. (1985). Fluvio-aeolian sedimentation – an example from Banks Island, NWT, Canada. *Geografiska Annaler*, 67A, 33–46.

Goodier, R. & Ball, D.F. (1969). Recent ground pattern phenomena in the Rhinog Mountains, North Wales. *Geografiska Annaler*, 51A, 121–126.

Goodier, R. & Ball, D.F. (1975). Ward Hill, Orkney: patterned ground features and their origins. In *The Natural Environment of Orkney*, ed. R. Goodier, pp. 47–56. Edinburgh: Nature Conservancy Council.

Gordon, J.E. (1979). Reconstructed Pleistocene ice-sheet temperatures and glacial erosion in northern Scotland. *Journal of Glaciology*, 22, 331–344.

Gordon, J.E. (1981). Ice-scoured topography and its relationships to bedrock structure and ice movement in parts of northern Scotland and west Greenland. *Geografiska Annaler*, 63A, 55–65.

Goudie, A.S. & Piggott, N.R. (1981). Quartzite tors, stone stripes, and slopes at the Stiperstones, Shropshire, England. *Biuletyn Peryglacjalny*, 28, 47–56.

Grant, D.R. (1977). Altitudinal weathering zones and glacial limits in Western Newfoundland, with particular reference to Gros Morne National Park. *Geological Survey of Canada Paper*, 77–1A, 455–463.

Grawe, O.R. (1936). Ice as an agent of rock weathering. *Journal of Geology*, 44, 173–182.

Gray, J.M. (1974). The Main Rock Platform of the Firth of Lorn, western Scotland. *Transactions of the Institute of British Geographers*, 61, 81–99.

Gray, J.M. (1978). Low-level shore platforms in the south-west Scottish Highlands: altitude, age and correlation. *Transactions of the Institute of British Geographers, New Series*, 3, 151–164.

Gray, J.M. (1982). The last glaciers (Loch Lomond Advance) in Snowdonia, N Wales. *Geological Journal*, 17, 111–133.

Gray, J.M. (1989). Distribution and development of the Main Rock Platform, western Scotland: comment. *Scottish Journal of Geology*, 25, 227–231.

Gray, J.M. & Brooks, C.L. (1972). The Loch Lomond Readvance moraines of Mull and Menteith. *Scottish Journal of Geology*, 8, 95–103.

Gray, J.M. & Coxon, P. (1991). The Loch Lomond Stadial glaciation in Britain and Ireland. In *Glacial Deposits in Great Britain and Ireland*, eds. J. Ehlers, P.L. Gibbard & J. Rose, pp. 89–105. Rotterdam: Balkema.

Gray, J.M. & Ivanovich, M. (1988). Age of the Main Rock Platform, western Scotland. *Palaeogeography, Palaeoecology, Palaeoclimatology*, 68, 337–345.

Gray, J.M. & Lowe, J.J. (1977). The Scottish Lateglacial environment: a synthesis. In *Studies in the Scottish Lateglacial Environment*, eds. J.M. Gray & J.J. Lowe, pp. 163–181. Oxford: Pergamon.

Gray, J.T. (1972). Debris accretion on talus slopes in the central Yukon Territory. In *Mountain Geomorphology*, eds. H.O. Slaymaker & H.J. McPherson, pp. 75–84. Vancouver: Tantalus Press.

Gray, J.T. (1973). Geomorphic effects of avalanches and rock-falls on steep mountain slopes in the central Yukon Territory. In *Research in Polar and*

Alpine Geomorphology, eds. B.D. Fahey & R.D. Thomson, pp. 107–117. Norwich: Geobooks.

Green, C.P., Coope, G.R., Currant, A.P., Holyoak, D.T., Ivanovich, M., Jones, R.L., Keen, D.H., McGregor, D.F.M. & Robinson, J.E. (1984). Evidence of two temperate episodes in Late Pleistocene deposits at Marsworth, UK. *Nature,* 309, 778–781.

Green, C.P. & Eden, M.J. (1971). Gibbsite in the weathered Dartmoor granite. *Geoderma,* 6, 315–317.

Green, C.P. & Eden, M.J. (1973). Slope deposits on the weathered Dartmoor granite, England. *Zeitschrift für Geomorphologie,* 18, 26–37.

Green, C.P. & McGregor, D.F.M. (1980). Quaternary evolution of the River Thames. In *The Shaping of Southern England,* ed. D.K.C. Jones, pp. 177–202. London: Academic Press.

Green, C.P., McGregor, D.F.M. & Evans, A.H. (1982). Development of the Thames drainage system in Early and Middle Pleistocene times. *Geological Magazine,* 119, 281–290.

Gregory, J.W. (1930). Stone polygons beside Loch Lomond. *Geographical Journal,* 76, 415–418.

Gregory, K.J. (1965). Aspect and landforms in north-east Yorkshire. *Biuletyn Peryglacjalny,* 15, 115–120.

Gregory, K.J. & Walling, D.E. (1968). The variation of drainage density within a catchment. *Bulletin of the International Association for Scientific Hydrology,* 13, 61–68.

Grønlie, O.T. (1924). Contributions to the Quaternary geology of Novaya Zemlya. In *Report of the Scientific Results of the Norwegian Expedition to Novaya Zemlya, 1921,* ed. O. Holtedahl, 124 pp.

Grove, J.M. (1972). The incidence of landslides, avalanches and floods in western Norway during the Little Ice Age. *Arctic and Alpine Research,* 4, 131–138.

Gruhn, R. & Bryan, A.L. (1969). Fossil ice wedge polygons in southeast Essex, England. In *The Periglacial Environment: Past and Present,* ed. T.L.Péwé, pp. 351–363. Montreal: McGill-Queen's University Press.

Guilcher, A. (1950). Nivation, cryoplanation et solifluction quaternaires dans les collines de Bretagne occidentale et du nord du Devonshire. *Revue de Géomorphologie Dynamique,* 1, 53–78.

Guilcher, A. (1981). Cryoplanation littoral et cordons glaciels de basse mer dans la région de Rimouski, côté sud de l'estuaire du Saint-Laurent, Québec. *Géographie Physique et Quaternaire,* 35, 155–169.

Guillien, Y. (1951). Les grèzes litées de Charentes. *Revue de Géographie des Pyrénées et du Sud Ouest,* 22, 155–162.

Gullentops, F. & Paulisen, E. (1978). The drop soil of Eisden type. *Biuletyn Peryglacjalny,* 27, 105–115.

Guodong, C. (1983). The mechanism of repeated segregation for the formation of thick layered ground ice. *Cold Regions Research & Technology,* 8, 57–66.

Haeberli, W. (1985). Creep of mountain permafrost: internal structure and flow of alpine rock glaciers. *Mitteilungen der Versuchsanstalt für Wasserbau, Hydrologie und Glaziologie,* 77, 142 pp.

Hails, J.R. & White, P.C.S. (1970). Periglacial features at Walton-on-Naze, Essex. *Proceedings of the Geologists' Association,* 81, 205–219.

Hall, A.M. (1983). *Weathering and Landform Evolution in North-East Scotland.* Unpublished PhD Thesis, University of St Andrews.

Hall, A.M. (1984). Editor. *Buchan Field Guide.* Cambridge: Quaternary Research Association.

Hall, A.M. (1985). Cenozoic weathering covers in Buchan, Scotland, and their significance. *Nature,* 325, 392–395.

Hall, A.M. (1986a). Weathering and relief development in Buchan, Scotland. In *International Geomorphology, 1986,* ed. V. Gardiner, pp. 991–1005. Chichester: Wiley.

Hall, A.M. (1986b). Deep weathering patterns in north-east Scotland and their geomorphological significance. *Zeitschrift für Geomorphologie,* 30, 407–422.

Hall, A.M. & Bent, A.J.A. (1990). The limits of the last British ice sheet in northern Scotland and the adjacent shelf. *Quaternary Newsletter,* 61, 2–12.

Hall, A.M. & Connell, E.R. (1986). A preliminary report on the Quaternary sediments at Leys Gravel Pit, Buchan, Scotland. *Quaternary Newsletter,* 48, 17–28.

Hall, A.M. & Connell, E.R. (1991). The glacial deposits of Buchan, north-east Scotland. In *The Glacial Deposits in Great Britain and Ireland,* eds. J. Ehlers, P.L. Gibbard, & J. Rose, pp. 129–136. Rotterdam: Balkema.

Hall, A.M. & Mellor, A. (1988). The characteristics and significance of deep weathering in the Gaick area, Grampian Highlands, Scotland. *Geografiska Annaler,* 70A, 309–314.

Hall, A.M. & Sugden, D.E. (1987). Limited modification of mid-latitude landscapes by ice sheets: the case of northeast Scotland. *Earth Surface Processes and Landforms,* 12, 531–542.

Hall, A.M. & Whittington, G.W. (1989). Late Devensian glaciation of southern Caithness. *Scottish Journal of Geology,* 25, 307–324.

Hall, K. (1986). Rock moisture content in the field and the laboratory and its relationship to mechanical weathering studies. *Earth Surface Processes and Landforms,* 11, 131–142.

Hall, K. (1988). A laboratory simulation of rock breakdown due to freeze–thaw in a maritime Antarctic environment. *Earth Surface Processes and Landforms,* 13, 369–382.

Hallet, B. (1983). The breakdown of rock due to freezing: a theoretical model. *Proceedings of the 4th International Conference on Permafrost, Fairbanks, Alaska,* pp.433–438. Washington DC: National Academy Press.

Hallet, B & Prestrud, S. (1986). Dynamics of periglacial sorted circles in western Spitzbergen. *Quatery Research,* 26, 81–89.

Hallet, B., Prestrud Anderson, S., Stubbs, C.W. & Carrington Gregory, E. (1988). Surface soil displacements in sorted circles, western Spitsbergen. *Proceedings of the 5th International Conference on Permafrost, Trondheim, Norway,* pp. 770–775. Trondheim: Tapir Press.

Hallet, B., Walder, J.S. & Stubbs, C.W. (1991). Weathering by segregation ice growth in microcracks at sustained sub-zero temperatures: verification from an experimental study using acoustic emissions. *Permafrost and Periglacial Processes,* 2, 283–300.

Halstead, C.A. (1974). Soil freeze–thaw recording in the Southern Uplands of Scotland. *Weather,* 29, 261–265.

Hamilton, T.D., Ager, T.A. & Robinson, S.W. (1983). Late Holocene ice wedges near Fairbanks, Alaska, USA: environmental setting and history of growth. *Arctic and Alpine Research,* 15, 157–168.

Hamilton, T.D. & Obi, C.M. (1982). Pingos in the Brooks Range, northern Alaska, USA. *Arctic and Alpine Research,* 14, 13–20.

Hampton, M.H. (1975). Competence of fine-grained debris flows. *Journal of Sedimentary Petrology,* 45, 834–844.

Hampton, M.H. (1979). Buoyancy in debris flows. *Journal of Sedimentary Petrology,* 49, 753–758.

Handa, S. & Moore, P.D. (1976). Studies in the vegetational history of mid Wales No. IV: Pollen analysis of some pingo basins. *New Phytologist,* 77, 205–225.

Hansen, K. (1970). Geological and geographical investigations in King Frederik IX's Land. Morphology, sediments, periglacial processes and salt lakes, Sandflugtdalen. *Meddelelser om Grønland,* 188, 22–45.

Hansom, J.D. (1983). Shore-platform development in the South Shetland Islands, Antarctica. *Marine Geology,* 53, 211–229.

Harker, A. (1901). Ice erosion in the Cuillin Hills, Skye. *Transactions of the Royal Society of Edinburgh,* 40, 221–252.

Harkness, D.D. & Wilson, H.W. (1979). Scottish Universities Research and Reactor Centre radiocarbon measurements III. *Radiocarbon,* 21, 203–256.

Harris, C. (1973). Some factors affecting the rates and processes of periglacial mass movements. *Geografiska Annaler,* 55A, 24–28.

Harris, C. (1981a). Microstructures in solifluction sediments from South Wales and North Norway. *Biuletyn Peryglacjalny,* 28, 221–226.

Harris, C. (1981b). *Periglacial Mass Wasting: A Review of Research.* Norwich: Geo Abstracts.

Harris, C. (1982). The distribution and altitudinal zonation of periglacial landforms, Okstindan, North Norway. *Zeitschrift für Geomorphologie,* 26, 283–304.

Harris, C. (1983). Vesicles in thin sections of periglacial soils from north and south Norway. *Proceedings of the 4th International Conference on Permafrost, Fairbanks, Alaska,* pp. 445–449. Washington DC: National Academy Press.

Harris, C. (1985). Geomorphological applications of soil micromorphology with particular reference to periglacial sediments and processes. In *Geomorphology and Soils,* eds. K.S. Richards, R.R. Arnett & S. Ellis, pp. 219–232. London: Allen and Unwin.

Harris, C. (1986). Some observations concerning the morphology and sedimentology of a protalus rampart, Okstindan, Norway. *Earth Surface Processes and Landforms,* 11, 673–676.

Harris, C. (1987a). Solifluction and related periglacial deposits in England and Wales. In *Periglacial Processes and Landforms in Britain and Ireland,* ed. J. Boardman, pp. 209–224. Cambridge: Cambridge University Press.

Harris, C. (1987b). Mechanisms of mass movement in periglacial environments. In *Slope Stability,* eds. M.G. Anderson & K.S. Richards, pp. 531–559. Chichester: Wiley.

Harris, C. (1989). Some possible Devensian ice-wedge casts in Mercia Mudstone near Cardiff, South Wales. *Quaternary Newsletter,* 58, 11–13.

Harris, C. (1990). Micromorphology and microfabrics of sorted circles, Front Range, Colorado, USA. In *Proceedings of the 5th Canadian Permafrost Conference, Quebec, June 1990,* eds. M.M. Burgess, D.G. Harry & D.C. Sego, pp. 89–94. Ottawa: National Research Council of Canada.

Harris, C. (1991). Glacial deposits at Wylfa Head, Anglesey, North Wales: evidence for Late Devensian deposition in a non-marine environment. *Journal of Quaternary Science,* 6, 67–77.

Harris, C. & Cook, J.D. (1988). Micromorphology and microfabrics of sorted circles, Jotunheimen, Norway. *Proceedings of the 5th International Conference on Permafrost, Trondheim,* pp. 776–783. Trondheim: Tapir Press.

Harris, C. & Ellis, S. (1980). Micromorphology of soils in soliflucted materials, Okstindan, northern Norway. *Geoderma,* 23, 11–29.

Harris, C. & McCarroll, D. (1990). Glanllynnau. In *North Wales: Field Guide,* eds. K. Addison, M.J. Edge & R. Watkins, pp. 38–47. Cambridge: Quaternary Research Association.

Harris, C. & Wright, M.D. (1980). Some last glaciation drift deposits near Pontypridd, South Wales. *Geological Journal,* 15, 7–20.

Harris, S.A. (1988). The alpine periglacial zone. In *Advances in Periglacial Geomorphology,* ed. M.J.Clark, pp. 369–414. Chichester: Wiley.

Harris, S.A. & Brown, R.J.E. (1978). Plateau Mountain: a case study of alpine permafrost in the Canadian Rocky Mountains. *Proceedings of the 3rd International Conference on Permafrost, Edmonton, Alberta,* pp. 385–391. Ottawa: National Research Council of Canada.

Harris, S.A., French, H.M., Heginbottom, J.A., Johnston, G.H., Ladanyi, B., Sego, D.C. & van Everdingen, R.O. (1988). *Glossary of Permafrost and Related Ground-Ice Terms.* National Research Council of Canada Technical Memorandum 142, Ottawa, 156pp.

Harry, D.G. (1988). Ground ice and permafrost. In *Advances in Periglacial Geomorphology,* ed. M.J. Clark, pp. 113–149. Chichester: Wiley.

Harry, D.G. & French, H.M. (1983). The orientation and evolution of thaw lakes, southwest Banks Island, Canadian Arctic. *Proceedings of the 4th International Conference on Permafrost, Fairbanks, Alaska,* pp. 456–461. Washington: National Academy Press.

Harry, D.G., French, H.M. & Pollard, W.H. (1985). Ice wedges and permafrost conditions near King Point, Beaufort Sea coast, Yukon Territory. *Geological Survey of Canada,* Paper 85-1A, 111–116.

Harry, D.G. & Gozdzik, J.S. (1988). Ice wedges: growth, thaw transformation, and palaeoenvironmental significance. *Journal of Quaternary Science,* 3, 39–55.

Harvey, A.M. (1986). Geomorphic effects of a 100 year storm in the Howgill Fells, Northwest England. *Zeitschrift für Geomorphologie,* 30, 71–91.

Harvey, A.M., Alexander, R.W. & James, P.A. (1984). Lichens, soil development and the age of Holocene valley floor landforms: Howgill Fells, Cumbria. *Geografiska Annaler,* 66A, 353–366.

Harvey, A.M., Oldfield, F., Baron, A.F. & Pearson, G.W. (1981). Dating of postglacial landforms in the central Howgills. *Earth Surface Processes and Landforms,* 6, 401–412.

Harvey, A.M. & Renwick, W.H. (1987). Holocene alluvial fan and terrace formation in the Bowland Fells, northwest England. *Earth Surface Processes and Landforms,* 12, 249–257.

Haugen, R.K., Outcalt, S.I. & Harle, J.C. (1983). Relationships between estimated mean annual air and permafrost temperatures in north-central Alaska. *Proceedings of the 4th International Conference on Permafrost, Fairbanks, Alaska,* pp. 462–467. Washington, DC: National Academy Press.

Hawkins, A.B. & Privett, K.D. (1981). A building site on cambered ground at Radstock, Avon. *Quarterly Journal of Engineering Geology,* 14, 151–167.

Hawkins, H.L. (1952). A pinnacle of Chalk penetrating the Eocene on the floor of a buried river channel at Ashford Hill, near Newbury, Berkshire. *Quarterly Journal of the Geological Society of London,* 108, 233–260.

Hay, T. (1936). Stone stripes. *Geographical Journal,* 87, 47–50.

Hay, T. (1937). Physiographical notes on the Ullswater area. *Geographical Journal,* 90, 426–445.

Hay, T. (1943). Notes on glacial erosion and stone stripes. *Geographical Journal,* 102, 13–20.

Henry, A. (1984a). *The Lithostratigraphy, Biostratigraphy and Chronostratigraphy of Coastal Pleistocene Deposits in Gower, South Wales.* Unpublished PhD Thesis, University of Wales, 517pp.

Henry, A. (1984b). Gower. In *Quaternary Research Association Field Guide, Wales: Gower, Preseli, Fforest Fawr,* ed. D.Q. Bowen, pp. 19–32. Cambridge: Quaternary Research Association.

Hey, R.W. (1965). Highly quartzose pebble gravels in the London Basin. *Proceedings of the Geologists' Association,* 76, 403–520.

Hey, R.W. (1991). Pre-Anglian glacial deposits and glaciations in Great Britain. In *Glacial Deposits in Great Britain and Ireland,* eds. J. Ehlers, P.L. Gibbard & J. Rose, pp. 13–16. Rotterdam: Balkema.

Higashi, A. & Corte, A.E. (1972). Growth and development of perturbations on the soil surface due to repetition of freezing and thawing. *Ice Research Laboratory, Department of Applied Physics, Faculty of Engineering, Hokkaido University, Sapporo, Japan, Research Papers,* 23.

Higginbottom, I.E. & Fookes, P.G. (1971). Engineering aspects of periglacial features in Britain. *Quarterly Journal of Engineering Geology,* 3, 85–117.

Hills, R.C. (1969). Comparative weathering of granite and quartzite in a periglacial environment. *Geografiska Annaler,* 51A, 46–47.

Hodgson, J.M. (1964). The low-level Pleistocene marine sands and gravels of the West Sussex coastal plain. *Proceedings of the Geologists' Association,* 75, 547–561.

Hoffman, R.S. (1974). Terrestrial vertebrates. In *Arctic and Alpine Environments,* eds. J.D. Ives, & R.G. Barry, pp. 475–568. London: Methuen.

Holdgate, M.W., Allen, S.E. & Chambers, M.J.G. (1967). A preliminary investigation of the soils of Signy Island, South Orkney Islands. *British Antarctic Survey Bulletin,* 12, 53–71.

Holdsworth, G. & Bull, C. (1970). The flow of cold ice: investigations on Meserve Glacier, Antarctica. *International Association of Scientific Hydrology Publication,* 86, 204–216.

Höllermann, P.W. (1964). Rezente Verwitterung, Abtragung und Formenschatz in den Zentralalpen am Beispiel des oberen Suldentales (Ortlergruppe). *Zeitschrift für Geomorphologie, Supplementband,* 4, 257 pp.

Höllermann, P.W. (1967). Zur Verbreitung rezenter periglazialer Kleinformen in den Pyrenäen und Ostalpen. *Göttinger Geographische Abhandlungen,* 40, 198 pp.

Hollingworth, S.E. (1934). Some solifluction phenomena in the northern part of the English Lake District. *Proceedings of the Geologists' Association,* 2, 167–188.

Hollingworth. S.E., Taylor, J.H. & Kellaway, G.A. (1944). Large scale superficial structures in the Northampton Ironstone Field. *Quarterly Journal of the Geological Society,* 99–100, 1–44.

Holmes, C.D. & Colton, R.B. (1960). Patterned ground near Dundas (Thule Air Force Base), Greenland. *Meddelelser om Grønland,* 158 (6), 15 pp.

Holmes, G. (1984). *Rock Slope Failure in Parts of the Scottish Highlands.* Unpublished PhD Thesis, University of Edinburgh.

Holmes, G.W., Hopkins, D.M. & Foster, H.L. (1968). Pingos in central Alaska. *United States Geological Survey Bulletin,* 1241-H, 40 pp.

Hopkins, D.M. & Sigafoos, R.S. (1951). Frost action and vegetation patterns on the Seward Peninsula, Alaska. *United States Geological Survey Bulletin,* 974-C, 100pp.

Horswill, P. & Horton, A. (1976). Cambering and valley bulging in the Gwash Valley at Empingham, Rutland. *Philosophical Transactions of the Royal Society,* A 283, 427–451.

Howarth, P.J. & Bones, J.G. (1972). Relationships between process and geometric form on high arctic debris slopes. *Institute of British Geographers, Special Publication,* 4, 139–153.

Huang, S.L., Aughenbaugh, N.B. & Wu, M-C. (1986). Stability study of the CRREL permafrost tunnel. *Journal of Geotechnical Engineering,* 112, 777–790.

Huber, T.P. (1982). The geomorphology of subalpine snow avalanche runout zones: San Juan Mountains, Colorado. *Earth Surface Processes and Landforms,* 7, 109–116.

Huddart, D. (1991). The glacial deposits and glacial history of the North and West Cumbrian lowlands. In *Glacial Deposits in Great Britain and Ireland,* eds. J. Ehlers, P.L. Gibbard & J. Rose, pp. 151–167. Rotterdam: Balkema.

Hudec, P.P. (1973). Weathering of rocks in arctic and subarctic environments. In *Proceedings, Symposium on the Geology of the Canadian Arctic,* eds. J.D. Aitken & D.J. Glass, pp. 313–335. Ottawa: Geological Association of Canada.

Hughes, O.L. (1972). Surficial geology and land classification, Mackenzie Valley transportation corridor. *Proceedings, Canadian Northern Pipeline Research Conference*, National Research Council of Canada, Technical Memorandum 104, 17–24.

Hunt, C.B. & Washburn, A.L. (1966). Patterned ground. *United States Geological Survey Professional Paper*, 494-B, 104–133.

Hustich, I. (1966). On the forest tundra and the northern treelines. *Report of the Kevo Subarctic Research Station*, 3, 7–47.

Hutchinson, J.N. (1967). The free degradation of London Clay cliffs. *Proceedings of the Geotechnical Conference, Oslo*, 1, 113–118.

Hutchinson, J.N. (1974). Periglacial solifluxion: an approximate mechanism for clayey soils. *Géotechnique*, 24, 438–443.

Hutchinson, J.N. (1980). Possible Late Quaternary pingo remnants in Central London. *Nature*, 284, 253–255.

Hutchinson, J.N. (1991). Periglacial and slope processes. In *Quaternary Engineering Geology*, eds. A. Forster, M.G. Culshaw, J.C. Cripps, J.A. Little & C.F. Moon, pp. 283–331. London: Geological Society Engineering Geology Special Publication No. 7, 283–331.

Hutchinson, J.N. & Hight, D.W. (1987). Strongly folded structures associated with permafrost degradation and solifluction at Lyme Regis, Dorset. In *Periglacial Processes and Landforms in Britain and Ireland*, ed. J. Boardman, pp. 245–256. Cambridge: Cambridge University Press.

Hutchinson, J.N., Somerville, S.H. & Petley, D.J. (1973). A landslide in periglacially disturbed Etruria Marl at Bury Hill, Staffordshire. *Quarterly Journal of Engineering Geology*, 6, 177–404.

Hutchinson, J.N. & Thomas-Betts, A. (1990). Extent of permafrost in southern Britain in relation to geothermal heat flux. *Quarterly Journal of Engineering Geology*, 23, 387–390.

Ingram, M. (1958). The ecology of the Cairngorms. IV. The *Juncus* zone: *Juncus triffidus* communities. *Journal of Ecology*, 46, 707–737.

Innes, J.L. (1982). *Debris Flow Activity in the Scottish Highlands*. Unpublished PhD thesis, University of Cambridge. 2 volumes.

Innes, J.L. (1983a). Lichenometric dating of debris flow deposits in the Scottish Highlands. *Earth Surface Processes and Landforms*, 8, 579–588.

Innes, J.L. (1983b). Debris flows. *Progress in Physical Geography*, 7, 469–501.

Innes, J.L. (1983c). Stratigraphic evidence of episodic talus accumulation on the Isle of Skye. *Earth Surface Processes and Landforms*, 8, 399–403.

Innes, J.L. (1985). Magnitude–frequency relations of debris flows in north-west Europe. *Geografiska Annaler*, 67A, 23–32.

Innes, J.L. (1986). Textural properties of regolith on vegetated steep slopes in upland regions, Scotland. *Transactions of the Royal Society of Edinburgh: Earth Sciences*, 77, 241–250.

Ives, J.D. (1957). Glaciation of the Torngat Mountains, northern Labrador. *Arctic*, 10, 66–87.

Ives, J.D. (1958). Glacial geomorphology of the Torngat Mountains, northern Labrador. *Geographical Bulletin*, 12, 47–75.

Ives, J.D. (1966). Blockfields, associated weathering forms on mountain tops and the nunatak hypothesis. *Geografiska Annaler*, 48A, 220–223.

Ives, J.D. (1975). Delimitation of surface weathering zones in eastern Baffin Island, northern Labrador and arctic Norway: a discussion. *Geological Society of America, Bulletin*, 86, 1096–1100.

Ives, J.D. (1978). The maximum extent of the Laurentide ice sheet along the east coast of North America during the last glaciation. *Arctic*, 31, 24–53.

Ives, J.D. & Fahey, B.D. (1971). Permafrost occurrence in the Front Range, Colorado Rocky Mountains, USA. *Journal of Glaciology*, 10, 105–111.

Iwata, S. (1983). Physiographic conditions for the rubble slope formation on Mount Shirouma-dake, the Japan Alps. *Geographical Reports of the Tokyo Metropolitan University*, 18, 51 pp.

Jackson, L.E., MacDonald, G.M. & Wilson, M.C. (1982). Paraglacial origin for terraced river sediments in Bow Valley, Alberta. *Canadian Journal of Earth Sciences*, 19, 2219–2231.

Jahn, A, (1961). Quantitative analysis of some periglacial processes in Spitsbergen. *Uniwersytet Wraclawski in Poleslawa Bieruta Zesvytynankowe, Nauki Przyrodnicze, Series B*, 5, 1–34.

Jahn, A. (1962). Geneza skalek Granitowych. *Czasopismo Geograficzne*, 23, 19–44.

Jahn, A. (1972). Niveo-aeolian processes in the Sudetes Mountains. *Geographica Polonica*, 23, 93–110.

Jahn, A. (1974). Granite tors in the Sudeten Mountains. *Institute of British Geographers, Special Publication*, 7, 53–61.

Jahn, A. (1975). *Problems of the Periglacial Zone*. Warsaw: Polish Scientific Publishers.

Jahn, A. (1976). Contemporaneous geomorphological processes in Longyeardalen, Vestspitsbergen (Svalbard). *Biuletyn Peryglacjalny*, 26, 255–268.

Jahn, A. (1983). Soil wedges on Spitzbergen. In *Proceedings of the 4th International Conference on Permafrost, Fairbanks, Alaska*, pp. 525–530. Washington DC: National Academy Press.

Jamieson, T.F. (1863). On the parallel roads of Glen Roy, and their place in the history of the glacial period. *Quarterly Journal of the Geological Society of London*, 19, 235–249.

Jamieson, T.F. (1892). Supplementary remarks on Glen Roy. *Journal of the Geological Society of London*, 48, 5–28.

Jenkins, A., Ashworth, P.J., Ferguson, R.I., Grieve, I.C., Rowling, P. & Stott, T.A. (1988). Slope failures in the Ochil Hills, Scotland, November 1984. *Earth Surface Processes and Landforms*, 13, 69–76.

Jennings, J.N. & Costin, A.B. (1978). Stone movement through snow creep, 1963-1975, Mount Twynam, Snowy Mountains, Australia. *Earth Surface Processes*, 3, 3–22.

John, B.S. (1970). Pembrokeshire. In *The Glaciations of Wales and Adjoining Regions*, ed. C.A. Lewis, pp. 229–265. London: Longman.

John, B.S. (1973). Vistulian periglacial phenomena in south-west Wales. *Biuletyn Peryglacjalny*, 22, 185–213.

Johnson, A.M. & Rodine, J.R. (1984). Debris flow. In *Slope Instability*, eds. D. Brunsden & D.B. Prior, pp. 257–361. Chichester: Wiley.

Johnson, G.A.L. & Dunham, K.C. (1963). The geology of Moor House. *Nature Conservancy Monographs*, 2, 182 pp.

Johnson, P. (1975). Snowmelt. In *Proceedings of the Flood Studies Conference, London, 1975*, ed. T.J. Derwent, pp. 5–21. London: Institute of Civil Engineers.

Johnson, P.G. (1983). Rock glaciers: a case for a change in nomenclature. *Geografiska Annaler*, 65A, 27–34.

Johnson, R.H. (1975). Some late Pleistocene involutions at Dalton-in-Furness, northern England. *Geological Journal*, 10, 23–34.

Johnsson, G. (1959). True and false ice-wedges in southern Sweden. *Geografiska Annaler*, 41, 15–33.

Johnston, G.H., Ladanyi, B., Morgenstern, N.R. & Penner, E. (1981). Engineering characteristics of frozen and thawing soils. In *Permafrost: Design and Engineering*, ed. G.H. Johnston, pp. 73–147. Toronto: Wiley.

Jones, D.K.C. (1981). *The Geomorphology of the British Isles: Southeast and Southern England*. London: Methuen.

Jones, P.F. & Derbyshire, E. (1983). Late Pleistocene periglacial degradation of lowland Britain: implications for civil engineering. *Quarterly Journal of Engineering Geology*, 16, 197–210.

Jones, R.C.B., Tonks, L.R. & Wright, W.B. (1938). Wigan District. *Memoir of the Geological Survey of the United Kingdom*. London: HMSO.

Jones, T.R. (1859). Notes on some granite tors. *The Geologist*, 2, 301–312.

Kear, B.S. (1968). *An Investigation into Soils Developed on the Shirdley Hill Sand in South-West Lancashire*. Unpublished MSc Thesis, University of Manchester.

Kear, B.S. (1977). Shirdley Hill Sand Formation. In *The Isle of Man, Lancashire Coast and Lake District. X INQUA Congress Excursion Guide*, pp. 11–12. Norwich: Geobooks.

Keeble, A.B. (1971). Freeze–thaw cycles and rock weathering in Alberta. *Albertan Geographer*, 7, 34–42.

Kellaway, G.A. & Taylor, J.H. (1953). Early stages in the physiographic evolution of a portion of the East Midlands. *Quarterly Journal of the Geological Society of London*, 108, 343–375.

Kelletat, D. (1970a). Rezente Periglazialerscheinungen im Schottischen Hochland. *Göttinger Geographische Abhandlungen*, 51, 67–140.

Kelletat, D. (1970b). Zum Problem der Verbreitung, des Alters und der Bildungsdauer alter (inaktiver) Periglazialerscheinungen im Schottischen Hochland. *Zeitschrift für Geomorphologie*, 14, 510–519.

Kerfoot, D.E. (1969). *The Geomorphology and Permafrost Conditions of Garry Island NWT*. Unpublished PhD Thesis, University of British Columbia, 308pp.

Kerfoot, D.E. & Mackay, J.R. (1972). Geomorphological process studies, Garry Island, NWT. In *Mackenzie Delta Area Monograph: 22nd International Geographical Congress, Montreal 1972*, ed. D.E. Kerfoot, pp. 115–130. St Catherines, Ontario: Brock University Press.

Kerney, M.P. (1963). Late-glacial deposits on the chalk of south-east England. *Philosophical Transactions of the Royal Society of London*, B 246, 203–254.

Kerney, M.P. (1965). Weichselian deposits on the Isle of Thanet, east Kent. *Proceedings of the Geologists' Association*, 76, 269–274.

Kerney, M.P., Brown, E.H. & Chandler, T.J. (1964). The late-glacial history of the Chalk escarpment near Brook, Kent. *Philosophical Transactions of the Royal Society*, B 248, 135–204.

Kerschner, H. (1978). Palaeoclimatic inferences from late Würm rock glaciers, Eastern Central Alps, Western Tyrol, Austria. *Arctic and Alpine Research,* 10, 635–644.

Kidson, C. (1971). The Quaternary history of the coasts of South West England, with special reference to the Bristol Channel coast. In *Exeter Essays in Geography*, eds. K.J. Gregory & W.L.D. Ravenhill, pp. 1–22. Exeter: Exeter University Press.

Kidson, C. (1977). The coast of southwest England. In *Quaternary history of the Irish Sea*, ed. C. Kidson & M.J. Tooley, pp. 257–298. Liverpool: Seel House.

King, L. (1983). High mountain permafrost in Scandinavia. *Proceedings of the 4th International Conference on Permafrost, Fairbanks, Alaska*, pp. 612- 617. Washington: National Academy Press.

King, L. (1986). Zonation and ecology of high mountain permafrost in Scandinavia. *Geografiska Annaler*, 68A, 131–138.

King, L.C. (1948). A theory of bornhardts. *Geographical Journal,* 112, 83–87.

King, L.C. (1958). The problem of tors. *Geographical Journal*, 124, 289–291.

King, R.B. (1968). *Periglacial Features in the Cairngorm Mountains.* Unpublished PhD Thesis, University of Edinburgh, 360 pp.

King, R.B. (1971a). Vegetation destruction in the sub-alpine and alpine zones of the Cairngorm Mountains. *Scottish Geographical Magazine,* 87, 103–115.

King, R.B. (1971b). Boulder polygons and stripes in the Cairngorm Mountains, Scotland. *Journal of Glaciology*, 10, 375–386.

King, R.B. (1972). Lobes in the Cairngorm Mountains, Scotland. *Biuletyn Peryglacjalny*, 21, 153–167.

Kirby, R.P. (1967) The fabric of head deposits in South Devon. *Proceedings of the Ussher Society*, 1, 288–290.

Kirkby, M.J. (1967). Measurement and theory of soil creep. *Journal of Geology*, 75, 359–378.

Kirkby, M.J. (1984). Modelling cliff development in South Wales: Savigear re-viewed. *Zeitschrift für Geomorphologie*, 28, 405–426.

Kirkby, M.J. (1987). General models of long-term slope evolution through mass movement. In *Slope Stability*, eds. M.G. Anderson & K.S. Richards, pp. 359–380. Chichester: John Wiley & Sons.

Kirkby, M.J. & Statham, I. (1975). Surface stone movement and scree formation. *Journal of Geology*, 83, 349–362.

Klatka, T. (1961). Indices de structure et de texture des champs de pierres de Lysogory. *Bulletin de la Société des Sciences et des Lettres de Lódz,* 12, 1–21.

Klatka, T. (1962). Geneza i wiek globorzy Lysogórskich. *Acta Geographica Lodziendzia*, 12, 129 pp.

Konischev, V.N. (1982). Characteristics of cryogenic weathering in the permafrost zone of the European USSR. *Arctic and Alpine Research,* 14, 261–265.

Konrad, J-M. & Morgenstern, N.R. (1982). Prediction of frost heave in the laboratory during transient freezing. *Canadian Geotechnical Journal*, 19, 250–259.

Koster, E.A. (1988). Ancient and modern cold-climate aeolian sand deposition: a review. *Journal of Quaternary Science*, 3, 69–84.

Koster, E.A. & Dijkmans, J.W.A. (1988). Niveo-aeolian deposits and denivation forms, with special reference to the Great Kobuk sand dunes, northwestern Alaska. *Earth Surface Processes and Landforms,* 13, 153–170.

Kostyaev, A.G. (1973). Some rare varieties of stone circles. *Biuletyn Peryglacjalny*, 22, 347–352.

Kotarba, A. (1972). Comparison of physical weathering and chemical denudation in the Polish Tatra Mountains. In *Processus Périglaciaires Étudies sûr le Terrain*, eds. P. Macar & A. Pissart, pp.205–216. Liège: Université Liège.

Kotarba, A. (1976). Morphodynamic characteristics of debris slopes in calcareous West Tatra Mountains. *Studia Geomorphologica Carpatho-Balcanica*, 10, 63–77.

Kotarba, A. (1984). Slope features in areas of high relief in maritime climate (with the Isle of Rhum as example). *Studia Geomorphologica Carpatho- Balcanica*, 17, 77–88.

Kotarba, A. (1987). Glacial cirques transformation under differentiated maritime climate. *Studia Geomorphologica Carpatho-Balcanica*, 21, 77–92.

Kotarba, A. & Strömquist, L. (1984). Transport, sorting and depositional

processes of alpine debris slope deposits in the Polish Tatra Mountains. *Geografiska Annaler*, 66A, 285–294.

Kutzbach, J.E. & Wright, H.E. (1985). Simulation of the climate of 18 000 years BP: results for the North American/North Atlantic/European sector and comparison with the geologic record of North America. *Quaternary Science Reviews*, 4, 147–187.

Lachenbruch, A.H. (1962). Mechanics of thermal contraction cracks and ice-wedge polygons in permafrost. *Geological Society of America Special Paper*, 70, 69pp.

Lagarec, D. (1982). Cryogenic mounds as indicators of permafrost conditions, northern Québec. In *The Roger J.E. Brown Memorial Volume, Proceedings of the 4th Canadian Permafrost Conference*, ed. H.M. French, pp. 43–48. Ottawa: National Research Council of Canada.

Lagerbäck, R. & Rodhe, L. (1985). Pingos in northernmost Sweden. *Geografiska Annaler*, 67A, 239–245.

Lagerbäck, R. & Rodhe, L. (1986). Pingos and palsas in northernmost Sweden – preliminary notes on recent investigations. *Geografiska Annaler*, 68A, 149–154.

Lake, R.D., Young, B., Wood, C.J. & Mortimore, R.N. (1987). Geology of the country around Lewes. *British Geological Survey Memoir*. London: HMSO.

Lamb, H.H. (1977). *Climate: Past Present and Future*. Volume 2: *Climatic History and the Future*. London: Methuen.

Lamb, H.H. (1979). Climatic variation and changes in the wind and ocean circulation: the Little Ice Age in the Northeast Atlantic. *Quaternary Research,* 11, 1–20.

Lamb, H.H. (1981). An approach to the study of the development of climate and its impact on human affairs. In *Climate and History*, eds. T.M.L. Wigley, M.J. Ingram & G. Farmer, pp. 291–309. Cambridge: Cambridge University Press.

Lamb, H.H. (1984). Some studies of the Little Ice Age of recent centuries and its great storms. In *Climatic Changes on a Yearly to Millenial Basis*, eds. N.-A. Mörner, & W. Karlén. Dordrecht: Reidel.

Lamb, H.H. (1985). The Little Ice Age period and the great storms within it (central England). In *The Climatic Scene*, ed. M.J. Tooley & G.M. Sheail, pp. 104–131. London: Allen and Unwin.

Lamb, H.H., Lewis, R.P.W. & Woodroffe, A. (1966). Atmospheric circulation and the main climatic variables between 8000 and 0 BC: meteorological evidence. In *World Climate, 8000–0 BC*, ed. J.S. Sawyer, pp. 174–217. London: Royal Meteorological Society.

Lapworth, H. (1911). The geology of dam trenches. *Transactions of the Institution of Water Engineers*, 16, 25–66.

Larsen, E. & Holtedahl, H. (1985). The Norwegian strandflat: a reconsideration of its age and origin. *Norsk Geologisk Tidsskrift*, 65, 247–254.

Larsen, J.A. (1974). Ecology of the northern continental forest border. In *Arctic and Alpine Environments*, eds. J.D.Ives & R.G. Barry, pp. 341–369. London: Methuen.

Larsson S. (1982). Geomorphological effects on the slopes of Longyear Valley, Spitsbergen, after a heavy rainfall in July 1972. *Geografiska Annaler*, 64A, 105–125.

Lautridou, J-P. (1988). Recent advances in cryogenic weathering. In *Advances in Periglacial Geomorphology*. ed. M.J. Clark, pp. 33–47. Chichester: Wiley.

Lautridou, J-P. & Ozouf, J.C. (1982). Experimental frost shattering. 15 years of research at the Centre de Géomorphologie du CNRS. *Progress in Physical Geography*, 6, 217–232.

Lautridou, J-P., Sommé, J. & Jamagne, M. (1984). Sedimentological, mineralogical and geochemical characteristics of the loesses of north-west France. In *Lithology and Stratigraphy of Loess and Palaeosols*, ed. M.Pécsi, pp. 121–132. Budapest: Hungarian Academy of Sciences.

Ledger, D.C., Lovell, J.P.B. & McDonald, A.T. (1974). Sediment yield studies in upland catchment areas in south-east Scotland. *Journal of Applied Ecology*, 11, 201–206.

Lee, M.P. (1979). Loess from the Pleistocene of the Wirral Peninsula, Merseyside. *Proceedings of the Geologists' Association*, 90, 21–26.

Lee, M.P. & Vincent, P.J. (1981). The first recognition of loess from North Wales. *Manchester Geographer*, 2, 45–53.

Leeder, M. (1972). Periglacially modified chalk and chalk-ridge diapirs from Norwich, Norfolk. *Transactions of the Norfolk and Norwich Naturalists' Society*, 22, 229–233.

Leffingwell, E. de K. (1915). Ground-ice wedges; the dominant form of ground-ice on the north coast of Alaska. *Journal of Geology*, 23, 635–654.

Lewin, J. (1966). Fossil ice wedges in Hampshire. *Nature,* 211, 728.

Lewin, J. (1969). The Yorkshire Wolds: a study in geomorphology. *University of Hull Occasional Papers in Geography*, No. 11, 88pp.

Lewin, J., Cryer, R. & Harrison, D.I. (1974). Sources for sediments and solutes in mid-Wales. *Institute of British Geographers Special Publication*, 6, 73–85.

Lewis, C.A. (1970). The glaciations of the Brecknock Beacons, Wales. *Brycheiniog*, 14, 97–120.

Lewis, C.A. & Lass, G.M. (1965). The drift terraces of Slaettartatindur, the Faroes. *Geographical Journal*, 131, 247–253.

Lewkowicz, A.G. (1983). Erosion by overland flow, central Banks Island, western Canadian arctic. In *Proceedings of the 4th International Conference on Permafrost, Fairbanks, Alaska*, pp. 701–706. Washington: National Academy Press.

Lewkowicz, A.G. (1987). Nature and importance of thermokarst processes, Sand Hills moraine, Banks Island, Canada. *Geografiska Annaler*, 69A, 321–327.

Lewkowicz, A.G. (1988). Slope processes. In *Advances in Periglacial Geomorphology*, ed. M.J. Clark, pp. 325–370. Chichester: Wiley.

Lewkowicz, A.G. (1989). Periglacial systems. In *Fundamentals of Physical Geography* (Canadian Edition), eds. D, Briggs, P. Smithson and T. Ball, pp. 363–397. Toronto: Copp Clark Pitman.

Lewkowicz, A.G. (1990). Morphology, frequency and magnitude of active-layer detachment slides, Fosheim Peninsula, Ellesmere Island, NWT. In *Proceedings of the 5th Canadian Permafrost Conference*, eds. M.M.Burgess, D.G. Harry & D.C. Sego, pp. 111–118. Québec: Université Laval.

Lewkowicz, A.G. & French, H.M. (1982). The hydrology of small runoff plots in an area of continuous permafrost. In *Proceedings of the 4th Canadian Permafrost Conference, Calgary, Alberta, 1981*, pp. 151–162. Ottawa: National Research Council of Canada.

Lewkowicz, A.G. & Young, K.L. (1990). Hydrology of a perennial snow-bank in the continuous permafrost zone, Melville Island, Canada. *Geografiska Annaler*, 72A, 13–21.

Liestøl, O. (1977). Pingos, springs and permafrost in Spitsbergen. *Norsk Polarinstitutt, Årbok 1975*, 7–29.

Lill, G.O. & Smalley, I.J. (1978). Distribution of loess in Britain. *Proceedings of the Geologists' Association*, 89, 57–65.

Lindner, L. & Marks, L. (1985). Types of debris slope accumulations and rock glaciers in south Spitsbergen. *Boreas*, 14, 139–153.

Linton, D.L. (1949). Unglaciated areas in Scandinavia and Great Britain. *Irish Geography*, 2, 25–33.

Linton, D.L. (1951). Unglaciated enclaves in glaciated regions. *Journal of Glaciology*, 1, 451–452.

Linton, D.L. (1955). The problem of tors. *Geographical Journal*, 121, 470–487.

Linton, D.L. (1959). Morphological contrasts of eastern and western Scotland. In *Geographical Essays in Memory of A.G. Ogilvie*, eds. R. Miller & J.W. Watson pp 16–45. Edinburgh: Thomas Nelson & Sons Ltd.

Linton, D.L. (1964). The origin of the Pennine tors: an essay in analysis. *Zeitschrift für Geomorphologie*, 8, 5–23.

Loch, J.P.G. & Kay, B.D. (1978). Water redistribution in partially frozen silt under several temperature gradients and overburden loads. *Journal of the Soil Science Society of America*, 42, 400–406.

Lockwood, J.G. (1979). Water balance of Britain 50 000 to the present day. *Quaternary Research*, 12, 297–310.

Løken, O. (1962). On the vertical extent of glaciation in north-eastern Labrador- Ungava. *Canadian Geographer*, 6, 106–119.

Long, D. (1991). The identification of features due to former permafrost in the North Sea. In *Quaternary Engineering Geology*, eds. A. Forster, M.G. Culshaw, J.C. Cripps, J.A. Little & C.F. Moon, pp. 369–372. London: Geological Society Engineering Geology Special Publication No. 7.

Lowe, J.J. & Walker, M.J.C. (1980). Problems associated with radiocarbon dating the close of the Lateglacial period in the Rannoch Moor area. In *Studies in the Lateglacial of North-West Europe*, eds. J.J. Lowe, J.M. Gray & J.E. Robinson, pp. 123–137. Oxford: Pergamon.

Lowe, J.J. & Walker, M.J.C. (1984). *Reconstructing Quaternary Environments*. London: Longman.

Luckman, B.H. (1970). The Hereford Basin. In *The Glaciations of Wales and Adjoining Regions*, ed. C.A. Lewis, pp. 175–196. London: Longman.

Luckman, B.H. (1971). The role of snow avalanches in the evolution of alpine talus slopes. *Institute of British Geographers Special Publication*, 3, 93–110.

Luckman, B.H. (1976). Rockfalls and rockfall inventory data: some observations from Surprise Valley, Jasper National Park, Canada. *Earth Surface Processes*, 1, 287–298.

Luckman, B.H. (1977). The geomorphic activity of snow avalanches. *Geografiska Annaler*, 59A, 31–48.

Luckman, B.H. (1978). Geomorphic work of snow avalanches in the Canadian Rocky Mountains. *Arctic and Alpine Research*, 10, 261–276.

Lundqvist, J. (1962). Patterned ground and related frost phenomena in Sweden. *Sveriges Geologiske Undersökning*, 55, 101 pp.

Mackay, J.R. (1972). The world of underground ice. *Annals of the Association of American Geographers*, 62, 1–22.

Mackay, J.R. (1974). Ice-wedge cracks, Garry Island, Northwest Territories. *Canadian Journal of Earth Sciences*, 11, 1336–1383.

Mackay, J.R. (1975). The closing of ice-wedge cracks in permafrost, Garry Island, Northwest Territories. *Canadian Journal of Earth Sciences*, 12, 1668–1674.

Mackay, J.R. (1977a). The widths of ice wedges. *Geological Survey of Canada Paper*, 77–1A, 43–44.

Mackay, J.R. (1977b). Pulsating pingos, Tuktoyaktuk Peninsula, NWT. *Canadian Journal of Earth Sciences*, 14, 209–222.

Mackay, J.R. (1977c). Changes in the active layer from 1968 to 1976 as a result of the Inuvik fire. *Geological Survey of Canada Paper*, 77–1B, 273–275.

Mackay, J.R. (1978). Sub-pingo water lenses, Tuktoyaktuk Peninsula, Northwest Territories. *Canadian Journal of Earth Sciences*, 15, 1219–1227.

Mackay, J.R. (1979a). Pingos of the Tuktoyaktuk Peninsula area, Northwest Territories. *Géographie Physique et Quaternaire*, 33, 3–61.

Mackay, J.R. (1979b). An equilibrium model for hummocks (non-sorted circles), Garry Island, North West Territories. *Geological Survey of Canada Paper*, 79–1A, 165–167.

Mackay, J.R. (1980). The origin of hummocks, western Arctic coast, Canada. *Canadian Journal of Earth Sciences*, 17, 996–1006.

Mackay, J.R. (1981a). Aklisuktuk (Growing Fast) Pingo, Tuktoyaktuk Peninsula, Northwest Territories, Canada. *Arctic*, 34, 270–273.

Mackay, J.R. (1981b). Active layer slope movement in a continuous permafrost environment, Garry Island, Northwest Territories, Canada. *Canadian Journal of Earth Sciences*, 18, 1666–1680.

Mackay, J.R. (1983a). Downward water movement into frozen ground, western Arctic coast, Canada. *Canadian Journal of Earth Sciences*, 20, 120–134.

Mackay, J.R. (1983b). Pingo growth and subpingo water lenses, western Arctic coast, Canada. *Proceedings of the 4th International Conference on Permafrost, Fairbanks, Alaska*, pp. 762–766. Washington DC: National Academy Press.

Mackay, J.R. (1984). The frost heave of stones in the active layer above permafrost, with downward and upward freezing. *Arctic and Alpine Research*, 16, 439–446.

Mackay, J.R. (1986a). The first 7 years (1978–1985) of ice-wedge growth, Illisarvik experimental drained lake site, western Arctic coast. *Canadian Journal of Earth Sciences*, 23, 1782–1795.

Mackay, J.R. (1986b). Growth of Ibyuk Pingo, western Arctic coast, Canada, and some implications for environmental reconstructions. *Quaternary Research*, 26, 68–80.

Mackay, J.R. (1988a). Ice wedge growth in newly aggrading permafrost, western Arctic coast, Canada. *Proceedings of the 5th International Conference on Permafrost, Trondheim*, pp. 809–814. Trondheim: Tapir Press.

Mackay, J.R. (1988b). Pingo collapse and palaeoclimatic reconstruction. *Canadian Journal of Earth Sciences*, 25, 495–511.

Mackay, J.R. & Mackay, D.K. (1976). Cryostatic pressures in nonsorted circles (mud hummocks), Inuvik, Northwest Territories. *Canadian Journal of Earth Sciences*, 13, 889–897.

Mackay, J.R. & Mathews, W.H. (1973). Geomorphology and Quaternary history of the Mackenzie River Valley near Fort Good Hope, NWT. *Canadian Journal of Earth Sciences*, 10, 26–41.

Mackay, J.R. & Mathews, W.H. (1974). Movement of sorted stripes, the Cinder Cone, Garibaldi Park, BC, Canada. *Arctic and Alpine Research*, 6, 347–359.

Maclean, A.F. (1991). *The Formation of Valley-Wall Rock Glaciers*. Unpublished Ph.D. Thesis, University of St Andrews.

Macpherson, J.B. (1980). Environmental change during the Loch Lomond Stadial: evidence from a site in the upper Spey Valley, Scotland. In *Studies in the Lateglacial of North-West Europe*, eds. J.J. Lowe, J.M. Gray & J.E. Robinson, pp. 89–102. Oxford: Pergamon.

Madgett, P.A. & Catt, J.A. (1978). Petrology, stratigraphy and weathering of Late Pleistocene tills in east Yorkshire, Lincolnshire and north Norfolk. *Proceedings of the Yorkshire Geological Society*, 42, 55–108.

Maizels, J.K. (1986). Frequency of relic frost-fissure structures and prediction of polygon patterns: a quantitative approach. *Biuletyn Peryglacjalny*, 30, 67–89.

Manley, G. (1949). The snowline in Britain. *Geografiska Annaler*, 13, 179–193.

Manley, G. (1952). *Climate and the British Scene*. London: Collins.

Manley, G. (1971a). Scotland's semi-permanent snows. *Weather*, 26, 458–471.

Manley, G. (1971b). The mountain snows of Britain. *Weather*, 26, 192–200.

Mantell, G. (1833). *The Geology of the South-East of England*. London.

Marr, J.E. (1916). *The Geology of the Lake District*. Cambridge: Cambridge University Press.

Marsh, P. & Woo, M (1981). Snowmelt, glacier melt and high arctic streamflow regimes. *Canadian Journal of Earth Sciences*, 18, 1380–1384.

Martini, A. (1969). Sudetic tors formed under periglacial conditions. *Biuletyn Peryglacjalny*, 19, 351–369.

Mathews, W.H. & Mackay, J.R. (1963). Snowcreep studies, Mount Seymour, BC: preliminary field investigations. *Geographical Bulletin*, 20, 58–75.

Matsuoka, N. (1990). The rate of bedrock weathering by frost action: field measurements and a predictive model. *Earth Surface Processes and Landforms*, 15, 73–90.

Matsuoka, N. (1991). A model of the rate of frost shattering: application to field data from Japan, Svalbard and Antarctica. *Permafrost and Periglacial Processes*, 2, 271–281.

Matthews, B. (1970). Age and origin of aeolian sand in the Vale of York. *Nature*, 227, 1234–1236.

Matthews, J.A., Dawson, A.G. & Shakesby, R.A. (1986). Lake shoreline development, frost weathering and rock platform erosion in an alpine periglacial environment, Jotunheimen, southern Norway. *Boreas*, 15, 33–50.

Matthews, J.A., Harris, C. & Ballantyne, C.K. (1986). Studies on a gelifluction lobe, Jotunheimen, Norway: ¹⁴C chronology, stratigraphy, sedimentology and palaeoenvironment. *Geografiska Annaler*, 68A, 345–360.

McArthur, J.L. (1975). Some observations on periglacial morphogenesis in the Southern Alps, New Zealand. *Geografiska Annaler*, 57A, 213–224.

McArthur, J.L. (1981). Periglacial slope planations in the southern Pennines, England. *Biuletyn Peryglacjalny*, 28, 85–97.

McCann, S.B. (1966). The Main Post-glacial Raised Shoreline of western Scotland from the Firth of Lorne to Loch Broom. *Transactions of the Institute of British Geographers*, 39, 87–99.

McCann, S.B. & Carlisle, R.J. (1972). The nature of the ice foot on the beaches of Radstock Bay, south-west Devon Island, NWT, Canada, in the spring and summer of 1970. *Institute of British Geographers, Special Publication*, 4, 175–186.

McCann, S.B. & Cogley, J.G. (1973). The geomorphic significance of fluvial activity at high latitudes. In *Research in Polar and Alpine Geomorphology*, eds. B.D. Fahey & R.D. Thompson, pp. 118–135. Norwich: Geobooks.

McCann, S.B., Howarth, P.J. & Cogley, J.G. (1972). Fluvial processes in a periglacial environment, Queen Elizabeth Islands, NWT, Canada. *Transactions of the Institute of British Geographers*, 55, 69–82.

McCarroll, D. (1991). Ice directions in western Lleyn and the status of the Gwynedd readvance of the last Irish Sea glacier. *Geological Journal*, 26, 137–143.

McConnell, D. (1988). The Ben Nevis Observatory log books. *Weather*, 43, 356–362 and 396–401.

McEwen, L.J. & Werritty, A. (1988). The hydrology and long-term geomorphic significance of a flash flood in the Cairngorm Mountains, Scotland. *Catena*, 15, 361–377.

McGreevy, J.P. (1981). Perspectives on frost shattering. *Progress in Physical Geography*, 5, 56–75.

McIntyre, A., Ruddiman, W.F. & Jantzen, R. (1972). Southward penetration of the North Atlantic Polar Front and faunal and floral evidence of large scale surface water mass movements over the past 225,000 years. *Deep Sea Research*, 19, 61–77.

McManus, J. (1966). An ice-wedge and associated phenomena in the Lower Limestone Series of Fife. *Scottish Journal of Geology*, 2, 259–264.

McRoberts, E.C. (1973). *The Stability of Slopes in Permafrost*. Unpublished PhD Thesis, University of Alberta, Edmonton.

McRoberts, E.C. (1978). Slope stability in cold regions. In *Geotechnical Engineering for Cold Regions*, eds. O.B. Andersland & D.M. Anderson, pp. 363–404. New York: McGraw-Hill.

McRoberts, E.C. & Morgenstern, N.R. (1974). The stability of thawing slopes. *Canadian Geotechnical Journal*, 11, 447–467.

Mellor, A. & Wilson, M.J. (1989). Origin and significance of gibbsitic montane soils in Scotland, UK. *Arctic and Alpine Research*, 21, 417–424.

Mellor, M. (1978). Dynamics of snow avalanches. In *Rockslides and Avalanches*, ed. B Voight, pp. 753–792. Amsterdam: Elsevier.

Metcalfe, G. (1950). The ecology of the Cairngorms. II. The mountain Callunetum. *Journal of Ecology*, 38, 46–74.

Miall, A.D. (1977). A review of the braided river depositional environment. *Earth Science Reviews*, 13, 1–62.

Michaud, Y., Dionne, J-C. & Dyke, L.D. (1989). Frost bursting: a violent expression of frost action in rock. *Canadian Journal of Earth Science*, 26, 2075–2080.

Miller, D.J. (1990). *Relict Periglacial Phenomena within the Tamar Basin, West Devon and East Cornwall: Their Significance with Regard to Climatic Reconstruction*. Unpublished PhD Thesis, University of Exeter, 2 volumes, 679 pp.

Miller, H. (1887). The country around Otterburn and Elsdon. *Memoir of the Geological Survey of Great Britain (England and Wales)*. London: HMSO.

Miller, R., Common, R. & Galloway, R.W. (1954). Stone stripes and other surface features of Tinto Hills. *Geographical Journal*, 120, 216–219.

Miller, R.D. (1980). Freezing phenomena in soils. In *Applications of Soil Physics*, ed. D. Hillel. New York: Academic Press.

Mitchell, G.F., Penny, L.F., Shotton, F.W. & West, R.G. (1973). A correlation of Quaternary deposits in the British Isles. *Geological Society of London, Special Report*, 4, 99 pp.

Mitchell, W.A. (1991). Loch Lomond Stadial landforms and palaeoglaciological reconstruction. In *Western Pennines: Field Guide*, ed. W.A. Mitchell, pp. 43–54. Cambridge: Quaternary Research Association.

Moign, A. (1974). Géomorphologie du strandflat au Svalbard: problems (age, origine, processus), methods de travail. *Inter-Nord*, 13–14, 57–72.

Moign, A. (1976). L'action des glaces flottantes sur le littoral et les fonds marins du Spitzberg central et nord-occidental. *Revue Géographique de Montréal*, 30, 51–64.

Moran, M. (1988). *Scotland's Winter Mountains*. Newton Abbot: David and Charles.

Morgan, A.V. (1971a). Polygonal patterned ground of Late Weichselian age in the area north and west of Wolverhampton, England. *Geografiska Annaler*, 53A, 146–156.

Morgan, A.V. (1971b). Engineering problems caused by fossil permafrost features in the English Midlands. *Quarterly Journal of Engineering Geology*, 4, 111–114.

Morgan, A.V. (1973). Pleistocene geology around Wolverhampton. *Philosophical Transactions of the Royal Society of London*, B 265, 233–297.

Morgan, R.P.C. (1971). A morphometric study of some valley systems in the English Chalklands. *Transactions of the Institute of British Geographers*, 54, 33–44.

Morgenstern, N.R. (1981). Geotechnical engineering and frontier resource development. *Géotechnique*, 31, 305–365.

Morgenstern, N.R. (1985). Recent observations on the deformation of ice and ice-rich permafrost. In *Field and Theory: Lectures in Geocryology*, eds. M. Church & O. Slaymaker, pp. 133–153. Vancouver: University of British Columbia Press.

Morgenstern, N.R. & Nixon, J.F. (1971). One-dimensional consolidation of thawing soils. *Canadian Geotechnical Journal*, 11, 447–469.

Mottershead, D.N. (1967). The evolution of the Valley of the Rocks and its landforms. *Exmoor Review*, 69–72.

Mottershead, D.N. (1971). Coastal head deposits between Start Point and Hope Cove, Devon. *Field Studies* 3, 433–453.

Mottershead, D.N. (1976). Quantitative aspects of periglacial slope deposits in Southwest England. *Biuletyn Peryglacjalny*, 25, 35–57.

Mottershead, D.N. (1977). The Quaternary evolution of the south coast of England. In *The Quaternary History of the Irish Sea*, eds. C. Kidson & M.J. Tooley, Geological Journal Special Issue No. 7, 299–320.

Mottershead, D.N. (1978). High altitude solifluction and Postglacial vegetation, Arkle, Sutherland. *Transactions of the Botanical Society of Edinburgh*, 43, 17–24.

Mottershead, D.N. (1982). Some sources of systematic variation in the main head deposits of Southwest England, *Biuletyn Peryglacjalny*, 29, 117–128.

Mottershead, D.N. & White, I.D. (1969). Some solifluction terraces in Sutherland. *Transactions of the Botanical Society of Edinburgh,* 40, 604–620.

Mugridge, S-J. & Young, H.R. (1982). Disintegration of shale by cyclic wetting and drying and frost action. *Canadian Journal of Earth Sciences,* 20, 568–576.

Müller, F. (1959), Beobachtungen über Pingos. *Meddelelser om Grønland,* 153, 127pp. (Translated by D.A. Sinclair: Observations on Pingos, National Research Council of Canada Translation 1073, 117 pp. (1963).)

Murchison, R.I. (1851). On the distribution of the flint drift of the south-east of England, on the flanks of the Weald, and over the surface of the South and North Downs. *Quarterly Journal of the Geological Society,* 7, 349–358.

Nansen, F. (1904). The bathymetrical features of the North Polar seas. In *The Norwegian North Polar Expedition, 1893–1896. Scientific Results, volume IV,* ed. F. Nansen, pp. 1–232. Kristiania: Dybwad.

Nansen, F. (1922). The strandflat and isostasy. *Videnskapsselskapet Skrifter 1921. I. Matematisk-Naturhistorisk Klasse,* 2. Kristiania: Norges Videnskaps Akademie, 313 pp.

Nelson, F.E. (1985). A preliminary investigation of solifluction macrofabrics. *Catena,* 12, 23–33.

Nesje, A. (1989). The geographical and altitudinal distribution of block fields in southern Norway and its significance to the Pleistocene ice sheets. *Zeitschrift für Geomorphologie Supplementband,* 72, 41–53.

Nesje, A., Anda, E., Rye, N., Lien, R., Hole, P.A. & Blikra, L.H. (1987). The vertical extent of the Late Weichselian ice sheet in the Nordfjord-Møre area, western Norway. *Norsk Geologisk Tidsskrift,* 67, 125–141.

Nesje, A., Dahl, S.O., Anda, E. & Rye, N. (1988). Block fields in southern Norway: significance for the Late Weichselian ice sheet. *Norsk Geologisk Tidsskrift,* 68, 149–169.

Nesje, A. & Sejrup, H. P. (1988). Late Weichselian/Devensian ice sheets in the North Sea and adjacent land areas. *Boreas,* 17, 371–384.

Newson, M.D. (1980). The geomorphological effectiveness of floods, a contribution stimulated by two recent events in mid-Wales. *Earth Surface Processes,* 5, 275–290.

Newson, M.D. (1981). Mountain streams. In *British Rivers,* ed. J. Lewin, pp. 59–89. London: George Allen and Unwin.

Newson, M.D. & Leeks, G.J. (1985). Mountain bedload yields in the United Kingdom: further information from undisturbed fluvial environments. *Earth Surface Processes and Landforms,* 10, 413–416.

Nichols, R.L. (1963). Miniature nivation cirques near Marble Point, McMurdo Sound, Antarctica. *Journal of Glaciology,* 4, 477–479.

Nicholson, F.H. (1969). *An Investigation of Patterned Ground.* Unpublished PhD Thesis, University of Bristol, 238pp.

Nicholson, F.H. (1976). Patterned ground formation and description as suggested by low arctic and subarctic examples. *Arctic and Alpine Research,* 8, 329–342.

Nicholson, F.H. & Granberg, H.B. (1973). Permafrost and snow cover relationships near Schefferville. *North American Contribution, 2nd International Conference on Permafrost, Yakutsk,* pp. 151–158. Washington DC: National Academy of Science.

Nickling, W.G. (1978). Eolian sediment transport during dust storms: Slims River Valley, Yukon Territory. *Canadian Journal of Earth Science,* 15, 1069–1084.

Nixon, J.F. & Ladanyi, B. (1978). Thaw consolidation. In *Geotechnical Engineering for Cold Regions,* eds. O.B. Andersland & D.M. Anderson, pp. 164–215. New York: McGraw-Hill.

Okuda S., Suwa, H., Okunishi, K., Yokoyama, K. & Nakano, M. (1980). Observations on the motion of a debris flow and its geomorphological effects. *Zeitschrift für Geomorphologie, Supplementband,* 35, 142–163.

Ollier, C.D. & Thomasson, A.J. (1957). Asymmetrical valleys of the Chiltern Hills. *Geographical Journal,* 123, 71–80.

Olyphant, G.A. (1983). Analysis of the factors controlling cliff burial by talus within Blanca Massif, southern Colorado, USA. *Arctic and Alpine Research,* 15, 65–75.

Onesti, L.J. & Walti, S.A. (1983). Hydrologic characteristics of small arctic-alpine watersheds, Central Brooks Range, Alaska. *Proceedings of the 4th International Permafrost Conference, Fairbanks, Alaska,* pp. 957–961. Washington DC: National Academy Press.

Ono, Y. & Watanabe, T. (1986). A protalus rampart related to alpine debris flows in the Kuranosuke Cirque, Northern Japanese Alps. *Geografiska Annaler,* 68A, 213–223.

Orme, A. R. (1964). The geomorphology of southern Dartmoor and the adjacent area. In *Dartmoor Essays,* pp. 31–72. Devonshire Association for the Advancement of Science, Literature and Art.

Oxford, S.P. (1985). Protalus ramparts, protalus rock glaciers and soliflucted till in the northwest part of the English Lake District. In *Field Guide to the Periglacial Landforms of Northern England,* ed. J. Boardman, pp. 38–46. Cambridge: Quaternary Research Association.

Paine, A.D.M. (1982). *Origin and Development of Blockfields in the Cairngorm Mountains.* Unpublished BA Dissertation, University of Cambridge, 90 pp.

Palm, E. & Tveitereid, M. (1977). On patterned ground and free convection. *Norsk Geografisk Tidsskrift,* 31, 145–148.

Palmer, J.A. (1956). Tor formation at the Bridestones in NE Yorkshire and its significance in relation to problems of valley-side development and regional glaciation. *Transactions of the Institute of British Geographers,* 22, 55–71.

Palmer, J.A. (1967). Landforms. In *Leeds and its Region,* eds. M.W. Beresford & G.R.J. Jones, pp. 16–29. Leeds: British Association for the Advancement of Science.

Palmer, J.A. & Neilson, R.A. (1962). The origin of granite tors on Dartmoor, Devonshire. *Proceedings of the Yorkshire Geological Society,* 33, 315–339.

Palmer, J.A. & Radley, J. (1961). Gritstone tors of the English Pennines. *Zeitschrift für Geomorphologie,* 5, 37–52.

Palmer, L.S. & Cooke, J.H. (1923). The Pleistocene deposits of the Portsmouth district and their relation to man. *Proceedings of the Geologists' Association,* 34, 253–282.

Parks, C.D. (1991). A review of the possible mechanisms of cambering and valley bulging. In *Quaternary Engineering Geology,* eds. A. Forster, M.G. Culshaw, J.C. Cripps, J.A. Little & C.F. Moon, pp. 373–380. London: Geological Society Engineering Geology Special Publication No. 7.

Parks, D.A. & Rendell, H.M. (1988). TL dating of brickearths from SE England. *Quaternary Science Reviews,* 7, 305–308.

Parks, D.A. & Rendell, H.M. (1992). Thermoluminescence dating and geochemistry of loessic deposits in southeast England. *Journal of Quaternary Science,* 7, 99–107.

Paterson, K. (1971). Weichselian deposits and fossil periglacial structures in north Berkshire. *Proceedings of the Geologists' Association,* 82, 455–468

Paterson, T.T. (1940). The effects of frost action and solifluction around Baffin Bay and the Cambridge district. *Quarterly Journal of the Geological Society of London,* 96, 99–130.

Paterson, W.S.B. (1981). *The Physics of Glaciers.* 2nd Edition. Oxford: Pergamon.

Payette, S., Gauthier, L. & Grenière, I. (1986). Dating ice wedge growth in subarctic peatlands. *Nature,* 322, 724–727.

Payette, S., Samson, H. & Lagarec, D. (1976). The evolution of permafrost in the taïga and in the forest-tundra, western Québec-Labrador Peninsula. *Canadian Journal of Forest Research,* 6, 203–220.

Peach, B.N., Gunn, W., Clough, C.T., Hinxman, L.W., Crampton, C.B. & Anderson, E.M. (1912). The Geology of Ben Wyvis, Carn Chuinneag, Inchbae and the surrounding country. *Memoirs of the Geological Survey, Scotland.* Edinburgh.

Peach, B.N., Horne, J., Gunn, W., Clough, C.T., Hinxman, L.T., Cadell, H.M., Greenly, E., Pocock, T.I. & Crampton, C.B. (1913). The geology of the Fannich Mountains and the country around upper Loch Maree and Strath Broom. *Memoirs of the Geological Survey, Scotland.* Edinburgh.

Peacock, J.D. (1975). Quaternary of Scotland – discussion. *Scottish Journal of Geology,* 11, 174–175.

Peacock, J.D. (1981). Scottish Late-glacial marine deposits and their environmental significance. In *The Quaternary in Britain,* eds. J. Neale & J. Flenley, pp. 222–236. Oxford: Pergamon Press.

Peacock, J.D. (1986). Alluvial fans and an outwash fan in upper Glen Roy, Lochaber. *Scottish Journal of Geology,* 22, 347–366.

Peacock, J.D. (1991). Glacial deposits of the Hebridean region. In *Glacial Deposits of Great Britain and Ireland,* eds. J. Ehlers, P.L. Gibbard & J. Rose, pp. 109–120. Rotterdam: Balkema.

Peacock, J.D., Graham, D.K. & Wilkinson, I.P. (1978). Late-glacial and post-glacial marine environments at Ardyne, Scotland, and their significance in the interpretation of the history of the Clyde sea area. *Reports of the Institute of Geological Sciences,* 78/17, 25 pp.

Peacock, J.D. & Harkness, D.D. (1990). Radiocarbon ages and the full-glacial to Holocene transition in seas adjacent to Scotland and southern Scandinavia: a review. *Transactions of the Royal Society of Edinburgh: Earth Science,* 81, 385–396.

Peacock, J.D., Harkness, D.D., Housley, R.A., Little, J.A. & Paul, M.A.

(1989). Radiocarbon ages for a glaciomarine bed associated with the maximum of the Loch Lomond Readvance in west Benderloch, Argyll. *Scottish Journal of Geology*, 25, 69–79.

Pears, N.V. (1975a). Radiocarbon dating of peat macrofossils in the Cairngorm Mountains, Scotland. *Transactions of the Botanical Society of Edinburgh*, 42, 255–260.

Pears, N.V. (1975b). The growth rates of hill peats in Scotland. *Geologiska Föreningens Stockholm Förhandlingar*, 97, 265–270.

Pearsall, W.H. (1968). *Mountains and Moorlands*. 2nd Edition. Glasgow: Collins.

Pearson, M.G. (1976). Snowstorms in Scotland, 1729–1830. *Weather*, 31, 390–393.

Peev, C.D. (1966). Geomorphic activity of snow avalanches. *International Association of Hydrological Sciences Publication*, 69, 357–368.

Peltier, L.C. (1950). The geomorphic cycle in periglacial regions as it is related to climatic geomorphology. *Annals of the Association of American Geographers*, 40, 214–236.

Pemberton, M. (1980). Earth hummocks at low elevation in the Vale of Eden, Cumbria. *Transactions of the Institute of British Geographers, New Series*, 5, 487–501.

Penn, S., Royce, C.J. & Evans, C.J. (1983). The periglacial modification of the Lincoln Scarp. *Quarterly Journal of Engineering Geology*, 16, 309–318.

Pennington, W. (1974). *The History of British Vegetation*. 2nd Edition. London: English Universities Press.

Pennington, W. (1975). A chronostratigraphic comparison of Late-Weichselian and Late-Devensian subdivisions, illustrated by two radiocarbon-dated profiles from western Britain. *Boreas*, 4, 157–171.

Pennington, W. (1977a). The Late Devensian flora and vegetation of Britain. *Philosophical Transactions of the Royal Society of London*, B 280, 247–271.

Pennington, W. (1977b). Lake sediments and the Lateglacial environment in Northern Scotland. In *Studies in the Scottish Lateglacial Environment*, eds. J.M. Gray & J.J. Lowe, pp. 119–141. Oxford: Pergamon.

Pennington, W. (1978). Quaternary Geology. In *Geology of the Lake District*, ed. F. Moseley, pp. 207–225. Leeds: Yorkshire Geological Society.

Penny, L.F., Coope, G.R. & Catt, J.A. (1969). Age and insect fauna of the Dimlington Silts, East Yorkshire. *Nature*, 224, 65–67.

Perrin, R.M.S., Davies, H. & Fysh M.D. (1974). Distribution of late Pleistocene aeolian deposits in eastern and southern England. *Nature*, 248, 320–324.

Perrin, R.M.S., Rose, J. & Davies, H. (1979). The distribution, variation and origins of pre-Devensian tills in eastern England. *Philosophical Transactions of the Royal Society of London*, B 287, 535–570.

Péwé, T.L. (1959). Sand wedge polygons (tesselations) in the McMurdo Sound region, Antarctica. *American Journal of Science*, 257, 542–552.

Péwé, T.L. (1966a). Ice-wedges in Alaska: classification, distribution and climatic significance. *1st International Conference on Permafrost, Washington*, pp. 76–81. Washington DC: National Academy of Science, National Research Council, Publication No. 1287.

Péwé, T.L. (1966b). Palaeoclimatic significance of fossil ice wedges. *Biuletyn Peryglacjalny*, 15, 65–73.

Péwé, T.L. (1968). Loess deposits of Alaska. *International Geological Congress, Report of the 23rd Session, Czechslovakia, Proceedings of Section 8, Genesis and Classification of Sedimentary Rocks*, pp. 297–309. Prague: Academia.

Péwé, T.L. (1969). Editor. *The Periglacial Environment*. Montreal: McGill-Queens University Press.

Péwé, T.L. (1974). Geomorphic processes in polar deserts. In *Polar Deserts and Modern Man*, eds. T.L. Smiley & J.H. Zumberge, pp. 33–52. Tucson: University of Arizona Press.

Péwé, T.L. (1979). Permafrost and its effects on human activities in arctic and subarctic regions. *Geo Journal*, 3, 333–344.

Péwé, T.L. (1983). Alpine permafrost in the contiguous United States: a review. *Arctic and Alpine Research*, 15, 145–156.

Phillips, L.M. (1976). Pleistocene vegetational history and geology in Norfolk. *Philosophical Transactions of the Royal Society, London*, B 275, 215–286.

Pierson, T.C. (1980). Erosion and deposition by debris flows at Mt Thomas, North Canterbury, New Zealand. *Earth Surface Processes*, 5, 227–247.

Pierson, T.C. (1981). Dominant particle support mechanisms in debris flows at Mt Thomas, New Zealand, and implications for flow mobility. *Sedimentology*, 28, 49–60.

Pissart, A. (1956). L'origine des viviers des Hautes Fagnes. *Annales de la Société Géologique de Belgique*, 79B, 119–131.

Pissart, A. (1963a). Les traces du 'pingos' du Pays de Galles (Grande Bretagne) et du Plateau des Hautes Fagnes (Belgique). *Zeitschrift für Geomorphologie*, 7, 147–175.

Pissart, A. (1963b). Les replats de cryoturbation au Pays de Galles (une variété géante des sols en guirlandes). *Biuletyn Peryglacjalny*, 12, 119–135.

Pissart, A. (1963c). Origine periglaciaire d'une variété géante des sols en guirlandes, découverte au Pays de Galles. *Comptes Rendues de l'Academie des Sciences Belgique*, 256, 222–224.

Pissart, A. (1965). Les pingos des Hautes Fagnes: les problèmes de leur genèse. *Annales de la Société Géologique de Belgique*, 88B, 277–289.

Pissart, A. (1966). Le rôle géomorphologique du vent dans la région de Mould Bay (île Prince Patrick, NWT, Canada). *Zeitschrift für Geomorphologie*, 10, 226–236.

Pissart, A. (1967). Les pingos de l'île Prince Patrick (76° N, 120° W). *Geographical Bulletin*, 9, 189–217.

Pissart, A. (1974). Les viviers des Hautes Fagnes sont des traces de buttes periglaciaires. Mais s'agissait-il réellement de pingos? *Annales de la Société Géologique de Belgique*, 9T, 357–381.

Pissart, A. (1983). Pingos et palses: un essai de synthèse de connaissance actuelles. In *Mesoformen des Reliefs in heutigen Periglazialraum*, eds. H. Poser & E. Schunke, pp. 48–69. Abhandlungen der Akademie der Wissenschaften in Göttingen, Mathematisch-Physikalische Klasse, 35. Göttingen: Vandenhoek & Ruprecht.

Pissart, A. (1985). L'origine des cryoturbations. In *Recent Trends in Physical Geography in Belgium*, ed. M. Van Molle, pp. 9–29. Brussels: Vrije Universiteit.

Pissart, A. (1988). Pingos: an overview of the present state of knowledge. In *Advances in Periglacial Geomorphology*, ed. M.J.Clark, pp. 279–297. Chichester: Wiley.

Pissart, A. & French, H.M. (1976). Pingo investigations: north central Banks Island, Canadian Arctic. *Canadian Journal of Earth Sciences*, 13, 937–946.

Pissart, A. & Gangloff, P. (1984). Les palses minérales et organiques de la vallée de l'Aveneau près de Kuujjuaq (Fort Chimo), Québec subarctique. *Géographie Physique et Quaternaire*, 38, 217–228.

Pissart, A. & Juvigne, E. (1980). Genèse et age d'une trace de butte périglaciaire (pingo ou palse) de la Konnerzvenn (Hautes Fagnes, Belgique). *Annales de la Société Géologique de Belgique*, 103, 73–86.

Pissart, A., Van Vliet-Lanoë, B. & Juvigne, E. (1988). Traces of ice in caves: evidence of former permafrost. *Proceedings of the 5th International Conference on Permafrost, Trondheim*, pp. 840–845. Trondheim: Tapir Press.

Pissart, A., Vincent, J-S. & Edlund, S.A. (1977). Dépôts et phenomènes éoliens sur l'île de Banks, territoires du Nord-Ouest, Canada. *Canadian Journal of Earth Sciences*, 14, 2462–2480.

Pitcher, W.S., Shearman, D.J. & Pugh, D.C. (1954). The loess of Pegwell Bay and its associated frost soils. *Geological Magazine*, 91, 308–314.

Pointon, W.K. (1978). The Pleistocene succession at Corton, Suffolk. *Bulletin of the Geological Society of Norfolk*, 30, 55–76.

Pollard, W.H. (1988). Seasonal frost mounds. In *Advances in Periglacial Geomorphology*, ed. M.J. Clark, pp. 201–229. Chichester: Wiley.

Pollard, W.H. & French, H.M. (1980). First approximation of the volume of ground ice, Richards Island, Pleistocene Mackenzie Delta, Northwest Territories, Canada. *Canadian Geotechnical Journal*, 17, 509–516.

Pollard, W.H. & French, H.M. (1984). The groundwater hydraulics of seasonal frost mounds, Northern Yukon. *Canadian Journal of Earth Sciences*, 21, 1073–1081.

Pollard, W.H. & French, H.M. (1985).The internal structure and ice crystallography of seasonal frost mounds. *Journal of Glaciology*, 31, 157–162.

Poser, H. (1954). Die Periglazial-Erscheinungen in der Umgebung der Gletscher des Zemmgrundes (Zillentaler Alpen). *Göttinger Geographische Abhandlungen*, 15, 125–180.

Potter, N. & Moss, J.H. (1968). Origin of the Blue Rocks blockfield and adjacent deposits, Berks County, Pennsylvania. *Geological Society of America, Bulletin*, 79, 255–262.

Potts, A.S. (1970). Frost action in rocks: some experimental data. *Transactions of the Institute of British Geographers*, 49, 109–124.

Potts, A.S. (1971). Fossil cryonival features in central Wales. *Geografiska Annaler*, 53A, 39–51.

Pounder, E.J. (1985). Periglacial river activity in upper Swaledale, north Yorkshire. In *Field Guide to the Periglacial Landforms of Northern*

England, ed. J. Boardman, 15–22. Cambridge: Quaternary Research Association.

Powers, M.C. (1958). A new roundness scale for sedimentary particles. *Journal of Sedimentary Petrology,* 23, 117–119.

Prestwich, J. (1892). The raised beaches and 'head' or rubble drift of the south-east of England; their relation to the valley drifts and to the glacial period; and on a postglacial submergence. *Quarterly Journal of the Geological Society of London,* 48, 263–343.

Price, L.W. (1973). Rates of mass-wasting in the Ruby Range, Yukon Territory. *North American Contribution, 2nd International Conference on Permafrost, Yakutsk,* pp. 235–245. Washington DC: National Academy Press.

Priesnitz, K. (1988). Cryoplanation. In *Advances in Periglacial Geomorphology,* ed. M.J. Clark, pp. 49–67. Chichester: Wiley.

Pullan, R.A. (1959). Tors. *Scottish Geographical Magazine,* 75, 51–55.

Pye, K. (1984). Loess. *Progress in Physical Geography,* 8, 176–217.

Pye, K. & Paine, A.D.M. (1983). Nature and source of aeolian deposits near the summit of Ben Arkle, Northwest Scotland. *Geologie en Mijnbouw,* 63, 13–18.

Ragg, J.M. & Bibby, J.S. (1966). Frost weathering and solifluction products in southern Scotland. *Geografiska Annaler,* 48, 12–23.

Ransom C.E. (1968). Ice wedge casts in the Corton Beds. *Geological Magazine,* 105, 74–75.

Rapp, A. (1959). Avalanche boulder tongues in Lappland. *Geografiska Annaler,* 41, 34–48.

Rapp, A. (1960a). Recent development of mountain slopes in Karkevagge and surroundings, northern Scandinavia. *Geografiska Annaler,* 42, 65–200.

Rapp, A. (1960b). Talus slopes and mountain walls at Tempelfjorden, Spitsbergen: a geomorphological study of the denudation of slopes in an arctic locality. *Norsk Polarinstitutt Skrifter,* 119, 96 pp.

Rapp, A. (1962). Karkevagge: some recordings of mass movements in the northern Scandinavian mountains. *Biuletyn Peryglacjalny,* 11, 287–309.

Rapp, A. (1983). Impact of nivation on steep slopes in Lappland and Scania, Sweden. *Abhandlungen der Akademie der Wissenschaften in Göttingen. Mathematische-Physikalische Klasse, Dritte Folge,* 35, 96–115.

Rapp, A. & Nyberg, R. (1981). Alpine debris flows in northern Scandinavia. *Geografiska Annaler,* 63A, 183–196.

Rapp, A., Nyberg, R. & Lindh, L. (1986). Nivation and local glaciation in N and S Sweden. A progress report. *Geografiska Annaler,* 68A, 197–205.

Rapson, S.C. (1985). Minimum age of corrie moraine ridges in the Cairngorm Mountains, Scotland. *Boreas,* 14, 155–159.

Rasmussen, A. (1981). The deglaciation of the coastal area NW of Svartisen, northern Norway. *Norges Geologiske Undersøkelse,* 369, 1–31.

Rasmussen, A. (1984). Late Weichselian moraine chronology of the Vesterålen islands, North Norway. *Norsk Geologisk Tidsskift,* 64, 193–219.

Ray, R.J., Krantz, W.B., Caine, N. & Gunn, R.D. (1983). A model for patterned ground regularity. *Journal of Glaciology,* 29, 317–337.

Reanier, R.E. & Ugolini, F.C. (1983). Gelifluction deposits as sources of palaeoenvironmental information. *Proceedings of the 4th International Conference on Permafrost, Fairbanks, Alaska,* pp. 1042–1047. Washington: National Academy Press.

Reed, W.J. (1988). *The Vertical Dimensions of the Last Ice Sheet and Late Quaternary Glacial Events in Northern Ross-shire, Scotland.* Unpublished PhD Thesis, University of St. Andrews.

Reger, R.D. & Péwé, T.L. (1976). Cryoplanation terraces: indicators of a permafrost environment. *Quaternary Research,* 6, 99–109.

Reid, C. (1887). On the origin of the dry chalk valleys and of the coombe rock. *Quarterly Journal of the Geological Society,* 43, 364–373.

Reid, C. (1892). The Pleistocene deposits of the Sussex coasts and their equivalent in other districts. *Quarterly Journal of the Geological Society,* 48, 344–361.

Reid, J.M., MacLeod, D.A. & Cresser, M.S. (1981). The assessment of chemical weathering rates within an upland catchment in North-East Scotland. *Earth Surface Processes and Landforms,* 6, 447–457.

Reid, J.R. & Nesje, A. (1988). A giant ploughing boulder, Finse, southern Norway. *Geografiska Annaler,* 70A, 27–33.

Rekstad, J. (1915). Om strandlinjer og strandlinjedannelse. *Norsk Geologisk Tidsskrift,* 3, 3–18.

Rendell, H.M., Worsley, P., Green, F. & Parks, D.A. (1991). Thermoluminescence dating of the Chelford Interstadial. *Earth and Planetary Science Letters,* 103, 182–189.

Reynolds, B. (1986). A comparison of element outputs in solution, suspended sediments and bedload for a small upland catchment. *Earth Surface Processes and Landforms,* 11, 217–222.

Reynolds, G. (1969). Rainfall, runoff and evaporation on a catchment in west Scotland. *Weather,* 24, 90–98.

Reynolds, R.C. & Johnson, N.M. (1972). Chemical weathering in the temperate glacial environment of the northern Cascade Mountains. *Geochimica et Cosmochimica Acta,* 36, 537–554.

Rice, R.J. (1959). A frost wedge in Angus. *Scottish Geographical Magazine,* 75, 50–51

Rice, R.J. & Douglas, T.D. (1991). Wolstonian glacial deposits and glaciation in Great Britain. In *Glacial Deposits in Great Britain and Ireland,* ed. J. Ehlers, P.L. Gibbard & J. Rose, pp. 25–36. Rotterdam: Balkema.

Richards, K.S. (1982). *Rivers.* London: Methuen

Richards, K.S. & McCaig, M. (1985). A medium-term estimate of bedload yield in Allt a'Mhuillin, Ben Nevis, Scotland. *Earth Surface Processes and Landforms,* 10, 407–411.

Richmond, G.M. (1962). Quaternary stratigraphy of the La Sal Mountains, Utah. *United States Geological Survey Professional Paper,* 324, 135 pp.

Rieger, S. (1974). Alpine Soils. In *Arctic and Alpine Environments.* eds. J.D. Ives & R.G. Barry, pp. 749–769. London: Methuen.

Riezebos, P.A., Boulton, G.S., van der Meer, J.J.M., Ruegg, G.H.J., Beets, D.J. Castel, I.I.Y., Hart, J., Quinn, I., Thornton, M. & van der Wateren, F.M. (1986). Products and effects of modern eolian activity on a nineteenth-century glacier-pushed ridge in West Spitsbergen, Svalbard. *Arctic and Alpine Research,* 18, 389–396.

Robertson-Rintoul, M.S.E. (1986). A quantitative soil-stratigraphic approach to the correlation and dating of post-glacial river terraces in Glen Feshie, western Cairngorms. *Earth Surface Processes and Landforms,* 11, 605–617.

Robinson, D.A. & Williams, R.B.G. (1976). Aspects of the geomorphology of the sandstone cliffs of the central Weald. *Proceedings of the Geologists' Association,* 87, 93–99.

Robinson, M. & Ballantyne, C.K. (1979). Evidence for a glacial readvance pre-dating the Loch Lomond Advance in Wester Ross. *Scottish Journal of Geology,* 15, 271–277.

Rochette, J-C. & Cailleux, A. (1971). Dépôts nivéo-éoliens annuels à Porte-de-la- Baleine, nouveau Québec. *Revue Géographique de Montréal,* 255, 35–41.

Rodine, J.D. & Johnson, A.M. (1976). The ability of debris, heavily freighted with coarse clastic materials, to flow on gentle slopes. *Sedimentology,* 23, 213–234.

Romanovskij, N.N. (1977). Formirovaniye poligonal'no-zhil'nykh struktur. *Akademie Nauk SSSR Sibirskoye Otdeliniye, Novosibirsk Izdatel'stvo Nauka,* 215pp.

Romanovskij, N.N. (1985). Distribution of recently active ice and soil wedges in the USSR. In *Field and Theory: Lectures in Geocryology,* eds. M. Church & O.Slaymaker, pp. 154–165. Vancouver: University of British Columbia Press.

Romans, J.C.C. & Robertson, L. (1974). Some aspects of the genesis of alpine upland soils in the British Isles. In *Soil Microscopy,* ed. G.K. Rutherford, pp. 498–510. Kingston (Ontario): Limestone Press.

Romans, J.C.C., Stevens, J.H. & Robertson, L. (1966). Alpine soils of north-east Scotland. *Journal of Soil Science,* 17, 184–199.

Rose, J. (1975). Raised beach gravels and ice wedge casts at Old Kilpatrick, near Glasgow. *Scottish Journal of Geology,* 11, 15–21.

Rose, J. (1980). In *Glasgow Region – Field Guide,* ed. W.G. Jardine, p. 37. Cambridge: Quaternary Research Association.

Rose, J. (1985). The Dimlington Stadial/Dimlington Chronozone: a proposal for naming the main glacial episode of the Late Devensian in Britain. *Boreas,* 14, 225–230.

Rose, J. (1987). Status of the Wolstonian glaciation in the British Quaternary. *Quaternary Newsletter,* 53, 1–9.

Rose, J. (1988). Stratigraphic nomenclature for the British Middle Pleistocene – procedural dogma or stratigraphic common sense? *Quaternary Newsletter,* 54, 15–20.

Rose, J. (1989a). Stadial type sections in the British Quaternary. In *Quaternary Type Sections: Imagination or Reality?* eds. J. Rose & C. Schlüchter, pp. 45- 67. Rotterdam: Balkema.

Rose, J. (1989b). Tracing the Baginton-Lillington Sands and Gravels from the West Midlands to East Anglia. In *West Midlands Field Guide,* ed. D.H. Keen, pp. 102–122. Cambridge: Quaternary Research Association.

Rose, J. & Allen, P. (1977). Middle Pleistocene stratigraphy of south-east Suffolk. *Journal of the Geological Society of London,* 133, 83–102.

Rose, J., Allen, P. & Hey, R.W. (1976). Middle Pleistocene stratigraphy in southern East Anglia. *Nature*, 263, 492–494.

Rose, J., Allen, P., Kemp, R.A., Whiteman, C.A. & Owen, N. (1985a). The Early Anglian Barham Soil of Eastern England. In *Soils and Quaternary Landscape Evolution*, ed. J. Boardman, pp. 197–229. Chichester: Wiley.

Rose, J., Boardman, J., Kemp, R.A. & Whiteman, C.A. (1985b). Palaeosols and the interpretation of the British Quaternary stratigraphy. In *Geomorphology and Soils*, eds. K.S. Richards, R.R. Arnett & S. Ellis, pp. 348–375. London: Allen and Unwin.

Rose, J., Lowe, J.J. & Switsur, R. (1988). A radiocarbon date on plant detritus beneath type till from the type area of the Loch Lomond Readvance. *Scottish Journal of Geology*, 24, 113–124.

Rose, J., Sturdy, R.G., Allen, P. & Whiteman, C.A. (1978). Middle Pleistocene sediments and palaeosols near Chelmsford, Essex. *Proceedings of the Geologists' Association*, 89, 91–96.

Rose, J., Turner, C., Coope, G.R. & Bryan, M.D. (1980). Channel changes in a lowland river catchment over the past 13,000 years. In *Timescales in Geomorphology*, eds. R.A. Cullingford, D.A. Davidson & J. Lewin, pp. 159–175. Chichester: Wiley.

Rowlands, B.M. (1971). Radiocarbon evidence of the age of an Irish Sea glaciation in the Vale of Clwyd. *Nature*, 230, 9–11.

Rowlands, P.H. & Shotton, F.W. (1971). Pleistocene deposits of Church Stretton (Shropshire) and its neighbourhood. *Journal of the Geological Society*, 127, 599–622.

Rudberg, S. (1962). A report on some field observations concerning periglacial geomorphology and mass movement in Sweden. *Biuletyn Peryglacjalny*, 11, 311–323.

Rudberg, S. (1963). Geomorphological processes in a cold semi-arid region. In *Axel Heiberg Preliminary Report*, ed. F. Müller. Montreal: McGill University.

Rudberg, S. (1964). Slow mass movement processes and slope development in the Norra Storfjäll area, south Swedish Lappland. *Zeitschrift für Geomorphologie, Supplementband*, 5, 192–203.

Rudberg, S. (1974). Some observations concerning nivation and snowmelt in Swedish Lapland. In *Geomorphologische Prozess und Prozesskombinationen in der Gegenwart unter verschiedenen Klimabedingungen*, ed. H. Poser, pp. 263–273. Göttingen: Vandenhoek und Ruprecht.

Ruddiman, W.F. & McIntyre, A. (1973). Time-transgressive deglacial retreat of polar waters from the North Atlantic. *Quaternary Research*, 3, 117–130.

Ruddiman, W.F. & McIntyre, A. (1981). The North Atlantic during the last deglaciation. *Palaeogeography, Palaeoclimatology, Palaeoecology*, 35, 145–214.

Ruddiman, W.F., Sancetta, C.D. & McIntyre, A. (1977). Glacial/interglacial response of subpolar North Atlantic water to climatic change: the record in ocean sediments. *Philosophical Transactions of the Royal Society of London*, B 280, 119–142.

Ruddiman, W.F., Sancetta, C.D., Niebler-Hunt, V. & Durazzi, J.T. (1980). Glacial/interglacial response rate of subpolar North Atlantic water to climatic change: the record in ocean sediments. *Quaternary Research*, 13, 33–64.

Rune, O. (1965). The mountain regions of Lappland. *Acta Phytogeographica Suedica*, 50, 64–77.

Russell, R.J. (1944). Lower Mississippi Valley loess. *Geological Society of America, Bulletin*, 55, 1–40.

Ryder, J.M. (1971). The stratigraphy and morphology of paraglacial alluvial fans in south-central British Columbia. *Canadian Journal of Earth Sciences*, 8, 279–298.

Ryder, R.H. & McCann, S.B. (1971). Periglacial phenomena on the Island of Rhum in the Inner Hebrides. *Scottish Journal of Geology*, 7, 293–303.

Samuelsson, C. (1926). Studien über die Wirkungen des Windes in den kalten und gemässigten Erdteilen. *Bulletin of the Geological Institute of Uppsala*, 20, 57–230.

Saunders, G.E. (1973). Vistulian periglacial environments in the Lleyn Peninsula. *Biuletyn Peryglacjalny*, 22, 257–269.

Saunders, I. & Young, A. (1983). Rates of surface processes on slopes, slope retreat and denudation. *Earth Surface Processes and Landforms*, 8, 473–501.

Savigear, R.A.G. (1952). Some observations on slope development in South Wales. *Transactions of the Institute of British Geographers*, 18, 31–51.

Savigny, K.W. (1980). *In Situ Analysis of Naturally Occurring Creep in Ice-Rich Permafrost*. Unpublished PhD Thesis, University of Alberta.

Schumm, S. (1963). A tentative classification of alluvial river channels. *US Geological Survey Circular*, 477.

Schumm, S. (1968). Speculations concerning paleohydraulic controls of terrestrial sedimentation. *Geological Society of America, Bulletin*, 79, 1573–1588.

Schunke, E. (1975). Die Periglazialerscheinungen Islands in Abhängigkeit von Klima und Substrat. *Abhandlungen der Akademie der Wissenschaften in Göttingen. Mathematisch-Physikalische Klasse, Dritte Folge*, 30, 273 pp.

Schunke, E. (1977). Zur Genese der Thufur Islands und öst-Grönlands. *Erdkunde*, 31, 279–287.

Schunke, E. & Zoltai, S.C. (1988). Earth hummocks (thufur). In *Advances in Periglacial Geomorphology*, ed. M.J. Clark, pp. 231–245. Chichester: Wiley.

Scott, K.M. (1978). Effects of permafrost on stream channel behaviour in Arctic Alaska. *United States Geological Survey Professional Paper*, 1068, 19pp.

Scotter, G.W. & Zoltai, S.C. (1982). Earth hummocks in the Sunshine area of the Rocky Mountains, Alberta and British Columbia. *Arctic*, 35, 411–416.

Scourse, J.D. (1985). *Late Pleistocene Stratigraphy of the the Isles of Scilly and Adjoining Regions*. Unpublished PhD Thesis, University of Cambridge.

Scourse, J.D. (1987). Periglacial sediments and landforms in the Isles of Scilly and West Cornwall. In *Periglacial Processes and Landforms in Britain and Ireland*, ed. J. Boardman, pp. 225–236. Cambridge: Cambridge University Press.

Seddon. M.B. (1984). *Pleistocene Permafrost Environments: Comparisons of Arctic Ice-Wedges and Ice-Wedge Casts in Britain*. Unpublished PhD Thesis, University of Reading. 342pp.

Seddon, M.B. & Holyoak, D.T. (1985). Evidence of sustained regional permafrost during deposition of fossiliferous Late Pleistocene sediments at Stanton Harcourt (Oxfordshire, England). *Proceedings of the Geologists' Association*, 96, 53–71.

Sekyra, J. (1961). Periglacial phenomena. *Instytut Geologiiszny (Warszawa), Prace*, 34, 199–207.

Selby, M.J. (1972). Antarctic tors. *Zeitschrift für Geomorphologie Supplementband*, 13, 73–86.

Seppälä, M. (1972). The term 'palsa'. *Zeitschrift für Geomorphologie*, 16, 463.

Seppälä, M. (1982). Present-day periglacial phenomena in northern Finland. *Biuletyn Peryglacjalny*, 29, 231–243.

Seppälä, M. (1986). The origin of palsas. *Geografiska Annaler*, 68A, 141–147.

Seppälä, M. (1987). Periglacial phenomena in northern Fennoscandia. In *Periglacial Processes and Landforms in Britain and Ireland*, ed. J.Boardman, pp. 45–55. Cambridge: Cambridge University Press.

Seppälä, M. (1988). Palsas and related forms. In *Advances in Periglacial Geomorphology*, ed, M.J. Clark, pp. 247–278. Chichester: Wiley.

Shackleton, N.J. (1987). Oxygen isotopes, ice volume and sea level. *Quaternary Science Reviews*, 6, 183–190.

Shakesby, R.A. (1975). An investigation into the origin of the deposits in a Chalk dry valley of the South Downs, southern England. *University of Edinburgh Geography Research Discussion Paper*, 5, 25pp.

Shakesby, R.A. (1981). Periglacial origin of slope deposits near Woolhope in the Welsh Borderland. *Cambria*, 8, 1–16.

Shakesby, R.A. & Stephens, N. (1984). The Pleistocene gravels of the Axe Valley, Devon. *Reports and Transactions of the Devonshire Association for the Advancement of Science*, 116, 77–88.

Sharp, R.P. (1942). Mudflow levées. *Journal of Geomorphology*, 5, 274–301.

Shaw, R. (1977). *Periglacial Features in Part of the South-East Grampian Highlands of Scotland*. Unpublished PhD Thesis, University of Edinburgh.

Shephard-Thorn, E.R. (1987). Quaternary. In *Geology of the Country around Hastings and Dungeness*, by R.D. Lake & E.R. Shephard-Thorn, British Geological Survey Memoir, pp. 46–56. London: HMSO.

Sheppard, S.M.F. (1977). The Cornubian batholith of SW England: D/H and O_{18}/O_{16} studies of kaolinite and other alteration minerals. *Quarterly Journal of the Geological Society of London*, 133, 573–592.

Shilston, D.T. (1986). Some possible injection structures observed near Stanstead Abbots, Herts. *Quaternary Newsletter*, 49, 7–13.

Shilts, W.W. (1978). Nature and genesis of mudboils, central Keewatin, Canada. *Canadian Journal of Earth Sciences*, 15, 1053–1063.

Shotton, F.W. (1953). Pleistocene deposits of the area between Coventry, Rugby and Leamington and their bearing on the topographic development of the Midlands. *Philosophical Transactions of the Royal Society of London*, B 237, 209–260.

Shotton, F.W. (1960). Large scale patterned ground in the valley of the Worcestershire Avon. *Geological Magazine*, 97, 404–408.

Shotton, F.W. (1977). *The English Midlands*. International Quaternary Association Guidebook A2, 10th INQUA Congress, Birmingham, 51pp.

Shotton, F.W. (1983). The Wolstonian stage of the British Pleistocene in and around its type area of the English Midlands. *Quaternary Science Reviews*, 2, 261–280.

Shotton, F.W. & Wilcockson, W.H. (1950). Superficial valley folds in an opencast working of the Barnsley Coal. *Proceedings of the Yorkshire Geological Society*, 28, 102–111.

Simpson, I.M. & West, R.G. (1958). On the stratigraphy and palaeobotany of a Late Pleistocene organic deposit at Chelford, Cheshire. *New Phytologist*, 57, 239–250.

Simpson, J.B. (1932). Stone polygons on Scottish mountains. *Scottish Geographical Magazine*, 48, 37.

Sissons, J.B. (1967). *The Evolution of Scotland's Scenery*. Edinburgh: Oliver and Boyd.

Sissons, J.B. (1969). Drift stratigraphy and buried morphological features in the Grangemouth–Falkirk–Airth area, central Scotland. *Transactions of the Institute of British Geographers*, 48, 19–50.

Sissons, J.B. (1972). The last glaciers in part of the south-east Grampians. *Scottish Geographical Magazine*, 88, 168–181.

Sissons, J.B. (1974a). Late-glacial marine erosion in Scotland. *Boreas*, 3, 41–48.

Sissons, J.B. (1974b). A lateglacial ice-cap in the Central Grampians, Scotland. *Transactions of the Institute of British Geographers*, 62, 95–114.

Sissons, J.B. (1975). A fossil rock glacier in Wester Ross. *Scottish Journal of Geology*, 11, 83–86.

Sissons, J.B. (1976a). A remarkable protalus rampart in Wester Ross. *Scottish Geographical Magazine*, 92, 182–190.

Sissons, J.B. (1976b). A fossil rock glacier in Wester Ross. *Scottish Journal of Geology*, 12, 178.

Sissons, J.B. (1976c). *The Geomorphology of the British Isles: Scotland*. London: Methuen.

Sissons, J.B. (1977a). Former ice-dammed lakes in Glen Moriston, Inverness-shire, and their significance in upland Britain. *Transactions of the Institute of British Geographers, New Series*, 2, 224–242.

Sissons, J.B. (1977b). The Loch Lomond Readvance in the northern mainland of Scotland. In *Studies in the Scottish Lateglacial Environment*, eds. J.M. Gray & J.J.Lowe, pp. 45–59. Oxford: Pergamon.

Sissons, J.B. (1978). The parallel roads of Glen Roy and adjacent glens, Scotland. *Boreas*, 7, 229–244.

Sissons, J.B. (1979a). The Loch Lomond Advance in the Cairngorm Mountains. *Scottish Geographical Magazine*, 95, 66–82.

Sissons, J.B. (1979b). The limit of the Loch Lomond Advance in Glen Roy and vicinity. *Scottish Journal of Geology*, 15, 31–42.

Sissons, J.B. (1979c). Palaeoclimatic inferences from former glaciers in Scotland and the Lake District. *Nature*, 278, 518–521.

Sissons, J.B. (1980a). The Loch Lomond Advance in the Lake District, northern England. *Transactions of the Royal Society of Edinburgh: Earth Sciences*, 71, 13–27.

Sissons, J.B. (1980b). Palaeoclimatic inferences from Loch Lomond Advance glaciers. In *Studies in the Lateglacial of North-West Europe*, eds. J.J. Lowe, J.M. Gray, & J.E. Robinson, pp. 31–44. Oxford: Pergamon.

Sissons, J.B. (1981a). British shore platforms and ice sheets. *Nature*, 291, 473–475.

Sissons, J.B. (1981b). The last Scottish ice sheet: facts and speculative discussion. *Boreas*, 10, 1–17.

Sissons, J.B. (1982). The so-called high 'interglacial' rock shoreline of western Scotland. *Transactions of the Institute of British Geographers, New Series*, 7, 205–216.

Sissons, J.B. & Cornish, R. (1983). Fluvial landforms associated with ice-dammed lake drainage in upper Glen Roy, Scotland. *Proceedings of the Geologists' Association*, 94, 45–52.

Sissons, J.B. & Dawson, A.G. (1981). Former sea levels and ice limits in part of Wester Ross, North-west Scotland. *Proceedings of the Geologists' Association*, 92, 115–124.

Sissons, J.B. & Grant, A.J.H. (1972). The last glaciers of the Lochnagar area, Aberdeenshire. *Scottish Journal of Geology*, 8, 85–93.

Sissons, J.B. & Sutherland, D.G. (1976). Climatic inferences from former glaciers in the south-east Grampian Highlands, Scotland. *Journal of Glaciology*, 17, 325–346.

Skempton, A.W. (1988). Geotechnical aspects of the Carsington dam failure. *Proceedings of the 11th International Conference on Soil Mechanics*, 5, pp. 2581–2591. Rotterdam: Balkema.

Skempton, A.W., Norbury, D. & Petley, D.J. (1991). Solifluction shears at Carsington, Derbyshire. In *Quaternary Engineering Geology*, eds. A. Forster, M.G. Culshaw, J.C. Cripps, J.A. Little & C.F. Moon, pp. 281–387. London: Geological Society Engineering Geology Special Publication No. 7.

Skempton, A.W. & Weeks, A.G. (1976) The Quaternary history of the Lower Greensand escarpment and Weald Clay vale near Sevenoaks, Kent. *Philosophical Transactions of the Royal Society of London*, A 283, 493–526.

Small, R.J. (1961), The morphology of Chalk escarpments: a critical discussion. *Transactions of the Institute of British Geographers*, 29, 71–90.

Small, R.J. (1964). The escarpment dry valleys of the Wiltshire Chalk. *Transactions of the Institute of British Geographers*, 34, 33–52.

Small, R.J. (1965) The role of spring sapping in the formation of Chalk escarpment dry valleys. *Southampton Research Series in Geography*, 1, 3–30.

Small, R.J., Clark, M.J. & Lewin,J. (1970) The periglacial rock-stream at Clatford Bottom, Marlborough Downs, Wiltshire. *Proceedings of the Geologists' Association*, 81, 87–98.

Smith, D.I. (1972). The solution of limestone in an arctic environment. *Institute of British Geographers Special Publication*, 4, 187–200.

Smith, D.J. (1988). Rates and controls of soil movement on a solifluction slope in the Mt Rae area, Canadian Rocky Mountains. *Zeitschrift für Geomorphologie, Supplementband*, 71, 25–44.

Smith, H.T.U. (1968). 'Piping' in relation to periglacial boulder concentrations. *Biuletyn Peryglacjalny*, 17, 195–204.

Smithson, F. (1953). The micro-mineralogy of North Wales soils. *Journal of Soil Science*, 4, 194–210.

Söderman, G. (1980). Slope processes in cold environments of Northern Finland. *Fennia*, 158, 83–152.

Sollid, J.L. & Reite, A.J. (1983). The last glaciation and deglaciation of central Norway. In *Glacial Deposits of North-West Europe*, ed. J. Ehlers, pp. 41–59. Rotterdam: Balkema.

Sollid, J.L. & Sørbel, L. (1979). Deglaciation of west-central Norway. *Boreas*, 8, 233–239.

Sparks, B.W. (1949). The denudation chronology of the dip-slope of the South Downs. *Proceedings of the Geologists' Association*, 60, 165–215.

Sparks, B.W. & Lewis, W.V. (1957). Escarpment dry valleys near Pegsdon, Hertfordshire. *Proceedings of the Geologists' Association*, 68, 26–38.

Sparks, B.W., Williams, R.B.G. & Bell, F.G. (1972). Presumed ground-ice depressions in East Anglia. *Proceedings of the Royal Society*, A 327, 329–343.

Spence, D.H.N. (1957). Studies on the serpentine vegetation of Shetland, 1: the serpentine vegetation on Unst. *Journal of Ecology*, 45, 917–945.

Spink, T.W. (1991). Periglacial discontinuities in Eocene Clays near Denham, Buckinghamshire. In *Quaternary Engineering Geology*, eds. A. Forster, M.G. Culshaw, J.C. Cripps, J.A. Little & C.F. Moon, pp. 389–396. London: Geological Society Engineering Geology Special Publication No. 7.

Spurrell, F.C.J. (1886). A sketch of the history of the rivers and denudation of West Kent. *Report of the West Kent Natural History Society*.

Statham, I. (1973). Scree slope development under conditions of surface particle movement. *Transactions of the Institute of British Geographers*, 59, 41–53.

Statham, I. (1976a). A scree slope rockfall model. *Earth Surface Processes*, 1, 43–62.

Statham, I. (1976b). Debris flows on vegetated screes in the Black Mountains, Carmarthenshire. *Earth Surface Processes*, 1, 173–180.

Statham, I. (1977). *Earth Surface Sediment Transport*. Oxford: Clarendon.

Statham, I. & Francis, S.C. (1986). Influence of scree accumulation and weathering on the development of steep mountain slopes. In *Hillslope Processes*, ed. A.D. Abrahams, pp. 245–267. London: Allen and Unwin.

Steers, J.A. (1966). *The English Coast and the Coast of Wales*. London: Collins.

Stephens, N. (1961). Pleistocene events in North Devon. *Proceedings of the Geologists' Association*, 72, 469–472.

Stephens, N. (1970). The West Country and Southern Ireland. In *The Glaciations of Wales and Adjoining Regions*, ed. C.A. Lewis, pp. 267–312. Longman: London.

Stevens, J.H. & Wilson, M.J. (1970). Alpine podzol soils on the Ben Lawers Massif, Perthshire. *Journal of Soil Science*, 21, 85–95.

Stewart, V.L. (1961). A permafrost horizon in the soils of Cardiganshire. *Welsh Soils Discussion Group Report*, 2, 19–22.

Stoker, M.S., Long, D. & Fyfe, J.A. (1985). The Quaternary succession in the central North Sea. *Newsletters in Stratigraphy*, 14, 119–128.

St Onge, D.A. (1965). La géomorphologie de l'île Ellef Ringnes, Territoires du Nord-ouest, Canada. *Étude Géographique, Direction de la Géographie*, 38, 46pp.

St Onge, D.A. (1969). Nivation landforms. *Geological Survey of Canada Paper*, 69–30, 12 pp.

Stott, T.A., Ferguson, R.I., Johnson, R.C. & Newson, M.D. (1986). Sediment budgets in forested and unforested basins in upland Scotland. *International Association of Hydrological Sciences Publication*, 159, 57–68.

Strachan, G.J. (1976). *Debris flows in Wester Ross*. Unpublished MA Dissertation, University of Aberdeen.

Straw, A. (1963). Some observations on the 'cover sands' of north Lincolnshire. *Transactions of the Lincolnshire Naturalists' Union*, 15, 260–269.

Straw, A. (1979). Eastern England. In *Eastern and Central England*, by A. Straw & K. Clayton, pp. 3–139. London: Methuen.

Straw, A. (1980). The age and geomorphological context of a Norfolk palaeosol. In *Timescales in Geomorphology*, eds. R.A. Cullingford, D.A. Davidson & J. Lewin, pp. 305–315. Chichester: Wiley.

Straw, A. (1983). Pre-Devensian glaciation of Lincolnshire (eastern England) and adjacent areas. *Quaternary Science Reviews*, 2, 239–260.

Straw, A. (1991). Glacial deposits of Lincolnshire and adjoining areas. In *Glacial deposits in Great Britain and Ireland*, eds. J. Ehlers, P.L. Gibbard & J. Rose, 213–221. Rotterdam: Balkema.

Stringer, C.B., Currant, P., Schwarcz, H.P. & Collcutt, S.M. (1986). Age of Pleistocene faunas from Bacon Hole, Wales. *Nature*, 320, 59–62.

Strömqvist, L. (1973). Geomorfologiska studier av blockhav och blockfält i Norra Skandinavien. *Uppsala University, Department of Physical Geography Reports*, 22, 161 pp.

Strömqvist, L. (1983). Gelifluction and surface wash, their importance and interaction on a periglacial slope. *Geografiska Annaler*, 65A, 245–254.

Stuart, A.J. (1977). The vertebrates of the last cold stage in Britain and Ireland. *Philosophical Transactions of the Royal Society of London*, B 289, 87–97.

Stuart, H. (1984). *A Comparative Study of Lateglacial and Holocene Talus Slopes in Snowdonia, North Wales*. Unpublished BSc Dissertation, University of St Andrews.

Sugden, D.E. (1968). The selectivity of glacial erosion in the Cairngorm Mountains, Scotland. *Transactions of the Institute of British Geographers*, 45, 79–92.

Sugden, D.E. (1970). Landforms of deglaciation in the Cairngorm Mountains, Scotland. *Transactions of the Institute of British Geographers*, 51, 201–219.

Sugden, D.E. (1971). The significance of periglacial activity on some Scottish mountains. *Geographical Journal*, 137, 388–392.

Sugden, D.E. (1977a). Did glaciers form in the Cairngorms in the 17–19th centuries? *Cairngorm Club Journal*, 97, 189–201.

Sugden, D.E. (1977b). Reconstruction of the morphology, dynamics and thermal characteristics of the Laurentide ice sheet at its maximum. *Arctic and Alpine Research*, 9, 21–47.

Sugden, D.E. (1982). *Arctic and Antarctic: A Modern Geographical Synthesis*, 472 pp. Oxford: Blackwell.

Sugden, D.E. & Watts, S.H. (1977). Tors, felsenmeer and glaciation in northern Cumberland Peninsula, Baffin Island. *Canadian Journal of Earth Sciences*, 14, 2817–2823.

Sukhodrovskiy, V.L. (1975). Particularités de l'évolution des versants dans les regions de pergélisol. *Biuletyn Peryglacjalny*, 24, 73–80.

Summerfield, M.A. & Goudie, A.S. (1980). The sarsens of southern England: their palaeoenvironmental interpretation with reference to other silcretes. In *The Shaping of Southern England*, ed. D.K.C. Jones, pp. 71–100. London: Academic Press.

Sutherland, D.G. (1980). Problems of radiocarbon dating of deposits from newly-deglaciated terrain: examples from the Scottish Lateglacial. In *Studies in the Lateglacial of North-West Europe*, eds. J.J. Lowe, J.M. Gray & J.E. Robinson, pp. 139–149. Oxford: Pergamon.

Sutherland, D.G. (1981). The high-level marine shell beds of Scotland and the build-up of the last Scottish ice sheet. *Boreas*, 10, 247–254.

Sutherland, D.G. (1984). The Quaternary deposits and landforms of Scotland and the adjacent shelves. *Quaternary Science Reviews*, 3, 157–254.

Sutherland, D.G. (1991a). The glaciation of the Shetland and Orkney Islands. In *Glacial Deposits in Great Britain and Ireland*, eds. J. Ehlers, P.L. Gibbard & J. Rose, pp. 121–128. Rotterdam: Balkema.

Sutherland, D.G. (1991b). Late Devensian glacial deposits and glaciation in Scotland and the adjacent offshore region. In *Glacial Deposits in Great Britain and Ireland*, eds. J. Ehlers, P.L. Gibbard & J. Rose, pp. 53–60. Rotterdam: Balkema.

Sutherland, D.G., Ballantyne, C.K. & Walker, M.J.C. (1984). Late Quaternary glaciation and environmental change on St. Kilda, Scotland, and their palaeoclimatic significance. *Boreas*, 13, 261–272.

Sutherland, D.G. & Walker, M.J.C. (1984). A Late Devensian ice-free area and possible interglacial site on the Isle of Lewis, Scotland. *Nature*, 309, 701–703.

Svensson, H. (1967). A tetragon-patterned block field. *Lund Studies in Geography, Series A*, 40, 8–23.

Svensson, H. (1976). Iskaldar som klimatindikatorer. *Svensk Geografisk Årsbok*, 52, 46–57.

Svensson, H. (1988) Ice-wedge casts and relict polygonal patterns in Scandinavia. *Journal of Quaternary Science*, 3, 57–67.

Synge, F.M. & Stephens, N. (1966). Late- and Post-glacial shorelines and ice limits in Argyll and North-east Ulster. *Transactions of the Institute of British Geographers*, 39, 101–125.

Taber, S. (1943). Perennial frozen ground in Alaska: its origin and history. *Geological Society of America, Bulletin*, 54, 1433–1548.

Takahashi, T. (1981). Debris flow. *Annual Review of Fluid Mechanics*, 13, 57–77.

Tallis, J.H. & Kershaw, K.A. (1959). Stability of stone polygons in North Wales. *Nature*, 183, 485–6.

Tansley, A.G. (1968). *Britain's Green Mantle*. 2nd Edition. London: Allen and Unwin.

Tarnocai, C. & Zoltai, S.C. (1978). Earth hummocks of the Canadian Arctic and Subarctic. *Arctic and Alpine Research*, 10, 581–594.

Taylor, A. (1978). Thermokarst depressions in southern Cambridgeshire. *Quaternary Newsletter*, 26, 1–2.

Taylor, A.E., Burgess, M., Judge, A.S. & Allen, V.S. (1982). Canadian geothermal data collection – northern wells 1981. *Earth Physics Branch, Energy Mines and Resources* Geothermal series, 13. Ottawa, Canada, 153pp.

Taylor, J.A. (1976). Upland climates. In *The Climate of the British Isles*, ed. S. Gregory, pp. 264–287. London: Longman.

Te Punga, M.T. (1956). Altiplanation terraces in southern England. *Biuletyn Peryglacjalny*, 4, 331–338.

Te Punga, M.T. (1957). Periglaciation in southern England. *Tijdschrift van het Koninklijk Nederlandsch Aardrijkskundig Genootschap*, 74, 400–412.

Tedrow, J.C.F. (1977). *Soils of the Polar Landscapes*, 638pp. New Brunswick: Rutgers University Press.

Tedrow, J.F.C. & Krug, E.C. (1982). Weathered limestone accumulations in the high arctic. *Biuletyn Peryglacjalny*, 29, 143–146.

Terzaghi, K. & Peck, R.B. (1967). *Soil Mechanics in Engineering Practice*. New York: John Wiley & Sons.

Thom, A.S. & Ledger, D.C. (1976). Rainfall, runoff and climatic change. *Proceedings of the Institution of Civil Engineers*, 61, 633–652.

Thomas, G.S.P. (1976). The Quaternary stratigraphy of the Isle of Man. *Proceedings of the Geologists' Association*, 87, 307–323.

Thomas, G.S.P. (1985). The Quaternary of the northern Irish Sea basin. In *The Geomorphology of North-West England*, ed. R.H. Johnston, pp. 143–158. Manchester: Manchester University Press.

Thompson, W.F. (1962). Preliminary notes on the nature and distribution of rock glaciers relative to true glaciers and other effects of the climate on the ground in North America. *International Association of Scientific Hydrology Publication*, 58, 212–219.

Thomson, M.E. & Eden, R.A. (1977). Quaternary deposits of the central North Sea. 3. The Quaternary sequence in the west-central North Sea. *Reports of the Institute of Geological Sciences*, 77/12, 1–18.

Thorarinsson, S. (1951). Notes on patterned ground in Iceland with particular reference to the Icelandic flás. *Geografiska Annaler*, 33, 144–156.

Thorn, C.E. (1976a). Quantitative evaluation of nivation in the Colorado Front Range. *Geological Society of America, Bulletin*, 87, 1169–1178.

Thorn, C.E. (1976b). A model of stony earth circle development, Schefferville, Quebec. *Proceedings of the Association of American Geographers*, 8, 19–23.

Thorn, C.E. (1979a). Ground temperatures and surficial transport in colluvium during snowpatch meltout: Colorado Front Range. *Arctic and Alpine Research*, 11, 41–52.

Thorn, C.E. (1979b). Bedrock freeze–thaw weathering regime in an alpine environment, Colorado Front Range. *Earth Surface Processes*, 4, 211–228.

Thorn, C.E. (1988). Nivation: a geomorphic chimera. In *Advances in Periglacial Geomorphology*, ed. M.J. Clark, pp. 3–31. Chichester: Wiley.

Thorn, C.E. & Darmody, R.G. (1980). Contemporary eolian sediments in the alpine zone, Colorado Front Range. *Physical Geography*, 1, 162–171.

Thorn, C.E. & Darmody, R.G. (1985). Grain-size sampling and characterization of eolian lag surfaces within alpine tundra, Niwot Ridge, Front Range, Colorado, USA. *Arctic and Alpine Research*, 17, 443–450.

Thorn, C.E. & Hall, K. (1980). Nivation: an arctic–alpine comparison and reappraisal. *Journal of Glaciology*, 25, 109–124.

Thorp, P. (1981). A trimline method for defining the upper limit of Loch Lomond Advance glaciers: examples from the Loch Leven and Glencoe areas. *Scottish Journal of Geology*, 17, 49–64.

Thorp, P. (1986). A mountain icefield of Loch Lomond Stadial age, western Grampians, Scotland. *Boreas*, 15, 83–97.

Thorp, P. (1987). Late Devensian ice sheet in the western Grampians, Scotland. *Journal of Quaternary Science*, 2, 103–112.

Thorp, P. (1991). The glaciation and glacial deposits of the western Grampians. In *Glacial Deposits in Great Britain and Ireland*, eds. J. Ehlers, P.L. Gibbard, & J. Rose, pp. 137–150. Rotterdam: Balkema.

Tinkler, K.J. (1966). Slope profiles and scree in the Eglwyseg Valley, North Wales. *Geographical Journal*, 132, 379–385.

Tipping, R.M. (1985). Loch Lomond Stadial *Artemisia* pollen assemblages and Loch Lomond Readvance firn-line altitudes. *Quaternary Newsletter*, 46, 1–11.

Tivy, J. (1957). Influence des facteurs biologiques sur l'érosion dans les Southern Uplands Écossais. *Revue de Géomorphologie Dynamique*, 8, 9–19.

Tivy, J. (1962). An investigation of certain slope deposits in the Lowther Hills, Southern Uplands of Scotland. *Transactions of the Institute of British Geographers*, 30, 59–72.

Tooley, M.J. & Kear, B.S. (1977). Mere Sands Wood (Shirdley Hill Sand). In *The Isle of Man, Lancashire Coast and Lake District*. X INQUA Congress Field Guide, pp. 9–10. Norwich: Geo Abstracts.

Tranter, M., Brimblecombe, P., Davies, T.D., Vincent, C.E., Abrahams, P.W. & Blackwood, I. (1986). The composition of snowfall, snowpack and meltwater in the Scottish Highlands – evidence for preferential elution. *Atmospheric Environment*, 20, 517–525.

Tranter, M., Davies, T.D., Abrahams, P.W., Blackwood, I., Brimblecombe, P. & Vincent, C.E. (1987a). Spatial variability in the chemical composition of snowcover in a small, remote Scottish catchment. *Atmospheric Environment*, 21, 853–862.

Tranter, M., Davies, T.D., Brimblecombe, P. & Vincent, C.E. (1987b). The composition of acidic meltwaters during snowmelt in the Scottish Highlands. *Water, Air and Soil Pollution*, 36, 75–90.

Trechmann, C.T. (1920). On a deposit of interglacial loess, and some transported preglacial freshwater clays on the Durham coast. *Quarterly Journal of the Geological Society of London*, 75, 173–201.

Trenhaile, A.S. (1983). The development of shore platforms in high latitudes. In *Shorelines and Isostasy*, ed. D.E. Smith & A.G. Dawson, pp. 77–93. London: Academic Press.

Trenhaile, A.S. & Mercan, D.W. (1984). Frost weathering and the saturation of coastal rocks. *Earth Surface Processes and Landforms*, 9, 321–331.

Trenhaile, A.S. & Rudakas, P.A. (1981). Freeze–thaw and shore platform development in Gaspé, Québec. *Géographie Physique et Quaternaire*, 35, 171–181.

Tricart, J. (1956). Étude expérimentale du problème de la gélivation. *Biuletyn Peryglacjalny*, 4, 285–318.

Tricart, J. (1970). *Geomorphology of Cold Environments*. New York: St Martin's Press.

Troll, C. (1944). Strukturböden, solifluktion und frostklimate der Erde. *Geologische Rundschau*, 34, 545–694.

Troll, C. (1958). Structure soils, solifluction and frost climates of the world. *US Army Snow, Ice and Permafrost Research Establishment, Corps of Engineers*, Translation 43, 121pp.

Troll, C. (1973). Rasenabschälung (turf exfoliation) als periglaziales Phänomen der subpolaren Zonen und der Hochgebirge. *Zeitschrift für Geomorphologie, Supplementband*, 17, 1–32.

Tuckfield, C.G. (1986). A study of dells in the New Forest, Hampshire, England. *Earth Surface Processes and Landforms*, 11, 23–40.

Tufnell, L. (1969). The range of periglacial phenomena in Northern England. *Biuletyn Peryglacjalny*, 19, 291–323.

Tufnell, L. (1971). Erosion by snow patches in the northern Pennines. *Weather*, 26, 492–498.

Tufnell, L. (1972). Ploughing blocks with special reference to north-west England. *Biuletyn Peryglacjalny*, 21, 237–270.

Tufnell, L. (1975). Hummocky microrelief in the Moor House area of the northern Pennines, England. *Biuletyn Peryglacjalny*, 24, 353–368.

Tufnell, L. (1976). Ploughing block movements on the Moor House Reserve (England), 1965–1975. *Biuletyn Peryglacjalny*, 26, 313–317.

Tufnell, L. (1985). Periglacial landforms in the Cross Fell–Knock Fell area of the North Pennines. In *Field Guide to the Periglacial Landforms of Northern England*, ed. J. Boardman, pp 4–14. Cambridge: Quaternary Research Association.

Unwin, D.J. (1969). The areal extension of rainfall records: an alternative model. *Journal of Hydrology*, 7, 404–414.

Unwin, D.J. (1975). The nature and origin of the corrie moraines of Snowdonia. *Cambria*, 2, 20–33.

Ussher, W.A.E. (1879). *The Post-Tertiary Geology of Cornwall*. Hertford.

Van Steijn, H. (1986). The interpretation of stratified slope deposits and laboratory simulation of transport mechanisms. In *International Geomorphology 1986*, ed. V. Gardiner, pp. 499–511. Chichester: Wiley.

Van Steijn, H. (1988). Debris flows involved in the development of Pleistocene stratified slope deposits. *Zeitschrift für Geomorphologie, Supplementband*, 71, 45–58.

Van Steijn, H., de Ruig, J. & Hoozemans, F. (1988). Morphological and mechanical aspects of debris flows in parts of the French Alps. *Zeitschrift für Geomorphologie*, 32, 143–161.

Van Steijn, H., van Brederode, L.E. & Goedheer, G.J. (1984). Stratified slope deposits of the grèze litée type in the Ardeche region in the south of France. *Geografiska Annaler*, 66A, 295–305.

Van Vliet-Lanoë, B. (1982). Structures et microstructures associées à la formation de glace de ségrégation: leur conséquences. In *The Roger Brown Memorial Volume, Proceedings of the 4th Canadian Permafrost Conference*, ed. H.M.French, pp. 116–122. Ottawa: National Research Council of Canada.

Van Vliet-Lanoë, B. (1985). Frost effects in soils. In *Soils and Quaternary Landscape Evolution*, ed. J. Boardman, pp. 117–158. Chichester: Wiley

Van Vliet-Lanoë, B. (1988a). The significance of cryoturbation phenomena in environmental reconstruction. *Journal of Quaternary Science*, 3, 85–96.

Van Vliet-Lanoë, B. (1988b). The origin of patterned ground in NW Svalbard. *Proceedings of the 5th International Conference on Permafrost, Trondheim*, pp 1008–1013. Trondheim: Tapir Press.

Van Vliet-Lanoë, B. (1988c). *Le Rôle de la Glace de Ségrégation dans les Formations Superficielles de l'Europe de l'Ouest*. Thèse de Doctorat d'Etat, Université de Paris, Sorbonne, 2 volumes, 854pp.

Van Vliet-Lanoë, B. (1991). Differential frost heave, load casting and convection: converging mechanisms; a discussion of the origin of cryoturbations. *Permafrost and Periglacial Processes*, 2, 123–139.

Van Vliet-Lanoë, B, Coutard, J-P. & Pissart, A. (1984). Structures caused by repeated freezing and thawing in various loamy sediments: a comparison of active, fossil and experimental data. *Earth Surface Processes and Landforms*, 9, 553–556.

Van Vliet-Lanoë, B. & Langhor, R. (1981). Correlation between fragipans and permafrost with special reference to Weichsel silty deposits in Belgium and northern France. *Catena*, 8, 137–154.

Van Vliet-Lanoë, B & Valadas, B. (1983). A propos des formations déplacées de versant cristallins des massifs anciens: le rôle de la glace de ségrégation dans le dynamique. *Bulletin de l'Association Française pour L'Étude du Quaternaire*, 4, 153–160.

Vandenberghe, J. (1983a). Ice-wedge casts and involutions as permafrost indicators, and their stratigraphic position in the Weichselian. *Proceedings of the 4th International Conference on Permafrost, Fairbanks, Alaska*, pp. 1289–1302. Washington DC: National Academy Press.

Vandenberghe, J. (1983b). Some periglacial phenomena and their stratigraphical position in the Weichselian. *Polarförschung*, 53, 97–107.

Vandenberghe, J. (1988). Cryoturbations. In *Advances in Periglacial Geomorphology*, ed. M.J.Clark, pp. 179–198. Chichester: Wiley.

Vandenberghe, J. & Van den Broek, P. (1982). Weichselian convolution phenomena and processes in fine sediments. *Boreas*, 11, 299–315.

Vaughan, P.R. (1976). The deformation of the Empingham valley slope. Appendix to Horswill, P. & Horton, A., 1976. Cambering and valley bulging in the Gwash Valley at Empingham, Rutland. *Philosophical Transactions of the Royal Society,* A 283, 451–462.

Veyret, Y. & Coque-Delhuille, B. (1989). Les versants à banquettes de Ronas Hill: Essai de définition d'une province périglaciaire aux îles Shetland. *Hommes et Terres du Nord,* 3, 171–178.

Vick, S.G. (1981). Morphology and the role of landsliding in the formation of some rock glaciers in the Mosquito Range, Colorado. *Geological Society of America Bulletin,* 92, 75–84.

Vierek, L.A. & Lev, D.J. (1983). Long-term use of frost tubes to monitor the annual freeze–thaw cycle in the active layer. *Proceedings of the 4th International Conference on Permafrost, Fairbanks, Alaska,* pp. 1309–1314. Washington: National Academy Press.

Vincent, P.J. (1976). Some periglacial deposits near Aberystwyth, Wales, as seen with a scanning electron microscope. *Biuletyn Peryglacjalny,* 25, 59–64.

Vincent, P.J. & Lee, M.P. (1981). Some observations on the loess around Morecambe Bay. *Proceedings of the Yorkshire Geological Society,* 43, 281–294.

Vincent, P.J. & Lee, M.P. (1982). Snow patches on Farleton Fell, South-East Cumbria. *Geographical Journal,* 148, 337–342.

Vitek, J.D. & Giardino, J.R. (1987). Rock glaciers: a review of the knowledge base. In *Rock Glaciers,* ed. J.R. Giardino, J.F., Shroder & J.D. Vitek, pp. 1–26. London: Allen and Unwin.

Vogt, T (1918). Om recente og gamle strandlinjer i fast fjell. *Norsk Geologisk Tidsskrift,* 4, 107–127.

Wahrhaftig, C. & Cox, A. (1959). Rock glaciers in the Alaska Range. *Geological Society of America Bulletin,* 70, 383–436.

Walder, J.S. & Hallet, B. (1985). A theoretical model of the fracture of rock during freezing. *Geological Society of America, Bulletin,* 96, 336–346.

Walder, J.S. & Hallet, B. (1986). The physical basis of frost weathering: toward a more fundamental and unified perspective. *Arctic and Alpine Research,* 18, 27–32.

Walker, H.J. (1983). *Colville River Delta, Alaska. Guidebook 2, Fourth International Conference on Permafrost, Fairbanks, Alaska.* Division of Geological and Geophysical Surveys, Department of Natural Resources, State of Alaska, Fairbanks, 34pp.

Walker, M.J.C. (1980). Late-Glacial history of the Brecon Beacons, South Wales. *Nature,* 287, 133–135.

Walker, M.J.C. (1984). Pollen analysis and Quaternary research in Scotland. *Quaternary Science Reviews,* 3, 369–404.

Walton, D.W.H. & Heilbronn, T.D. (1983). Periglacial activity on the sub-arctic island of South Georgia. *Proceedings of the 4th International Conference on Permafrost, Fairbanks, Alaska,* pp. 1356–1361. Washington: National Academy Press.

Warburton, J. (1985). Contemporary patterned ground (sorted stripes) in the Lake District. In *Field Guide to the Periglacial Landforms of Northern England,* ed. J. Boardman, pp. 54–62. Cambridge: Quaternary Research Association.

Warburton, J. (1987). Characteristic ratios of width to depth-of-sorting for sorted stripes in the English Lake District. In *Periglacial Processes and Landforms in Britain and Ireland,* ed. J. Boardman, pp. 163–171. Cambridge: Cambridge University Press.

Ward, J.C. (1873). The glaciation of the northern part of the Lake District. *Quarterly Journal of the Geological Society,* 29, 422–441.

Ward, J.C. (1896). The geology of the northern part of the English Lake District. *Memoirs of the Geological Survey.* London.

Ward, R.G.W. (1980). Avalanche hazard in the Cairngorm Mountains, Scotland. *Journal of Glaciology,* 26, 31–41.

Ward, R.G.W. (1984). Avalanche prediction in Scotland: 1. A survey of avalanche activity. *Applied Geography,* 4, 43–62.

Ward, R.G.W. (1985). Geomorphological evidence of avalanche activity in Scotland. *Geografiska Annaler,* 67A, 247–256.

Ward, R.G.W., Langmuir, E.D.G. & Beattie, B. (1985). Snow profiles and avalanche activity in the Cairngorm Mountains, Scotland. *Journal of Glaciology,* 31, 18–27.

Wardle, P. (1974). Alpine timberlines. In *Arctic and Alpine Environments,* eds. J.D. Ives & R.G. Barry, pp. 371–402. London: Methuen.

Washburn, A.L. (1947). Reconnaissance geology of portions of Victoria Island and adjacent regions, Arctic Canada. *Geological Society of America Memoir,* 22, 142 pp.

Washburn, A.L. (1956a). Unusual patterned ground in Greenland. *Geological Society of America, Bulletin,* 67, 807–810.

Washburn, A.L. (1956b). Classification of patterned ground and review of suggested origins. *Geological Society of America, Bulletin,* 67, 823–865.

Washburn, A.L. (1967). Instrumental observations of mass-wasting in the Mesters Vig district, NE Greenland. *Meddelelser öm Grønland,* 166, 1–297.

Washburn, A.L. (1969). Weathering, frost action and patterned ground in the Mesters Vig district, Northeast Greenland. *Meddelelser øm Grønland,* 176, 303 pp.

Washburn, A.L. (1973). *Periglacial Processes and Environments.* London: Edward Arnold, 320pp.

Washburn, A.L. (1979). *Geocryology: A Survey of Periglacial Processes and Environments.* London: Edward Arnold.

Washburn, A.L. (1980). Permafrost features as evidence of climatic change. *Earth Science Reviews,* 15, 327–402.

Washburn, A.L. (1985). Periglacial problems. In *Field and Theory: Lectures in Geocryology,* eds. M. Church & O. Slaymaker, pp. 167–202. Vancouver: University of British Columbia Press.

Washburn, A.L., Smith, D.D. & Goddard, R.H. (1963). Frost cracking in a middle-latitude climate. *Biuletyn Peryglacjalny,* 12, 175–189.

Waters, R.S. (1960). Pre-Würm periglacial phenomena in Britain. *Biuletyn Peryglacjalny,* 9, 163–176.

Waters, R.S. (1961). Involutions and ice-wedges in Devon. *Nature,* 189, 389–390.

Waters, R.S. (1962). Altiplanation terraces and slope development in Vest-Spitsbergen and South-West England. *Biuletyn Peryglacjalny,* 11, 89–101.

Waters, R.S. (1964a). The Pleistocene legacy to the geomorphology of Dartmoor. In *Dartmoor Essays,* ed. I.G. Simmons, pp. 73–96. Devonshire Association for the Advancement of Science, Literature and Art.

Waters, R.S. (1964b). Reports of the Commission on Periglacial Morphology: Great Britain. *Biuletyn Peryglacjalny,* 14, 109–110.

Waters, R.S. (1965). The geomorphological significance of Pleistocene frost action in south-west England. In *Essays in Geography for Austin Miller,* eds. J.B. Whittow & P.D. Wood, pp. 39–47. Reading: University of Reading.

Waters, R.S. (1971). The significance of Quaternary events for the landform of South-West England. In *Exeter Essays in Geography,* eds. K.J. Gregory & W. Ravenhill, pp. 23–31. Exeter: University of Exeter.

Watson, E. (1965a). Periglacial structures in the Aberystwyth region of central Wales. *Proceedings of the Geologists' Association,* 76, 443–462.

Watson, E. (1965b). Grèzes litées ou éboulis ordonnées tardiglaciaires dans la région d'Aberystwyth. *Bulletin, Association de Geographes Francaise,* 338, 16–25.

Watson, E. (1966). Two nivation cirques near Aberystwyth, Wales. *Biuletyn Peryglacjalny,* 15, 79–101.

Watson, E. (1969a). The slope deposits in the Nant Iago Valley, near Cader Idris, Wales. *Biuletyn Peryglacjany,* 18, 95–113.

Watson, E. (1969b). The periglacial landscape of the Aberystwyth region. In *Geography at Aberystwyth,* eds. E.G. Bowen, H. Carter & J.A. Taylor, pp. 35–49. Cardiff: University of Wales Press.

Watson, E. (1970). The Cardigan Bay area. In *The Glaciations of Wales,* ed. C.A. Lewis, pp. 125–145. London: Longman.

Watson, E. (1971). Remnants of pingos in Wales and the Isle of Man. *Geological Journal,* 7, 381–392.

Watson, E. (1972). Pingos of Cardiganshire and the latest ice limit. *Nature,* 238, 343–344.

Watson, E. (1976). Field excursions in the Aberystwyth region, 1–10 July, 1975. *Biuletyn Peryglacjalny,* 26, 79–112.

Watson, E. (1977a). Editor. Mid and North Wales. *Excursion Guide for the 10th INQUA Congress,* 48 pp. Norwich: Geobooks.

Watson, E. (1977b). The periglacial environment of Great Britain during the Devensian. *Philosophical Transactions of the Royal Society,* B 280, 183–198.

Watson, E. (1981). Characteristics of ice-wedge casts in west central Wales. *Biuletyn Peryglacjalny,* 28, 164–177.

Watson, E. & Watson, S. (1967). The periglacial origin of the drifts at Morfa Bychan, near Aberystwyth. *Geological Journal,* 5, 419–440.

Watson, E. & Watson, S. (1971). Vertical stones and analogous structures. *Geografiska Annaler,* 53A, 107–114.

Watson, E. & Watson, S. (1972). Investigation of some pingo basins near Aberystwyth, Wales. *Proceedings of the 24th International Geological Congress, Montreal,* Section 12, 212–223.

Watson, E. & Watson, S. (1974). Remains of pingos in the Cletwr Basin, southwest Wales. *Geografiska Annaler,* 56A, 213–225.

Watson, E. & Watson, S. (1977). The mid-Wales uplands. In *Mid and North Wales: Excursion Guide for the 10th INQUA Congress,* ed. E. Watson, pp. 21–27. Norwich: Geobooks.

Watt, A.S. (1947). Pattern and process in the plant community. *Journal of Ecology,* 35, 1–22.

Watt, A.S. (1955). Stone stripes in Breckland. *Geological Magazine,* 92, 173–174.

Watt, A.S. & Jones, E.W. (1948). The ecology of the Cairngorms. I. The environment and the altitudinal zonation of the vegetation. *Journal of Ecology,* 36, 283–304.

Watt, A.S., Perrin, R.M.S. & West, R.G. (1966). Patterned ground in Breckland: structure and composition. *Journal of Ecology,* 54, 239–258.

Watts, S.H. (1983). Weathering processes and products under arid arctic conditions. *Geografiska Annaler,* 65A, 85–98.

Watts, W. (1905). Geological notes on sinking Langsett and Underbank concrete trenches in the Little Don Valley. *Transactions of the Institution of Mining Engineers,* 31, 668–680.

Wayne, W.J. (1981). Ice segregation as an origin for lenses of non-glacial ice in 'ice-cemented' rock glaciers. *Journal of Glaciology,* 27, 506–510.

Webber, P.J. (1974). Tundra primary productivity, In *Arctic and Alpine Environments,* eds. J.D. Ives & R.G. Barry, pp. 445–476. London: Methuen.

Weeks, A.G. (1969). The stability of slopes in south-east England as affected by periglacial activity. *Quarterly Journal of Engineering Geology,* 5, 223–241.

Wells, S.G. & Harvey, A.M. (1987). Sedimentologic and geomorphic variations in storm-generated alluvial fans, Howgill Fells, northwest England. *Geological Society of America, Bulletin,* 98, 182–198.

West, R.G. (1968). Evidence for pre-Cromerian permafrost in East Anglia. *Biuletyn Peryglacjalny,* 17, 303–304.

West, R.G. (1977). *Pleistocene Geology and Biology.* 2nd Edition. London: Longman.

West, R.G. (1980a). *The Pre-glacial Pleistocene of the Norfolk and Suffolk Coasts.* Cambridge: Cambridge University Press.

West, R.G. (1980b). Pleistocene forest history in East Anglia. *New Phytologist,* 85, 571–622.

West, R.G. (1987). Origin of small hollows in Norfolk. In *Periglacial Processes and Landforms in Britain and Ireland,* ed. J. Boardman, pp. 191–194. Cambridge: Cambridge University Press.

West, R.G. (1991). On the origin of Grunty Fen and other landforms in southern Fenland, Cambridgeshire. *Geological Magazine,* 128, 257–262.

West, R.G., Dickson, C.A., Catt, J.A., Weir, A.H. & Sparks, B.W. (1974), Late Pleistocene deposits at Wretton, Norfolk: II Devensian deposits. *Philosophical Transactions of the Royal Society, London.* B 267, 337–420.

West, R.G., Funnell, B.M. & Norton, P.E.P. (1980). An Early Pleistocene cold marine episode in the North Sea: pollen and faunal assemblages at Coverhithe, Suffolk, England. *Boreas,* 9, 1–10.

Whalley, W.B. (1976). A rock glacier in Wester Ross. *Scottish Journal of Geology,* 12, 175–179.

Whalley, W.B., Gordon, J.E. & Thompson, D.L. (1981). Periglacial features on the margin of a receding plateau icecap, Lyngen, North Norway. *Journal of Glaciology,* 27, 492–496.

White, E., Starkey, R.S. & Saunders, M. (1971). An assessment of the relative importance of several chemical sources to the waters of a small upland catchment. *Journal of Applied Ecology,* 8, 743–749.

White, I.D. & Mottershead, D.N. (1972). Past and present vegetation in relation to solifluction on Ben Arkle, Sutherland. *Transactions of the Botanical Society of Edinburgh,* 41, 475–489.

White, S.E. (1976a). Is frost action really only hydration shattering? *Arctic and Alpine Research,* 8, 1–6.

White, S.E. (1976b). Rock glaciers and block-fields, review and new data. *Quaternary Research,* 6, 77–97.

White, S.E. (1981). Alpine mass movement forms (noncatastrophic): classification, description and significance. *Arctic and Alpine Research,* 13, 127–137.

White, S.E. (1987). Differential movement across transverse ridges on Arapaho rock glacier, Colorado Front Range, U.S.A. In *Rock Glaciers,* eds. J.R. Giardino, J.F. Shroder & J.D. Vitek, pp. 145–149. London: Allen and Unwin.

White, S.E., Clark, G.M. & Rapp, A. (1969). Palsa localities in Padjelanta National Park, Swedish Lappland. *Geografiska Annaler,* 51A, 97–103.

Whitehouse I.E. & McSaveney M.J. (1983). Diachronous talus surfaces in the Southern Alps, New Zealand, and their implications to talus accumulation. *Arctic and Alpine Research,* 15, 53–64.

Whiteman, C.A. & Kemp, R.A. (1990). Pleistocene sediments, soils and landscape evolution at Stebbing, Essex. *Journal of Quaternary Science,* 5, 145–161.

Whittaker, A. (1972). Geology of Bredon Hill, Worcestershire. *Bulletin of the Geological Survey of Great Britain,* 42, 1–30.

Whittington, G.W. (1985). The Little Ice Age and Scotland's weather. *Scottish Geographical Magazine,* 101, 174–178.

Whittow, J.B. & Ball, D.F. (1970). North-west Wales. In *The Glaciations of Wales,* ed. C.A. Lewis, pp. 21–58. London: Longman.

Whyte, I.D. (1970). *Periglacial Formation and Mass-Movement in the Mamore Forest Area.* Unpublished MA Dissertation, University of Edinburgh, 70 pp.

Wilkinson, T.J. & Bunting, B.T. (1975). Overland transport of sediment by rill water in a periglacial environment in the Canadian high arctic. *Geografiska Annaler,* 57A, 105–116.

Williams, P.J. (1958). The development and significance of stony earth circles. *Skrifter Utgitt an det Norske Videnskaps Akademi i Oslo I. Matematisk- Naturhistorisk Klasse,* 3, 14pp.

Williams, P.J. (1959). An investigation into processes occurring in solifluction. *American Journal of Science,* 257, 42–58.

Williams, P.J. (1966). Downslope soil movement at a sub-arctic location, with regard to variations in depth. *Canadian Geotechnical Journal,* 3, 191–203.

Williams, P.J. (1988). Thermodynamic and mechanical conditions within frozen soils and their effects. *Proceedings of the 5th International Conference on Permafrost, Trondheim,* pp. 493–498. Trondheim: Tapir Press.

Williams, P.J. & Smith, M.W. (1989). *The Frozen Earth: Fundamentals of Geocryology.* Cambridge: Cambridge University Press.

Williams, R.B.G. (1964). Fossil patterned ground in eastern England. *Biuletyn Peryglacjalny,* 14, 337–349.

Williams, R.B.G. (1965). Permafrost in England during the last glacial period. *Nature,* 205, 1304–1305.

Williams, R.B.G. (1968). Some estimates of periglacial erosion in Southern and Eastern England. *Biuletyn Peryglacjalny,* 17, 311–335.

Williams, R.B.G. (1969). Permafrost and temperature conditions in England during the last glacial period. In *The Periglacial Environment,* ed. T.L. Péwé, pp. 399–410. Montreal: McGill-Queen's University Press.

Williams, R.B.G. (1971). Aspects of the geomorphology of the South Downs. In *Guide to Sussex Excursions,* ed. R.B.G. Williams, pp. 35–42. Institute of British Geographers.

Williams, R.B.G. (1973). Frost and the works of man. *Antiquity,* 47, 19–31.

Williams, R.B.G. (1975). The British climate during the last glaciation: an interpretation based on periglacial phenomena. In *Ice Ages Ancient and Modern,* eds. A.E. Wright & F. Moseley, pp. 95–120. Liverpool: Seel House.

Williams, R.B.G. (1980). The weathering and erosion of Chalk under periglacial conditions. In *The Shaping of Southern England,* ed. D.K.C. Jones, pp. 225- 248. London: Academic Press.

Williams, R.B.G. (1987). Periglacial phenomena in the South Downs. In *The Scientific Study of Flint and Chert: Proceedings of the 4th International Flint Symposium,* eds. G. de G.Sieveking & M.B. Hart, pp. 161–167. Brighton: Brighton Polytechnic.

Williams, R.B.G. & Robinson, D.A. (1991). Frost weathering of rocks in the presence of salts – a review. *Permafrost and Periglacial Processes,* 2, 347- 353.

Wilson, J.M. (1873). *The Imperial Gazetteer of Scotland,* Volume 2. Edinburgh: Fullarton.

Wilson, J.W. (1952). Vegetation patterns associated with soil movement on Jan Mayen Island. *Journal of Ecology,* 40, 249–264.

Wilson, M.J. (1985). The mineralogy and weathering history of Scottish soils. In *Geomorphology and Soils,* eds. K.S. Richards, R.R. Arnett & S. Ellis, pp. 233–244. London: Allen and Unwin.

Wilson, M.J. & Bown, C.J. (1976). The pedogenesis of some gibbsite soils from the Southern Uplands of Scotland. *Journal of Soil Science,* 27, 513–522.

Wilson, P. (1981). Periglacial valley-fill sediments at Edale, north Derbyshire. *East Midland Geographer,* 7, 263–271.

Wilson, P. (1990). Clast size variations on talus: some observations from northwest Ireland. *Earth Surface Processes and Landforms,* 15, 183–188.

Wilson, P., Bateman, R.M. & Catt, J.A. (1981). Petrology, origin and environment of deposition of the Shirdley Hill Sand of southwest Lancashire, England. *Proceedings of the Geologists' Association,* 92, 211–229.

Wingfield, R.T.R. (1987) Giant sand waves and relict periglacial features on the sea bed west of Anglesey. *Proceedings of the Geologists' Association*, 98, 401–404.

Wintle, A.G. (1981). Thermoluminescence dating of Late Devensian loesses in southern England. *Nature*, 289, 479–480.

Wintle, A.G. (1990). A review of current research on TL dating of loess. *Quaternary Science Reviews*, 9, 385–397.

Wintle, A.G. & Catt, J.A. (1985a). Thermoluminescence dating of soils developed in Late Devensian loess at Pegwell Bay, Kent. *Journal of Soil Science*, 36, 293–298.

Wintle, A.G. & Catt, J.A. (1985b). Thermoluminescence dating of Dimlington Stadial deposits in Eastern England. *Boreas*, 14, 231–234.

Woo, M-k. (1983). Hydrology of a drainage basin in the Canadian High Arctic. *Annals of the Association of American Geographers*, 73, 577–596.

Woo, M-k. (1986). Permafrost hydrology in North America. *Atmosphere-Ocean*, 24, 201–234.

Woo, M-k. (1990). Permafrost hydrology. In *Northern Hydrology: Canadian Perspectives*, eds. T.D. Prowse & C.S.L. Ommanney, National Hydrological Research Institute Science Report No 1, pp. 63–75. Saskatoon: Environment Canada

Woo, M-k. & Marsh, P. (1977). Effect of vegetation on limestone solution in a small arctic basin. *Canadian Journal of Earth Sciences*, 14, 571–581.

Woo, M-k., Marsh, P. & Steer, P. (1983). Basin water balance in a continuous permafrost environment. *Proceedings of the 4th International Permafrost Conference, Fairbanks, Alaska*, pp. 1407–1411. Washington DC: National Academy Press.

Woo, M-k. & Steer, P. (1982). Occurrence of surface flow on arctic slopes, southwestern Cornwallis Island. *Canadian Journal of Earth Sciences*, 19, 2368–2377.

Woo, M-k. & Steer, P. (1983). Slope hydrology as influenced by thawing of the active layer, Resolute, NWT. *Canadian Journal of Earth Sciences*, 20, 978–986.

Wood, B.L. (1969). Periglacial tor topography in southern New Zealand. *New Zealand Journal of Geology and Geophysics*, 12, 361–375.

Wood, S.V. (1882). The Pliocene period in England. *Quarterly Journal of the Geological Society*, 38, 667–745.

Worsley, P. (1966a). Some Weichselian fossil frost wedges from East Cheshire. *Mercian Geologist*, 1, 357–365.

Worsley, P. (1966b). Fossil frost wedge polygons at Congleton, Cheshire, England. *Geografiska Annaler*, 48A, 211–219.

Worsley, P. (1970). The Cheshire–Shropshire Lowlands. In *The Glaciations of Wales*, ed. C.A. Lewis, pp. 83–106. London: Longman.

Worsley, P. (1977). Periglaciation. In *British Quaternary Studies*, ed. F.W. Shotton, pp. 205–219. Oxford: Clarendon.

Worsley, P. (1984). Periglacial Environment. *Progress in Physical Geography*, 8, 270–276.

Worsley, P. (1985). Pleistocene history of the Cheshire–Shropshire Plain. In *Geomorphology of North-West England*, ed. R.H. Johnson, pp. 201–221. Manchester: Manchester University Press.

Worsley, P. (1986). Periglacial environment. *Progress in Physical Geography*, 10, 265–274.

Worsley, P. (1987). Permafrost stratigraphy in Britain – a first approximation. In *Periglacial Processes and Landforms in Britain and Ireland*, ed. J. Boardman, pp. 89–99. Cambridge: Cambridge University Press.

Worsley, P. (1991a). Possible Early Devensian glacial deposits in the British Isles. In *Glacial Deposits in Great Britain and Ireland*, eds. J. Ehlers, P.L. Gibbard & J. Rose, pp. 47–52. Rotterdam: Balkema.

Worsley, P. (1991b). Glacial deposits of the lowlands between the Mersey and Severn Rivers. In *Glacial Deposits of Great Britain and Ireland*, eds. J. Ehlers, P.L. Gibbard & J. Rose, pp. 203–212. Rotterdam: Balkema.

Wrammer, P. (1973). Palsmyrar i Taavavuoma, Lappland. *Göteborgs Universitet Naturgeografisk Institut Rapport*, 3, 140 pp.

Wright, M.D. (1983). The distribution and engineering significance of superficial deposits in the Upper Clydach Valley, South Wales. *Quarterly Journal of Engineering Geology*, 16, 319–330.

Wright, M.D. (1991). Pleistocene deposits of the South Wales Coalfield and their engineering significance. In *Quaternary Engineering Geology*, eds. A. Forster, M.G. Culshaw, J.C. Cripps, J.A. Little & C.F. Moon. Geological Society, Engineering Geology Special Publication 7, 441–448.

Wright, M.D. & Harris, C. (1980). Superficial deposits in the South Wales Coalfield. In *Cliff and Slope Stability*, ed. J.W. Perkins, pp. 193–205. Cardiff: Dept. of Extra-Mural Studies, University College Cardiff.

Wright, W.B. (1914). *The Quaternary Ice Age*. London: Macmillan.

Wymer, J.J. (1977). The archaeology of man in the British Quaternary. In *British Quaternary Studies*, ed. F.W. Shotton, pp. 93–106. Oxford: Clarendon.

Wymer, J.J. (1988). Palaeolithic archaeology and the British Quaternary sequence. *Quaternary Science Reviews*, 7, 79–98.

Young, A. (1972). *Slopes*. Edinburgh: Oliver and Boyd.

Young, A. & Saunders, I. (1986). Rates of surface processes and denudation. In *Hillslope Processes*, ed. A.D. Abrahams, pp. 3–27. Boston: Allen & Unwin.

Zenkovich, V.P. (1967). *Processes of Coastal Development*. Edinburgh: Oliver & Boyd.

Index